AF768398

Master Airline Pilot

Master Airline Pilot offers a process for improving pilots' skills in risk management, situational awareness building, decision making, communications, and crew management. It links aviation human factors with practical airline operations to promote the development of master-level aviation skills across the full range of pilot experience.

Serving as a practical guide for operational aviation challenges, the book discusses exceptional events such as operations under marginal condition, intervening to interdict an unsafe operation, and resolving crew conflicts. It also provides techniques for handling more common airline flying challenges like delays, holding, diverting, and continuing versus aborting a deteriorating game plan.

The book is intended for airline pilots, training captains, simulator instructors, and aviation students taking courses in flight safety and crew management to improve their skillset, proficiency, and expertise toward peak performance.

Master Airline Pilot

Applying Human Factors to Reach Peak Performance and Operational Resilience

Steve Swauger

CRC Press
Taylor & Francis Group
Boca Raton London New York

CRC Press is an imprint of the
Taylor & Francis Group, an **informa** business

Designed cover image: Stephen Gay

First edition published 2023
by CRC Press
6000 Broken Sound Parkway NW, Suite 300, Boca Raton, FL 33487-2742

and by CRC Press
4 Park Square, Milton Park, Abingdon, Oxon, OX14 4RN

CRC Press is an imprint of Taylor & Francis Group, LLC

© 2023 Steve Swauger

Reasonable efforts have been made to publish reliable data and information, but the author and publisher cannot assume responsibility for the validity of all materials or the consequences of their use. The authors and publishers have attempted to trace the copyright holders of all material reproduced in this publication and apologize to copyright holders if permission to publish in this form has not been obtained. If any copyright material has not been acknowledged please write and let us know so we may rectify in any future reprint.

Except as permitted under U.S. Copyright Law, no part of this book may be reprinted, reproduced, transmitted, or utilized in any form by any electronic, mechanical, or other means, now known or hereafter invented, including photocopying, microfilming, and recording, or in any information storage or retrieval system, without written permission from the publishers.

For permission to photocopy or use material electronically from this work, access www.copyright.com or contact the Copyright Clearance Center, Inc. (CCC), 222 Rosewood Drive, Danvers, MA 01923, 978-750-8400. For works that are not available on CCC please contact mpkbookspermissions@tandf.co.uk

Trademark notice: Product or corporate names may be trademarks or registered trademarks and are used only for identification and explanation without intent to infringe.

Library of Congress Cataloging-in-Publication Data
Names: Swauger, Steve, author.
Title: Master airline pilot: applying human factors to reach peak performance and operational resilience/Steve Swauger.
Description: First edition. | Boca Raton, FL: CRC Press/Taylor & Francis Group, LLC, 2023. |
Includes bibliographical references and index. |
Identifiers: LCCN 2022046915 (print) | LCCN 2022046916 (ebook) |
ISBN 9781032383446 (hardback) | ISBN 9781032383453 (paperback) | ISBN 9781003344575 (ebook)
Subjects: LCSH: Aeronautics–Human factors–Study and teaching. |
Airplanes–Piloting–Human factors. | Air pilots–Training of. | Aeronautics–Safety measures.
Classification: LCC TL553.6.S93 2023 (print) | LCC TL553.6 (ebook) |
DDC 629.130071—dc23/eng/20221013
LC record available at https://lccn.loc.gov/2022046915
LC ebook record available at https://lccn.loc.gov/2022046916

ISBN: 9781032383446 (hbk)
ISBN: 9781032383453 (pbk)
ISBN: 9781003344575 (ebk)

DOI: 10.1201/9781003344575

Typeset in Times
by codeMantra

Contents

SECTION III Introduction to Challenging and Non-Normal Operations

Acknowledgments

I would like to acknowledge the invaluable contribution of the following people:

- The many Master Class pilots I have had the honor to fly with and observe during my aviation career. This book is about you and the many pearls of wisdom that you have shared.
- The many Human Factors scientists who helped this simple stick-and-rudder pilot understand the complexities of this field of study. I am especially grateful for the kind patience and wisdom of R. Key Dismukes, Immanuel Barshi, Barbara Burian, and Loukia Loukopoulos.
- My friends and colleagues from the Human Factors Roundtable, the Active Pilot Monitoring Working Group, and the industry experts from the ISAP and PACDEFF conferences.
- Captain Stephen Gay for the use of his photograph on this book's cover.
- My family, Joline and Zack – for their patience, encouragement, support, and graphics expertise.

About the Author

Steve Swauger is an Aviation Human Factors Consultant with Human Factors Excellence in Aviation, LLC. He holds a BA in Human Factors from the USAF Academy and a MA in National Security Studies from CSU San Bernardino. Over his 40-year pilot career, he flew over 18,000 hours in military and air carrier aircraft. His airline career spanned over 26 years, where he served over 20 years as a Captain on a variety of Boeing 737 aircraft. While flying for the airline, he served for 23 years in a variety of pilots' union safety committee roles including Safety Committee chair, Human Factors subcommittee lead, GO-Team accident investigator, Professionalism Council lead, and safety investigator.

During his tenure, he authored over 200 articles under his column, "The Human Factor", which were published in union member publications. He has presented and led workshops at the International Symposium of Aviation Psychology and the PACDEFF Human Factors/CRM Conference. He is an original member of the Human Factors Roundtable (a group of Human Factors and Safety department professionals from U.S. Major Airlines and industry). He served on the Active Pilot Monitoring Working Group, which published "A Practical Guide for Improving Flight Path Monitoring" through the Flight Safety Foundation. Additionally, he has served on several workgroups tasked with researching and redesigning airline flight-deck procedures and operating manuals.

Introduction

Safe airline aviation is one of the most successful achievements of modern society. Proactive safety management systems, effective CRM programs, and proven training methods have combined to deliver a nearly perfect safety record. Despite our progress, some stubborn safety problems still linger. We still commit avoidable errors. We still make questionable decisions. Bad events still happen. Our nearly perfect safety record is not good enough.

When a mishap event occurs, we wonder why. We study the aviation environment and see how conditions rise and fall, appear and disappear, and mix and match to generate unpredictable situations. Positioned in the middle of this turbulent, uncertain environment is us – pilots trying our best to fly our aircraft and achieve safe outcomes. When we don't handle aviation challenges skillfully, latent vulnerabilities surface. Problems slip through.

Surveying across our pilot group, we identify one subgroup that does very well almost all of the time. However, when particular conditions combine in unfavorable ways, this group encounters problems that they sometimes handle unskillfully. We identify another subgroup that consistently demonstrates superior levels of awareness, skill, and judgment to successfully handle every challenge they encounter. These Master Class pilots employ reliable mindsets, techniques, and practices that generate their success. The purpose of this book is to identify these skills and attributes, understand how they work, and guide the process of weaving them into our personal aviation wisdom.

BUILDING MASTER CLASS PILOTS

Whether we are new to the airline industry or approaching retirement, we share an extraordinary journey. It begins at the first moment when we choose to pursue a career in professional aviation. It concludes when we respond to the last item of the last checklist after our last flight. Every day marks another step along our Master Class journey.

We have worked hard, studied, and practiced to get to where we are today. When we were students, every step followed a singular training path defined by the curriculum. When our formal training ends, available paths diverge. We have a choice with how we wish to proceed. One path leads toward settling into our comfort zone, relaxing, and enjoying the ride. Another path leads toward Master Class growth, continuous learning, and enjoying the ride. With either path, we get to enjoy the ride because aviation is an awesome career. What a great life we have chosen for ourselves.

Some may ask, what is wrong with relaxing into our comfort zone? Reaching it means that we have finally developed the skills to fly easily and confidently. It is a natural outcome after achieving proficiency. In truth, there is nothing inherently wrong with settling into our comfort zone. Vulnerabilities only arise because our personal awareness and learning begin to subside *after* settling into our comfort zone. In this relaxed state, we can erroneously conclude that we have reached the summit of the mountain and that there is nothing left to learn.

The truth is that the path continues. There is no summit. There is always a new level of learning to explore. For perspective, consider the simple act of eating food. We eat every day. Each of us amasses enough experience to become comfortable with our current level of food knowledge. We know what we like, we know what we don't like, and we know how different foods affect us. Yet, no matter how proficient we are with eating, we can't advance our understanding of nutrition. Nutrition represents a deeper level of knowledge beyond the familiarity acquired by simply consuming food. If eating food leads to dietary proficiency, understanding nutrition leads to dietary mastery.

Carrying this example a step further, imagine that we each make a commitment to learn more about nutrition. We might start by reading an informative book or surfing relevant websites. We might discover that our current diet, however enjoyable and tasty, is actually endangering our health. We might even discover that we aren't even chewing our food sufficiently. By delving into nutrition, we discover that we have unknowingly fostered latent health vulnerabilities that may surface later in life. Until we make the commitment to study the science of nutrition, we can't move beyond the barrier created by just eating proficiently. What's more, knowledge of nutrition alone is not enough. There are expert nutritionists who still choose to eat poorly. It is only by balancing science (knowledge of nutrition), awareness (psychology), and intention (committed practice) that we can achieve the mastery with our diets.

This same logic follows for aviation. As we amass flying experience, we become proficient and comfortable. Settling for only proficiency and comfort, we reach a barrier that limits our progress toward mastery. It keeps us vulnerable to committing errors and making questionable decisions. To overcome this barrier, we each need to pursue the Master Class path. Just as nutrition reveals a deeper level of food knowledge, aviation human factors uncovers a deeper level of aviation wisdom.

AVIATION HUMAN FACTORS (HF)

Many of us have a sense of what HF entails, but may not be able to clearly define it. One reason for this is that the breadth of HF is so extensive. It spans across how we interact with our aircraft, automation, each other, procedures, operational systems, and our own psychology. Some specific areas include:

- *Human/machine interface:* The field of ergonomics covers the physical design of everything we can see, touch, reach, or move on the flightdeck. From the shape and movement of switches and dials, to the size and presentation of screens and displays, to the adjustments of our flightdeck seat, ergonomics engineering plays a guiding role.
- *Automation:* Automation encompasses how we interact with aircraft systems, computers, and displays.
- *CRM:* Interacting with each other delves into team psychology, leadership, communications, shared mental models, and interacting with outside workgroups and agencies.

- *Procedures:* Procedures cover the range of flightdeck tasks, checklists, planning, and workload. These include the execution, sequence, and interdependence between hundreds of individual aviation tasks.
- *Aviation environment:* HF studies how we interact with the many systems that surround us including airline management, flight operations, dispatch, local station operations, passenger cabin operations, aircraft maintenance, air traffic control, and the local system created by pilots working together on a flightdeck.
- *Psychology:* At the center of all of these interactions is the human mind. Psychology illuminates the subconscious processes, biases, and limitations that affect how we interact with our aviation environment.

FOR EXPERIENCED AIRLINE PILOTS

Even the greatest athletes have coaches – not because they aren't performing at the top of their game, but because they believe that their pursuit of mastery benefits from an outside perspective. Coaches help them to both improve the strengths and eliminate the weaknesses in their performance. They detect nuances that the athletes don't perceive themselves. They identify subtle habits that open vulnerabilities and guide techniques to avoid errors and improve success. As line-flying pilots, we don't have coaches flying with us every day. The next best options are coaching each other and becoming our own coach. Using the ideas in this book, we will discover ways to strengthen our performance and eliminate our weaknesses. At this point in our careers, most of us have become skilled airline pilots. We successfully handle every operational challenge that we encounter. Still, every once in a while, proficient, capable pilots like us find ourselves in difficult situations that we are ill-equipped to handle. It is along this turbulent edge of complexity, unpredictability, uncertainty, and risk that we sometimes fail. This book reveals concepts, skills, and techniques that we can use to become better risk managers, decision makers, error mitigators, and team leaders. Along the way, we will develop the foresight, resilience, and aviation mastery to navigate along this turbulent edge.

FOR PILOTS PURSUING AN AIRLINE CAREER

As aspiring airline pilots, we are still developing the skills that will serve us for the rest of our careers. We'll make mistakes and learn lessons. There is an old adage in aviation that the best way to avoid making mistakes is to gain experience – and the best way to gain experience is by making mistakes. Rather than learning the hard way ourselves, we will study mishaps, incidents, and accidents to recreate the perspectives of the involved pilots. By examining their events, we will discover ways to avoid falling into similar traps. We will use their experiences to build our own experience. Many of the concepts and techniques may seem unusable for us at this early stage of our career. Some ideas will be immediately useful for us and others will become useful later. Along the way, we'll shape our perspective to maximize our personal growth of aviation wisdom and mastery. We'll plant the seeds today and enjoy the harvest tomorrow.

PERSPECTIVE AND PRESENTATION OF THIS BOOK

The concepts and techniques presented in this book should prove useful for all pilots, especially those operating in crew environments. It uses the perspective of passenger-carrying airlines because this environment fully encompasses the range of complications created by passenger service, crew interaction, and system complexity. Other aviation professionals including military and corporate pilots encounter different challenges, but in the end, we are all pilots flying aircraft together within the same aviation environment. The lessons and techniques presented on these pages still apply.

I refer to the left seat pilot as Captain (CA). This position is sometimes referenced as pilot, first pilot, or aircraft commander. For the right seat pilot, I use First Officer (FO). This position is sometimes referenced as co-pilot or second pilot. For continuity, assume that the left seat pilot (Captain) is serving as the aircraft commander and that the right seat pilot (FO) is assuming the second-in-command role.

AUTHOR'S AVIATION BACKGROUND AND PERSPECTIVE

My perspective endeavors to join the two different worlds of professional aviation and HF science. I have devoted decades toward both perfecting my craft as a professional pilot and studying how HF science helps us reach the highest levels of aviation mastery. As a professional pilot of 40 years of experience, I flew the line at a major airline for over 26 years – 20 years as Captain.[1] I was also deeply involved with safety work as an accident/incident investigator, flightdeck procedures development project pilot, union safety committee chairman, and HF subcommittee lead. I am actively involved with airline industry HF organizations and have presented my work at many conferences. For over 20 years, I authored hundreds of articles applying HF science to airline aviation. My writing explores the hidden forces underlying aviation to understanding why procedures and techniques work the way they do, how we unintentionally fall short, and how we can do better. My writing builds upon the brilliant work of many giants within the HF community. I hope that these chapters will kindle your desire to explore the wisdom of their writings.

My unique perspective comes as a bridge builder and as a translator. I have explored both sides of the gap dividing HF scientists and pilots. On one side are people with answers to questions that the people on the other side ponder – and vice versa. Unfortunately, each side speaks different languages and views aviation from slightly different perspectives. In this book, I will interpret aviation HF science from a line pilot's perspective.

WHAT FOLLOWS

The early chapters of this book will lay the framework for our understanding of commonly used HF terms. The later sections will apply the HF perspective across many line-flying situations to evaluate techniques toward becoming wise Master Class pilots. Thank you for joining me on this journey.

NOTE

1 "Flying the line" refers to the daily schedule of flying of passengers, freight, and flight missions. Refer to the Glossary of Commonly Used Airline Terms for this and other airline references.

1 The Master Class of Pilots

When we begin our aviation career, we are satisfied with learning our basic flight maneuvers and completing them without instructor intervention. Since our initial aviation journey is controlled by the training syllabus, we focus on *learning*. As we acquire experience and hours, we transition from learning to *doing*. Sure, we still need to learn about new aircraft, procedures, and operations, but the scales measuring our priorities and mental focus tip decidedly toward *doing the job*. As we settle into the daily routine of line-flying, we level off, slide our seat back, and let the autopilot take over. The urgency to learn fades away. If some new procedure arises, we shift our mental focus long enough to learn it and then return to our comfortable routine. The consequence of *doing* is that it unintentionally creates a barrier to further *learning*. As Master Class pilots, we choose to maintain our focus on learning even after we reach proficiency. To understand our learning process, let's examine how our mindset changes as we gain experience.

1.1 HOW WE LEARN

Let's begin with how we learn to perform a psychomotor task – any task that involves thinking and physical movement. A simplistic view of flying is that we perceive a series of indications from a variety of sources, form an understanding of what they mean, decide what action needs to be taken, and perform that action to a flight control or aircraft system. We guide the aircraft along a desired path by continuously repeating this process. This sounds simple enough, but in practice, it can become complicated by the volume of information and our ability to perceive and process it. It is further complicated by our decision-making process, training, repetition, environmental factors, quality of feedback, personal capabilities, and so much more. Consider the following process for completing a task.

- Perceive sensory inputs from our environment.
- Sort through these inputs and match them to recognized patterns.
- Apply meaning to the patterns.
- Choose a game plan that will move us from our present position toward our goal.
- Select decisions and actions that enact that game plan.
- Apply appropriate control inputs to the aircraft.
- Assess the effectiveness of our actions on the aircraft's trajectory.
- Repeat the process for each follow-on adjustment.

The stream of thinking and flying happens so quickly that it appears to flow seamlessly from one aviation task to the next. It is much like a motion picture where each individual frame is followed by the next so quickly that it creates the experience of

DOI: 10.1201/9781003344575-1

smooth motion. The mental effort and attention focus that we use to fly changes as we gain experience. To understand this better, consider pilots divided between three categories – novice, proficient, and Master Class. Each category engenders unique sets of abilities, motivations, perspectives, and behaviors.

1.1.1 How Novice Pilots Learn

Let's skip ahead along our career path to a point where we begin training with our first airline as a newly hired FO. Everything is new and different. Whether our journey was through the traditional flight school track or through the military track, reaching our first airline job feels a bit like starting over. Even as fully certified pilots, starting with a new aircraft or in a new organization imposes a unique set of policies, procedures, and expectations. Everything feels unfamiliar.

At first, we experience information overload. We don't know how to skillfully rank-order everything that is happening. It feels like there are too many demands on our time and attention. We struggle to filter and prioritize. Every piece of information needs to be detected, considered, and judged for relevance. Then, we need to choose whether to act on it, defer it, or ignore it. As we struggle to process everything, we find ourselves falling further and further behind the pace of the operation. This adversely affects our flying. We over-control or under-control the aircraft. All the while, new inputs arrive that demand our attention. More time is lost and we fall further behind. To control the flood, we start narrowing our focus. This promotes tunnel vision. We make lots of mistakes. As frustrating as these early days may seem, each repetition gives us valuable experience and practice. We rapidly improve.

This early learning phase requires a great deal of mental effort. The good news is that this isn't our first flying experience. We apply skills and techniques that we learned with our previous aircraft to handle these new challenges. We have successfully accomplished novice learning many times in our past. We will succeed again. The novice phase is fast-paced, but our learning curve is steep.

1.1.2 How Proficient Pilots Learn

Over time, we continue to gain experience and knowledge through repetition and practice. We learn the tricks of the trade. What once was rough and challenging becomes smooth and easy. We fly the aircraft effortlessly. That same stream of sensory inputs that completely overwhelmed us as novices now seems easy to process. As proficient pilots, we have seen-it-all and done-it-all. We learn to combine inputs and tasks. Instead of perceiving and evaluating each sensory input, we recognize large blocks of information by how they match with familiar patterns. For example, consider a pilot cleared for a visual approach from the downwind while following another aircraft that is currently on base. The major decisions are managing energy (timing configuration changes, thrust, and speed to arrive stabilized on final), ensuring spacing (maintaining an acceptable distance behind the preceding aircraft), and judging the turn to final (smoothly transitioning from a level downwind through a descending turn to roll out on a stabilized final approach segment). Within each of these sub-tasks are dozens of embedded decisions requiring our skillful judgment.

We need to factor considerations like aircraft weight, speed, altitude, the preceding aircraft type, typical approach characteristics, weather parameters, runway layout, and many more. Novices see each of these as separate tasks to manage. Proficient pilots see the situation as one continuous flow with each portion connected to the next.

Another characteristic of proficient pilots is subconscious pattern recognition and decision making. If we ask a proficient pilot how they fly the base-to-final turn, we might get a vague answer like, "I know where the aircraft needs to be, and I just fly it there." We know what looks right and fly the aircraft to reproduce that mental picture. Using this visualization technique, we not only interpret parameters accurately and quickly, but we anticipate future trends to stay ahead of changing conditions. This frees time and attention to monitor indications that are more subtle. This is why proficient pilots notice so much more than novices. While novices struggle to expand their situational awareness beyond the immediate confines of the flightdeck, proficient pilots easily extend their situational awareness to anticipate future conditions.

With repetition, we discover our comfort zone. We become more relaxed and confident. Proficiency is the level reached by the vast majority of us – good, capable, successful, and skillful pilots. As our comfort zone deepens, we form habits and routines to make each phase of the flight familiar and easy. Over time, these routines become automatic. We reach a point where we can do them without consciously thinking about them. Like our aircraft, we can complete many aviation tasks while our brains run on autopilot.

Proficient pilots who are settled in their comfort zones develop an expectation of normalcy – that this event will go just like it did the last 10, 100, or 1,000 times. While our pace of learning in the novice phase is fast, learning in the proficient phase slows. To be clear, the proficient level does not necessarily reflect capability or skill. Some proficient pilots are considered to be the finest pilots within their organizations. Other proficient pilots are content with mediocrity. Instead, the proficient level is characterized by the mental stasis we experience as we settle into our comfort zone. The key distinction is that proficient pilots reach their comfort zone and are content to remain there.

1.1.3 How Master Class Pilots Learn

Master Class pilots commit themselves to move beyond the comfort zone accepted by proficient pilots. Pursuing mastery, we strive to deepen our knowledge and understanding – both of our aviation profession and of ourselves. This requires sustained effort. Instead of settling into a comfort zone of automatic, subconscious, habitual routine, we actively pursue higher levels of understanding, awareness, and skill-refinement. This requires commitments to practice mindfulness, pursue knowledge, and contemplate ourselves. Not content to simply complete each task successfully, we strive to understand the many conditions, indications, interactions, and underlying processes that influence an effect or task. Even when we fly well, we ask ourselves questions. Why did this work out so well this time? What were the subtle indications that I need to incorporate into my monitoring routines? How can I improve my personal techniques to do this task better in the future? While proficient pilots are content to settle for successful performance, Master Class pilots strive to steadily raise their flying to higher levels of precision. Why go through the extra effort to pursue the Master Class level? There are three main benefits.

- Master Class pilots are better prepared to handle complex and non-standard situations. We are better prepared to handle the unexpected.
- Master Class pilots make better instructors and mentors. Like the wise elders of the tribe, we collect techniques to pass along to others.
- Master Class pilots find self-satisfaction through their pursuit of mastery. By approaching each task as a learning experience, every flight remains fresh and interesting.

1.2 HOW WE LEARN TO MANAGE AVIATION TASKS

Let's examine some aspects of learning to better understanding the transition between the novice, proficient, and Master Class levels.

1.2.1 Heuristic Development – Learning the Tricks of the Trade

A useful way to understand the learning process is by how we each develop our personal heuristics. Heuristics are the mental techniques that we use for detecting inputs, interpreting them, and applying procedures to successfully complete tasks. Simply stated, heuristics are habits and techniques we use to complete aviation tasks. Our personal heuristics may not be the best methods or even the procedures we are supposed to use, but they work for us. On the information-processing side, they help us sift through the flood of sensory inputs. They help us combine indications and organize groupings into patterns that make sense. They also guide where to focus our attention. We learn which parameters warrant our attention and which can be ignored. The process of heuristic development evolves and changes as we gain experience and knowledge.

1.2.2 Novice Pilot Heuristics

In the classroom, we are taught how to complete a common aviation task like an engine start using a standardized process. The instructor ensures that we can repeat the steps of the task sequentially, consistently, and accurately. We adopt a heuristic that views the engine start as ordered steps of equal importance. For example:

1. Check for system preconditions (bleed air pressure).
2. Rotate that engine start switch to the START position.
3. Look at the duct pressure gauge for a pressure drop, and so on.

As we perform engine starts over and over again, our engine start heuristic begins to change. The ordered sequence blends into one continuous flow where each step naturally and logically leads to the next. The lines between each step become blurred. We begin to clump certain steps together into a series of mini-flows where each step triggers or creates preconditions for the next. Look at that, do this, wait for that, move on to the next step, and so on. We also develop a sense of the pacing for how long each task typically takes and whether the task is a quantitative (does the light turn on, or not) or qualitative (does the engine core temperature rise at a normal rate, or not). Returning to our engine start example, our heuristic becomes a much more detailed and blended flow.

1. Start by checking the pressurization panel to ensure that we have adequate start pressure. Typically, we develop a mini-flow to ensure proper system configuration (APU bleed air ON, air conditioner packs OFF, duct pressure adequate for engine start). That mini-flow becomes a precondition to…
2. … Engage the engine starter motor. We turn a switch or depress a button and determine whether it engaged successfully (a yes/no, quantitative determination). This leads to …
3. … Verify that the start valve has opened by observing a change in the bleed air duct pressure gauge (a yes/no, quantitative determination). We check to see that the engine is beginning to spin by looking at the appropriate gauge (also a quantitative determination). This cues us to …
4. … Switch from quantitative assessments to several qualitative assessments like checking the rate of engine acceleration and the appropriate rise of oil pressure. Each parameter is evaluated against what we have come to accept as normal – does this parameter react like it normally does or not. Following this, we …
5. … Check that the engine reaches the appropriate speed. Then, we engage fuel flow. This, in turn, cues us to …
6. … Monitor engine light-off and the rise in turbine speed and core temperature (both qualitative) while monitoring for start failure lights (quantitative), and so on.

Until we establish our proficient heuristic, we tend to over-process information. We expend a great deal of effort and time detecting, deciding, understanding, and acting. At the novice level, we frequently become task-overloaded. Like driving on narrow roads in an unfamiliar country, we struggle to stay in our lane, keep up with the flow of traffic, interpret the signs, and find our way. We miss important cues while trying to filter through many unimportant ones. Additionally, we lack skill with processing abnormalities. When something happens out of the ordinary, we are slow to process the new conditions and take appropriate actions.

1.2.3 Proficient Pilot Heuristics

Over time, we acquire rules-of-thumb, simplifications, or tricks-of-the-trade that help us to easily complete the engine start. We learn to devote more attention to relevant indications and less attention to unimportant ones. We also mentally combine information into useful groupings or baskets. Numbers gain contextual meaning. The rise on the fuel flow gauge indicates that the fuel flow valve has opened, but the value on the gauge allows us to assess whether it is appropriate for normal start (whether it is a typical initial fuel flow value) or an early indication of an impending problem (like an initial fuel flow that is abnormally high). A non-typical indication might signal a condition that may lead to a hung start, slow start, or an engine over-temperature. Our understanding of these nuances grows with repetition and experience. Proficient pilot heuristics evolve from the novice's lists of steps into interdependent flows blending one task to the next. The more individual steps we combine, the more we can handle. In addition to visual indications, we begin to notice sounds and vibrations that the novice pilot wouldn't. We also refine our sense of pacing. This sub-task should take this long to complete. The next one should take slightly longer. When the pacing of an event doesn't match what we expect, we investigate the cause. Each successful repetition reinforces our routine and makes it more familiar, solid, and easy.

1.2.4 Master Class Pilot Heuristics

As proficient pilots solidify their habit patterns, their levels of attention and effort decrease. As Master Class pilots, however, we don't let familiarity influence our attention or effort. Like proficient pilots, we expect tasks to go normally, but we remain diligent for exceptions. Each engine start is not just another engine start. Each remains important because we might experience that one hung start of our entire career. We don't *expect* something to go wrong, but we elevate our vigilance *in case* something goes wrong.

This improves our anticipation. We become experts on the many ways (however rare) that an otherwise normal event can go wrong. For comparison, if you asked a proficient pilot to list the ways that an engine start can go awry, they might give you a basic list – start switch failure, starter shaft failure, no light-off, hung start, engine fire, and over-temperature. A Master Class pilot could present a much longer list with many subtle cases. They might list all of the standard starting abnormalities plus, first start of the day in very cold weather, middle of the day start in very hot conditions, high altitude engine start, external start cart failure during engine start, strong tailwind, APU failure during start, and interrupted start. Additionally, Master Class pilots could cite specific examples that they had experienced, read about, and heard about.

Master Class pilot heuristics are richer and deeper as we continue to add more nuance and detail. Viewed as woven cloth, the novice's understanding is like a simple cross-weave, the proficient pilot's understanding is like a sturdy canvas, and the Master Class pilot's understanding is like an ornate tapestry – interwoven with many different colors and types of threads – each one representing a different experience and piece of knowledge integrated within the whole.

BIBLIOGRAPHY

Mosier, K. (2010). The Human in Flight: From Kinesthetic Sense to Cognitive Sensibility. In E. Salas & D. Maurino (eds), *Human Factors in Aviation - 2nd Edition* (pp. 147–174). Burlington: Academic Press.

Section I

Introduction to Core Concepts

DOI: 10.1201/9781003344575-2

I.1 THE LIMITATIONS OF COMMONLY USED TERMS AND CONCEPTS

An aircraft mishap occurs and an investigation follows. The report attributes the cause as complacency, inattention, fatigue, non-compliance, or unprofessional behavior. We use these labels because they provide simple cause-and-effect explanations using "if-then" logic. "If they weren't complacent, then the accident wouldn't have happened." "If they weren't acting unprofessionally, then they would have made better decisions." Another convenience of these labels is that they imply defects of motivation, attention level, ability, priorities, due diligence, and professionalism. This connects deficient behaviors with unfavorable outcomes.

When challenged to accurately define these terms, however, we struggle. What is complacency? Where does it come from? What does it look like when it starts? How does it evolve? The underlying truth is that most of us don't really understand these terms despite how freely we use them. They are convenient generalizations that allow us to distill complex, variable processes into seemingly solid concepts. Further, they allow us to dismiss what happened as detached and separate from anything that we would actually do ourselves. They were unprofessional. We are not. They made poor decisions. We do not. We use these terms to reinforce an illusion of a well-ordered world with clear borders. We confidently conclude, "we aren't like those pilots."

These same flaws arise with concepts like risk management, complexity, decision making, situational awareness, error, distraction, safety, and time management. None of these terms are new to us. We think that we understand them. We use them freely. Envisioning a concept like risk management, we know what good risk management looks like and what bad risk management looks like. We know when we have managed risk well. We know when we have managed risk poorly.

As we dig deeper, we realize that our understanding of many of these terms only achieves clarity when we view them from a hindsight perspective. Our hindsight perspective hides the underlying processes and vulnerabilities that affect risk management. Less clear is what poor risk management looks like to mishap crews while they are flying in-the-moment as the situation is unfolding around them. What does poor risk management look like before anything starts going wrong? Before we can practice Master Class analysis, we need to expand our understanding of these foundational concepts.

I.2 HINDSIGHT BIAS IN MISHAP INVESTIGATIONS

Hindsight bias allows us to label the motivations, mindsets, and actions of the mishap crew. We review the event and notice that the pieces match our preconception of poor risk management, so we apply the label. This is a wrong step down the wrong path. Clearly, these pilots did not intend to be poor risk managers. They didn't intend to make bad choices, fail, bend metal, or injure passengers. More accurately, they started their flight with a crystal-clear intention to do everything right and arrive safely at their destination. Something happened along the way to deflect that event trajectory. Certain conditions, together with their choices and actions, combined to produce an unfavorable outcome. In hindsight, we choose to label the cause, "poor risk management". Once we apply the label, our logic, perceptions, and conclusions fall conveniently into place. We can examine the sequence of events and locate the points where the crew selected "bad" choices, applied "poor" actions, or "missed" important indications. The labels guide our explanation. The explanation uses other labels to support the first label. This process is convenient, but it doesn't advance our aviation profession. It doesn't give us Master Class insight into the mishap. If, instead, we resist applying labels until we fully understand how the event unfolded to the crew, we gain useful insight into the forces and dynamics that combined to generate the event.

I.3 EVALUATING MISHAPS USING AN IN-THE-MOMENT PERSPECTIVE

A better approach is to examine the event from the in-the-moment perspective of the mishap crew.

- What was happening before the mishap event?
- What were their preconceptions and assumptions?
- What were their expectations of how the flight would unfold?
- What information did they have available? What was missing?
- What indications did they actually notice at the time?
- If they missed important information, what contributed to their oversight?
- What factors shaped their mindset?
- How did their mindset change as the event progressed?
- How did workload, complexity, and other factors affect the progression of events?

When we adopt an in-the-moment perspective, we discover that many of the labels we previously used begin to lose clarity. For example, what is complexity? What makes it rise? What makes it fall? What does it look like before it begins to directly affect us? How does it affect our decision making? How do we accommodate rising complexity? When we trade our hindsight perspective for an in-the-moment perspective, we see that outcomes are constantly being pulled and pushed in many directions. We see how latent vulnerabilities emerge from seemingly unrelated forces to generate undesirable outcomes. The path of a mishap is rarely simple.

We also gain a deeper understanding of ourselves. Instead of judging a crew on what they did and should have done, we project ourselves into their environment to see how we might have experienced it. By seeing ourselves in their situation, we discover our own latent vulnerabilities. We evaluate the resilience of our personal techniques and habit patterns to test whether they would have guided us toward a successful outcome.

I.4 UNDERSTANDING CONCEPTS ON A MASTER CLASS LEVEL

As Master Class pilots, we move beyond a cursory understanding of core concepts and dig deeply into the underlying "why" and "how". Throughout this section, we will review many pilot reports. Some will end favorably and some won't. In each case, we should endeavor to project ourselves into the situation. Strive to understand how the pilots experienced their event at the time. Look closely to discover some familiar shadows lurking within our own decision-making processes. We can test our personal countermeasures to see how they would have served us with that event. Some questions we can ask ourselves are:

- When was the earliest indication that the event was deflecting away from their intended path?
- How would we recognize those earliest indications using our own monitoring techniques?
- What features or combinations of conditions have we encountered in similar events?
- How did the pacing of the event change?
- How did their workload rise or fall?
- How effective was their crew resource management (CRM)? How would our CRM techniques have handled the event?

We'll make three passes as we review these cases. First, we'll scan through the event report to survey when each condition or event arose. We'll apply our hindsight perspective to imagine how conditions and indications would have unfolded in real time. Second, we'll review the event as witnesses, as if we were actually sitting on the jumpseat. We'll watch the crew for signs of confusion, deterioration of active risk management, how they fell behind with their workload, the quickening of the operational pace, and their attempts to recover. Third, we'll reimagine the event as if we were in the pilot seat. Would our techniques, decision making, balancing of priorities, and CRM have prevented the mishap? The first two passes help us understand the in-the-moment flow of the event. The third offers us an opportunity to refine our own Master Class skills.

Organization of the Core Aviation Concepts Section: The following chapters will delve into some core concepts of aviation human factors (risk management, complexity, decision making, situational awareness, error, distraction, safety, and time management). It is important that we develop a shared understanding of the concepts before we examine Master Class techniques, pilot roles, non-normal operations, and professionalism.

2 Risk Management

2.1 THERE IS ALWAYS SOME RISK

Aviation involves risk. First, we reduce and mitigate risk as well as we can. Then, we manage what remains. On balance, we do a commendable job. When risk is low, we easily follow familiar game plans. As risk rises, we reach a point where we can no longer justify continuing with that original game plan. We switch to a backup game plan or suspend operations. Between those extremes, we apply policy guidance and our experience to assess and balance priorities.

Risk is not a static, predictable factor. It varies with conditions, vulnerabilities, capabilities, and limitations. All of these forces interact – sometimes adding risk, sometimes reducing risk, and sometimes spawning unpredictable situations that seem unrelated to any of the underlying forces. Even with seemingly identical conditions, different situations can emerge. Lacking predictability, we can't definitively judge when a risky situation becomes unsafe. We rely on our judgment.

Despite these challenges, we can succeed as long as we preserve a safe escape option. We don't have to land from an unstabilized approach. We can go around. We don't have to depart into a thunderstorm. We can wait until the storm passes. Our challenge is to accurately detect the risk factors, weigh how they are affecting our situation, predict the outcomes from the interactions between conditions, and balance options to choose the best game plan. It is this balancing act, plus knowing when to change the plan, that generates resilience and safety.

2.2 OPERATIONS AND RISK MANAGEMENT

Commercial airlines balance their operations between profitability and safety. While every company strives to avoid risk, market pressures steadily push them toward riskier choices. Resources are limited and expensive. Squeezing more out of available resources improves profitability. Actions that reduce revenue include cancelling flights, hiring extra personnel, grounding aircraft with multiple mechanical deferrals (MELs), reducing arrivals for anticipated marginal weather, and increasing discretionary fuel loads. Conversely, actions that enhance profitability include increasing the utilization rate of each aircraft, boosting employee productivity, pushing arrivals to beat approaching marginal weather, flying aircraft with multiple MELs, and reducing discretionary fuel loads. A compromise evolves with efficiency pushing the limits toward riskier choices and safety pushing back.

This dynamic is also affected by past successes. When risky operations succeed, they encourage repetition. "It worked last time, let's do it again." With each success, companies become more confident in their ability to manage risk even more tightly. The risk boundary becomes less scary and more approachable. Some organizations even pride themselves with how well they walk this tightrope. The same goes for pilots. The more risk-tolerant we are, the more aggressively we will explore ways to

DOI: 10.1201/9781003344575-3

complete our flights under increasingly risky conditions. Every time we succeed, it encourages us to do it again.

Along with market pressures, we have legality limits set by the regulator. These govern a wide range of qualifications, currencies, operations specifications, and operating limits. By their very nature, the regulator's limitations define absolute boundaries. Companies are allowed the discretion to operate inside of these boundaries even when a combination of conditions becomes quite risky. Weather limits are instructive examples. The regulator defines the point of deteriorating weather when operations must cease, but nothing prevents a company from operating up to the very edge of that limit.

Viewed individually, each regulatory limit makes sense. Combined with other factors, risk becomes unacceptable. Conceivably, an airline can pair their least-experienced Captain with their least-experienced FO, both flying at the limits of their operational duty day, on a MEL-degraded aircraft, in weather that is approaching regulatory limits, and at the performance limits of the aircraft. While no individual regulatory limit is exceeded, few of us would find this accumulation of risk factors to be acceptable. Unless airline leadership halts the operation, it falls to us pilots to accurately judge whether to continue the flight, or not. We are the final judges of acceptable risk. We are the final barrier.

2.3 PILOTS AS THE LAST DEFENSIVE BARRIER

No matter how risky or safe an operation is, pilots remain as the last barrier of defense against an accident. Recall the "Swiss Cheese" accident model we studied during ground school (Figure 2.1).

Developed by James Reason, it depicts the latent vulnerabilities that exist within each layer of defense as holes.[1] In an accident scenario, these holes align to allow a hazardous trajectory to pass through every layer of defense. We design safety programs and procedures to locate and shrink these vulnerabilities, thereby reducing accident probability. Normally, at least one defensive layer intercepts each hazardous

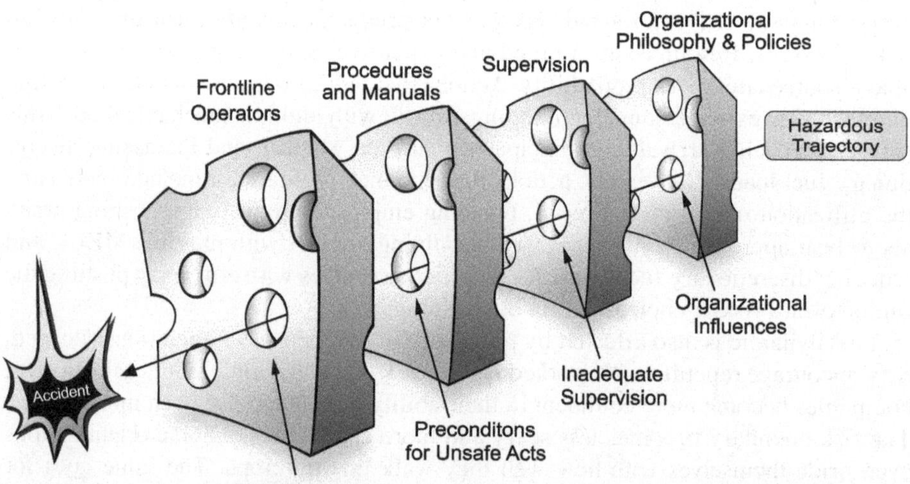

FIGURE 2.1 An aviation version of Reason's Swiss Cheese accident model.

trajectory and stops it before it can culminate in an accident, but no matter how careful an organization is, some holes always remain. We rely on resilient defenses-in-depth through a series of detection and mitigation strategies. Organizations erect early barriers through leadership, philosophies, and policies. Supervision, manuals, and procedures form subsequent barriers. Frontline operators form the final barriers. A risk-tolerant, profit-driven operation is modeled with larger holes. A risk-adverse, safety-driven organization is modeled with smaller holes.

Organizational and operational barriers compose *strategic* protections that address vulnerabilities across the entire flight operation. Leadership intervenes by making systemic decisions whenever risk becomes unacceptable. An example would be suspending operations at an airport ahead of an approaching ice storm. Pilots compose *tactical* barriers. We are well positioned to assess and balance local conditions. Strategic countermeasures tend to respond slowly and generally, while pilot defenses tend respond immediately and specifically. To use military or fire-fighting conventions, we are the on-scene commanders.

Since we represent this final layer of defense, most accident summaries list some reference to pilot error in their list of causes. Hindsight easily identifies actions that the pilots should have taken to prevent the accident. In the Swiss Cheese model, this is labeled as "Unsafe Acts" by the frontline operators. The inference is that if the pilots hadn't performed a particular unsafe act (or failed to perform a safe act), the accident would not have happened.

We are that final layer of defense between mitigated incidents and unmitigated accidents. Why is this so important? It's important because the earlier strategic defenses cannot manage the subtle assessments and agile corrections that flying in-the-moment requires. We are the only ones who can detect and assess the rapidly changing or complex conditions that might line up the vulnerability holes. By quickly detecting and shrinking vulnerabilities, we block almost every accident trajectory. Master Class skills and techniques shrink the holes even further.

2.4 PILOTS AS SEPARATE BARRIERS

To expand on the model, Figure 2.2 shows how the final frontline operator layer is actually three distinct layers – the pilot flying (PF), the pilot monitoring (PM), and the flight crew as a team. Given their separate responsibilities, each of these three barriers have their own strengths (solid barriers) and weaknesses (vulnerability holes). For standardization, we will consider a two-pilot crew.

- *PF strengths:* Since PFs are actively controlling the aircraft, they know what they intend to do at each moment. They continuously see, decide, and act. When their game plans and abilities appropriately match the conditions and objectives, risk is effectively managed.
- *PF weaknesses:* PFs can be poor judges of their own level of task saturation. When they become overloaded, they may succumb to mental biases and habitual behaviors that negatively affect risk management. As stress and time-pressure rise, the effects of these biases increase.
- *PM strengths:* Since they aren't actively controlling the aircraft, PMs are well positioned to assess the appropriateness and effectiveness of their PFs' actions. From their detached perspectives, they aren't locked into the same

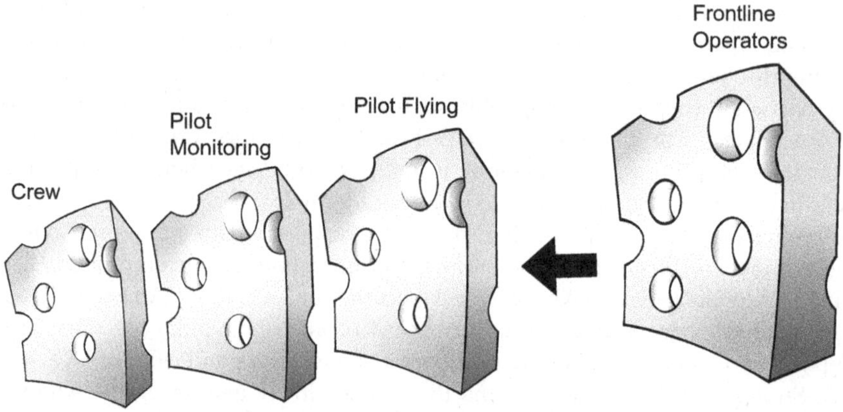

FIGURE 2.2 The three layers of frontline operator accident prevention.

assumptions and habits that may be adversely influencing the PFs' choices and actions. PMs can perform independent assessments and form different conclusions. This provides greater resistance against deceptive biases and habits that arise under stress. Because of this, PMs are often the first to detect adverse trends in a failing game plan.

- *PM weaknesses:* It is difficult for PMs to assess what PFs are thinking, especially since the quality of communication often degrades under stress and time pressure. PMs infer what their PFs must be thinking based on their actions. They silently observe and form assumptions that seem to match what the PFs must be planning.

2.4.1 The Flight Crew Team as a Third Barrier

Our policies, procedures, and guidance direct PFs and PMs to fulfill their duties independently, but also to coordinate their actions as a crew.

- *Flight crew strengths:* With active communication, pilots can share ideas, divide workload, detect adverse trends, and evaluate the viability of backup game plans. Numerous HF research studies confirm that effective crews consistently outperform disjointed crews.
- *Flight crew weaknesses:* There are times when confusion adversely undermines PF and PM roles.

2.4.2 Additional Considerations to Safety Barriers

A number of other factors affect our ability to interdict accident trajectories.

- *Familiar operations:* The vast majority of airline operations are standardized and familiar. Over time, success fosters a degree of laxity.[2] The more familiar and normal a situation, the greater our human tendency to relax our attention and focus. When crews recognize a predictable and familiar sequence during a flight, they believe events will unfold as they always have in the past. PFs can lower their vigilance and fly on mental autopilot. PMs can lower their vigilance, divert their attention away from flightpath monitoring, and become distracted with discretionary tasks.

- *Short time-to-fail phases of flight:* During certain phases of aircraft movement (like taxi operations or landing), the time-to-fail is very short. Safe operations require both pilots to remain fully engaged while controlling or monitoring the aircraft's path. What looks like a normal taxi path one moment can quickly become a hazardous deviation toward an obstacle. While momentary deviations are safely managed, longer deviations rapidly elevate risk. As FOs, we monitor the aircraft's taxi path, but we also subjectively monitor how well the Captain is controlling that path. Are their corrections accurate? Are they smooth? Is the Captain looking outside or distracted by something inside? While we wouldn't think of diverting our attention from monitoring the aircraft path while landing, we sometimes divert our attention during taxi operations because the slower speeds feel less risky.
- *Unexpected changes:* Fast-changing events, especially when they are unfamiliar, unexpected, or novel may cause a crew's coordination and communication process to break down. For example, consider wingtip strikes during taxi movement. A crew expecting to taxi to their gate is directed to hold out until the gate clears. Maneuvering on an unfamiliar holding pad, both pilots may divert their attention inside referencing their charts. Entering unfamiliar areas, Captains may focus their attention on their intended direction of turn (looking right while turning right or looking left while turning left). While turning one way, they swing the opposite wingtip into obstacles. FOs are equally challenged. Even if they are trying to clear the threatened wingtip, they are not controlling the aircraft. They cannot assess the track of a wingtip without knowing how tightly the Captain intends to turn (Captain turning left and looking left with the FO monitoring the right wingtip as it swings toward obstacles, for example).
- *Failed anticipation:* In some events, PMs sacrifice reaction time while trying to anticipate what their PFs are thinking or doing. Instead of speaking up, PMs assume that their PFs must be making appropriate choices. Lacking the full picture, they feel confusion. They effectively become spectators of the event, not active participants. This is an especially powerful effect when FOs are inexperienced and assume that their experienced Captains must know what they are doing.
- *Misaligned perspectives:* The PF/PM system is designed to keep both pilots independently assessing the aircraft's path. This separation makes it easier to detect emerging errors. When the PF and PM become perfectly aligned in their thinking, they risk a case where both pilots miss the same warning signs while making the same mistakes. It would be like purposely aligning the holes of both Swiss cheese slices to allow an error to pass through. This unintentional alignment is common in many accidents and incidents.

2.5 ASSESSING RISK

Our ability to manage risk depends on how well we assess it. This is affected by pilot personalities and risk assessment strategies. First, some assumptions:

- Pilots do not intentionally choose to commit serious errors or cause mishaps.
- Most mishap events involve pilots with no history of past risky behavior.
- Most pilots are sufficiently proficient, experienced, and knowledgeable before their mishap events.
- Pilots choose every decision and action that they take.
- Pilots had previously performed the intended flight maneuver successfully many times in the past.

If these assumptions feel familiar, they should. In many ways, mishap pilots are just like the rest of us. They are skilled, proficient pilots who found themselves in unfavorable situations. This is why we will focus on what happens to deflect pilots toward unfavorable event trajectories as a means toward improving our own Master Class skills.

2.5.1 Risk Management Model

All of our aviation choices require us to manage risk. While we strive to account for all conditions, many remain unknown. Most of the time, we accept the existing level of risk and manage our game plan to avoid adverse consequences. Our choices range from conservative to hazardous. Consider Figure 2.3.

Starting at the left, risk conditions steadily increase as we move toward the right. All of our risk management strategies fall somewhere between these extremes. From left to right, this graphic depicts three zones and two lines.

- *Lower risk zone – conservative:* Risk management strategies at the far left area reflect a conservative bias. Adverse to risk, these pilots add their own safety margins on top of what the company, aircraft manufacturer, or regulator already impose. Pilot decisions in this regime might include carrying

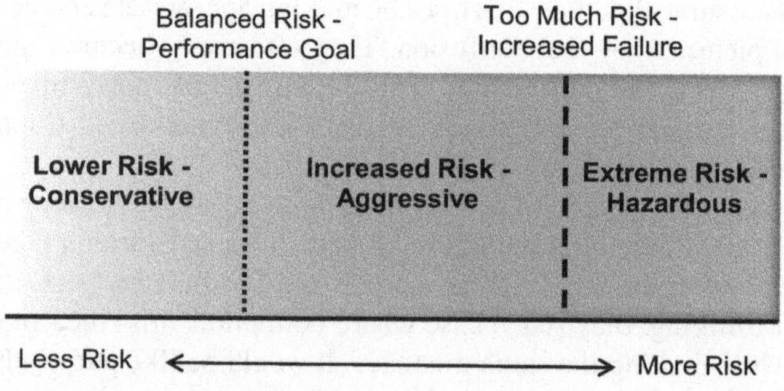

FIGURE 2.3 Risk continuum with pilot risk personalities.

more fuel than is deemed appropriate or customary. Pilots might slow and configure for landing earlier than their contemporaries. Since adding an additional safety margin wastes time and resources, organizations generally do not encourage or train to these conservative standards.

- *Balanced risk line – performance goal:* Organizations train to a standard that balances efficiency and risk. This reflects a safety margin that matches their corporate goals. A risk-adverse organization promotes choices that would push the line further to the left. A risk-tolerant organization would encourage choices that push the line further to the right. The risk strategy of a company is guided both through their published/trained standards and through their organizational culture. Sometimes, these conflict. An organization that looks risk adverse on paper might quietly promote a risk-tolerant frontline culture. Ideally, the company aligns their written and cultural standards to promote a consistent standard for all operations. Fully compliant and professionally motivated pilots strive to select choices that fall near this Balanced Risk – Performance Goal line.
- *Increased risk zone – aggressive:* When pilots routinely choose riskier options, their choices fall in this center zone. There are several reasons why. First, pilot personalities may bias them toward riskier choices. Examples might include continuing approaches with convective weather affecting the landing runway or taxiing at higher speeds on wet taxiways. Second, riskier conditions may be imposed by outside agencies. Examples include dispatch trimming fuel loads down to regulatory minimums to reduce weight or ATC assigning tight vectors and higher airspeeds before clearing pilots for visual approaches. Third, riskier conditions may arise from unforeseen or unanticipated conditions. Examples might include an unexpected marginal weather or unreported reduced stopping performance on an icy runway.
- *Too much risk line – increased failure:* As risk increases, we reach a point where the probability of failure begins rising steeply. Choices to the left of this line may result in close calls and incidents, while choices to the right risk mishaps or accidents. While shown as a vertical line on the model, it marks a range where the probability of failure begins to rise sharply. In later graphics, we will remove this line and consider this as a range of increasing risk shown by darker shades of gray.
- *Extreme risk zone – hazardous:* The final hazardous risk zone indicates a range of high probability of failure. While failure is not certain, more events will result in failure. For perspective, at the increased failure line, maybe 1% of pilots might experience failure. At the far right of the shaded area, 99% might experience failure.

2.5.2 Classification of Pilots by Risk Tolerance

To compare pilot risk tolerance, let's remove everything but the balanced risk line and a deepening gray zone of rising risk and probability of failure. Pilot risk management strategies center along their particular bias resulting in a skewed bell curve.

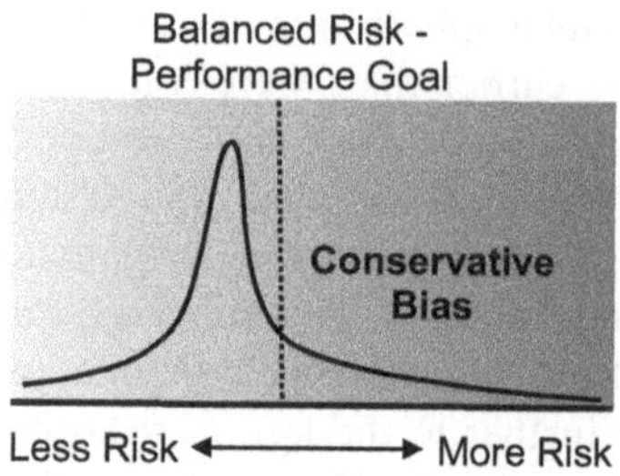

FIGURE 2.4 Conservative pilot's risk management strategy.

- *Conservative pilots:* In Figure 2.4, pilots who favor conservative choices skew their choices to the left of the balanced risk performance goal line. In trying to avoid risk, they choose more conservative options that preserve their personal safety margins. Rarely do they find themselves in the deepening gray area of riskier operations. For example, they may slow and configure earlier, divert more often, add more discretionary fuel, and refuse mechanically degraded aircraft more often than their contemporaries.

  As a consequence of their conservative shift, these pilots rarely encounter risky events. Ironically, this lack of exposure also results in lack of practice. As an unintended consequence, they may be less skillful at managing more-challenging situations. On the rare occasions when they may find themselves faced with difficult profiles, they may struggle.
- *Standardized pilots:* Standardized pilots (Figure 2.5) attempt to follow company guidance that balance priorities of efficiency and risk. They balance their risk management strategies between conservative choices and aggressive choices, as the situation dictates. As compliant pilots, they accept the need to find a way to conduct their flights even when conditions push them toward riskier options (marginal weather, increased loads, tighter turn times). Compared with conservative pilots, a higher percentage of their choices end up flying profiles that fall within the gray zone of rising risk.

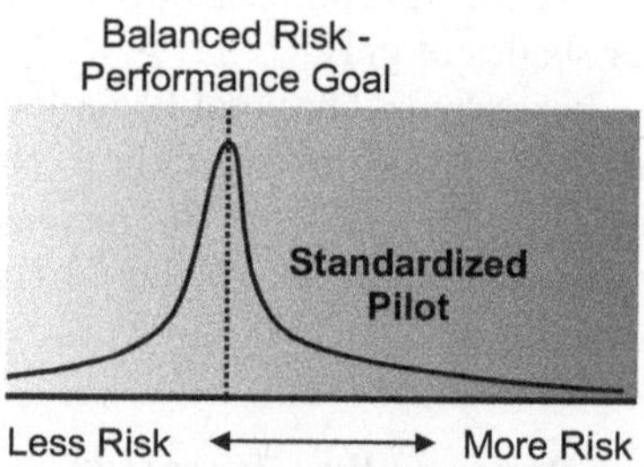

FIGURE 2.5 Standardized pilot's risk management strategy.

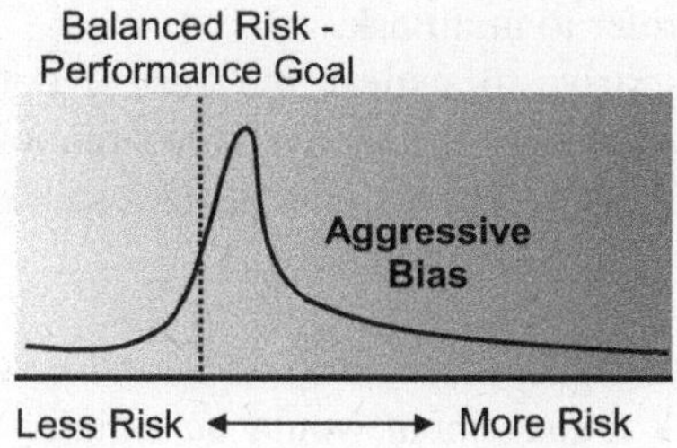

FIGURE 2.6 Aggressive pilot's risk management strategy.

- *Aggressive pilots:* Some pilots tend to favor riskier choices (Figure 2.6). They know the performance goal but believe that they can squeeze more efficiency from the operation. They may find existing safety margins to be overly restrictive. They trust in their personal ability to operate safely. Sometimes their motivations are company-friendly, like reducing discretionary fuel loads to lighten aircraft weight, improve efficiency, and increase profits. Sometimes their motivations are personal. By making their flight profiles more challenging, flying them feels more exciting and personally rewarding. While these aggressive pilots are often judged to be superior aviators by their peers, they are unsuitable role models since their choices often rationalize company standards to justify their bias toward riskier choices. In fairness, they are not inherently reckless or unsafe. Many aggressive pilots fly their entire careers free of mishaps.

2.5.3 The Gray Zone of Increasing Risk

On each model, pilots encounter some scenarios that fall within the gray zone of increasing risk. This is unavoidable. Even conservative pilots can find themselves operating in increased risk, make errors, and fail. The tightness of each pilot's bell curve tends to reflect their experience level. Less-skilled pilots might exhibit flatter, wider bell curves since they lack the refined judgment to accurately match their choices with the conditions. Highly skilled pilots might exhibit steeper, tighter bell curves since they skillfully match their choices to the existing conditions.

The bottom line is probability. The further to the right we choose to operate, the higher our probability of failure. Here are some additional characteristics of pilots operating across the gray zone.

- Conservative, low-risk operations tend to be easy, slow, and methodical.
- Aggressive, high-risk operations seem more time-pressured and rushed.
- Conservative pilots prefer to process information piece by piece and do one thing at a time.

- Aggressive pilots prefer to multitask.
- Standardized pilots strive to achieve the best of both sides. They match their attention and effort with the needs of the situation. They ramp up their aggressiveness when conditions demand faster response and favor conservative choices when workload is low.

If we could gauge where the gray starts deepening and how quickly the probability of failure is growing, risk management would be simple. We would simply select choices that stay clear of the deepening gray. The truth is that the gray zone is actually a variable, turbulent area that proves to be difficult to measure as we are flying. A number of factors affect the grayness of the risky failure zone.

- *Complexity:* Risk accumulates and compounds. Risk factors combine in unpredictable ways. We may successfully handle two smaller risk factors but fail when a third is added.
- *Novelty:* A novel situation carries greater risk because we lack applicable knowledge. The more unfamiliar the situation, the less direct experience we have to draw upon. We might err by applying inaccurate assumptions between past experiences and the current one. Treating them equally can prove to be an unreliable strategy.
- *Time assessment:* We may misjudge the time needed to handle the workload with an unfamiliar event. Risky scenarios seem to move faster and take longer than we expect. We find ourselves falling behind which further increases time pressure.
- *Fatigue:* Fatigue impairs our ability and motivation to accurately assess risk and weigh options. When we are tired, complex decision making takes more effort than we feel like expending. Choosing an expedient, familiar game plan seems quicker and easier. Even when we recognize our inadequate decision making, we may lack the motivation to evaluate better options.
- *Startle and surprise:* Unexpected events startle or surprise us. While we are startled, our risk assessment process shifts from conscious assessment to gut-felt reactions.
- *Expectations:* Expectations strongly influence our assessment. The stronger our expectation, the less willing we are to notice or accept information that contradicts it. Expectations encourage rationalization over rational thinking.
- *Outlook or mindset:* If we are too optimistic about our decisions in a situation, we might be less willing to accept indications that our game plan is failing. If we are too pessimistic, we might abandon an otherwise viable game plan.
- *Assessment skill:* Some people are better-skilled at assessing risk than others. Pilots who believe they are good risk assessors are more confident with their decisions. Coupled with an aggressive bias, this may lead these pilots to make quick decisions based on incomplete information. Less-skilled risk assessors, coupled with a conservative bias may employ prolonged analysis and have difficulty choosing a game plan.

2.5.4 Changing Conditions within the Gray Zone

Real-world conditions make the location of the failure zone fluid, variable, and ultimately, unpredictable. We can sense when we are approaching it, but can't reliably gauge how deeply into the gray zone that we have moved. The more that conditions are changing, the more difficult our assessment becomes. Consider an example of deteriorating arrival weather. Conservative pilots may choose to avoid the risk by entering holding. While this gives them time to assess options, waiting too long can drive their fuel so low that they end up diverting. Aggressive pilots may try to rush their approach to beat the deteriorating weather and encounter windshear or breaking action hazards. Even good, standardized pilots can find themselves challenged trying to select the best choice.

2.5.5 The Proficient Pilot's Path to Failure While in the Gray Zone

The path to success and the path to failure begin at the same point. Since we pursue and achieve success every day, we rarely see or practice failing events during line-flying. If all we ever do is manage successful outcomes, how can we learn what failing scenarios look like, feel like, or how they evolve? In training, we routinely practice failing scenarios. While useful, simulator training scenarios don't match the feel of real-world line flying. No matter how experienced we are, encountering a failing game plan during line-flying is a rare and surprising event. Consider the following story.

BOX 2.1 CLEM'S BAD DAY – A VISUAL APPROACH GONE WRONG

Clem is a proficient, typical line-flying Captain. One day, he was on vectors for an airport with a fairly short runway. Earlier than he expected, ATC cleared him for a visual approach. Free to maneuver on his own and running behind schedule, he chose to press closer the field and delay configuring the aircraft. When he finally began slowing for the approach, he realized that he had misjudged his energy state. He was closer and faster than he wanted to be. He took steps to shed the excess energy. He pulled the thrust levers back to the idle stops, directed his FO to lower the landing gear, and extend maneuvering flaps. This helped. He intercepted glideslope and started down. The approach was tight, but he felt like he could still salvage it. Then, he noticed that his speed was not decreasing as expected. He shallowed his glidepath and slowed enough to extend full flaps. At this point, he was fully configured, still fast, and trending high on glidepath. He reduced pitch to rejoin glidepath, but this caused his speed to stagnate about 20 knots fast. Running out of time and ideas, he decided that his parameters were close enough to continue. He chose to land fast and deplete the excess speed during rollout. He relied on a ATIS-reported 10-knot headwind to help with slowing. Unfortunately, the winds had switched to a light tailwind by the time he reached the flare. He overcontrolled his flare.

He felt that it was not bad enough to warrant a go around, so he continued even though he overflew his planned touchdown point. Increasingly frustrated, he eased the wheels down for a smooth touchdown about 3000′ down the runway. He lowered the nose, initiated reverse thrust, and engaged wheel braking. He felt normal deceleration which restored his confidence that he would successfully stop. Unfortunately, there was a latent hazard that he had not anticipated. A light rain had fallen on the airport just before his arrival. This was the first rainfall following many weeks of dry, dusty weather. While the touchdown portion of the runway appeared mostly dry, a slippery layer of water, reverted rubber deposits, and dust contaminants covered the last 2000′ of the runway. Clem's first indication of this hazard began when his brakes started antiskid cycling. Next, the aircraft began sliding. He maintained aircraft alignment, but slid past the end and stopped well into the overrun. Until this event, Clem had enjoyed an unblemished safety record.

With this example, we recreate a fairly typical sequence of events leading up to a mishap. Using our event review process, let's avoid judging or labeling this event exclusively using hindsight. Sure, Clem clearly exceeded his company's stabilized approach criteria and should have gone around either because of the unstabilized approach or the long landing. In reality, many of us typically make these same kinds of snap judgments as Clem did. So, while not condoning his decision to continue, let's endeavor to understand it. Our first review of the event reveals a typical visual approach with a crew that becomes rushed.

With our second review of the event (imagined while sitting on the jumpseat), we sense the point when Clem first realized that he was behind. We understand his sense of urgency to deplete excess energy, that he decided to land anyway when he thought he had handled the landing successfully, his surprise when the autobrakes begin cycling, and his disappointment with causing the mishap.

With our third review of the incident, we join Clem in the Captain seat and mentally recreate the in-the-moment experience of flying this approach. We become focused on solving problems as we detect them. Noticing that we are too fast, we increase drag by lowering gear and extending flaps. These are normal corrections we do anytime we need to deplete excess energy. This drives us too high on glideslope, so we decrease our pitch to rejoin glidepath. At each point along the approach and landing, we detect problems and take reasonable measures to mitigate them – normal pilot stuff. What is different with this event is that as soon as we solve one problem, another one seems to crop up. We never fully solve our approach energy problem. On short final, we make a crucial decision. We decide that we have done as much as we can to stabilize the approach and choose to land. Even this is not an uncommon decision. It happens every day. What is unknown, and highly consequential, is a latent vulnerability of slick pavement on the last third of the runway. The long and fast landing carries us into a portion of the runway where no aircraft had previously ventured. They all landed normally and turned off at the high-speed exit, so no one experienced or reported the reduced braking hazard at the far end. As we roll onto this pavement, we encounter a condition that is fundamentally different than the rest of the runway and is beyond our aircraft's stopping capability.

In hindsight, we know that all of Clem's problems stemmed from the same cause – an unsalvageable, unstabilized approach. Clem did not realize this at the time. Every action he took felt like the correct steps to solve familiar approach problems. He detected and responded to problems like we all do. The process should have worked, but it didn't. This is because, while his corrections were appropriate, the unstabilized approach generated conditions that exceeded the effectiveness of all of his combined efforts. He fell into a Recognition Trap.

2.6 THE RECOGNITION TRAP

How do good pilots like Clem make mistakes like this? When we examine this scenario in hindsight, we see all the places where he should have recognized his energy problem. While he was flying in-the-moment, he didn't foresee the adverse parameters as warning signs. He saw them as correctable deviations. In the end, he pursued an unachievable goal to an unfavorable outcome. Why would an experienced, capable pilot like Clem choose an unsuccessful decision path over safer alternatives?

2.6.1 Recognition Primed Decision Making

Gary Klein's model of Recognition Primed Decision Making (RPDM) models the decision-making behavior of experts. RPDM describes how experts view a situation and quickly assess whether it matches a similar event from their past experience. In this model, pilots like us follow this following thought process[3]:

- Detect the indications generated by the situation.
- Recognize the situation as something we have experienced before.
- Select a game plan to move from our current position toward our goal.
- Run a mental simulation to see if the game plan fits.
- Select the decisions needed to match the game plan and simulation.
- Identify the action steps needed.
- Follow this familiar set of steps and decisions to match that simulation.

In Clem's case, being cleared for a visual approach was something he had experienced many times before. Moreover, it was a scenario he had routinely and successfully handled. When he *recognized* the situation, it *primed* him to use the same *decision making* that worked for him in the past. Conversely, if he faced a scenario that he didn't recognize, he would have known not to apply his past decisions to what appeared to be a novel situation.

As airline pilots, we typically operate in familiar environments. A seasoned airline pilot has effectively, successfully, and reliably performed the same task over and over again. In fact, many airline Captains operate for years without ever experiencing a failing approach like the one that Clem encountered. We are highly confident in our ability to adapt to changing conditions to achieve consistent, successful results.

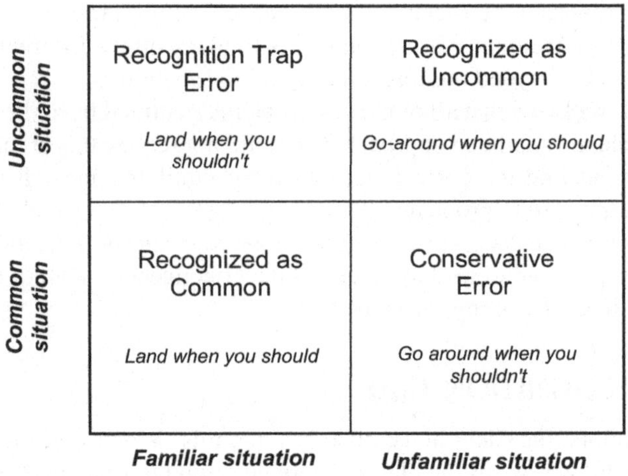

FIGURE 2.7 Situation recognition options.

On the other hand, because we are so rarely exposed to failure, many of us have little practice with detecting and rejecting failed decision paths. Our daily exercise of good judgment is demonstrated by our ability to deal with changing conditions and to achieve success, not by our ability to detect and respond to failing decision paths. It is precisely this inexperience, combined with our confidence in our judgment which triggers the Recognition Trap. Consider Figure 2.7.

The horizontal axis reflects whether we are familiar with the situation or not. The vertical axis depicts whether the situation is commonly encountered or not. The four blocks break down as follows:

- *Recognized as common:* This lower-left block represents most common line-flying events – pilots experiencing expected parameters and achieving consistent success. Pilots recognize these situations and follow proven, reliable game plans. The example cited in the block is a stabilized approach where the pilots choose to continue and land.
- *Recognized as uncommon:* This upper-right block represents situations that the pilots recognize as uncommon and are actually uncommon. Knowing that the parameters don't match a stabilized approach or favor a landing, pilots make the appropriate go around decision. While this isn't the original goal of the approach, it is still considered a successful outcome.
- *Conservative error:* This lower-right block reflects the fairly rare case where pilots detect an unfamiliar situation that is actually common. In the cited example, they choose to go around even though parameters favorably support a normal landing. Technically, this is an error since the pilot misidentifies the situation. This error is wasteful, but rarely results in serious consequences.
- *Recognition Trap error:* This upper-left block is the most consequential. It depicts an uncommon/unfavorable situation which the pilot incorrectly

identifies as common. This is the kind of error that Clem made. He thought he recognized an approach that was "good enough" to be salvaged. Unfortunately, the parameters were too extreme. His failed recognition trapped him into a pursuing an unachievable game plan. Recognition Trap errors can be highly consequential. Clem was able to stop the aircraft in the overrun with minimal damage, but he conceivably could have departed the pavement, damaged the aircraft, and injured passengers.

2.6.2 Recognition Trap and Rationalization

Referring to Figure 2.3 earlier in this chapter, Clem's Recognition Trap profile fell within the deep-gray portion of the risk zone. He didn't realize it at the time. His only option to avoid the mishap would have been to abort the approach and go around. Instead, he continued with his familiar game plan for a visual approach to landing. To be clear, Clem did not have any intention to place his aircraft in a hazardous position. He was a good pilot who mistakenly recognized this situation as something familiar and manageable. Perhaps, his internal dialog went something like this:

- *Cleared visual. I'll keep the speed up a bit longer.*
- *Too much. Better slow it down. Thrust to idle.* "Landing gear down, flaps [intermediate setting]."
- *Intercepting glidepath. Still too fast. I'll shallow my glidepath a bit to get more flaps out.*
- *Good. Now, more flaps.* "Flaps [full landing position]."
- *Now, I'm too steep. I just need to push over a bit and rejoin the glidepath.*
- *Good. I'm on glidepath, but still too fast. The speed should bleed off before landing.*
- *Why isn't the speed dropping off like it should?*
- *That didn't work. Too late, now. I'll land and get on the brakes during rollout.*
- *I over-flared. Now, I'm floating long. I'll just ease it down.*
- *Nice touchdown. I landed long, but I still have good reversers and braking.*
- *Good. The aircraft is slowing. I'll make the last turnoff at the end of the runway.*
- *Whoa! Brakes aren't working. I'm skidding! I've got to keep it tracking straight!*
- *[Aircraft slides into the overrun]. I can't believe that just happened!*

Notice that with each problem, Clem had a solution. Each solution appeared to help, but never quite enough. He recognized that he was high and fast on a visual approach final, but he had been there before and knew what to do to fix it. He thought that he could duplicate successful experiences from his past and land the aircraft safely. He applied normal corrective inputs and expected them to solve his energy problem. They worked before, so they should work again. But that day, they didn't.

We can see from Clem's mental dialog that he continued to receive positive reinforcements that his corrections were working. He responded to each individual approach challenge and remained confident that it would all work out in the end. Notice his rationalization. Instead of detecting, assessing, and choosing, he became surprised, tunnel-focused, and reactive. The successful game plan from his mental simulation didn't match what was happening. He became consumed with working faster and faster to try and stay ahead of the deteriorating situation. At no point did he step back and challenge the validity of his original game plan. No parameter was strong enough to snap his attention away from continuing to land. Instead of looking for indications that his approach wasn't working, he looked for (and found) positive reinforcement to continue.

Under the rising stress of this accelerating situation, his mind rationalized good reasons to keep going. When the aircraft stopped in the overrun, Clem looked back in disbelief. With the clarity of hindsight, he realized that his approach was unstabilized by company standards. He should have gone around. He should have anticipated the stopping problems on the slick pavement on the far end of the runway. He wondered, "What was I thinking?" In truth, he wasn't thinking. He was reacting. He was flying the aircraft and solving approach problems in-the-moment.

2.6.3 Questioning Our Judgment Replaces Situational Assessment

Clem is a good pilot. He thinks like a good pilot. Unfortunately, even good pilots can use flawed risk assessment. Consider Figure 2.8.

The first block follows the RPDM process. The good pilot assesses the situation, matches it to familiar experiences, selects a game plan, validates it through mental simulation, and assesses whether it is working. If it is, then great. If it isn't, they start

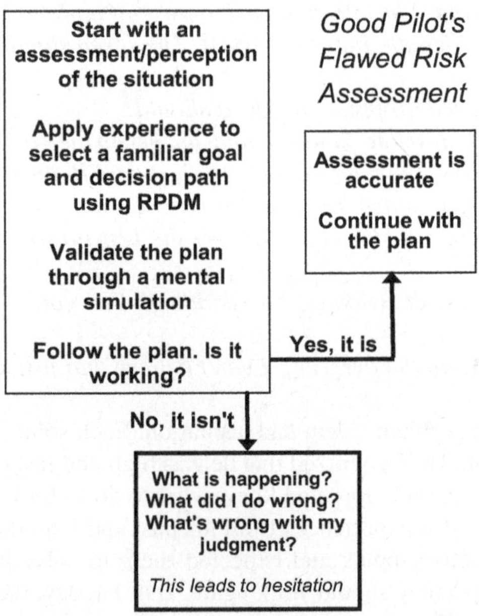

FIGURE 2.8 Good pilot's flawed risk assessment.

by questioning themselves. What is happening? What did I do wrong? What did I miss? Valuable time is lost to confusion, internal dialog, and self-doubt. Our mental focus is diverted from actively solving the problem to self-assessment. Time pressure leads to more rationalization. We handle the problem using split-second reactions instead of rationally choosing our actions. We fall behind the aircraft. Our situational awareness vanishes.

2.6.4 Pilot Recollection of Risk during Mishap Investigations

When assessing their event in hindsight, few mishap pilots could accurately recall what was happening at the time. Instead of specific aircraft parameters, they remembered general impressions like being "fast" or "high on glidepath". When asked to estimate specific parameters, they typically under-assessed the severity. Their recollection tended to bias toward what their parameters should have been. The values they could recall tended to be limited to one or two particular parameters to the exclusion of all others. In the following event, the crew mishandled the approach, executed a go around, but then became task-saturated, mismanaged their automation, and found themselves in an undesirable attitude. Notice how many of the details of the event were completely missed or confused. Also notice how the Captain seemed to devote a great deal of time and effort trying to understand where they went wrong and how they contributed to their problems.

BOX 2.2 CAPTAIN LOSES SA, THEN EXPENDS UNNECESSARY EFFORT TRYING TO UNDERSTAND WHY

Captain's report: … So, with the interrupted calls for flaps and gear, I had to ask for the flaps to 3 again, maybe twice (I really can't remember) because I got stepped on by the ATC controller, and then gear up, and then a frequency change call, which we didn't get the frequency – *because so much was happening so fast and with 4 ATC calls on a real go around, gear up call again, I was really losing SA (Situational Awareness).* … As we acquired "SPEED ALT" I noticed that we also had the "TOGA LOCK" FMA which totally confused me – where did that come from? *Did we get an Alpha floor event? I'm thinking, no way.* If we did, it must have been in the rotation to GA [go around] attitude because neither of us had any idea where an alpha floor event happened. *I'm thinking to myself as all this messed up go around is happening, "I must be really screwed up and fatigued if I just caused an alpha floor event and I didn't even notice it."*

In our debrief we both agreed that with the intrusive ATC calls, the interruption in the go around procedures, the missed frequency change, the TOGA lock mystery, the uncommanded (or unintended) descent and turn, the extra ATC calls, [the] frequency change, and confusing instructions on go around clearance resulted in us both being overwhelmed and stressed and apparently getting the airplane in an undesired state, all starting with a missed glideslope intercept. *We discussed, in detail, all the events that happened and realize that we most likely missed some things. [I] couldn't even recall most of it. It was a very stressful event and very disturbing to both of us because of all the things that happened that we didn't expect, didn't understand and couldn't even explain afterward.*[4]

2.7 HANDLING AVIATION THREATS

How well we manage risk depends on how well we detect threats. As professional aviators, we expect a high level of performance from ourselves. Likewise, society expects us to always deliver safe travel regardless of conditions. When we fall short of our expectations or theirs, we feel disappointment and disbelief. Consider the following report from an FO returning to flying following an extended absence.

Notice the self-recrimination and self-doubt in the FO's report. For perspective, the error was rather insignificant. The FO missed a required call to a ramp control agency which was easily remedied. The rest of the pilot's report outlined the underlying reason for missing the ramp call to include an unexpected taxi routing and distraction.

BOX 2.3 FO STRUGGLES TO REGAIN SKILLS FOLLOWING LONG ABSENCE FROM FLYING

FO's report: Upon arrival to Gate X in ZZZ we were asked to call OPS. The Captain called them and we were reminded that we should call Ramp Control before entering the ramp next time. *I couldn't believe that I had made such a mistake after over XXXXX hours on the fleet and almost XX years of flying at the company with many, many ZZZ arrivals. I looked at the Captain and I said, "I am sorry I have no excuse," which is how I felt. I then called the ramp and apologized. ... As we debriefed, I started questioning the why of my mistake and I realize that I am taking for granted my lack of situational awareness as a result of my 7 month absence from flying. ...* I was still very conscious of my weaknesses because of my lack of recency, but now as I am finishing my second trip after OE I feel that "I am back" which obviously I am not.[5]

In their excellent book, *The Limits of Expertise: Rethinking Pilot Error and the Causes of Airline Accidents*, the authors examine 19 airline accidents and mishaps with a particular focus on the human factors influencing the pilots.[6] They document the clear difference in pilot psychology between normal, familiar situations and stressful, unfamiliar situations. Time and again, we witness deliberate, rational, engaged pilots reduced to task-saturated, reactive, confused pilots when confronted by unique, complex, and time-pressured situations. As professional airline pilots, we expect ourselves to rise above every challenge, and in most cases, we do. Still, every situation contains vulnerabilities. Our inherent biases, limitations, and weaknesses remain. Stress brings them to the surface. Just because we have successfully overcome all aviation obstacles for years doesn't mean we are ready for whatever may come today. The complex interactions between conditions create completely unpredictable outcomes. We can be good pilots our entire career and still become completely surprised by the exceptional combinations of events.

2.7.1 Levels of Knowing

When detecting threats, we need to understand what events to look for and what they mean. This assumes that we possess specific knowledge and recognition. To understand this better, we'll start with a review of four classifications of "knowing":

- *We know what we know:* This is, by far, the most common classification of events for experienced pilots. As novices, this category starts out small because everything is new. It quickly grows as we gain experience. As experienced pilots, most of what we see are events that we recognize. We've been there. We've done that. We have a full inventory of past successful game plans to draw upon. Whether we tend toward conservative, standardized, or aggressive risk management strategies, we have clear ideas of what works for us.
- *We know what we don't know:* This category is much smaller. As we gain experience, we encounter fewer new and unique events. Even if we haven't directly experienced something, chances are that we have heard about it from other pilots. We remember what worked for them and what didn't. If we do encounter something completely new and accurately recognize that it is completely new, we conclude that it is a unique situation that may require a unique solution. This gives us a starting point for innovating new game plans.
- *We don't know what we don't know:* This is where we can begin to run into trouble. Encountering a unique event, we feel uncomfortable with not understanding what is happening. We pride ourselves with being knowledgeable, decisive, and capable. When we encounter a new event, many of us tend to simplify our assessment of the situation to make it more manageable. Rather than innovate a plan to fit the new case, we assume that the current situation is close enough to a familiar case that we successfully handled in the past. Like a worker with limited tools, we try to make do with the few familiar tools we have instead of acquiring an appropriate tool. This may lead to adopting flawed assumptions and taking shortcuts.
- *We think we know what we actually don't know:* This is Recognition Trap territory. We misidentify a unique situation as something we recognize and confidently select a familiar game plan to solve it. We don't even recognize that this event is something new. We confidently select the familiar (flawed)

plan and force it to work. Unfortunately, it doesn't. The harder we force the failing plan, the more difficult and time-compressed the event becomes. Instead of questioning or replacing the flawed plan, we question ourselves and our judgment. "Why isn't this working?"

2.7.2 Threat Detection, Game Plans, and Confidence Level

As risk rises, situations become more unpredictable. The more unpredictable a situation, the less confident we should feel about it. Some pilots miss this. They assert that they should possess high confidence in all situations. Ego adversely affects honest assessment. Misplaced confidence actually promotes biases which encourage us to push failing game plans too far and too long.

- *Measuring risk versus assessing how it is trending:* While flying the aircraft in-the-moment, it is difficult to consistently rate the level of risk. Even on the same flightdeck, each pilot may report widely different assessments. This can become problematic since we usually don't have time to discuss mismatched assessments during time-pressured events.

  A more-consistent and reliable assessment of risk is whether it is rising or falling. Logically, we expect our confidence level to fall as we perceive that risk is rising. It's important to distinguish that we are not talking about self-confidence. We are referring to the confidence we have in our risk assessment – confidence in how well our game plan fits the conditions. If we have a simple, common situation (great weather, daylight, fully stabilized approach) and accurately identify it as a simple, common situation, then we are highly confident with our game plan. As we sense that risk is rising, we need to acknowledge that our confidence in our plan should drop accordingly.

  Referring back to our risk model (Figure 2.3), as risk rises, our confidence drops, and the further into the gray zone we find ourselves. We find it difficult to know how gray or risky a situation is, but we can always sense our *movement* to the left or right within the gray zone.
- *Adding "buts" to our game plan:* As our confidence in our game plan drops, we compensate by amending it with qualifiers. For example, "We have a great weather, daylight, and a fully stabilized approach, *but* we're 5 miles behind a heavy aircraft landing on our same runway." The possibility of wake turbulence increases the risk and reduces our confidence in the game plan. We compensate by adding contingency options, "We'll still fly the approach, *but* we'll look for warning signs of wake turbulence and be ready to go around." We might choose to mitigate the threat by using countermeasures such as offsetting slightly upwind or flying slightly above the heavy's flightpath. This also opens a CRM opportunity for PFs to share which indications they will use to trigger a go around.

2.7.3 Warning Signs

Small divergences from our game plan are normal. As long as they remain small, we make appropriate corrections and continue. Perfect stability is unachievable, so we constantly apply corrections to our path. When divergences become more intense or when our corrections become ineffective, something larger is at play. We recognize these increased deviations as warning signs that our game plan may be failing. Some of these warning signs are:

- *Increasing magnitude of corrections:* Imagine a situation where we are on glidepath and we fail to notice that the relative wind is shifting from a 10-knot tailwind to a 15-knot headwind. As the wind shifts, it causes our indicated airspeed to increase. We reduce thrust. We probably also experience increased turbulence as we pass through the transition layer. This affects our runway aimpoint, so we redirect our attention to restore our path. When we return our attention to our airspeed, we notice that it is still high – maybe even climbing higher. We reduce thrust further. Failing to detect the wind shift, we might perceive this as a warning sign.

  Now, imagine the same scenario except that we detect the wind shift. The thrust corrections and the turbulence would now seem completely normal. We wouldn't perceive our airspeed problem as a warning sign. Knowing the cause of the indication changes how we assess it. Sensing a warning sign means that something is happening that we may not fully understand. Time allowing, warning signs encourage us to investigate. When time is short and the deviation is significant, it may signal a deep excursion into the gray, high-risk zone. We might consider aborting the game plan.
- *Quickening:* Increase the intensity of the previous wind shift example. Imagine that the wind shift was actually a windshear event caused by a developing microburst. Anticipating a 25-knot tailwind-to-headwind shift (from a 10-knot tailwind to a 15-knot headwind), we would form a game plan for how much thrust correction will be required. What if, instead, the microburst causes a 50-knot wind shift? Entering the shear, we make our planned thrust reduction. We notice that the airspeed is continuing to rise, so we reduce the thrust levers to the idle stops. Next, we notice that the airspeed is continuing to rise and that we are drifting high on glideslope. The warning signs continue to escalate in frequency and intensity. We make our corrections faster and larger. This quickening is a sign of a failing game plan. We may not know the cause (windshear versus wind shift), but we do recognize that something is wrong. Again, while we may inconsistently judge how risky our problem is, we are very good at recognizing a worsening trend. When we start making corrections faster and faster while our problems continue to mount, it is a warning sign of a failing game plan.
- *Running out of time:* A co-factor of quickening is time-compression. Events feel like they are moving faster. Pilots sometimes rationalize this effect because, even in normal situations, events like landing naturally quicken as we sequence from configuring, intercepting final, short final, flare, and

touchdown. We use our experience to guide how to gauge this pacing. When the cascade of problems exceeds our expected pacing (quickening), it is a warning sign.

- *Reaching the stops:* Another warning sign is reaching the stops of our available corrections. To continue with our windshear example, if we reduced power to the idle thrust stops for a sustained period and it doesn't solve our problem, it is a warning sign. We rarely operate against the stops in the airline-category aircraft. Aircraft are engineered to maintain operational margins "around the middle". Take landing flaps, for example. While they give us the benefit of reducing our approach pitch angle and reducing our landing speed, they also require a "middle" thrust setting during a stabilized approach – not too high and not too low. This gives us an ample range for making corrections requiring more or less thrust. Whenever conditions force us out of this middle range, it feels strange.
- *Trading one problem for another:* Pilots encountering failing game plans sometimes choose strategies that trade one problem for another. In our scenario with Clem's unstable approach, his airspeed was too high to select full flaps. He increased his pitch angle to slow. This traded his airspeed problem for a glidepath problem. After setting full flaps, he tried to correct his glidepath problem by decreasing pitch, which renewed his airspeed problem. Whenever we trade problems from one parameter to another, it is a warning sign.
- *Changing the measure of success – goal shifting:* Pilots on failed profiles often shift from failing goals to achievable goals. Referring to Clem's approach, he started with a goal of a normal stabilized approach and landing. When he misjudged his energy management and jammed the approach, he discarded the glidepath goal and concentrated on configuring to get full landing flaps. When that caused a glidepath problem, regaining glidepath became his new goal. Realizing that he wasn't going to satisfy stabilized approach standards, he ultimately accepted the excessive parameters and switched his goal to landing. Floating the landing, his next goal became easing the aircraft down for a smooth touchdown. Sliding toward the overrun, his final goal became holding aircraft alignment and stopping. Anytime we detect ourselves switching between goals, it is a warning sign.

2.7.4 The Difference between Normal Path Deviations and Warning Signs

There is a thin line dividing manageable deviations that we can handle and extreme deviations that are uncorrectable. While we are flying in-the-moment, this distinction blurs because we use the same corrective actions regardless of the intensity. When we are slow, we advance the thrust levers. When we are really slow, we advance them further and quicker. As long as we know the cause of the deviation and it remains manageable, we make the correction and continue. Let's intensify the deviation and see what changes.

- *Surprise:* The first effect we may notice is that the severity of the deviation surprises us. Our past experience guides our expectation. For example, if a previous crew reported moderate turbulence on final, we immediately recall our past experiences with moderate turbulence on final. We form an expectation of what we are likely to experience. Given that expectation, we are ready when we encounter bumpy air jostling us down final. If, instead, we encounter an extreme bump that lifts our right wingtip way up and pushes our nose way left, that would surprise us. Anytime we are surprised by events beyond our expectations, it is a warning sign.
- *Strong or abrupt disruptions in the flightpath:* Even if we aren't surprised, if the disruption is especially strong, we treat it as a warning sign. For example, if we are on final behind a large aircraft and expect possible wake turbulence, we probably wouldn't be surprised by a strong wake encounter. Still, when we encounter that dramatic jolt that upsets the flightpath, we treat is as a warning sign.
- *Unexpected:* If we encounter something that is unexpected, we treat it as highly important. Mishap pilots often treat anomalous events as "interesting" or "something to talk about later" rather than something that should immediately concern them. As we gain experience, we discover more nuances that surround effects. So, we encounter fewer unexpected events. If we recall that a left-quartering headwind to a particular runway at a particular airport usually generates a noticeable sink event in the flare (due to the way that the wind rolls over some large hangars near the runway), we expect it and prepare for it. Someone new to that combination of runway and wind wouldn't know this and would not expect it.
- *Anomalous effects of unknown origin:* If an anomalous event occurs and we don't know what caused it, we treat it as a warning sign. It might be caused by something hazardous. We'll need to investigate the cause before proceeding with our original game plan. The temptation is to accept the anomaly as a "one-off" indication and continue with the game plan.
- *Exceeding normal control inputs:* After hundreds and thousands of flights, we develop a sense for what is typical and what isn't. Even when we can't identify the cause, something in our gut-feel senses the wrongness of a deviation. For example, consider an event where the crew noticed that it took excessive force to rotate during takeoff. They knew that the trim was set correctly, so they treated the heavy nose effect as a warning sign. They compensated by slowing their rotation rate and holding a normal takeoff pitch attitude. After a few more knots of acceleration, the aircraft broke ground and started to climb. After reaching a safe altitude, the crew investigated further and discovered that operations had failed to include 8,000 pounds of freight to their performance computation. During rotation, the crew didn't know the cause of the heavy nose. It could have been from several factors, but the safest course of action was to slow the rotation rate and let the aircraft fly when it reached sufficient airspeed. Had the crew continued with their typical rotation rate, they might have induced a tailstrike.

- *Multiple anomalies:* When several anomalous indications or events occur together, their cause can be especially difficult to identify. It is prudent to treat this as a warning sign. Partial electrical failures are good examples since they may manifest in seemingly unrelated system failures. Some system losses may be immediately apparent, while others may not manifest until later.
- *Initial indications warn of future problems:* Many failure modes are preceded by warning signs. In some turboprop engines, chip lights warn of possible impending engine failure. The same is true for loss of oil indications on turbojet engines. Crews need to decide whether to keep the engine running at low thrust or shut it down as a precaution. Another example is a steadily dropping hydraulic system quantity after lowering landing gear. Anticipating that this is caused by a leak in the hydraulic pressure line, we can prepare for the loss of other systems (spoilers and thrust reversers, for example) during landing rollout.
- *Gut feeling – something instinctively warns us:* The more experienced we become, the greater the connection between our mental assessment and our intuition. Many mishap pilots wish they had trusted their gut feeling when they first sensed that a situation wasn't working out. Trusting our gut feeling may work in our favor or not, but the prudent course of action is to trust it, investigate it if we have time, or abandon the game plan if we don't.

NOTES

1 A summary of the evolution of the Swiss Cheese model is available in: Reason (2008), pp. 95–103.
2 Our safety culture often labels this as *complacency*. The term is inaccurate. The label "complacency" implies *lack of caring*. I have never interviewed a mishap pilot that didn't genuinely care about the success of their flight. Their actions didn't reflect a lack of caring. For this reason, I prefer the term *laxity* because it accurately reflects how otherwise well-intentioned crews can inappropriately relax their attention and level of diligence during flight phases when they should have remained engaged and focused.
3 Klein (1999), p. 27, the listed steps summarize Klein's model.
4 Edited for brevity. Italics added. NASA ASRS report #1580868. An alpha floor event is an Airbus automated protection feature that advances aircraft thrust if an approach-to-stall is detected during a low-speed condition.
5 Edited for brevity. Italics added. NASA ASRS report #1771957. Note: NASA ASRS deidentifies this report by changing the airport identifier to ZZZ and the flight hours to XXXXX. This convention is commonly used in ASRS reports. OE is Operational Experience, which are recertification flights with a training Captain before being release to normal line-flying.
6 Dismukes, Berman, and Loukopoulos (2007).

BIBLIOGRAPHY

ASRS – Aviation Safety and Reporting System – National Aeronautics and Space Administration (NASA) Online Database. (n.d.). Retrieved from Aviation Safety and Reporting System: https://asrs.arc.nasa.gov/search/database.html.

Dekker, S. (2015). *Safety Differently: Human Factors for a New Era.* Boca Raton, FL: CRC Press.

Dismukes, R. K., Berman, B. A., & Loukopoulos, L. D. (2007). *The Limits of Expertise: Rethinking Pilot Error and the Causes of Airline Accidents.* Burlington: Ashgate Publishing Company.

Klein, G. (1999). *Sources of Power: How People Make Decisions.* Cambridge, MA: The MIT Press.

Reason, J. (2008). *The Human Contribution: Unsafe Acts, Accidents, and Heroic Recoveries.* Burlington: Ashgate Publishing Company.

3 Complexity

Complexity adversely affects us. Because it emerges from so many sources and creates such wide-ranging effects, it challenges our ability to control the flight. As we struggle with it, we can become task-overloaded. We narrow our focus to process the immediate indications of what is happening and fail to recognize the underlying causes.

Complexity is not simply additive. The synergistic and emergent properties interact in unpredictable ways to generate unanticipated situations. We have more complex aircraft, ATC environments, arrival and departure procedures, and airports pressed against their operational capacity limits. The more sources of complexity, the more combinations and unpredictable outcomes arise. Airlines, ATC, and regulators attempt to contain these wild effects with policies, procedures, and regulations. Their efforts to bound undesirable consequences in one area can cause vulnerabilities to emerge in another.

Using optimization programs, airlines tighten their operating margins to squeeze more productivity out of each asset – aircraft, employees, gates, and slots. While this efficiency increases profits and reduces waste, it makes the system more vulnerable to disruptions. A loosely built schedule contains slack that we can use to dampen disruptions. Reeling in that slack keeps the schedule running on time. With tightly optimized operations, we lose recovery options. A single delayed departure spreads across the rest of the system with misconnections, crew time-outs, and cancellations. Other complexity delays include traffic saturation, gate availability, local transfer of passengers/freight, misconnected flight crewmembers, disruptions of airport support services, security issues, runway closures from disabled aircraft, ATC radio failures, computer outages, power failures in the airport terminal, mechanical failures of the jetway, customer ticket scanning breakdowns, and more. While everyone strives to reduce complexity, it remains unavoidable. As Master Class pilots, we strive to deeply understand the underlying sources of complexity, what makes them rise and fall, how they manifest on the flightdeck, and how they affect us.

3.1 COMPLEXITY THEORY

Let's start by understanding how complexity arises and how it affects the airline aviation system.

3.1.1 Six Features of Complex Systems

Sidney Dekker offers the following list of complex system features.[1]

1. *Unbounded systems are affected by environmental factors:* Our flight operations are highly affected by many outside forces. On a macro-level, we are influenced by the Department of Transportation (DOT), FAA, NTSB,

DOI: 10.1201/9781003344575-4

ICAO, aircraft manufacturer, fuel prices, our competitors, the consumer market, and more. On the micro-level, each flight is affected by weather, ATC, cities/stations, gate availability, and other employee groups, to name a few. The boundaries between these groups blur and overlap as they each pursue their own objectives.

Complex systems are highly dynamic and affected by unseen forces. Envision floating on a raft on a fast-flowing river. We use our oars to control our course over water that is constantly buffeted by unseen rocks below. We read the flow of the river and steer away from the hazards. Likewise, we respond to forces affecting our flight. We anticipate how these forces might affect our game plan and take measures to keep our flights moving safely. The main point is that we have little influence over the forces generating the complexity in our aviation environment, but we benefit by understanding them and improve our ability to predict future situations and mange consequences.

2. *Complex interactions:* The effects from our decisions are not isolated. Every action and choice that we make propagates throughout the system. Whether we push on time or delay our push for a maintenance issue, we trigger ripples of consequence – many of which we can't predict or control. Our situation creates both local effects and broader systemic effects. Imagine arriving on-time to our destination only to discover that the flight before us took the last available gate. This was caused by a separate aircraft that experienced a mechanical delay on a different gate. We landed on time, but we are the ones incurring a delay. As the ripples spread, other connecting flights become delayed as they wait for our transferring passengers. This wave of local and systemic effects propagated from a single mechanical delay on one aircraft.

   Through repetition and experience, we improve our decision-making skills. By monitoring the success of others and sharing information from our local perspective, we improve the system's resilience. For example, seeing that fog is developing at our current location, we alert our dispatchers. They begin adding fuel and assigning takeoff alternates to future flights. Our timely report helps them respond more quickly than waiting for official weather reports.

3. *System complexity and local influence:* While Dekker is correct in warning that, "each component is ignorant of the behavior of the system as a whole" (Dekker, 2011, p. 13), we can still improve our awareness of how our local actions affect the system and vice versa. Even in a complex system, we can find ease and simplicity within our portion of the operation. Complexity within the system does not mean we need to select complex choices. Our gate hold may be due to departure traffic overload, weather blocking departure corridors, thunderstorms blocking jet routes, airspace saturation, marginal arrival weather, or our company's decision to hold our flight. Regardless of the cause, we are simply "on gate hold". The causes of systemic complexity that created our delay, while important for our understanding, don't necessarily affect the choices we make within our aircraft

or for our flight. The daily operation is a flow of work. We contribute to that systemic flow with our contribution. Others around us contribute to that same flow. We benefit by keeping our portion simple and safe.

4. *Operating near the turbulent edge of chaos:* Aviation, like driving and walking, is an inherently unstable undertaking. The moment we stop actively controlling the process, the aircraft will crash, the car will drift into a ditch, or we will stumble and fall. It is only through our ceaseless monitoring and continuous control inputs that we maintain our safe trajectory.

   Aviation systems also require constant monitoring and control inputs to maintain a safe course. Market pressures encourage systems to push against their boundaries. The most profit is gained from the last passengers boarding a full aircraft, the final boxes of freight loaded, flying with the minimum required fuel, and having exactly enough workers. While operating on this operational edge, one overbooked passenger, uploading 1,000 pounds less fuel than we need, or one ramp agent calling in sick destabilizes the system. Optimization makes the system fragile. Despite this fragility, fully optimized systems actively strive to balance along this edge.

   While the system works to maximize profit, it cannot manage it within each individual flight. It is up to us to evaluate the unique characteristics affecting our flight. Standard arrival fuel may be appropriate for one flight, but may be too low for another. Only we can predict where we think our edge lies and employ appropriate measures to ensure safe operations. We are uniquely positioned to balance safety and efficiency for our particular flight.

5. *Path dependency:* Dekker concludes that most interactions are fairly localized. We are most affected by our immediate contacts – our dispatcher, our gate agent, our local station agent, and our ramp agents. The late-arriving jet we receive during our aircraft swap was caused by events that occurred well before our involvement. Disruptive events that occurred hours before we received the jet continue to affect us and our choices.

   Disruptive events have a way of amplifying delays unless we take proactive steps to dampen them down. A late-arriving flight motivates an operational effort to get the schedule back on time. We increase the pace of many individual operations to regain lost minutes. Some choices prove appropriate. Other choices don't. Ultimately, the flight crew remains the final authority in the decision-making process and the final barrier blocking an unsafe event trajectory. If the pace needs to be slowed, we need to be the ones to slow it. We can still choose to move quickly and accurately, but we must avoid rushing.

6. *The butterfly effect:* One of the more significant features of complex systems is that the consequence of an event can either dampen or magnify as it propagates through the operation. The same type of event can result in vastly different consequences from one occurrence to the next. An ice storm bearing down on our intended destination can result in our flight being cancelled while another flight is encouraged to hurry. Large local effects can also be affected by small changes in conditions. One flight can land

successfully on an icy runway while a slight temperature drop can cause the next flight to slide uncontrollably into the overrun.

As we operate closer to the edge, the probability of adverse consequences rises. When we back away from the edge, adverse consequences subside. For this reason alone, it is incumbent upon us to detect the edge and back away from it whenever we can. If we push against the edge, we risk losing control over the outcome. If we land on an icy runway and experience poor braking, we need to share this information with others, even if it causes subsequent aircraft to go around and divert. The fact that we successfully operated on the edge doesn't mean the next crew will succeed. If the local conditions or system complexity push the operation toward the edge, someone needs to pull it back. Many components of our system are designed to push the operation forward, even during deteriorating conditions. We stand in the best position to detect the edge and pull the operation back.

3.1.2 The Marble Analogy

These six features seem to depict our aviation environment as being unpredictable and chaotic. On the other hand, we've proven that we handle system complexity rather well. We succeed because each one of us actively manages our flights within our piece of the system. Envision walking while carrying a bowl. In the bottom of the bowl rolls a single marble. As long as we keep the marble from pitching over the edge of the bowl, we succeed. When we pay attention and the environment is unchallenging, the task is easy. As the environment becomes more chaotic and complex, the task becomes more difficult. Operational demands force us to divide our attention between the marble and the many other tasks that fill our day. The marble rolls around – sometimes racing perilously toward the edge. When this happens, we return our attention to the task and actively dampen the trajectory of the marble until it settles safely back into the center of the bowl. As long as we devote an appropriate level of attention toward managing the task, we succeed.

The shape of each bowl is governed by the management philosophy of the airline. An airline that chooses to routinely push against limits creates an environment that is like controlling the marble in a shallow-sided dish. Each crew needs to devote extra time and attention to keep their marbles safely contained. An example would be an airline that only follows regulatory limits and transfers local operating decisions to line pilots. Each pilot monitors their own gray zone and manages their own local operational decisions. On the other extreme, a proactive, resilient airline creates an environment like a bowl with steep sides. This airline monitors complexity and adverse trends to actively manage risk exposure across the entire operation. Their pilots find it quite easy to keep their marbles safely contained because the airline operation doesn't expose them to risky situations.

3.2 SOURCES OF COMPLEXITY

With our foundational understanding of complexity in the aviation environment, let's examine the external and internal sources that generate it.

3.2.1 External Sources of Complexity

Complexity is affected by a number of sources outside of ourselves and our immediate flightdeck environment.

- *Aircraft and automation:* The trend in automation has been to design complex aircraft components to ease pilot workload. Simply stated, smart automation takes care of many complex tasks so we don't have to. With most innovations, this works well – very well. In fact, many pilots completely relegate responsibility for task management to the automation. They don't even try to understand how the system works. The problem is that if we don't know the design assumptions underlying the system, we probably won't understand the causes behind a malfunction. The more complex the system, the more failure modes are theoretically possible and the less likely we will understand what is causing them.

  The problems and subsequent grounding of the Boeing 737 MAX due to MCAS confusion is a prominent example. Crews were informed that the aircraft had a system that automatically compensated for the pitch effects associated with large thrust changes, but not what would cause it to fail or how the failure modes would affect aircraft handling. Trust in the automation led crews to assume that any malfunctions would be automatically corrected by the system's software programming. Pilots remained ignorant of a latent vulnerability created by an associated system failure. When that system failed on two accident flights, the crews lacked the understanding to identify the cause, disable the malfunctioning automation, and recover the aircraft using a backup mode.

  Complex automation affects how we discover and handle malfunctions. Imagine a series of gauges that measure hydraulic fluid volume, pressure, and temperature. If a leak displayed a steady decline on a volume gauge, we could detect the problem well before it caused a rise in temperature or drop in pressure. Now imagine that engineers replace all three of these gauges with a single warning light designed to trigger for either a loss of volume, low pressure, or overtemp. Granted, when the light illuminated, we would draw the same conclusion, "I've lost that hydraulic system – probably a leak." Still, by the time we get the warning light, we will have lost the ability to plan ahead and form a mitigation plan.

  The point is that automation may reduce operational complexity, but it can also introduce latent vulnerabilities that arise when conditions combine in rare or unexpected ways. We understand that pushing this button transfers the operation of certain aircraft systems to an automated controller, but we should strive to understand the underlying conditions that the automation targets.

  - What is the intended function of the automated system?
  - Why did the engineers think it was important to automate this function?
  - What failure modes does the automation protect against?
  - What failure modes are not covered by the automation?

  - What associated failures might disable the automation protections?
  - How would these failures present themselves through aircraft indications?
- *Procedures:* Procedures are intended to guide how we complete necessary tasks. When we need to modify a process, procedures are rewritten. Later, when something undesirable happens, procedures are amended to address that newly discovered vulnerability. Over time, this patchwork approach can make procedures unintentionally complex. This complexity often spreads into interdependent areas. A new procedure designed to address a particular problem in one area might generate unanticipated vulnerabilities in another. Pilots can become confused with how to apply the guidance. Which procedure applies here? How do we adapt that procedure to our specific case? If two separate procedures seem to fit, which one should we use?

  Unique interpretations proliferate on both the airline level and local pilot level. Consider an airline that provides minimal guidance and relies on their Captains to modify procedures as they see fit. This improves versatility, but increases variability. The opposite extreme is an airline that overly controls operational procedures. This reduces both variability and versatility. This may adversely affect the crews' resilience with responding to unscripted variations encountered while line flying.
- *Regulatory:* Complexity may increase when companies translate regulations into procedural guidance. Years ago, a tripped circuit breaker (CB) on the flightdeck was treated as a simple procedure. We touched the CB to feel if it was warm. If not, we could reset it. If it remained engaged, we continued normally. Following the TWA 800 accident (inflight fuel tank explosion – July 17, 1996), new regulations prohibited the resetting of any fuel system CBs while in flight. Later procedural changes added a logbook documentation requirement for any tripped CB. Following that, the prohibition against resetting CBs was expanded to include any systems not required to complete the flight. Each procedural change increased the complexity.

  As regulatory complexity rises, more details and nuances are added. This often increases our reliance on automation and software. For example, the previous regulations for flight time and duty hours limits were typically managed by pilots using their personal logbooks. Fatigue-mitigating improvements incorporated a range of additional limits that applied look-back histories and look-forward projections. This raised the complexity and effective forced us to use software applications to monitor duty day legality. Additional company policies, union contract modifications, and the option of waiving some contractual limits made it even more complex.
- *Interface devices:* The main computer for aircraft guidance is the flight management computer (FMC or equivalent). It combines inputs from aircraft sensors, accesses databases, and incorporates crew inputs to guide the flight director and autopilot. Most of us acquire a user's understanding of the system. We know how to make inputs to get the aircraft to do what we want it to do. Few of us understand the underlying assumptions involved in

the internal computations. Very few of us possess the deeper level understanding of how each embedded function works. As complexity rises, our level of understanding falls. We trust the FMS to perform in ways that we really don't understand. This is reflected in the common pilot declaration, "What's it doing now?"

Another interface platform is our electronic flight bag (EFB). Typically supplied as a touchscreen tablet, it has become the repository for all of our aviation charts, operations manuals, systems manuals, regulations, company guidance, and more. Companies can quickly issue, update, and change any document in the EFB. Pilots updating their EFBs before a flight may discover a host of new changes. Some may be flagged requiring the pilots to open and review them before the flag is removed. As a minimum, pilots need only acknowledge receiving the change.

- *ATC environment:* The limiting bottlenecks in the ATC system are arrivals and departures. More accurately, it is airspace and aircraft separation. These limitations are reflected by the airport's acceptance rate. The arrival flow challenge is optimizing the movement of dissimilar aircraft from different directions while requiring different spacing toward multiple runways. Similar challenges for departures include accommodating ATC departure corridor spacing (miles in trail) and destination acceptance rates (resulting in metered departure times and gate holds). These optimization challenges push ATC systems against their limits. Any disruptions quickly spread. If an airport with an arrival rate of 60 aircraft per hour unexpectedly goes IFR, the acceptance rate may drop to 30 aircraft per hour. Subsequent aircraft get slowed down until the number of aircraft is reduced. Any operation planned against these capacity limits becomes extremely vulnerable to disruptions.
- *Airline operations:* Like ATC, every airline strives to optimize their operation. This translates into how they manage available gates, slots, and workers. Operating at their capacity limit relies on everything going right. When a single aircraft fails to vacate its gate while working a mechanical problem, it forces the inbound flight for that gate to hold out. This delays the arrival and transfer of those passengers and crew, which spreads delays to other flights, and so on. Any operation planned against its operating limit becomes extremely vulnerable to complexity.

3.2.2 Internal Sources of Complexity

Complexity can arise from our personal limitations. These can vary from unavoidable to self-generated.

- *Lack of experience:* The less experience we have with a situation, the more complex it appears. As novices, many new experiences feel complex. As we gain experience, flying becomes easier. With experience, we assemble a toolbox of successful game plans that we lacked as novices.

- *Lack of exposure:* Experience in our aircraft type and total flying time may not translate into skills needed to cope with a complex situation. Compare a highly experienced Captain flying for the first time into a particular congested hub airport with a new-hire FO who had been based at that airport with a previous employer. The FO's extensive experience makes the operation seem easy for them. Conversely, the novelty and complexity of the operation can make the flight quite stressful for the Captain.
- *Lack of currency:* Pilots returning to flying following a long absence (like a medical grounding) may have accumulated significant time and experience on paper, but their lack of recent practice may slow their pacing, especially as complexity rises. Even recalling and applying normal procedures may take extra effort. Many airlines recognize this vulnerability and schedule pilots returning from long absences to fly with training pilots until they regain their proficiency.
- *Lack of recency:* For airlines with extensive route structures, years may pass between a pilot's exposure to a particular airport. Long gaps and high complexity increase their vulnerability to errors. This is especially evident when an airport employs unique local procedures. Local operators and ATC controllers, who use those procedures every day, find them to be familiar and easy. This promotes an expectation bias where they assume that everyone must know their provincial quirks.
- *Stress from outside of the flightdeck:* Stress from outside endeavors or home life follow us onto the flightdeck. Our personality profile assumes that we are skilled compartmentalizers. This means that we can mentally put aside concerns unrelated to our current flying task. Still, we have our limits. Showing up to work stressed from outside concerns weakens our ability to process operational complexity. A disturbing phone call between flights can adversely affect our attention focus during the next flight.
- *Lack of knowledge:* A growing concern is the rising volume of required knowledge. Most of us are responsible for thousands of pages of information. The sheer volume has shifted our strategy from memorizing material toward learning how to locate it in our EFBs. We are still expected to remember the numbers and parameters that we use every day. For exceptional and rarely encountered situations, we are expected to know that the particular guidance exists and how to locate it in our EFBs. For particularly unique situations, airlines provide experienced resources (typically at a centralized operations facility) to assist us.

  Automation has replaced some requirements for specific knowledge. For example, the previous requirement to remember system overtemp parameters has been replaced by an overheat warning light. The automation may even save a record of the overtemp event to aid us with making our maintenance discrepancy writeup.
- *Lack of preparation:* We know that preparation reduces the adverse effects of complexity. Still, we find cases where high-time, high-currency pilots encounter problems when they fail to adequately prepare. Their sense of

familiarity sometimes desensitizes them to common risks. Imagine flying back and forth from Chicago O'Hare (ORD) every day for years. Despite the inherent complexity of operating at this busy hub airport, it can feel repetitious. Success and familiarity can promote a sense of invulnerability to error. If it always worked out every time in the past, why go through the repetitive effort of planning and briefing?

- *CRM breakdowns:* Complexity magnifies latent CRM weaknesses within a crew. If all crewmembers cannot agree on the nature of the problem, then developing a shared mental model becomes difficult. Complexity also obscures team roles. When crewmembers aren't clear about their roles, then effective communication suffers. Finally, maintaining a shared mental model is disrupted by emerging counterfactuals (indications that conflict with what we expect to see). Is a conflicting indication just a minor anomaly or is it a sign that the game plan is failing? For example, there are cases where an FO called for a go around, but the Captain overrode the directive and continued to land. The FO clearly felt that parameters indicated a need for a go around. The Captain felt that those same parameters reflected acceptable deviations. In addition to major CRM breakdowns, there are minor CRM problems. These include reluctance to voice concerns, withholding knowledge of counterfactuals, self-doubt, and silence (which is usually interpreted as consent to continue).
- *Self-initiated complications:* A difficult issue is borderline non-compliance. For example, if one pilot voices a conversational observation during taxi-out, is that an act of sterile flightdeck non-compliance or harmless flightdeck conversation? Technically, it does violate sterile flightdeck protocols since it may adversely affect monitoring, planning, and task accomplishment. It could lead to an unnecessary distraction, especially in complex environments. On the other hand, we can dismiss it as the kind of brief comment that rarely affects the crew's workflow.

  Another aspect of self-initiated complication is purposefully increasing the difficulty level of required tasks. For example, consider a Captain who intentionally arrives to the flightdeck mere minutes from pushback time and rushes their preparation, planning, and briefing. FOs are generally inclined to adapt. While these Captains may be highly skilled at completing "their stuff" in record time, this practice doesn't allow their FOs a choice in the matter. Latent errors may be missed because the Captain fails to verify the FO's work.
- *Resignation:* When a situation becomes "too complex", some pilots accept the complexity as inevitable and resolve to push forward. "Yes, there is a storm cell on final, but the last flight made it through. We are running late and we need to get these people to the gate. We'll be fine." We accept that there is a distinction between a complex situation and one that becomes hazardous, but that line is unclear. Complexity can encourage us to rationalize the level of risk.
- *Get-home-itis:* A significant percentage of accidents occur on the last scheduled flight of a crew's pairing. Pilots know that they just need to finish

that one last flight and they can go home. Also, they are particularly familiar with their home airport, so complex conditions are easier to dismiss. This combination seems to promote a higher willingness to overlook warning signs and continue failing game plans.

3.3 HOW COMPLEXITY AFFECTS OUR GAME PLAN

In his RPDM model, Klein explains the process we use to make operational decisions. First, we view the situation and compare it with our past experiences. We identify familiar aspects and select a previous game plan that succeeded under similar conditions. We test it by running a mental simulation that imagines how our flight will track from our present position toward the desired conclusion. This mental simulation is a bit like watching a video of how the scenario will unfold. If the video seems plausible, we use that game plan. The game plan then guides our SA-building and decision making. In this way, the mental simulation is predictive (foresees the outcome) and descriptive (describes how we will accomplish each step of the game plan).[1]

3.3.1 Complexity Limits Familiar Game Plans

Complexity makes selecting a game plan more challenging. The more complex the situation, the more nuanced and numerous the possible outcomes. Perhaps, none of our past game plans exactly match the situation. This means that we may need to adapt or innovate one of our promising options. This immediately pushes us into the gray zone of increased risk. We can't freeze time to sift through the indications, balance priorities, and make well-reasoned decisions. Time pressures force us to choose quickly. To keep up with the pace of events, we find ourselves doing everything at the same time – detecting, balancing, selecting game plans, running mental simulations, making decisions, and communicating, all while flying the aircraft. Managing an evolving game plan becomes an iterative process. We may start with a promising game plan only to discover that it needs to be modified or replaced.

3.3.2 Unique Situations

When a situation is unique, we won't have any convenient past game plans that match. One strategy is to simplify our goals and innovate the follow-on steps. Take an example of a complex flight control malfunction made even more difficult by marginal weather. The final goal is a safe landing – simple enough. How to achieve this requires innovation and balancing. Maybe we start by deciding where to land. If an alternate airport is more favorable than our planned destination, then we divert. This reduces complexity. Next comes the process of preparing the aircraft for landing. Our procedures will help but may fall short of addressing complicated aspects of our particular flight control malfunction. Time allowing, we can access outside expertise from company operations or our aircraft manufacturer – again, reducing complexity. Eventually, we construct a game plan focused on establishing a safe configuration for landing. We strive to keep all conditions as manageable as possible. We stack the cards in our favor to create additional safety margin against unforeseen effects.

3.3.3 Monitoring the Trend of Complexity

If our game plan is valid, it should unfold simply and predictably. If it is inaccurate, invalid, or unachievable, the situation will become more complex or difficult. The key indicator is the trend of the level of complexity. Is it getting worse or is it getting better? If it is becoming worse, unpredictable outcomes can emerge. We can try to compensate, but if our game plan is fundamentally flawed, new problems appear or past problems reemerge. Despite our best efforts to hold it together, the game plan crumbles. As it falls apart, our in-the-moment perception is that the situation feels too complex or unmanageable. In many cases, we won't have time to investigate why. We only need to detect the rising complexity and switch to a safer backup plan to reverse the trend.

3.4 RISING COMPLEXITY MODEL

Consider Figure 3.1. Using the RPDM process, we select our game plan. If the plan is valid, the situation resolves as expected in the lower box (Simple Situation). Everything matches our mental simulation. As the scenario unfolds, it remains simple and complexity doesn't rise.

The middle box (Difficult Situation) represents a more complex scenario. It will be more challenging to manage, but our corrections will work. Complexity remains steady and the game plan works.

The upper box (Failing Situation) reflects a deteriorating situation. To a point, a failing scenario looks exactly like a difficult one. That is how the Recognition Trap disguises itself. There are five important distinctions.

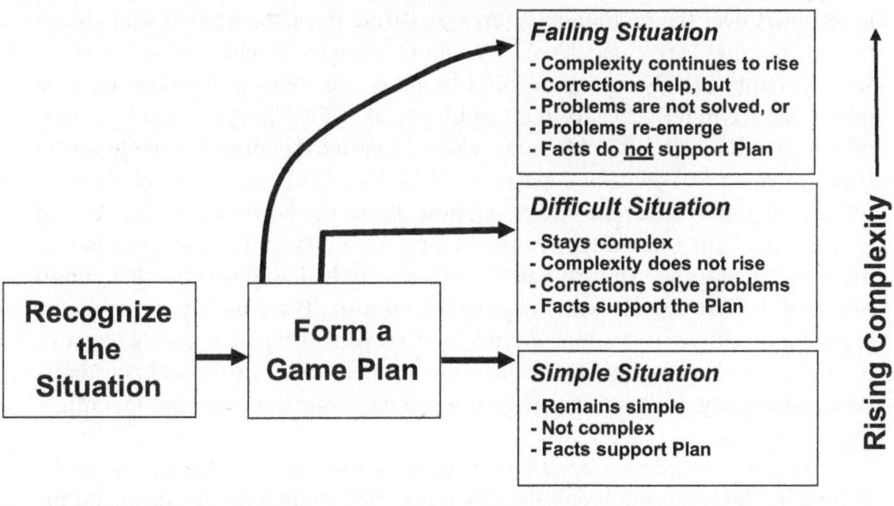

FIGURE 3.1 Game plan responses to rising complexity.

1. *The level of complexity increases.* Despite our best efforts to resolve our problem, it continues to worsen.
2. *Our corrections seem to help, but they are not sufficient to solve our problems.* The illusion of progress is the hook that encourages many mishap pilots to continue their failing game plans.
3. *Problems are not solved by our corrections.* They persist or worsen.
4. *Problems that we thought we solved either reemerge or are replaced by new problems.* For example, during an unstabilized approach, we can shallow our descent rate to reduce excessive airspeed, but it creates a glidepath problem.
5. *Facts do not support the game plan.* This is often masked because tunnel vision encourages us to focus on limited parameters to the exclusion of the bigger picture.

3.5 THE PUSHES AND PULLS OF THE SYSTEM

Some aspects of the flight operation unintentionally push us toward the edge of hazardous risk. This is not because the system is inherently unsafe. It is because the system doesn't know where the edge is or that we are approaching it. This is especially evident during slowly decaying conditions. Consider two cases.

BOX 3.1 EXAMPLES: WHERE THE OPERATION PUSHES AGAINST THE EDGE OF RISK

Case one – Increasing winds and turbulence in the arrival area: A reoccurring condition occurs in the Los Angeles basin during Santa Ana wind flow. It manifests primarily at Burbank (BUR), Van Nyes (VNY), and Ontario (ONT). The prevailing west winds are replaced by a strong northerly flow. As the airmass spills over the mountains, it creates strong turbulence. Airlines monitor reports as crews arrive and depart. The point when the turbulence becomes too severe is subjectively measured pilot-by-pilot and moment-by-moment. One pilot may accurately classify the turbulence as MODERATE, while the next pilot may call it SEVERE. Moderate allows continued operations, while severe shuts down all arrivals and departures. Which call is accurate? Once a severe turbulence report shuts the operation, how do we know when it is safe to start it back up again? Logically, we should wait for a change in conditions before resuming operations. Unfortunately, surface winds do not reflect what conditions will be like for departing or arriving aircraft. There are strong company and pilot incentives to resume operations. Everyone is trying to locate the edge of unacceptable risk. This makes us vulnerable to supervisory and peer pressure that push or pull us forward even when our experience and gut-feel might urge us to wait.

Case two – Steady snowfall and dropping temperatures: Another situation is steadily deteriorating breaking action reports as temperatures drop and the snow continues to fall. We know that when the runway surface temperature approaches 0°C, our breaking effectiveness drops dramatically. Contaminated runways warmer than approximately +5°C or colder than −5°C typically provide braking action reports of MEDIUM (RCC 3) or POOR (RCC 1). Contaminated runways near 0°C, however, often generate braking reports of NIL (RCC 0). Consider a situation where the snow is falling and temperature is dropping. Pilot breaking action reports progress from GOOD (RCC 5), to MEDIUM (RCC 3), to POOR (RCC 1). The airport managers plow and treat the runway. The first crew to land after the treatment reports MEDIUM TO POOR (RCC 2). Minutes later, the next crew reports POOR (RCC 1). If we are the next aircraft to land, what do we do? Assuming that we are legal to land under POOR (RCC 1), we could rely on the last report and land. Conversely, we could consider that the runway temperature is trending toward the hazardous zone near 0°C. Is the snowfall continuing? What was the aircraft type from the last POOR report? The airport managers feel like they have done their job by treating the runway, so they expect us to land. If the temperature is nearing the danger zone, however, we might conclude that we won't get anything better than POOR, and will probably get closer to NIL. We might prudently wait until immediately after another plowing and treatment or divert to an airport with favorable conditions.

In both of these cases, we witness operations that are probing to find the edge while urging us to keep moving forward. Everyone acknowledges the legal limits, but until we actually reach them, we are permitted to operate at our discretion. Our companies will usually defer to us because we are in a position to judge the immediate conditions and trends. Unfortunately, this is exactly the situation that describes conditions preceding many mishaps.

3.5.1 Factors That Push Us Forward

Pushes are forces within the system that urge us to keep the operation moving forward. The main push in the airline system is the flight schedule. The entire organizational structure is designed to move passengers, freight, and aircraft efficiently and on time. When a flight becomes delayed, the company responds by channeling additional assets to help it. Pushes can come from dispatchers, station managers, local operations agents, or supervisors. They urge us forward even as we approach the edge of regulatory or prudent operating conditions. One reason is that companies base their decisions on reported information. With rapidly changing conditions, information may become stale or inaccurate. We are in better position to weigh the current effects of complex conditions by accessing immediate sources of information. What conditions did that last flight report? How fast is that storm cell moving? What are the current winds reported by ATC? How have they changed over the last 10 minutes?

Other pushes include ATC (biased to keep aircraft moving), traffic flows (pilots flying in a line of aircraft following one another), and limited fuel loads (which limit the time to either delay our arrival or divert).

3.5.2 Factors That Pull Us In

While pushes tend to be external components that urge us forward, pulls are internal and personal. Imagine that we are in IMC conditions on final approach. Suddenly, we break out of the weather and see the runway. Seeing the runway is one of the strongest pulls in aviation. It is like a green light clearing us to land. It tends to suppress the importance of warning signs surrounding the runway environment. What if we then notice evidence of a microburst in the airport environment? What if we see heavy rain falling on the runway past the touchdown zone? Do pulls urge us to discount these warning signs and continue?

Another pull is flight accomplishment. To draw upon military references, our "mission" is to move passengers and freight on time between locations. While the schedule pushes us forward, our personal commitment to fulfilling that mission pulls us along.

Another aspect is our ego or sense of professional pride. If we mismanage an approach and end up unstabilized on final, we'll probably feel professionally embarrassed. Good pilots don't mismanage their approach energy and find themselves unstabilized on final. We might feel tempted to continue the approach and rationalize our decision to land like Clem did.

NOTE

1 Dekker (2011, pp. 138–139), Dekker derives his list from Paul Cilliers (1998), *Complexity and Postmodernism: Understanding Complex Systems.* London: Routledge.

BIBLIOGRAPHY

Dekker, S. (2006). *The Field Guide to Understanding Human Error.* Burlington: Ashgate Publishing Company.

Dekker, S. (2011). *Drift into Failure: From Hunting Broken Components to Understanding Complex Systems.* Burlington: Ashgate Publishing Company.

Dekker, S. (2015). *Safety Differently: Human Factors for a New Era.* Boca Raton, FL: CRC Press.

Dismukes, R. K., Berman, B. A., & Loukopoulos, L. D. (2007). *The Limits of Expertise: Rethinking Pilot Error and the Causes of Airline Accidents.* Burlington: Ashgate Publishing Company.

Klein, G. (1999). *Sources of Power: How People Make Decisions.* Cambridge, MA: The MIT Press.

Woods, J., Dekker, S., Cook, R., Johannesen, L., & Sarter, N. (2010). *Beyond Human Error - 2nd Edition.* Burlington: Ashgate Publishing Company.

4 Decision Making

Aviation decision making affects every action we take. Our decisions feel less like individually considered choices than as a continuous stream of conscious and subconscious selections. Our decisions are also highly interdependent. Previous decisions affect our current choices which will then influence our follow-on choices. This chapter will examine aviation decision making with the goal of understanding how we can learn to choose skillfully and avoid decision-making errors.

4.1 AVIATION DECISION MAKING

Recall the decision-making models we learned in school. They typically followed a circular process of detection, interpretation, selection, and feedback. Those steps include:

1. Detect and identify the problem.
2. Establish and weigh decision-making criteria.
3. Generate and evaluate alternatives.
4. Choose one and act.
5. Use feedback to evaluate the decision's effectiveness.
6. Make adjustments and repeat the process.

This deliberative process works well for decisions that are simple and independent. In aviation, research indicates that our decision making is often complicated, the lines are hazy, and the options are highly interdependent with changing conditions. From her research, Dr. Kathleen Mosier observed that:

> …most crews did not wait until they had a complete understanding of the situation to make and implement decisions. Rather they seemed to make a recognitional, almost reflexive judgment, based upon a few, critical items of information; and then spent additional time and effort verifying its correctness though continued situational investigation. If later information changed situation assessment enough to prompt a change of decision, a second option was generated and implemented. Virtually no time was spent in any comparisons of options. In fact, the bulk of time was spent in situation assessment…
>
> Mosier (1991, pp. 266–271)

Her observations align with a branch of HF science called Naturalistic Decision Making (NDM). NDM strives to explain how experts make decisions in highly dynamic and complex environments. Compare the previous decision-making process with the following list of components and considerations from Gary Klein (1999, p. 288):

- Intuition
- Mental simulation

DOI: 10.1201/9781003344575-5

- Using leverage points to solve ill-defined problems
- Seeing the invisible
- Storytelling
- Analogical and metaphorical reasoning
- Reading peoples' minds
- Rational analysis
- Team mind
- Judging the typicality of a situation
- Judging typical goals
- Recognizing typical courses of action
- Judging the solvability of a problem
- Detecting anomalies
- Judging the urgency of a problem
- Detecting opportunities
- Making fine discriminations
- Detecting gaps in a plan
- Detecting barriers that are responsible for gaps in the plan

Notice how Klein replaces the step-by-step process with a list of considerations and strategies. These considerations reflect Recognition Primed Decision Making (RPDM). He proposes that our decision-making process follows these steps (Klein, 2003, p. 28):

1. Observed indications match recognized patterns.
2. Patterns activate action scripts.
3. Action scripts are assessed through mental simulation.
4. Mental simulation is driven by mental models.

As pilots, we may not think that we are choosing action scripts and applying mental models to drive mental simulations, but if we view this process from an aviation perspective, it does make sense. As we fly, we continuously monitor information from our instruments and surroundings to construct a mental picture that makes sense of our current position. We then visualize the flightpath we need to follow to continue moving toward our desired goal. We apply our past experiences to match this current picture with a game plan that promises to reach that desired goal. That game plan includes a sequence of previously validated decisions to guide our aircraft along.

Klein's research is centered on how experts use RPDM under stressful conditions.[1] We also use RPDM with common, everyday situations. Through repetition, it becomes our preferred method of decision making. The more experience we accumulate, the more situations we recognize, and the more we use recognition to guide our decision making. RPDM delivers quick, reliable decisions with minimal effort. Over many years and many thousands of flights, almost all scenarios become familiar. Nearly every flight matches some familiar action script from our past.

This provides us with a steady stream of ready-made decisions that have been flight-tested and validated.

While RPDM guides us when situations are familiar, the opposite is also useful. We recognize unfamiliar situations as situations where familiar game plans won't fit. When situations are unfamiliar and we recognize that they are unfamiliar, we need to change to a *rational choice strategy.*[2] Since we lack familiar action scripts, we draw upon intuition, inspiration, and crew coordination to form a workable strategy. Here, the steps might look like:

1. Indications don't match recognized patterns.
2. Patterns inspire innovation and/or modification of successful past action scripts.
3. New action scripts are assessed through mental simulation, discussion, and coordination.
4. The crew communicates their revised shared mental model.

In this way, RPDM seems to offer two distinct pathways that apply similar steps. Familiar situations are recognized as familiar and are matched to past successful game plans. Unfamiliar situations are recognized as unfamiliar and require innovative or modified game plans.

Our stream of aviation decisions is interdependent. Choices we previously made highly influence our perception of the present moment and choices we are likely to consider in the future. We can't extract a single decision from a flight, analyze it using our hindsight perspective, and then judge its worth. This is a fallacy we sometimes see during accident analysis. Knowing the unfavorable outcome of a mishap event encourages us to apply negative labels to past decisions. This is neither fair, nor useful, for our learning process.

Decisions emerge from our training, experience, bias, and mindset. Each of these interacts and constantly changes. Consider a typical training event. We conduct the training and document that the crew has completed it. We really don't know how well each pilot understood the material, how well they will retain over time, if they will accurately access it from their memory while flying, or whether it will ultimately aid or hinder their decision-making process. Claiming that a mishap crew "was trained" is a factual statement, but it doesn't help us understand their decision-making process during their event.

Next, we consider the effect of experience. Hopefully, experience improves our decision making over time and guides us to select optimal choices. Unfortunately, we encounter events where a pilot's experience actually hinders their effective decision making. For example, if they use a personal technique that unintentionally creates a latent vulnerability, an unfortunate combination of conditions may allow an error to surface.

Third, human biases form natural, low-resistance, decision-making pathways in our minds. Since our biases work on the subconscious level, they may be difficult for us to recognize, especially while we are under stress. Over time, some pilots drift toward habitually selecting familiar, biased choices over more appropriate options.

The fourth and most variable parameter is our mindset. This category attempts to encompass situational awareness, alertness, motivation, mood, predispositions, and social dynamics. Each of these is highly variable. When combined, they can create unpredictable and unrepeatable outcomes. Moreover, our mindset constantly changes and leaves little historical trace. Mishap pilots often cannot accurately recall what their mindset was following an event. They struggle to recreate the complex combinations of contributing factors that formed their mindset at the time. All of this may seem like a tangled ball of string, so let's untangle it.

4.2 THREE CATEGORIES OF AVIATION DECISION MAKING

To frame our discussion, let's examine decision making within three practical categories:

1. *Familiar – spontaneous and standardized:* Most of the time, our decision making flows spontaneously as a smooth stream of choices that match a familiar game plan. These decisions are guided by our habits or standardized procedures.
2. *Unfamiliar or nuanced – consciously considered:* On fewer occasions, something unfamiliar causes our decision path to generate two or more viable options. Typically, we pause and consider the available choices. We evaluate them, select one, and then move forward. We make these decisions either alone or as a crew.
3. *Unknown or rare – highly consequential – crew coordinated:* On rare occasions, the best option is not clear and conditions threaten adverse outcomes. We need to weigh our choices, coordinate with other team members, decide on a strategy, and innovate a detailed game plan.

4.3 SPONTANEOUS AND STANDARDIZED DECISION MAKING WITHIN FAMILIAR SITUATIONS

This category describes the vast majority of simple, scripted, or standardized decisions that we make as experienced pilots. As we gain experience, most decisions fall within this category. The more variations that we experience, the more familiar choices we have available to us. Having seen almost everything, we recognize almost everything.

4.3.1 Spontaneous, Unconscious Decisions

At the simplest level, we make decisions automatically. We select choices without deliberating or considering options. If turbulence bumps our wing up, we spontaneously move the flight controls to level the wing. While this qualifies as a decision (since we could choose not to correct for the turbulence), it is a decision we make almost subconsciously. Watching a pilot fly down final approach on a turbulent summer day, we see them making constant control inputs. Each action demonstrates that they detected the flightpath disruption and chose a response to restore their stabilized approach path.

4.3.2 Standard Procedures

A routine form of decision making is driven by procedure, experience, and habit. Procedures are written to guide the most commonly encountered cases or ideal scenarios. They guide a sequence of actions that all company pilots are expected to follow. Consider a repetitive action like securing our lap belt before operating the aircraft. As directed by procedure and reinforced by experience, we should always secure our lap belt before flight. With repetition, the decision becomes habitualized. We don't really give it much thought. Our hands automatically locate and secure the lap belt. We don't contemplate the choice of doing it versus not doing it. We wouldn't consider flying without securing our lap belt. It would feel wrong to leave it unfastened.

Decisions in this category are framed by normal, standardized procedures. When it is time to start an engine, we use the scripted engine start procedure. The decision path is already laid out for us, step by step. It still requires our attention, accurate compliance, and monitoring, but it follows a highly predictable sequence. To the inexperienced observer, performing an engine start might seem like a complex procedure where we simultaneously monitor many changing parameters and make many split-second decisions. To us, having started aircraft engines hundreds and thousands of times, it us just a familiar sequence that we follow.

Procedures provide a process that we can reliably repeat whenever we form a new crew. If we swap Captains for a crew change, we can confidently expect that they will complete the engine start steps exactly like the last Captain did. We know what we are expected to do and what to expect from each other. Flight standards give us both predictability and context. Decisions flow smoothly and organically.

4.3.3 Familiarity Breeds Contempt

As this old proverb warns, the more familiar we are with something, the less attention we pay to it.[3] An engine start is a highly practiced and standardized task that rarely goes wrong in operational line flying. After thousands of successful engine starts, many pilots unintentionally lower their vigilance.

The same goes for completing checklists. Ideally, the PM reads the checklist step. Both pilots verify the correct system status or switch position. Then, the PF recites the required response. Over time, our verification vigilance can regress while our verbal callouts remain accurate. We can become accurate checklist reciters and sloppy checklist verifiers. Some telltale signs are when PMs stop looking at switch positions as they read the steps from the checklist card or when PFs accurately recite the responses without looking at the switch positions. Whether we have allowed our minds to wander or we have diverted our attention to other considerations, familiarity can lead us to miss important steps. Consider the following account from an ATR Captain.

BOX 4.1 ROTE HABIT PATTERNS DEGRADE VERIFICATION

Captain's report: I called for the Before Start Checklist (Below the Line). As we worked through the checklist items, we passed through the "Cabin secure items and doors", and *unfortunately, I did not adequately think about the checklist items as we discussed them.* I gave the signal to release the prop brake to the ground crew, who replied that the engine was clear. …[After starting the engine] I then moved into the After Start flow, and glanced up to see the ground crew giving the shutdown signal for engine #1. I glanced up at the overhead panel and noticed that the passenger door was still open. I immediately shut down engine #1 and called the Flight Attendant, who said that she was just finishing up the cabin. She closed the door, and I then re-cleared the engine with the ground crew and started engine #1 again. …This was the result of inattention and complacency on my part. *Though I had called for the checklist items, I was not adequately paying attention to the responses I was giving, instead falling back on rote replies to the checklist challenge items and not giving each item the required attention.* I … will endeavor to pay closer attention to what each checklist item says.[4]

Notice that the Captain fell "back on rote replies". This kind of error occurs more often with highly experienced pilots as they settle into their comfort zones.

4.3.4 Latent Vulnerabilities Surface When Something Rare Happens

Standardized procedures can mask gaps in our knowledge. These latent vulnerabilities simmer below the surface until a particular set of conditions combine to allow an error to surface. Referring back to our Swiss Cheese model, this would happen when the holes in the cheese line up to allow the error to pass through. When something goes wrong, surprise and startle can delay our response. Consider the following account of a Boeing 737 engine start mishap.

BOX 4.2 LATENT VULNERABILITY EMERGES DURING AN ENGINE OVERHEAT

FO's report: Normal sequence up to push, cleared to start #2 engine. Turned packs off, … moved switch to GND start, hacked clock. I am not sure when I added fuel as I usually do it at 25% N_2, but today I cannot remember verifying that step. I know I did move the start lever to the idle detent, then saw EGT increasing past normal indications. I kept watching, thinking there is no way it was going to go past red line. N_2 was at 42%, never got any higher than that [indicating a hung start]. EGT went through red line and I should have cut it off but I was under the understanding it was supposed to cut off itself for a hot start. … I was slow to cut off the start lever and did so once prompted by Captain.

Captain's report: … Since the aircraft was approaching the [towbar] release point, I was not monitoring the engine start at all as I anticipated the "Set Parking Brake" instruction from the tug driver and was looking outside to ensure that the aircraft was no longer moving. Once the brakes were set, I thought to myself that I had not heard the starter-cutout and I glanced down at the N_2 [Note: The starter switch is solenoid-held which releases with an audible snap when N_2 passes 56%]. The N_2 was at 42% and not accelerating. … As I was orienting myself to what was happening with the engine start [when] the tug driver made his required calls and I informed him to remain in contact with me via the intercom. Once I saw the starter still engaged, I looked back at the N_2 which was still at 42% and I said to the FO, "It looks like a hung start." As soon as I uttered my thought that we had a hung start, I looked up at the EGT which was already approaching 800°C [overheat is 725°C] and I said, "Hot Start - you are going to have to abort the start" to the FO. The FO replied that the aircraft had an auto-abort function that should abort the start [Note: The procedure calls for the FO to anticipate the impending overtemp and manually close the start lever]. I commanded an aborted start again and told the FO that it made no difference what the book said that the engine start had to be aborted and I reached for the start lever. At this point the FO aborted the start.[5]

Notice how a repetitive, familiar event can mask knowledge gaps (latent vulnerabilities) and lead to flawed decision making. Imagine the FO's mindset at the time. They might have assumed that they never need to abort an engine start because the engine control computer will always do it automatically. This FO's latent misconception, combined with a rare hung start event, combined to cause this undesirable overtemp event.

4.3.5 When Procedures Are Disrupted

Standardized procedures serve many purposes. Along with providing a clear playbook to guide what to do during each phase of the flight, their predictability helps us detect errors. Consider the following cases:

- *When a pilot misses a standardized step*, it is easier for the other pilot to detect the omission, highlight it, and correct it.
- *When a pilot adds a standardized step that is not applicable*, it cues the other pilot of a mismatch in the expected sequence.
- *When a pilot mis-orders a standardized step*, it cues the other pilot that they may be inattentive, task-saturated, distracted, or fatigued.

In each case, the pilot detects the *difference* between the actions taken by the other pilot and what the procedure directs. Even if they don't detect the actual error at that moment, they sense a *wrongness* about it. Since we expect a smooth, predictable flow from task to task, we are particularly skillful at detecting deviations. Mismatches immediately cue us to investigate the reason and to restore order. Consider the task of notifying the flight attendants that takeoff is imminent.

1. The procedure starts with the Captain pressing the attendant call button (chime or ding).
2. This cues the attendants make their departure public address (PA) to the passengers.
3. The Captain monitors the PA audio to confirm that the attendants heard the chime/ding and are safely seated for takeoff.
4. The Captain calls for the Before Takeoff Checklist.
5. The FO reads the checklist which includes the challenge, "Attendant Notification".
6. The Captain responds, "Complete".

Now, let's look at some examples where Captains choose to make nonstandard responses to the checklist.

- *Altering the response:* Pilots sometimes state the response in a different way. Using our example, instead of "Complete", pilots may respond, "That's Complete". This is a minor deviation, but still nonstandard.
- *Adding to the response:* The responding pilot adds additional information. This may even be applicable or useful information, but still nonstandard. For example, for the challenge of, "Attendant Notification", they may respond, "Complete and the cabin is secure."
- *Completely changing the response:* Pilots may significantly alter the response. Perhaps they try to add a little humor to spice up an otherwise monotonous task sequence. To communicate that they have pressed the attendant call button and monitored the response, they might say, "Dung and Done".

These variations seem to emerge more often among highly experienced pilots who have become comfortable. These responses are well-intentioned. They try to impart useful information or keep the flightdeck environment casual and light-hearted. While seemingly harmless, each modification is initially perceived as a mismatch by the FO. They sense a wrongness from what is expected. They pause and process whether the reply is valid or not. This may only take a moment, but that momentary pause still interrupts an otherwise smooth flow. At a deeper level, it may alter how the FO views the Captain's perspective on standardization. "I guess standardization isn't very important to this Captain."

With pilots who fly together often (like pilots within smaller flight organizations), each pilot learns to accommodate each other's variations, so these deviations have less impact. Within large airlines, where frequent crew swaps are commonplace, these deviations can cause significant disruptions. Moreover, the effects of these disruptions are shouldered predominantly by FOs. Captains repeat their personal modifications day after day, while each new FO has to adapt to each Captain's unique quirks.

4.3.6 The Comfort Zone Trap in Decision Making

Decision making flows seamlessly during standardized, familiar, and spontaneous operations. As we gain experience, more of our decisions fall within this easy-going flow. To highly experienced pilots, entire flights and pairings flow smoothly along without encountering a single decision that strays from their standardized and familiar scripts. When everything is standardized and familiar, an atmosphere of ease and comfort surrounds us. We settle in. While not inherently bad, we can find our attention level dropping. Our decision-making process may become lax. This may set the stage for future errors.

Familiar scenario decision making uses recognition and matching. Unfamiliar scenario decision making prompts us to adapt and innovate. How do we handle situations that fall between these extremes? Imagine encountering a situation that is close to, but not quite the same as, past familiar cases. We can decide that it falls "close enough" to a past familiar game plan, so we use it. Because it doesn't match the conditions, we may need to force it a bit to get it to work. This still keeps us within our comfort zone. Next, consider a scenario that is a bit more unfamiliar – a bit more of a mismatch. We could judge that it is still close enough to our familiar game plan. We'll have to use more force, but we can still make it work. Notice how the practice of forcing situations to fit our familiar game plans can become a habit. When a situation doesn't match, we just need to push a bit harder to stay within our comfort zone.

Next, consider scenarios from the other extreme. Imagine a situation that is completely unfamiliar and unexpected. It would require analysis and intuition to innovate a workable solution. Forcing a familiar plan would be clearly inadvisable. Logically, we conclude that there must be a point along this continuum where we need to switch from using/forcing familiar solutions to innovating unique solutions. Misjudging this crossover point is a potential trap. This is one of the signs of the Recognition Trap error – pushing a familiar/common decision-making strategy onto an unfamiliar/uncommon type of situation.

4.3.7 Blended Innovation

If we accept that forcing our familiar game plans with unfamiliar scenarios can lead to failure, what should we do? The truth is that there is no clear point where we switch from recognition/matching to adapting/innovating. Consider Figure 4.1.

On one extreme, we have familiar, comfortable situations that we recognize and match to past game plans. On the other extreme, we have unfamiliar, unexpected situations requiring innovation. Between is a crossover zone where either force or adaptation is needed to make the plan work. At the far left, we really don't want adaptation or innovation. We want consistent, standardized procedures guided by

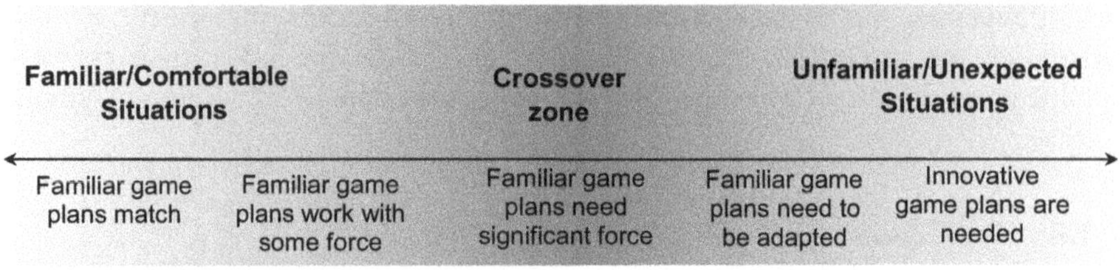

FIGURE 4.1 Game plan continuum.

consistent, standardized decision making. At the far right, we need to innovate solutions for unique problems. In between, we have situations requiring both sets of skills. There isn't a clear line where we switch from one form of decision making to the other.

How do we know where our current scenario fits on the continuum? In truth, we don't. Experience guides our decision making. Only with hindsight can we determine how well we chose. The defining parameter is the amount of *force* we need to execute our game plan. The more force we need to apply, the more unsuitable the game plan is. As Master Class pilots, we develop both a keen awareness of the current conditions and the ability to accurately select a game plan. This will become clearer as we examine the remaining two classifications: unfamiliar/nuanced and unknown/rare.

4.4 DECISION MAKING WITH UNFAMILIAR OR NUANCED SITUATIONS

As situations become more unfamiliar, we find ourselves faced with situations that lack clear directives. We need to apply our understanding of the policy intentions that underly written procedures. This category divides as follows:

- *Unfamiliar, but simple:* These are situations that include aspects that we haven't encountered before. They may be rare nuances or situations that are not implicitly covered in the manuals. While they remain simple and uncomplicated, we'll need to acquire more information and make a judgment call. For example, consider some detail on our dispatch release that we haven't seen before or don't recognize. We pause our preparation flow, access the expanded dispatch release section of our flight operations manual, and clarify the meaning of the unknown item. We could also call our dispatcher and ask them. Either way, we take the time to convert the unfamiliar situation into a familiar one.
- *Familiar, but requiring balancing:* Sometimes, we encounter conflicting priorities. The airline operation is driven by the schedule, so staying on-time is a strong motivator. Sometimes, making up for lost time pushes the pace of our operation faster than we wish to go. Delays can create problems with crew duty day limits, passenger connections, curfews, and slot times. As we look for ways to make up for lost time, we may explore shortcuts to

shave minutes. Shortcuts tend to conflict with typical priorities. Our decisions seek a balance between these conflicting priorities.

- *Familiar, but complex:* Some airline operations are inherently complex, especially with high-volume, tight-airspace airports. Flying to large hub airports is complicated. Deteriorating weather and arrival/departure saturation make it more challenging. If it also happens to be one of our airline's hub airports, most of our passengers will need to make connections and we may need to swap to another aircraft. The more systemic stress we add to a flight, the more consequential our decisions become.
- *Familiar, but with unpredictable risk factors:* Sometimes, the situation is familiar, but the conditions move us further into the crossover zone. We encounter higher risk and need to pay closer attention to maintaining the balance between conflicting goals. Unpredictability increases stress on our decision making.

4.4.1 Conscious Consideration

While decisions within the previous category of standardized/spontaneous situations were scripted, easy, and clear (see, recognize, match, and do), decisions within this unfamiliar and nuanced category require more conscious consideration (see, analyze, adapt, innovate, and do). Conscious consideration requires time and attention – resources often in short supply when we are late or stressed.

- *Human biases:* How we perceive, consider, and choose our decisions are subjects of extensive HF research. Psychologists have documented over 100 human cognitive biases.[6] We discover that we typically don't select our choices based on rational logic. Instead, we filter our choices through feelings, emotions, subconscious mindsets, and instincts. This is not necessarily bad. In fact, our biased, instinctual, gut-felt choices often serve us quite well. Sometimes, however, these choices prove to be poorly suited for time-pressured decision making. The more aware we are of our personal biases, the better we can control their undesirable side effects. In summary, we tend to choose successful choices that:
  - we find most *familiar*
  - we have used most *often*
  - we have used most *recently*
  - we have used in *similar conditions*
  - we have *used previously* at this location or airport

  Let's take a moment to reflect on some choices we have made recently in unfamiliar or nuanced situations. Compare what we chose to do with the biases of this list. Notice similarities? This list of biases tends to conform with our comfort zone. These are the well-worn tools in our toolbox – the ones we select most often. When we encounter an unfamiliar situation, we derive comfort from using familiar tools, even if they don't fit the job. Trying an unfamiliar new tool feels like adding uncertainty to an already difficult situation.

- *Uncertainty rising:* The significant variable here is uncertainty. Sources of uncertainty arise from[7]:
  - Missing information
  - Unreliable information
  - Ambiguous or conflicting information
  - Complex information

  In our minds, familiarity eases the uncertainty and supports the illusion that we are effectively compensating for the unknown. The greater the level of uncertainty, the more strongly we are attracted toward familiarity bias. Additionally, the stronger our affinity toward our favorite choices, the more likely we are to neglect potentially better options.

  NDM holds that we are biased toward quickly selecting an option that we view as *workable*. We assess a situation and select the first viable game plan that pops into our head. We do this even when conditions favor innovating a potentially better option. Moreover, our choices strongly align with our gut-feeling, not by how we logically evaluate options.

4.4.2 The Drift Towards "Strong but Wrong"

Over time, our decisions migrate from choices-by-thinking to choices-by-feeling. This happens very slowly – so slowly that we don't detect the gradual drift. We tend to select strongly felt choices that may be inappropriate for the given scenario – called a "strong-but-wrong" error (Reason, 2008, p. 17). We tend to see this kind of error more often with pilots who have become particularly set in their ways.

As Master Class pilots, we also have our favorites – our comfortable go-to choices. The difference is that we remain aware of our biases and accurately assess the suitability of our choices. This ensures that they are not just workable solutions, but that they effectively control risk.

4.4.3 Equal Choices

What if we have two or more choices that are essentially equal? Most experienced pilots do well with this dilemma. We recognize that since each choice delivers equal benefit, it doesn't really matter which one we pick. We default to our biases and pick the one that is most familiar or recent. An example is when Tower asks us, "Say requested runway." If runways are essentially equal, many of us would select the one we have used most often. We recall a mental picture of landing, the available runway turn-offs, and expected taxi route to our gate.

An undesirable effect can emerge when the severity of potential consequences rises. If faced with two choices offering similar benefit, but with each containing elevated risk, some pilots exhibit reluctance to select either. Their fear of making the wrong choice inhibits them from choosing.[8] Pilots with this bias tend to avoid making tough decisions. They may try to get another authority, like the chief pilot, dispatcher, or station manager, to decide for them. This transfers the responsibility for any undesired consequences onto the other decision maker.

4.4.4 Similar Choices with Conflicting Benefits/Penalties

Another problem arises when each choice contains different benefits and penalties. Take landing at the intended airport under marginal weather conditions versus diverting to an alternate airport. Pilots are drawn toward their original destination because it falls within their comfort zone (successful arrival at the scheduled destination airport), but repelled from it by the rising risk (possible windshear or landing performance issues). Compare this with a much safer choice of diverting to the alternate (undesirable, though with safer outcomes). While a much safer choice, it creates additional problems with the schedule, passenger connections, and duty day limits. Pilots can second-guess themselves about the unused option while experiencing stress over the chosen option.

4.4.5 Moving Toward One Choice as the Availability of the Other Decreases

In this case, as we move closer to our selected choice, we begin losing the availability of the alternate option. Consider the decision to continue in a holding pattern versus diverting to our alternate airport. Assume that we have a diversion plan, but before we reach our divert decision fuel, ATC announces that the airport is accepting arrivals. We accept the approach. As we descend to lower altitude, ATC starts assigning delay vectors and slowdowns. Our fuel reserves drop. At some point, we need to commit to either continuing with the approach and hoping that we make it in or exiting the traffic flow and diverting to our alternate with even less fuel. As we move closer to the landing option, our fuel reserves may drop below what we need to safely divert. The safer diversion option is lost.

4.5 THE LENS OF BIAS AND EXPERIENCE

Our decisions are biased by a lens formed by our biased mindset and past experiences.

4.5.1 How Our Experience Affects Familiar, Spontaneous, and Standardized Decisions

Consider Figure 4.2 of how our recognition of the situation is affected by our lens of experience and bias.

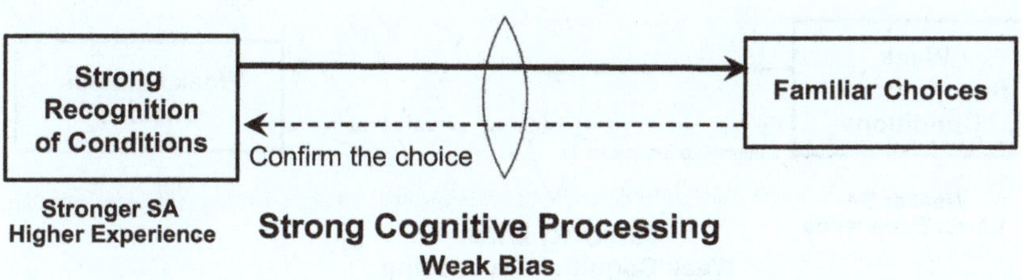

FIGURE 4.2 Strong cognitive processing with weak bias.

We start with strong recognition of conditions in a simple, uncomplicated situation. We see, hear, and sense what is happening. Our minds process information through a lens shaped mostly by our accurate thinking and to a lesser extent by our personal bias. When the situation is familiar and our situational awareness is strong, we think clearly. There is little distortion through the clear lens. After selecting a familiar choice, we reverse backward in a feedback loop which confirms the accuracy of our decision.

Passing through this lens both forward and backward models how our recognition affects our choice and how our choice affects our assessment. As long as the lens remains clear, the process works. If the lens becomes cloudy, we subconsciously focus our attention on parameters that support our selected choice and decrease our attention toward parameters that might contradict it. For example, if we are flying a normal visual approach on a clear, calm day, we feel confident with our decision to land. We rarely consider searching for reasons to go around. We remain aware of possible threats, but when everything supports continuing to land and nothing warns us that we should go around, we strongly favor our landing decision.

Subconsciously, we devote more attention to parameters that encourage us to continue. We devote less effort toward monitoring for warning signs. Why waste effort looking for something that doesn't exist? Instead, we focus our attention toward flying a smooth approach to a perfect landing. It is a beautiful, calm day with a stabilized approach. Everything is following the game plan. Over the years, this process works well for us. Our RPDM process is reinforced and confirmed.

4.5.2 How Our Bias Affects Unfamiliar and Nuanced Decisions

Now, let's examine our second category – situations that are unfamiliar or nuanced (Figure 4.3). Since our situational awareness is lower, we have less experience to draw upon, so our recognition is weaker. We need to consider our choices more carefully. Time pressures, complexity, and unpredictability tend to weaken our rational thinking. The lens becomes cloudy and allows biases to distort our decision-making process. As we are pushed outside of our comfort zone, we look for something familiar to latch onto. Our mind uses RPDM like a life preserver. The seemingly strongest choices are those familiar options that we use most often. Since familiar choices have worked hundreds and thousands of times in the past, we feel reassured by selecting one of them. This distorted, biased decision making imparts a false sense of legitimacy. Our

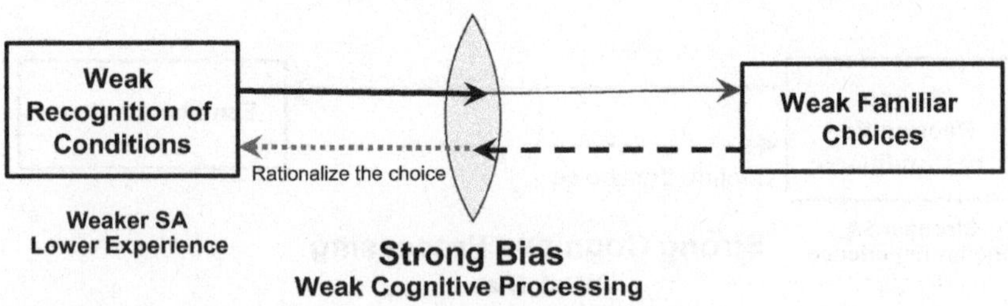

FIGURE 4.3 Strong bias with weak cognitive processing.

biased mind then feeds that selected choice back through the lens to confirm our recognition. Again, our biases kick in and encourage us to subconsciously elevate parameters that support our selected choice and minimize warning signs that might conflict against it. The higher our stress and greater the time pressure, the stronger this bias effect. In the end, the same mental processes that lead us to select familiar choices with familiar situations lead us to select familiar, but possibly inappropriate, choices under unfamiliar situations. Moreover, RPDM disguises these inappropriate choices by wrapping them in the secure feeling created by our comfort zone.

4.5.3 How the Bias Affects Unknown or Rare Situations

The third category of rare and unknown events offers two decision paths depicted by Figures 4.4 and 4.5.

- *How the biased process distorts the lens:* In Figure 4.4, we start at the left block (No/Low Recognition of Conditions). Flying in-the-moment, we feel time-pressured to select a game plan. Our analysis process is depicted by the solid black arrow pointing to the right (Analyze the situation). The absence of our familiar RPDM mode allows bias to influence our decision-making process (cloudy lens). Assuming that we are well-intentioned, experienced, comfortable pilots, almost every situation we typically encounter is one we have seen before and handled successfully. Suddenly, we are faced with a rare anomaly. The cloudy lens distorts our perception. We sense discomfort. Our stress rises (thin, gray arrow pointing to the right). Scanning the range of available options, our biases encourage us to pick a familiar game plan that appears to be marginally usable. It feels workable or close enough that we can force to conform with conditions. This preserves the secure feeling of our comfort zone. Our feedback and verification loop reverses back through the same cloudy, biased lens (black, dashed line pointing to the left). It again biases our perception. In our need to confirm the "rightness" of our decision (gray, dotted arrow), we allow rationalization to subconsciously deflect our attention toward indications that support our choice and suppress indications that contradict it. Our biased decision feels right.

  Applying an inflight example, consider a flap malfunction detected while configuring to land. The appropriate game plan is to break off the approach, complete the appropriate flap malfunction checklists, and return for what will probably be an emergency landing – all tasks outside of our comfort zone.

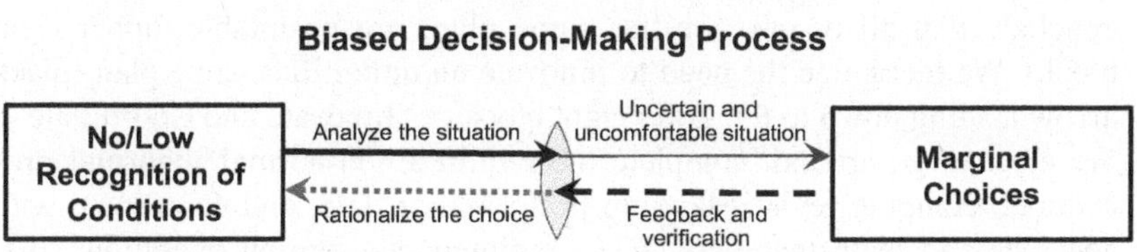

FIGURE 4.4 Biased decision-making process.

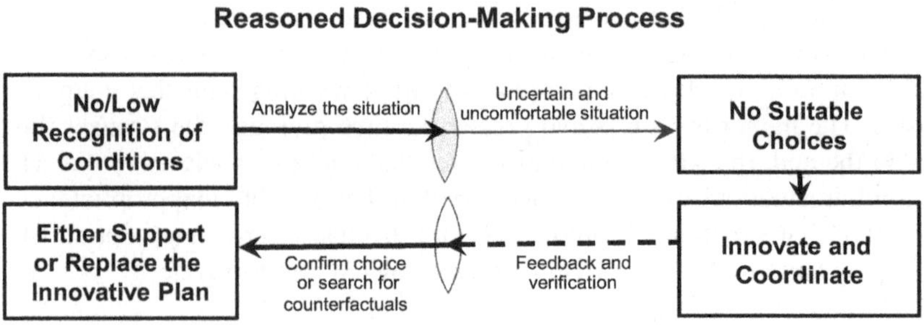

FIGURE 4.5 Reasoned decision-making process.

Following the Biased Decision-Making Process, our crew chooses a marginal, familiar game plan. They accept the flaps in their partially deployed state, add a few extra knots to their approach speed, and continue their approach. Rationalizing their decision, they evaluate aircraft handling and conclude that the aircraft is safely flyable. They judge that they have sufficient runway length for stopping margin. They further rationalize that this choice stays on profile, avoids a low fuel situation from a delayed landing, keeps them on schedule, and gets them to the gate where they can start coordinating for mechanical repair. Their familiar sequence of approach, configure, and land is retained. Safely on the ground, their decision feels validated. While they didn't follow procedure, their marginal game plan remained familiar and workable.

Reviewing this crew's actions, we may not agree with their decision-making process. We have the advantage of assessing their actions from our no-stress, no-time pressure, hindsight perspective. Instead, if we place ourselves within their scenario, we can appreciate how pilots, making decisions in-the-moment, might select this option. Again, our purpose here is not to judge right or wrong, but to begin to understand the mental process that generates biased decisions. We then use this awareness to clean our own lens and guide our own well-reasoned choices.

- *How the reasoned process cleans the lens:* The Reasoned Decision-Making Process is represented by Figure 4.5. The top half of the graphic is the same as for the Biased Decision-Making Process. The difference is that we recognize that we have a unique situation with our flap malfunction. We recall that our aircraft has different checklist procedures for no flap, partial flaps, and asymmetric flaps for both leading edge and trailing edge flaps. Combining the rarity of these malfunctions with operational pressures, weather, and fatigue makes this one of the more challenging scenarios in aviation. We conclude that all of our familiar game plans are unsuitable (upper right block). We recognize the need to innovate an unfamiliar game plan (black arrow leading down to the lower right block of "Innovate and Coordinate"). We elect to go around, complete the required non-normal/abnormal procedures, compute revised landing performance data, and coordinate with agencies and team members (ATC, maintenance, station operations, dispatch, and cabin crew). We assign duties, rehearse contingencies, and make sure everything is ready before commencing our approach. As our feedback loop reverses back through the clear lens (in the lower half of Figure 4.5), we guard against the adverse effects of bias. We recognize the need to monitor for both confirming indications that our game plan is working and for warning signs that it might not.

4.6 LENS DISTORTION AND RECOGNITION TRAP ERRORS

Examining the differences between the Biased Decision-Making Process (Figure 4.4) and the Reasoned Decision-Making Process (Figure 4.5), we start with the assumption that there is always some distortion in our lenses. The higher our level of stress, uncertainty, and time pressure, the more the distortion. The second assumption is that everyone wants to select the best decision. The difference lies with how each process influences our decision making.

The path that concerns us is the Biased Decision-Making Process (Figure 4.4). This is where we risk making Recognition Trap errors – where we try to force familiar game plans into uncommon situations where they won't fit. To avoid the Recognition Trap, we need to become aware of how our minds work. Our minds will always distort the view through the decision-making lens. None of us is immune. The lens will:

- Bend our perception to select the familiar choice, whether it is appropriate or not
- Bend our selection toward a game plan that supports our comfort zone
- Bend our assessment to magnify indications that our plan is working and suppress warning signs that imply that our game plan is failing

Knowing that we cannot fully trust the perspective we see through the lens, we compensate by raising our level of skepticism. Skepticism widens our perception and challenges us to accurately rate the effectiveness of our game plan. This assessment is not just a one-time evaluation made when we select our game plan. We continue reassessing it to correct the distortion from our optimism biases.

4.6.1 When Effort Replaces Reassessment – The Frozen Bolt

Recognition Trap errors are present in many aviation accidents and mishaps. Pilots become task-overloaded while trying to push their failing game plans across the finish line. They start with the best of intentions. They want to comply with procedures and safe practices, but as stress and time pressure increase, they become blinded to rising risk. Forcing their game plan becomes more important than making sure that it remains viable.

Imagine that we are mechanics working in an auto repair shop. We need to remove an engine bolt. We recognize this as a familiar task. We have successfully removed many engine bolts over the years. We position our wrench and try removing the bolt. It doesn't budge. If we paused to evaluate the situation, we would identify three possible outcomes: the bolt loosens, the bolt breaks, or we innovate another plan.

Our good intention is to successfully remove the bolt. In all of our past attempts, we succeeded. Since we are confident in our skills, our egos become involved. We decide to push harder on the wrench. Still, the bolt doesn't move. Now, we feel frustrated. As we redouble our efforts, we brace one foot against the workbench and push really hard. Snap! The bolt breaks, we tumble over the fender, and fall down. Alerted by the commotion, our boss walks over and asks us why we didn't stop and spray the frozen bolt with some penetrant to unfreeze it first. We are already berating ourselves for not doing exactly that. Encouraged by our past successes and pushed by ego and impatience, we thought the bolt would loosen if we only applied a bit more force. In the heat of that moment, we ignored the possibility that the bolt might actually break. We didn't consider an alternative game plan because, in-the-moment, pushing harder appeared to be the only viable option.

Let's examine this process with an example with an aviation story. Consider this report from a crew that continued an approach resulting in a hard landing and aircraft damage.

BOX 4.3 CREW PUSHES THEIR GAME PLAN IN ADVERSE CONDITIONS RESULTING IN AIRCRAFT DAMAGE

Captain's report: Taking vectors for ILS approach to RWY 20, we called field in sight at 10 miles turning final. Cleared to land at the 5-mile fix and Tower began to give us windshear alerts on approach end of RWY 02 [The opposite end of the landing runway]. Fully configured, we continued with the approach with no windshear indications through 1,000′. Descending through 500′, the First Officer clicked off autopilot and began to feel some wind gusts. Somewhere around 200′ we experienced a 20-knot gain followed by a quick high-speed clacker. *While correcting and beginning landing flare, we received windshear caution alert just prior to over the numbers. A go around was our trained procedure for these indications, yet I felt we were so close to touchdown that a transition to climb would have put us in a precarious state, so I stated, "continue" to First Officer.* Floating and trying to maintain airspeed, we dropped on the runway with a hard landing. The right main gear touching down first causing a whiplash effect on the left wing. We rolled out and exited runway as normal. Taxied to the gate and reported our windshear to Tower as a 20-knot gain at around 100′.

I asked First Officer to check [the gear] for any indication of the hard landing and our surprise, we found the left wing damaged at the trailing edge of the outboard flap and wing tip below the winglet. I reported the damage to Maintenance and called chief pilots.

Our aircraft arrived in the undesired state of flying in low level windshear. We continued an approach we had made many times before in very similar conditions and were looking for cues to discontinue approach. *We got to a point, perhaps so close to landing, that the threat of windshear was not going to keep us from landing. I, as Captain, felt that the sudden low level windshear*

> *we encountered was maybe too much to overcome and simply landed the plane.* Take note of this incident and realize that although we operate in windshear conditions regularly, it can become strong enough to alter flight at low levels to a point where no recovery is possible. It is our responsibility as pilots to see the windshear clues and either abort the maneuver or not attempt it. *We failed to do so with the untimely clues we received and resulted in aircraft damage. Teach not only escape guidance, but when to simply not attempt it. I operated cautiously [with respect] to weather all the time, yet this one I did not see coming.*[9]

Notice the similarities between our bolt-breaking mechanic and our winglet-damaging pilots. In both examples, they pushed a familiar game plan past the breaking point. They both detected the warning signs (termed "untimely cues" by the Captain in the report), but discarded the alternatives because they felt that it was too late. Each ignored the option of switching to an alternative game plan.[10] Both were fully immersed in their effort to push through their problem and achieve their desired outcome. In hindsight, the warning signs were clearly present, but they didn't seem clear while they were in-the-moment. Their emotional commitment (confidence, stress, determination, and preconception) blinded them to safer alternatives.

4.6.2 Feeling That the Plan Is Right Replaces Getting It Right

As we compare successful events with mishap events, we notice crews using similar RPDM processes, but arriving at noticeably different game plans and decisions. Why does one crew select a successful path, while another chooses one that leads to failure? How can experienced pilots, equally motivated to succeed, select very different game plans? The distinction may lie with mishap pilots choosing game plans that *feel right* instead of plans that *get it right.*

Our comfort zone is like sitting in our favorite chair at home. Suddenly, a fly enters the room and starts buzzing around. It irritates us. We feel a strong urge to remove that irritant and restore our comfort zone. Removing the fly becomes more important than anything else. We get out of our chair and begin chasing it around with a rolled-up magazine. Swatting at the fly, we knock over a lamp causing it to shatter on the floor. The fly zips away unharmed. Cleaning up our mess, we consider our decision. It all started with an irritating fly, but ended with a broken lamp – not because the fly was a threat, but because it disturbed our comfort zone.

Consider Clem's failed approach. Discovering that he was too fast and steep on final, he might have felt irritated by his situation. His decisions followed the quickest path to remove the irritant. He chopped his thrust levers to idle and pushed the nose over. This restored the desired glidepath, but his airspeed rose above stabilized approach limits. This was doubly irritating, so he chose to ignore the problem, accept his excess speed, and land. Because of a slippery runway, he slid into the overrun. He just broke the lamp. It happened because, in the heat of the moment, landing seemed like the best way to remove the irritant of being fast and steep.

The Master Class perspective recognizes that the right decision doesn't need to *feel* right. In fact, it can feel like failure. Going around from an approach that we botched feels embarrassing. It admits to everyone that we failed to manage our profile. Going around forces us to accept responsibility for the steps that we should have taken earlier on the approach to avoid the unstabilized final.

In the heat of the moment, we may experience irritation with events that don't go well. It will feel uncomfortable. How do we use this bad feeling? We use it as a warning sign. RPDM relies on experienced pilots recognizing their situations accurately. When a game plan is failing, it *should feel* like it is failing. Acknowledging actual parameters, we realize this too late to salvage our original game plan. Remember that a Recognition Trap error is only consequential if we continue it to its unfavorable conclusion. As soon as we detect that our game plan is failing (bad feeling), we switch to an alternative or innovative decision (try something else or start over).

4.6.3 Exceptional, Novel, and Uncertain Situations

Much has been written about "black swan" events.[11] These are rare and surprising events that can generate highly consequential outcomes. They describe events that are so unique that no procedures were developed and no one was specifically trained to mitigate them. The combination of our surprise along with our lack of preparedness magnifies the intensity of the shock and startle that we may experience. Ideally, when we lack familiar references for a solution, we should pause, examine the indications, solicit guidance, and innovate a viable solution. Instead, bias-vulnerable pilots may latch onto a familiar solution, even when it is clearly inappropriate. This is an example of *availability bias*, where we focus on the many past occasions when a chosen game plan succeeded. The surprise factor seems to amplify our tunnel vision toward supporting parameters and away from contradictory parameters.

4.7 The Five Stages of a Failing Decision in Recognition Trap Errors

A failing aviation decision often progresses through five stages (Figure 4.6).

1. It starts with *surprise*. This represents the sudden disruption of our comfort zone. Everything was going fine until events started unraveling.

Five Stages of a Recognition Trap Error	
Surprise	What's happening here?
Effort	I need to slow down!
Hope	Please slow down already!
Resignation	This isn't working.
Acceptance	Might as well just land.

FIGURE 4.6 The five stages of a Recognition Trap error.

2. The next phase is *effort*. We work harder to solve the problem. For an unstabilized approach, this would include chopping power, increasing drag, aggressively regaining the glidepath, and S-turning.
3. The third stage is *hope*. Having used every trick from our toolbox and still not solving our energy problem, we realize that we can't do anything more. We hope that our corrections will prove to be enough.
4. Next is *resignation*. We realize that we are just not going to achieve stabilized approach parameters. We admit that we've done all that we can. We resign ourselves to accepting the failed approach.
5. The final stage is *rationalization*. We shift to mitigating the consequence of the failure. Moving the goalposts, we rationalize that we've failed the approach, but that we are going to ensure that it ends safely. "It's a long runway." "I'm going to get on the brakes early."

The obvious question is, "Why don't pilots just go around from unstabilized approaches?" Statistics across the airline industry indicate that a relatively small percentage of unstabilized approaches end in a go around. Far more often, we land. During debriefs, we often agree that a go around would have been the right choice, but under the stress of the moment, something subconscious takes over. This instinctive reaction leads most pilots to land.

As Master Class pilots, we recognize this subconscious process and develop personal countermeasures to detect it in ourselves. We notice when it is happening, recognize it as a sign of possible failure, abandon our game plan, and select a new course that reverses the problem. Despite this strong psychological instinct pushing us to continue a failing game plan, self-awareness redirects us to switch to a contingency backup plan.

NOTES

1 Klein cited conditions of "greater time pressure, higher experience level, dynamic conditions, [and] ill-defined goals" to describe the professions that he studied – features common with career aviation professionals and the aviation environment (Klein, 1999, p. 95).
2 Klein cites conditions that include a "need for justification, conflict resolution, optimization, greater computational complexity." (Klein, 1999, p. 95).
3 Quote attributed to Geoffrey Chaucer, in his work, Tale of Melibee in the 1300s.
4 Narrative edited for clarity and brevity. Italics added. NASA ASRS report #1700977.
5 Both narratives edited for brevity and clarity. NASA ASRS report #1691492.
6 To explore the range of human biases, reference the "Cognitive Bias Codex" on Wikipedia. It is an interactive graphic that catalogs over 100 human biases. They are divided into four categories: too much information, not enough meaning, how we remember, and the need to act fast.
7 Klein (1999, p. 277), Previously published under: Schmitt and Klein (1996, August). "Fighting the Fog: Dealing with Battlefield Uncertainty". *Marine Corps Gazette 80,* pp. 62–69.
8 Known as the Buridan's Ass Paradox – It refers to a hypothetical situation where a donkey that is equally hungry and thirsty is placed precisely midway between a stack of hay and a pail of water. Since the paradox assumes the animal will always go to whichever is closer, it dies of both hunger and thirst since it remains torn between the equal desires of food and water (Wikipedia: https://en.wikipedia.org/wiki/Buridan%27s_ass).

9 Narrative edited for clarity and brevity. Italics added. NASA ASRS report #955213.
10 In fairness, this report dates from 2011. Training now encourages MAX thrust go arounds from windshear even if the aircraft momentarily touches down.
11 The theory was developed by Nassim Nicholas Taleb to explain: the disproportionate role of high-profile, hard-to-predict, and rare events that are beyond the realm of normal expectations…, and the psychological biases that blind people, both individually and collectively, to uncertainty... (Wikipedia: https://en.wikipedia.org/wiki/Black_swan_theory).

BIBLIOGRAPHY

ASRS – Aviation Safety and Reporting System – National Aeronautics and Space Administration (NASA) Online Database. (n.d.). Retrieved from Aviation Safety and Reporting System: https://asrs.arc.nasa.gov/search/database.html.
Klein, G. (1999). *Sources of Power: How People Make Decisions*. Cambridge, MA: The MIT Press.
Klein, G. (2003). *The Power of Intuition: How to Use Your Gut Feelings to Make Better Decisions at Work*. New York: Currency Books.
Mosier, K. L. (1991). Expert Decision Making Strategies. Proceedings of the Sixth International Symposium on Aviation Psychology, (pp. 266–271), Columbus, OH.
Reason, J. (2008). *The Human Contribution: Unsafe Acts, Accidents, and Heroic Recoveries*. Burlington: Ashgate Publishing Company.

5 Situational Awareness

Loss of situational awareness (SA) is routinely cited as a contributing factor in many accident and mishap reports. Across the industry, we find the term useful, but do we really know what it means?

- We describe as a quantity, but can't accurately measure it.
- We quickly build it, but can instantly lose it.
- We believe we have it, but don't realize that we have lost it.
- We might actually have it, but convince ourselves that we don't.
- We can build it to describe our current position, but it changes moments later.

It is a wide-ranging and flexible concept. This lack of clarity is reflected by the ways we describe SA.

- *All or nothing:* We generalize SA as an all-or-nothing commodity. We conclude that the mishap crew "didn't have SA", or that "they lost it." We use the label "zero SA" to describe pilots who seemed to have little awareness of what was going on. We use "total SA" to describe pilots who completely understood what was happening.
- *More or less:* We refer to SA in relative quantities. "They didn't have enough SA." "They needed more SA." This reflects our assumption that SA can be cumulative. The more SA we collect, the better we understand the situation.
- *Good or bad:* We label accurate SA as good and misguided SA as bad. We make these judgments in hindsight. This is because we know the outcome of the event, detect all of the relevant conditions, and discard all of the non-relevant conditions.
- *Noticed or unnoticed:* Hindsight judgment allows us to accurately separate useful pieces of information from irrelevant ones. "The crew should have noticed that." "If they had noticed that their airspeed, the mishap wouldn't have happened."

These common uses demonstrate how our concept of SA spans dimensions of presence, quantity, quality, time, judgment, and confidence. As Master Class pilots, we need to find ways to understand what SA is, how we form it, what degrades it, and what it looks like on the flightdeck.

DOI: 10.1201/9781003344575-6

5.1 UNDERSTANDING AND IMPROVING OUR SA

We use sensory perceptions, ideas, and impressions to build our SA. SA, in turn, shapes our understanding about what has happened, what is happening, and what is likely to happen.

- *Sensory perceptions* come primarily from what we see, but we also include our other senses. We constantly scan our instruments, environment, and flightpath to assess our current position and trajectory.
- *Ideas* are the mental concepts that we form from the indications that we perceive. We often express them as judgments. We are where we want to be, or not. Our path is either following our game plan, or it isn't. We are moving the way we want to move, or not.
- *Impressions* arise from our subconscious. They reflect the gut-felt judgments that emerge from our past experience. The situation feels right, or it doesn't.

All three of these SA components are fluid, interactive, and interdependent. Let's examine how we use them to skillfully build our own SA.

5.1.1 Studying How We Form Our SA

From a hindsight perspective, we can accurately rate the quality of the SA we had with any past event. This is because we know how that flight progressed and how it ended. We also know which parameters, decisions, and actions contributed positively or negatively to that end result. To improve our SA-building abilities, we need to translate this hindsight wisdom into skills that we can actively apply while flying.

Consider a mishap event. Examining the timeline, we identify an early period where everything appeared to be going well. As we trace the timeline forward, we identify a later portion where the event appeared to be going poorly. Between these two time segments, we locate the point of time when the game plan began deviating from the crew's desired trajectory. We focus our investigation at this point. Analyzing the scenario around this deviation point, we ask some questions.

- What was the first indication or event that led the crew to realize their problem?
- What were the warning signs that preceded that realization?
- Where was their attention focused as they missed these early warning signs?
- What was their mindset?
- How did their mindset affect which indications they were actively monitoring?
- How did their mindset affect missing those indications?
- How was their crew coordination?
- Was there a breakdown of crew roles and responsibilities?
- Was there a pivotal moment when their priorities or game plan began to change?

- When did they first notice that their SA was flawed?
- How did their realization of their failing SA evolve?
- What were their reactions to this realization?
- What outside or inside factors contributed to this realization?
- What planning steps could they have taken earlier in their scenario to preserve their SA?
- What could they have done to improve rebuilding their SA?

These questions analyze their background conditions, perceptions, and mindsets. We do this before we judge or rate their SA. By exploring these questions, we begin to form an understanding of how the event unfolded for them. In cases where the mishap crew successfully rebuilds their SA, the following questions are useful.

- How quickly did they rebuild their SA?
- What could they have done to improve the process?
- What were the steps they used?
- How did they prioritize their SA-building process with other competing flightdeck tasks?
- What crew factors improved or hindered their SA-building?

It doesn't really matter that we don't know what the crew was actually thinking during their event. Our motivation is to use their event to improve our own SA-building skills. As we contemplate these questions, we draw upon our own processes, biases, experience, and insight. We reorient our perspective away from them and toward ourselves. We learn to look less through a microscope at their SA-building process and more into a mirror toward ours.

- *The jumpseat perspective:* Imagine sitting on the jumpseat while the mishap crew handled their event. We sense the pacing and workload as the event unfolded. Notice the points where mounting workload caused their pace to quicken. Notice the times when conflicting information caused confusion.
- *The pilot seat perspective:* Next, rerun the scenario as if we are the actual pilots with the mishap event. Start at a point well before the event started to unravel, then press PLAY. Exercises like this reconstruct what HF scientists call the naturalistic environment. As we imagine the event transpiring in real time, we speculate how our personal SA-building practices might have handled the event.
    - o How would our SA-building practices have shaped our mindset during this event?
    - o What critical information was available?
    - o When and how would that information appear? How would it change?
    - o How would each new piece of information affect our SA in-the-moment?
    - o What parameters might have seemed important in-the-moment?
    - o Reflecting on our personal habit patterns, how do we typically allocate our attention toward detecting relevant indications?

- How would our shared mental model affect our crew performance with this event?
- How is our mindset typically affected by our mood? Fatigue level? Frustration level?
- How are we affected by environmental conditions (light level, sun position, turbulence)?

Consider how the mishap crew's SA changed during their event. Imagine how our SA would have changed if we were there. When we look into the mirror, we can evaluate:

- If we were the pilots involved in this event, how would we have reacted at critical decision points during this scenario?
- Would our personal techniques have helped us detect the warning signs and mitigate adverse consequences?
- Do we have past flights that feel familiar with this event? What did we do to avoid unfavorable outcomes?
- Would our answers change if we were tired, frustrated, or on the last leg of the trip?

5.1.2 Using Stories to Improve Our SA-Building Skills

We never have "zero SA". SA always exists in some form because our minds structure events as stories. No matter how disjointed individual facts may seem, our minds will strive to link them together along a logical sequence with a beginning, a middle, and an end. Likewise, SA-building links our recollection of the past with our perception of the present to envision the future. The beginning/middle/end of story writing mirrors how we link past/present/future to build SA. When our story transitions smoothly through these three time frames, it feels right.

Consider what happens when these three time frames don't flow smoothly. Consider an example of watching a magic show. As the magician moves around the stage manipulating the props, our minds follow their actions. We analyze what has happened and what is happening to build a story to predict how it will end. When the magician finally reveals the trick, we discover that we have been fooled. The story we wrote in our mind was wrong. Are we guilty of having zero SA? Insufficient SA? Bad SA? If asked to recall the details building up to the magician's reveal, we would accurately list all of the steps. Doesn't that sound like good SA? Our problem was not with perceiving the details. It was with missing the *relevant* details or the *hidden meaning* behind them. If we define good SA as simply knowing what transpired and what was happening, then our SA at this magic show would be considered as excellent. The fact that we failed to predict the outcome means that our SA failed to transition smoothly from present moment SA to future SA (in this case, because it was skillfully misdirected by the magician). While this disconnection between past, present, and future SA is entertaining at a magic show, we find it highly undesirable in aviation.

5.1.3 Understanding Failed SA Transitions in Aviation

We routinely use a story-building process in aviation. It guides our expectations for how events should unfold in the future. These expectations affect how we monitor and perceive details. Consider the following report from a crew that convinced themselves that they were cleared for takeoff.

BOX 5.1 CREW CONVINCES THEMSELVES THAT THEY WERE CLEARED FOR TAKEOFF

Captain's report: ... We departed the ramp from Spot X and Ground Control assigned XXR, which was a very short taxi from Spot X. I made the Takeoff PA, and, as I said I would, I taxied very slowly while the FO loaded the box with the new data, read the LEGS page aloud (from the new runway) as I followed along on the ZZZZZ 3 Dept on my iPad, and briefed the Takeoff Data. As we were in the middle of the [ramp] pad, northbound on taxiway, still fairly far from the departure end of the runway, Tower asked if we were up (on frequency). The FO said we were. *Tower then asked if we would be ready at the end. The FO and I briefly checked with each other and acknowledged that, yes, we would be ready at the end. In my recollection, I then heard "RNAV to ZZZZZ1, Cleared for Takeoff, XXR". I chimed the flight attendants and turned on my two Landing Light switches, which is something I purposefully only do once Cleared for Takeoff. The FO finished the checklist with plenty of time (everything was properly set up before she began it), and prior to crossing the hold short line onto XXR, she said something like, "I'm pretty sure I heard Cleared for Takeoff, but do you want me to verify with them?" I looked at my Landing Lights, saw that I had moved them to the ON position, which I "only do" once cleared for Takeoff, and thought I specifically heard "RNAV to ZZZZZ1, Cleared for Takeoff." So, I said to her, "No. I don't need you to verify. I heard Cleared for Takeoff, too... But please... verify if you want to! Do you want to?" She said something like, "No, I think we're good. We both heard the same thing."* I said something like, "I specifically remember hearing 'RNAV to ZZZZZ1', which they never say until they clear us for Takeoff. We're okay."

While on the Takeoff Roll and prior to V_1, Tower called, "Aircraft X?" We did not answer. They called again once we were airborne and told us to fly Runway Heading. I pretty much knew at that point that we either were NOT cleared for Takeoff... OR we WERE cleared for Takeoff and the Tower Controller forgot that he cleared us. Shortly after that call, the Tower Controller said something like, "Aircraft X, possible pilot deviation. Advise when ready to copy." And of course, at this point, I had a very sinking feeling.[1]

This Captain's report included a detailed analysis into how they made the error and ideas to prevent reoccurrence. There were several references to personal departure practices, a late departure change, concern over the FO's lack of currency, and personal habits that contributed to their flawed preconceptions. Of particular interest is that both pilots were unsure of the takeoff clearance, discussed it, but that neither was sufficiently moved to verify the clearance with ATC. All of the relevant facts were present, but were overlooked due to their flawed SA.

5.1.4 The SA Balloon Metaphor

Imagine SA using a balloon. Our task is to keep our SA balloon inflated. As we build our SA, we add air to the balloon. The larger it grows, the more SA we have. Unlike a reliable balloon, however, our balloon has three flaws. First, it leaks. After we stop adding air, it slowly deflates. This reflects how our SA doesn't endure over time. We repeatedly have to recharge it with updated conditions and changes. A related problem is that filling our SA balloon is not our only task. We still have to fly the aircraft and attend to many other demands on our time and attention. If we only work on these other tasks, our neglected SA balloon slowly deflates.

The second problem is that our balloon has a tricky valve. When we encounter unanticipated conditions, the valve can suddenly release and immediately deflate our SA balloon. We see this effect when something unexpected completely changes our game plan. For example, when a vehicle intrudes onto the landing runway and forces us to go around, it instantly deflates our SA balloon that projected an uneventful landing.

The third problem with our balloon is that we may need to maintain several different balloons at the same time – one for each game plan that may be needed. In a complex environment, we select and fill a separate balloon for each potential game plan. Filling these extra balloons keeps our contingency backup plans primed and ready for use. An example is flying an approach to weather minimums. We need to be ready to see the runway and land, but we also need to be ready to execute a missed approach.

5.2 THE TIME FRAMES OF SA

Using our balloon metaphor, we'll explore how SA changes along the timeline of our flight. Specifically, three different aspects of SA emerge.[2]

- *Past SA:* Our recall of what we have planned and what has previously transpired
- *Present SA:* Our perception of what is happening at the present moment
- *Future SA:* Our prediction of what will likely happen in the future

5.2.1 SA from Past Planning and Events

We start with flight planning. At the gate, we review all of our planning materials like the dispatch release, weather, NOTAMS, EFB, and ATC clearance. We construct a story of how we expect the flight to progress. We fill our past SA balloon by forming an expectation for the taxi-out, takeoff, and departure. How will the weather conditions affect us? Will we be restricted by traffic saturation? We also form an expectation of how the descent, arrival, and landing will unfold. What weather conditions will we encounter? What can we expect for an arrival flow? What is the planned arrival fuel? What contingency options will we need to prepare?

We form, brief, and discuss our game plan. Our game plan forms the foundation for future SA-building. It shapes our mindset which guides which parameters to monitor. It guides our strategy for completing procedures, running checklists, and managing aircraft automation. As a crew, we communicate a shared mental model of our game plan. This forms a common vision of how we expect the flight to unfold.

Fast forward to the cruise flight phase as we prepare for descent and arrival. We start with the past SA that we formed while planning and briefing at the gate. However, what we now recall has become mixed and modified by our memories from hundreds and thousands of past flights. This often distorts what we remember preparing at the gate. Also, what we planned back at the gate may not match current conditions and information. The air in the balloon has become stale.

We acquire new details that may change our expectations. We now know the current weather at the destination, the arrival STAR, and our current fuel state. The air in our past SA balloon is now a mixture of what we initially inflated at the gate plus the fresh information that we have recently added.

Let's continue forward to the approach and landing phase. Imagine that both the weather forecast and the ATIS indicated that the airport would be VMC. Because we don't expect visibility problems, we skip briefing an instrument approach. During the approach, however, we enter a cloud deck. Flying in unexpected IMC, we realize that our SA is mistaken. We are unprepared to fly a charted instrument approach, so we go around. This is an example of the SA balloon valve failing. Would we conclude from this example that we did a poor job building our SA? Probably not. We had great SA for the expected VMC conditions. We just lacked appropriate SA for the unexpected IMC conditions. Like our example with the magic act, our inaccurate SA resulted from unexpected and unnoticed details.

5.2.2 SA in the Present Moment

When we visualize SA, we typically envision our aircraft state at the present moment. How good is our SA *right now*? If we were in the simulator, we could demonstrate present moment SA by freezing the simulator and accurately describing our SA. We could describe what we expected to happen, how well our game plan is currently tracking, and each parameter displayed on our gauges. We don't have this option during a flight. While flying, our SA is more like a motion video formed by a stream of constantly changing moment-by-moment assessments. Each momentary assessment flows seamlessly into the next. Following this flow, we assess our position, aircraft systems status, and energy state. Does the current flow of events agree with what we planned, nor not? Is it progressing smoothly and predictably? Do we have the right balloon? Is it sufficiently inflated?

Present moment SA is our bridge to future SA. When the indications smoothly link the two, we conclude, "We have good SA." The story matches what we expected to see, what we are seeing, and what we predict to see next. Our game plan is validated. When parameters contradict our expectations, we either make corrections to restore our game plan or abandon it for a different one. These two cases indicate how our present moment SA can be either good (because nothing unexpected happened – a well-inflated balloon) or bad (something happened that we didn't anticipate – wrong balloon or valve failure).

To cite a practical example, imagine that we are flying an ILS approach. Just as we near localizer final, the ground transmitter suddenly fails. Our expectation is that we will see the localizer bar moving smoothly from the edge of our situation display toward our aircraft symbol where it will be captured by the flight director to join the final approach course. Instead, we see ILS OFF flags or wildly gyrating localizer and glideslope symbols. The valve fails and our balloon deflates. Our SA instantly flips from good to poor. At first, we may not even understand what is happening – only that we've lost our SA.

There is always more happening than our physical senses and minds can process. We learn to selectively filter the many sources of information by devising techniques and habit patterns to guide what to look for and what to ignore. Our techniques and habit patterns are biased to seek information that supports our SA. The more overloaded our workload is, the more we prioritize expected information and overlook contradictory information.

5.2.3 SA that Predicts the Future

Accurate prediction is the ultimate goal of SA. Past SA sets the stage, present SA verifies it, but future SA is the prize. The reason we put so much effort into planning, briefing, and monitoring is to increase the accuracy and resilience of our future prediction. It becomes our measure of success. If we planned and briefed for a particular outcome, monitored parameters to guide our aircraft along a smooth trajectory toward that outcome, and reached it as planned, then we will have achieved a desirable level of SA. We chose the right balloons and inflated them well.

A qualitative dimension of future SA reflects how far it projects into the future. Consider the divisions within Figure 5.1[3]. Viewed from an aviation perspective, the tactical level (situational assessment to reach situational awareness) measures our *awareness* of immediate parameters – like values displayed on flightdeck gauges. Applying the example of aircraft fuel gauges, this would reflect glancing at the gauges and seeing that we have 20,000 pounds of fuel and that the balance between the tanks is within limits. Expanding to the strategic level (sensemaking to reach understanding), we derive *meaning*. What does 20,000 pounds of fuel actually mean to us? Comparing it with our fuel burn plan, we could conclude that

Objective	Phase	
	Process	***Outcome***
Tactical (short-term)	situational assessment	situation awareness
Strategic (long-term)	sensemaking	understanding
Scientific (longer-term)	analysis	prediction

FIGURE 5.1 Matrix of tactical, strategic, and scientific SA categories.

we are tracking either above or below the expected burn for each fix along our route. Expanding our perspective further to the scientific level (analysis to reach prediction), we could conclude that our trend of fuel burn will have us arriving at our destination with 10,000 pounds of fuel. This would allow for 30 minutes of holding and a subsequent diversion to our planned alternate to arrive with 5,000 pounds of fuel. Additionally, it would mean that we have sufficient fuel to attempt two approaches and still have enough reserve to reach a planned alternate. As our SA progresses through these three levels, our balloon inflates a bit more and our future SA ranges further out.

To expand on this fuel example, this single parameter can assume different meanings depending on our needs. When assessing aircraft performance, fuel is primarily viewed from a *weight* perspective. For inflight contingencies, fuel is viewed from *time available* or *diversion range* perspectives. If we are in an aircraft that carries fuel for multiple flights, our current fuel load can be viewed from a *logistics* perspective as we determine where to schedule our next refueling.

5.3 EXPANDING OUR SA AT EACH LEVEL

We strive to expand our SA balloons at each level – past, present, and future (Figure 5.2).

5.3.1 Expanding Our Past SA

Using our experience, procedures, and knowledge, we begin filling our past SA balloon at the gate. Each pilot conducts their own review of the flight planning materials. Then, we coordinate together to identify the threats and challenges. The pilot designated by the company's procedures briefs the game plan. At some operations, the Captain conducts this briefing. At others, the PF does. However it's done, we construct our game plan and build a shared mental model. This establishes a common mindset of the conditions, threats, and challenges that we expect to face.

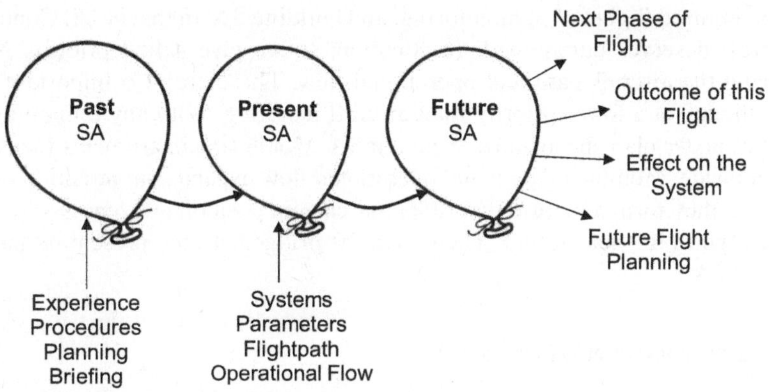

FIGURE 5.2 Balloon model of past, present, and future SA.

Envisioning the flow of the flight, we imagine how the story will unfold.[4] The richer and more detailed our story, the more we inflate our past SA balloon. We are like chess masters constructing tactical and strategic game plans before advancing our first piece. We accept that our plans will change to respond to unexpected moves. We plan our lines of attack while building defensive contingencies for whatever our opponent might attempt. A good aviation game plan includes these same features. We start with a plan to deal with clear threats like convective weather and low ceilings. Then, we look deeper for subtle threats like a tight temperature/dewpoint spread could threaten fog by our arrival time. We compensate by adding fuel and evaluating alternates. Our balloons fill nicely.

The enemies of good briefing and planning are distraction and time pressure. When our planning is interrupted, we divert our attention toward the distraction, devote time to deal with it, and then return to our planning. As we recover from the disruption, we may subconsciously combine thoughts that we had before the distraction with similar thoughts drawn from past flight experiences. This can distort our SA. Did I already review that, or am I remembering doing it from a previous flight? Resuming our planning, we may lose the detail that we had reached before the distraction occurred. Our SA suffers.

Planning and experience pre-charge our past SA balloon. The usefulness of this pre-charge depends on the time frame of the flight phase. The closer the time frame, the more useful our planning, and the better our SA. We base the near-term conditions on fresh observations, while distant conditions rely on forecasts. This allows us to plan quite accurately for takeoff/departure while still at the gate. Since approach/landing may be hours away, that planning will be less useful. That is why we schedule a second planning/briefing time block during cruise before beginning our descent and approach. In cruise, we update the initial game plan (past SA) with current information (present SA). This recharges our balloon for the next high-workload phase of flight.

5.3.2 Expanding Our Present SA

At any particular moment, we have limited time and attention available. We divide it between required flight tasks, monitoring, and building SA, in that order. Controlling the aircraft deserves our immediate attention, so we give it first priority. Next is monitoring the aircraft path and operational flow. These are also important flying tasks, although at a lower priority than aircraft handling. With any remaining time and attention, we plan ahead and expand our SA. Monitoring instruments focuses our attention inside. Monitoring path and operational flow expands our attention outside. Combined, they form a picture that links our current position awareness with where we expect go in the near future. They form the bridge between present moment SA and future SA.

5.3.3 Expanding Our Future SA

Each time we update our future SA, we extend our prediction of what will happen. At its simplest level, SA shapes our understanding of what is currently happening in our aircraft and how well our flightpath is tracking toward our objective. Expanding our SA further, we encompass the movement and actions of other operators around us.

With past SA, we draw primarily from outside sources – weather products, dispatch plan, and charts. With present SA, we draw primarily from immediate observations. With future SA, we project forward in time and space.

- How well is our flightpath tracking toward the next restriction?
- How about the restriction after that?
- Where do we fit into the traffic flow?
- How will arrival conditions affect our game plan?
- What can we expect on the ground after landing?
- Will the station be ready with a parking gate and ground support crew?
- Are there concerns about passenger or crew connections?

We build future SA by assessing two factors – how our current game plan is tracking (flightpath trajectory) and monitoring for adverse conditions that might signal that our game plan is failing (counterfactuals). Consider a scenario where we are being vectored for an approach to a particular runway. Under favorable conditions, our monitoring supports continuing. As we monitor for counterfactuals, we detect a storm cell that is developing along our expected flight track. This warns us to anticipate possible deviation vectors, windshear, go around, or a runway change to avoid the storm cell. In this way, building future SA cues us to divide our attention between continuing with the game plan and monitoring for threats that may jeopardize it.

When we skillfully combine present SA with future SA, we attend to all immediate flight-management tasks, effectively monitor our current position, and accurately predict both the path that our aircraft will follow and any threats that may upset our game plan. As we gain skill and experience, the first two tasks become easier. This frees up extra time and attention. We can either expand our future SA balloon or we can accept it at its current level, reduce our attention level, and relax.

As they gain proficiency, many pilots choose to relax. Since they have flown each particular flight many times before, they "know" how it will go. Expending extra effort to expand their future SA feels unnecessary to them. What seems to happen is that they begin subconsciously metering their attention level to match their perceived workload. When they have lots to do or when threats are immediately apparent, they elevate their attention level. When workload is low or threats are not apparent, they reduce their attention level. While this makes practical sense, it is exactly what sets them up for surprise, startle, task overload, and Recognition Trap errors.

As Master Class pilots, we consciously choose to modulate our attention based on the flight phase, not on perceived workload. For example, maneuvering for approach and landing always demands our full attention, even under benign conditions. We resist the comfort zone bias that might lead us to inappropriately relax.

5.4 FACTORS THAT DEGRADE SA

Using our understanding of the three time frames of past, present, and future SA, consider some factors that tend to degrade our SA.

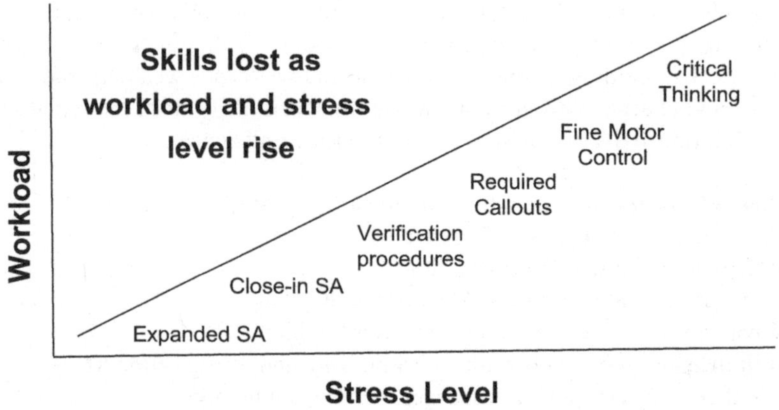

FIGURE 5.3 Skills lost as workload and stress level rise.

5.4.1 Excessive Workload

During the flight, we need extra time and attention to maintain our SA. This comes from whatever remains after controlling the aircraft and monitoring flight parameters (high-priority flight tasks). Figure 5.3 depicts how specific aviation skills erode as our workload and stress levels rise.

With a manageable workload, we have plenty of time to complete of all of our immediate flightdeck tasks, expand our SA, and assess the progress of our game plan. As workload and stress increase, the first aviation skill we lose is our expanded SA. It steadily shrinks until we can only pay attention to a few parameters. After this, we lose our immediate, close-in SA. We only have enough time and attention to attend to immediate flight tasks. Our crew duties suffer. We drop verification protocols such as announcing system changes and verifying system responses. We have too much to do and not enough time to confirm the accuracy of our actions. Next, our CRM communication degrades as we are too busy to share information back and forth. After this, our personal aviation skills begin to degrade. Our fine motor control worsens as we get rougher on the aircraft flight controls. At extreme levels of overload and stress, our minds become so befuddled that we effectively stop thinking. A wise aviator summarized these last three points as, "First, you stop talking. Then, you stop flying. Finally, you stop thinking." Following is an example from an EMB-175 crew that lost the race to keep up with their excessive workload.

BOX 5.2 CREW ALTITUDE BUST DUE TO TASK OVERLOAD

Captain's report: This is one incident and report that I'm not sure exactly where to begin. It was the First Officer's leg, and right off the bat on climbout from our departure airport, ATC gave us a reroute. The ATC communications, instructions, and clearances became increasingly frequent. ATC required multiple, and I repeat, multiple speed changes, heading changes, clearances to a fix, altitude changes, and entire reroutes multiple times! Then, when we would change the frequency, the new controller would change it all over again. Then, the approach controller changed the arrival runway, also. *It wasn't that the flight crew was inattentive, incompetent, or inexperienced, it got to the point it was literally overwhelming. It was excessive! Both crewmembers were working full-time to keep up with ATC demands.* One particular ATC clearance was to cross 30 miles south of Rockford VOR at 13,000′. It was put in the FMS; however, the Pilot Flying failed to change the altitude alert and the Pilot Monitoring failed to notice due to the extremely busy nature of this flight. Both pilots and ATC all noticed about the same time that the aircraft was high on the profile and ATC asked if we were starting down. We had just started down, said yes, and expedited the descent. We were a little high crossing the fix, asked ATC if there was a problem with not making the crossing restriction and he said no. If ATC could limit number of instructions to flight crew unless absolutely necessary, it would reduce crew saturation.[5]

From this report, we surmise that the crew began their flight with a good pre-charge of planning and briefing (past SA balloon). Unfortunately, ATC repeatedly changed their departure instructions. This deflated their SA and forced them to adjust repeatedly to a stream of changes. Every time they tried to rebuild their SA, ATC deflated it. They were devoting all of their attention toward managing immediate flight tasks. *Complying* and *flying* exhausted all of their available time and attention. *Verifying* and *monitoring* eroded until the crew failed to properly manage an assigned altitude restriction.

5.4.2 Complexity and Novelty

With experience, we acquire a relaxed feeling in our comfort zone. With some pilots, this can become a goal in itself. In a sense, we become overly comfortable with feeling comfortable. Our SA-building skills can deteriorate.

- *Comfort zones and complexity:* Consider a hypothetical situation where we have flown thousands of flights in a single aircraft type. We have used our seniority to limit our flying to a single city pair – between New York and London. We've become very comfortable with flying that route. One day, however, scheduling reassigns us to Mexico City. The flight goes poorly. Despite our extensive airline experience, settling into our comfort zone has caused us to lose our ability to handle new and complex situations. The lesson here is that while flying in our comfort zone seems attractive, we need to maintain the skills to adapt to all situations across the range of our flight operation.

  This problem also occurs with infrequently practiced flight skills. For example, hand-flying a CAT III approach using the HUD is fundamentally different than flying that same approach using a fully coupled autopilot and autoland. If we only fly coupled approaches month after month, we lose proficiency hand-flying with the HUD. When a situation forces us to use HGS procedures, we may lack the skills to succeed.

- *Complexity and emergence*[6]*:* Limiting ourselves to familiar situations, building future SA becomes easy. Once we experienced every nuance, we'll know how to handle deviations. As complexity and novelty increase, we'll have less experience to draw upon. There are just too many variables interacting in too many unpredictable ways. Unanticipated outcomes begin to emerge. With convective activity, for example, one flight might land with light turbulence while the next aircraft encounters severe turbulence. When emergence generates these wildly variable outcomes, we can't reliably predict how our current situation will play out.
- *Complexity coping skills:* Experience with a complex operation serves us across other complex environments. Consider an RJ crew based out of Chicago O'Hare (ORD). Flying in and out of that complex airport day after day, they become well-versed with both the specific ORD challenges and with the general challenges common to any complex airport. If these pilots are reassigned to LaGuardia (LGA), they will still need to learn the unique local procedures for LGA, but they will already have useful high-volume airport operating skills from ORD.
- *Task saturation increases error rates:* As task saturation mounts, we switch from a responsive mode (detect, consider, choose, respond) to a reactive mode (detect and react). Organizing, managing, choosing, and communicating become time-consuming luxuries. Since falling behind the aircraft feels uncomfortable, we engage each task as quickly as possible. We willingly sacrifice SA-building to stay ahead of mounting immediate tasks. It's much like the arcade game of Whac-A-Mole. In the game, the player uses a padded bat to wack each mole as it pops up. They score a point every time they successfully hit the mole while its head is exposed. The pace accelerates as the moles emerge and retract faster and faster. Soon, the pace exceeds their ability to keep up. Errors include missing the mole, hitting an empty hole after the mole has retracted, hitting a hole where no mole has yet emerged, missing a mole they are aiming for, and failing to detect a mole's appearance. As the pace increases, their "errors" increase and their "hits" decrease. While intended for arcade entertainment, it accurately demonstrates how our flightdeck performance can break down as we become task-saturated.
- *Deterioration of CRM:* As complexity weakens our crew coordination, our shared mental model and team SA also become degraded. To cope, crews temporarily suspend communications with the intention of catching back up later when task-loading eases. It starts when both pilots recognize that they need to complete many tasks quickly. They separate their efforts and individually dive into their half of the task list. When the intended aircraft state is restored and the time pressures ease, they circle back, communicate, verify each other's work, and rebuild SA. This CRM process of dividing the workload under trying conditions, then reforming the team, rebuilding SA, and restoring a shared mental model is an important team skill. Unfortunately, this recovery opportunity doesn't always happen. In quickening scenarios, SA shrinks and mistakes slip through unnoticed and

uncorrected. Consider the following pilot report resulting in an ATC low altitude warning. Notice how their communication, verification, and role assignments broke down as their task saturation increased.

BOX 5.3 CREW ROLE BREAKDOWN FROM TASK OVERLOAD

FO/PM report: … We asked and were given a vector so that we could accomplish the task. The Captain began to manipulate the FMS. As he was working on the changes, approach gave us a base turn and asked us to keep it tight. *This caused the Captain to become task-saturated. I asked if I could manipulate the FMS and free [time for] him to focus on flying the aircraft to which he agreed.* While I was reloading the approach, the controller cleared us direct LIFTT and for the LOC DME 15. I read back the clearance, however in my distraction, I did not notice if he gave us a crossing altitude at LIFTT, nor did I read one back. I was able to execute the direct to LIFTT and returned to monitoring our progress. We were VMC in a descent and almost at LIFTT when the controller instructed us to maintain 13,000′ until crossing LIFTT. At that time, I looked and saw that we were at 12,600′. We began to correct the altitude when the controller gave as a low altitude alert. We were VMC with adequate terrain separation. We corrected and continued the approach. Unfortunately, the fog bank had completely covered the airport and surrounding area. We went missed [approach] again and diverted.

It was a challenging day at a very challenging airport. *We allowed ourselves to be rushed into a second attempt at an approach without enough time to adequately prepare. In the rush, the PF attempted to manipulate the FMS during a critical phase of flight and lost situational awareness. When the PM took over FMS load responsibilities, he was rushed and also lost situational awareness. While the PM was loading the FMS, the PF mistakenly started a descent for 12,300′, the published altitude after LIFTT, without making proper challenge/response of altitude selection. As a result, the PM was unaware of the selected altitude change. To avoid recurrence, the flight crew need to follow SOP guidance, requiring PM to make all FMS changes during critical phases of flight, and requiring the PF to confirm all altitude changes in Alt Pre-select with the PM.* Above all, the crew must be more assertive in communicating with ATC their need to set up between approaches with such tight geographic and airspace constraints.[7]

In this event, the Captain/PF chose to reprogram the FMS while task-saturated with flying the aircraft. The FO/PM pitched in to help and both pilots divided their efforts to accommodate the workload. Neither pilot noticed the PF's descent altitude-setting error nor the PM's FMS programming error. Verification protocols broke down. They apparently perceived an advantage with dividing their workloads with the good intention of catching up when their workload eased. With too much to do, the priorities of crew coordination and verification were suspended.

This also highlights the thin line that we draw between success and failure. Imagine if this crew had handled this challenge successfully. Their actions, including discarding crew coordination and verification, might have served as an example of effectively handling complexity. Their decision to "disregard the manual and get the job done" would have been validated.

- *Reverting to autonomous personal techniques:* Another task overload strategy is for each pilot to handle their individual tasks using their own preferred techniques without crew coordination. Reviewing the previous NASA ASRS report, we note how each pilot became very busy with their efforts to keep up with the situation. We can imagine how each entered into their own autonomous mode. Each became so busy that CRM verification apparently broke down. The PF's mis-programming of LIFTT and the PM's failure to detect the error led to the altitude bust. Neither pilot detected the oversight until ATC queried their low altitude.

  When our workload increases, we speed up our pacing. We take shortcuts and skip steps that we deem are less important or might be completed later. As our pace increases even more, we abandon crew coordination. When the quickening pace prevents us from verifying each other's work, we depend on perfect performance by each pilot – a standard that becomes increasingly difficult to deliver under complexity and time pressure.

5.4.3 Unskillful Monitoring

Effective monitoring is essential for maintaining good SA. Two monitoring components are:

- *Monitoring the flightpath:* Tracking the flightpath to assure that it is proceeding as expected. This applies to any aircraft movement, including taxi. This uses close-in, future SA. Is the aircraft moving in the direction that we want it to go?
- *Monitoring the progress of the game plan:* We scan our environment to determine whether conditions either support or contradict our game plan. Is the aircraft proceeding along or straying away from our desired path? This is tracked through our long-range or expanded future SA. When indications and conditions support our game plan, our future SA is validated. When contradictory indications (counterfactuals) arise, our future SA becomes unreliable.

We monitor the transition from short-term, close-in future SA to expanded, long-term future SA. If we smoothly connect both perspectives, it validates our choices. As novelty, task-loading, fatigue, and complexity rise, our biases begin to emerge. Future SA shrinks and effective decision making suffers. Biases lead us to overweigh the parameters that confirm our game plan and minimize indications that suggest failure. Consider the following PM report of an altitude bust:

BOX 5.4 WHILE ATTENTION WAS DIVERTED, FO MISSES CAPTAIN'S ALTITUDE-SETTING ERROR

FO/PM report: We had just been handed over to Houston Center from Fort Worth and were on the DRLLR FOUR arrival. Houston Center gave us a descent to FL240. The altitude was set and we started our descent. Shortly thereafter, ATC gave us a [clearance] "descend via the DRLLR except maintain 310 KTS to SSTAX, then comply with the published speeds." I read back the clearance and the bottom altitude was set and confirmed. *The threat that was never discussed or caught was that the fix, OILLL, was supposed to be crossed at or above FL240. During the brief, the CA said he would fly the Vertical Path Indicator (VPI) on the arrival. I'm not sure what happened, but for some reason he didn't fly the VPI. I was heads-down for some reason, I believe I was looking in my Jeppesen binder to see where we were parking and to find the approach chart for the runway.* The next thing I know, he states that he is stopping the descent. I look up to see what's going on and he says that he was supposed to cross OILLL at or above FL240. We had descended almost 1,000′ below that when he stopped the descent. I'm disappointed in myself for not being a diligent PM. *I trusted that he was going execute the descent the way he had briefed and therefore didn't ever think we were going to bust through a crossing restriction. I'm normally very meticulous about following along on the arrival to make sure that we will meet all the restrictions.* This one seemed so benign because we weren't even in the challenging part of the arrival yet, i.e. there were no windows or speed restrictions that had to be met. … *The error was that the PF must have lost his situational awareness and didn't realize where the aircraft was in relation to the procedure. I, the PM, failed to monitor and crosscheck because of performing other tasks.* Be more diligent when performing challenging tasks in our aircraft. Don't get complacent and assume the PF is on top of things.[8]

Their shared mental model set the expectation for how the PF was going to manage the flightpath. This biased how the FO monitored and tracked their future SA. They weren't task-overloaded. The FO rated the complexity as "benign". Instead, their expectation bias influenced the PM verification and allowed the PF's error to slip through.

5.4.4 Rationalization and Goal Shifting

Even when we detect counterfactuals, time pressures may affect how we process them. Imagine that we detect a counterfactual. If we didn't have time to determine the cause or significance, we might quickly dismiss it as an anomaly. We could also detect it, accept it as accurate, but judge that it is unimportant. We may normally be quite skillful at detecting counterfactuals, but fail when fatigue, laxity, and misinterpretation conspire to weaken our vigilance.

Even when we accurately conclude that our game plan is failing, we can rationalize that the deviation is manageable by shifting our goal. Consider an approach where everything is falling apart like Clem's unstabilized approach and runway departure (from Chapter 2). The acceleration of events and the increasing severity of the counterfactuals reached a point where Clem had two choices – go around and start over or accept the unstable approach and land. He chose to land. Since he slid into the overrun, we agree that he made the wrong choice. Surveying reports from across the airline industry, we discover that the vast majority of unstabilized approaches like Clem's are continued and successfully stopped. We rarely hear about those. In many of those unstabilized approach landings, the pilots had the same SA as Clem – that the stabilized approach goal was no longer achievable. They switched their goal to successfully landing and stopping. Consider this NTSB report of a runway excursion:

> The Captain later stated that he had considered calling for a go-around before touchdown but the "moment had slipped past and it was too late." He said that "there was little time to verbalize it" and that he instructed the FO to get the airplane on the ground rather than call for a go-around. He reported that, in hindsight, he should have called for a go-around the moment that he recognized the airplane was floating in the flare.[9]

There are several interesting aspects with this mishap. For most airline aircraft, the latest point to begin a go around is when the thrust reverser levers are raised during rollout. This Captain recalled deciding that it was "too late" to go around even while they were still airborne and floating down the runway. They touched down over 4,200′ down with less than 2,800′ of runway remaining. Adding to the problem was a late deployment of speed brakes and wet pavement. They departed the end of the runway with minor aircraft damage and no injuries. In hindsight, the Captain acknowledged that a go around would have been the appropriate choice. In-the-moment, however, his mind formed a judgment that it was "too late" to go around from the flare. Another detail of the story is that they had Vice Presidential candidate Mike Pence on the aircraft while landing at New York LaGuardia airport (LGA). Everyone was watching and waiting for that jet to land. We can imagine that this created an extremely strong "mission priority" to land successfully versus what might have felt like a professionally embarrassing go around. The goals of a stabilized approach and successful landing in the touchdown zone shifted to landing and successfully stopping. This offers a glimpse into how strongly mental biases can influence our SA and how easy it is to shift from a procedural goal to an immediately attainable goal.

5.4.5 Distractions and Disruptions

As the aircraft moves along the operational flow, our SA moves along with it. Our briefed expectation (past SA) flows into what we see happening right now (present SA) which flows into what we predict will happen next (future SA). We value the efficient transition between these time frames. Nothing upsets this smooth flow more than distractions and disruptions. Distractions are momentary bumps that interrupt what we are doing or thinking. Disruptions are major obstacles that block or derail our desired path.

SA-building is a time-consuming task that is vulnerable to disruption. Consider the task of planning for taxi-in and parking after landing. We review the airport diagram (for probable runway exits, taxi hotspots, probable taxi routes, company operations guidance, and local procedures), and NOTAMS (for taxiway closures and construction). Now, imagine that we are in the middle of this review process when ATC assigns us a reroute. This interrupts our planning process. After we process the ATC reroute, we'll need to return and finish our ground operations planning. Our challenge is determining where we left off. The stronger the distraction/disruption, the greater the chance of misjudging what we did and what we had left to do. A number of errors are possible:

- We may mistakenly believe that we had finished our task and move on.
- We may confuse completing the task with past memories of completing that task.
- We may inaccurately recall what we had already checked the NOTAMS.
- We may inadvertently recall past occasions of reviewing the NOTAMS.
- We might remember a similar flight when nothing important was listed in the NOTAMS.

If this day, the airport had closed some taxiways for construction, we might overlook these important restrictions. Remembering what we have previously done doesn't always attach to the most recent occurrence. The most resilient technique is to back up our review process to a point well before the interruption and start over.

5.4.6 Disrupted SA and Crew Resource Management (CRM)

When conditions force a change in our game plan, we need to update our shared mental model. As we become task-saturated, CRM safeguards tend to deteriorate. Consider Figure 5.4 of degraded CRM features.

The first CRM feature to suffer is communicating our shared mental model. PFs may mentally update changes to their game plan, but may feel too busy to immediately brief the changes with their PMs. Next, lacking an accurate shared vision, the quality of PM's monitoring begins to degrade. Next, pilots stop verbalizing the

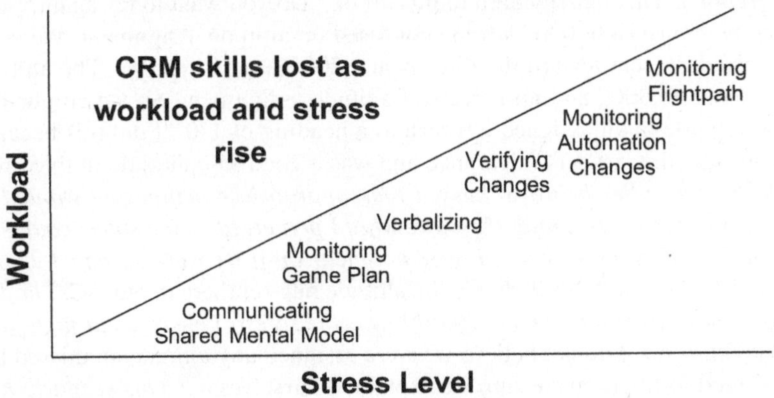

FIGURE 5.4 CRM skills lost as workload and stress rise.

changes that they make to aircraft systems. It's quicker to just make the changes and return to other pressing tasks. These first three aspects reflect desirable CRM functions that promote detecting and mitigating errors.

The remaining three functions on the upper right slope of the graph represent essential, foundational CRM duties. They must remain intact to prevent highly consequential errors. In order, they are failing to verify system changes, failing to monitor automation changes, and suspending our monitoring of the aircraft's flightpath. Our layered defenses of detection and mitigation disappear. We effectively become single-pilot operators.

Following is an account of an overloaded crew doing their best to keep up with an increasingly difficult situation. Notice how the rising workload adversely affects their performance. Imagine how it must have felt during the constant stream of ATC calls to them and other aircraft. On top of this, add the noise of the autopilot disengagement horn, rapid conversation snippets between the pilots as they tried to divide workload, and numerous ATC frequency changes.

BOX 5.5 CRM BREAKDOWN DURING TASK OVERLOAD

Captain's (PM, then PF) report: We were given the TELLR 2 RNAV Runway 34R ILS transition. DEN had just changed from south to north flow; we received a new clearance for the TELLR Runway 34 while still at cruise. Once talking to Approach, *the controller was very busy.* The arrival had a 40 to 30-knot tailwind all the way down. Approximately 10 miles from CRAGG we were vectored off the arrival and [were] told to maintain 210 knots. We were then on a shorter ground track, asked to slow, and had a 40-knot tailwind. We used all available drag devices to slow and descend but were still high. We leveled at 13,000′. *Due to communication congestion, we were not able to descend on schedule. All of these made us high on final.* We asked for turns on final to try to correct, but were unable to correct. I recollect that Approach asked, "Do you want to try again?" We understood that to mean go missed approach, and agreed with ATC. *This began the "threats" section of this [report].* The non-standard radio call of, "Do you want to try again?" followed by "Turn right 020" left me confused on altitude assignment. We were in a visual descent toward the runway at 9,200′ (approximately). The altitude bug was at 10,000′, and no perceived altitude assignment. So we climbed to 10,000′ while in an assigned left turn to a heading of 170°. I did this because 10,000′ was the last ATC clearance and was a good safe altitude in that area. *With the controller being so busy, I took appropriate action and wanted to ensure aircraft control and, therefore what I perceived as the safest course of action. The First Officer and I were not clear on if we were assigned 9,000′ or 10,000′.* I chose 10,000′ due to the altitude bug set there in our AC. *I had to make a quick decision and felt 10,000′ was the safest; I cleared our flightpath and we climbed.* I do not believe we were assigned any altitude on missed but we wanted to file to make sure we explained ourselves. … *This scenario had three turns in less than 30 seconds, with some hesitation that caused CRM to break down.*[10]

Combined with the FO's report, we know the following:

- ATC jammed the crew for the approach.
- Both pilots became task-saturated.
- The game plan switched from landing, to a go around, and back to landing.
- Task saturation led to SA and CRM breakdown.
- Confusion developed between the pilots regarding the flap selection. The FO even considered moving the flaps himself rather than directing the Captain to do it. This option is not allowed at most airlines. This implies that the FO considered switching to an expeditious, autonomous mode.
- The Captain determined that the FO was too overloaded to successfully complete the approach and assumed control of the aircraft. This reversed roles while both pilots were task-overloaded.
- Confusion developed over the altitude assignment. The Captain chose 10,000′ as the higher option (without ATC confirmation), despite FO's attempts to convince him otherwise.
- ATC, PF, and PM all on "different pages".

This event illustrates how individual and crew SA can break down as we become overloaded. It became so untenable that the Captain took control of the aircraft and climbed to a "safe" altitude of 10,000′. Imagine the moment just before the Captain took control of the aircraft. How was the Captain's SA at that point? How about the FO's?

5.4.7 Inadequate Preparation for Future Contingencies

Stress and complexity undermine contingency preparation. When we don't have a prepared backup plan, we tend to stick with our original plan, even if it is failing. The better we plan for future contingencies, the easier it is to make the change. Compare two crews with different preparation strategies.

BOX 5.6 CREWS ALPHA AND OMEGA - A COMPARISON OF PLANNING EFFECTIVENESS

Crew Alpha – effective contingency planning: Consider crew Alpha, an effectively communicating, high-SA team. They review their planning products for the expected conditions and select an appropriate game plan for the flight. They also discuss a number of possible contingency options. As the flight progresses, they monitor conditions for warning signs that may indicate that a contingency plan might be needed. They use SA-building options including contacting their dispatcher, calling ahead to the arrival station, and monitoring real-time weather through EFB apps. They notice a significant line of thunderstorms building between their scheduled destination and their assigned alternate. Contacting their dispatcher, they coordinate a more favorable alternate that will remain unaffected by the thunderstorms. During their arrival, they discover significant weather conditions building at their destination. They continue to fill both future SA balloons – the original one for landing at the scheduled destination and a second one for possible diversion. On final, they encounter moderate turbulence. Their windshear warning activates for a moment, then stops. They anticipated this contingency, so they aren't surprised by it. Expanding their scan to the runway environment, they notice signs of microburst activity. They execute a go around and divert to their alternate.

Crew Omega – ineffective contingency planning: Next, consider team Omega that doesn't update their SA or plan for contingencies. They review the flight planning products at the gate and conclude that they are unlikely to encounter significant weather. They rely on the dispatcher's assigned alternate. Approaching top of descent, they begin their standard arrival review process. They notice the deteriorating weather at the destination airport, but decide it must not be too bad since no one is diverting or being assigned holding. During descent, they see storm cells building near the airport. Quickly reviewing their options, they discover the line of thunderstorms that now blocks the route to their assigned alternate. Too busy to contact their dispatcher, they press ahead. The Captain comments that it is especially important for them to land since the assigned alternate is no longer viable. On final, they encounter moderate turbulence. They are startled when their windshear warning activates. While they are processing what to do, the windshear warning stops. They are still in moderate turbulence, but since they have the runway in sight, they continue. The turbulence continues to worsen as they cross the overrun. The FO calls for a go around, but the Captain overrides them and continues. In the flare, they experience a significant wind gust from the left and start drifting well right of centerline. The Captain dips the left wing to correct back to runway centerline and scrapes the left winglet.

5.4.8 Alpha and Omega Analysis

Team Alpha began building good SA during their planning back at the gate. They discussed game plans for both the intended flight and for a possible diversion. Since it was too early to evaluate the conditions at their destination, they ensured that they had enough fuel to support a range of options. As they neared top of descent for their destination, they anticipated both the problems with the destination weather and the thunderstorms blocking their route toward their assigned alternate. They coordinated a more favorable alternate with their dispatcher. Joining final, they could see that conditions were nearing the tipping point between continuing and diverting. When the windshear warning sounded and they detected signs of microburst, they easily switched to their contingency backup plan, executed a go around, and diverted to their revised alternate.

Team Omega started with a viable plan, but neglected to evaluate contingencies. They filled their past SA balloon with only the destination airport forecast. It wasn't until their descent that they realized that they had neglected to revise it for changing conditions. They detected the worsening weather, but they rationalized that the destination weather must be good enough because "everyone was getting in". At this point, they discovered the line of thunderstorms between them and their assigned alternate. Instead of triggering a desire to select a better alternate, it solidified their urgency to land. This may not have been a conscious decision, but we can imagine it subconsciously affecting their mindset. The sight of the runway pulled their attention forward. The hook was set. The turbulence worsened and the windshear warning briefly sounded. It quickly stopped. As the counterfactual (the windshear warning) disappeared, it felt like the threat stopped, also. It felt like a momentary red light followed by a steady green light. Even in the flare with a strong gust causing them to drift sideways, the Captain continued. Every bit of their attention was focused on completing the landing. The FO/PM directed a go around, which the Captain overruled. The FO didn't repeat the call or act to intervene.

When we become task-saturated, we tend to revert to established habits. This tends to undermine our SA-building and deliberative decision making. We need to build contingency game plans when we have adequate time and attention to interpret information and evaluate options. That is why we schedule blocks of time both at the gate and before top of descent for this purpose. We know that once we get busy, we are far more likely to continue with our current game plan than build an alternate plan.

NOTES

1 Edited for brevity. Italics added. NASA ASRS report #1769610.
2 Endsley (1995), These three time windows parallel a widely accepted SA model from Dr. Mica Endsley. Her model cites three levels:
   - *Level 1:* perception of relevant aspects of the environment.
   - *Level 2:* comprehension of these aspects relative to goals.
   - *Level 3*: projection of future states of the environment.
3 Fiore, S.M. (6 November 2007) from Psychology Wiki: https://psychology.wikia.org/wiki/Situation_awareness.
4 This reflects the *mental simulation* step from the RPDM process, (Klein, 1999), p. 27.
5 Italics added. Edited for brevity. NASA ASRS report #1689075.
6 Emergence is a statistical effect that describes how conditions interact in unpredictable ways to generate outcomes that vary widely.
7 Italics added. Edited for brevity and clarity. NASA ASRS report #1424221.
8 Italics added. Edited for brevity. NASA ASRS report #1277242.
9 Minor editorial changes were to maintain consistency with pilot designations. NTSB Final report – October 27, 2016, Eastern Airlines Flight 3452 – DCA17IA020 (NTSB, 2017).
10 Italics added. Edited for brevity. NASA ASRS report #1649453.

BIBLIOGRAPHY

ASRS – Aviation Safety and Reporting System – National Aeronautics and Space Administration (NASA) Online Database. (n.d.). Retrieved from Aviation Safety and Reporting System: https://asrs.arc.nasa.gov/search/database.html.

Endsley, M. R. (1995). Toward a theory of situation awareness in dynamic systems. *Human Factors*, 1(37), 32–64.
Endsley, M. R. (2022). *SA Technologies – Publications*. Retrieved from SA Technologies: https://satechnologies.com/publications/.
Klein, G. (1999). *Sources of Power: How People Make Decisions*. Cambridge, MA: The MIT Press.
NTSB. (2017 September 22). *NTSB Issues Final Report for Oct. 2016 Mike Pence Boeing 737-700 LaGuardia Runway Excursion*. Retrieved from Aviation Safety Network: An Exclusive Service of Flight Safety Foundation: https://reports.aviation-safety.net/2016/20161027_B737_N278EA.pdf.
Tremblay, S., & Banbury, S. (Eds.). (2004). *A Cognitive Approach to Situation Awareness: Theory and Application*. London: Routledge.

6 Error

As professional aviators, we cringe when we see "pilot error" listed in the causes of an accident. It seems definitive and conclusive because our society is predisposed to attribute bad outcomes to broken components or unwise decisions. In effect, this bias uses failures to explain failures.[1] The label "pilot error" creates the impression that we have broken pilots flying our aircraft. We know that this isn't true.

Society conditions us to view aviation mishaps the same way we view TV crime dramas. Start with the crime scene, collect clues, construct a timeline that tells a story, identify the offenders, establish motive or negligence, convict them, and declare that justice is served. The flaw in this logic is that mishap pilots don't conspire to crash aircraft. They aren't even complacent operators. In truth, pilots are career professionals who devote time, effort, training, and intention toward conducting every single flight safely. Society doesn't know what to do with this contradiction – good pilots, fully trained and certified, holding the best of intentions, who somehow find themselves at the center of mishap investigations. Using the same analysis process that we use to investigate crimes or analyze broken machines just doesn't fit. In this chapter, we will examine four main areas.

- The flaw with assuming that pilots are the broken parts in an accident event
- How complexity changes our perspective when exploring error
- The pilot contribution to error
- A better perspective for analyzing mishaps

6.1 SOCIETY'S PERCEPTION OF ERROR

Ever since the scientific revolution in the early 1500s, logical analysis has shaped society's mindset. Central to this thinking are two constructs – the perspective of linear time and mechanical deconstruction.[2] By linear time, we refer to the process of drawing a timeline that tracks the event's progression. When evaluating a mishap, investigators use this timeline to frame a feasible story to explain the sequence of events. Next, they imagine an alternative timeline that describes how the event would have unfolded if everything had gone right. Comparing the two, they locate the point where the ill-fated trajectory veered away from desired path. This identifies which decision, act, or effect caused the mishap. Like diagnosing a malfunctioning machine, they open the panel, examine the components, and locate the broken one. This identifies the "cause" of the accident.

The next step of assessing the blame easily follows. Whoever is most responsible for that action, inaction, or decision is the one to blame. The whole process is neatly packaged into an engaging story where all of the pieces fit neatly together. It all wraps up like a TV crime drama – bad event (accident), detective work (investigation),

DOI: 10.1201/9781003344575-7

constructed timeline (the story of what happened), guilty party (whoever or whatever failed that caused the accident), and blame (punishment).

- Cause: Because of what they did (or didn't do), an accident occurred.
- Supporting conclusion: Had they not made that bad decision or action, or made the one that they should have made, that bad outcome would not have happened.

The "guilty" parties are neatly bounded between the two bookends of this deductive logic. Guilt is assigned, the gavel falls, and court is adjourned.

In fairness, most professional investigations are conscientious and thorough. Many systemic contributing factors are identified which lead to constructive improvements within our industry. Aviation is safer because of this important investigative work. Still, in our sound-bite driven society, the headline, "pilot error" is all that most people remember. In one analysis, "...between 1999 and 2006, 96% of investigated U.S. aviation accidents were attributed in large part to the flight crew. In 81%, people were the *sole* reported cause."[3] Politicians and media look at these compelling numbers and demand that airlines and regulators act. The problem appears to be human-caused, so there must be deficiencies in training or performance that we can remedy to solve the problem. Unfortunately, we have been flying for over 100 years and applying "broken part" logic has not prevented aviation accidents.

6.2 FLAWS WITH THE LOGICAL ANALYSIS OF ERROR

The promise of scientific logic is that it describes complex events in simple terms. Unfortunately, simple explanations drive simplistic solutions that don't address complex problems in complex environments.

6.2.1 Flaw of the Bad Apple Theory

If we accept broken component logic, then we search for things that appear broken. Find the malfunctioning part and replace it. Retrain or remove the malfunctioning pilot and aviation problem should be solved. This is especially appealing when we have an accident where the pilots appear to have acted negligently or carelessly. Remove the bad apples and the remaining apples should be good to go. There are two problems with this logic. First, it didn't work, doesn't work, and won't work. After removing or retraining the mishap pilots, the same problems resurface with other pilots in similar situations. Like our Whac-a-Mole metaphor, the same problems continued to pop up. Whacking them down didn't prevent them from emerging again. Second, as we constructed the story from the pilot perspective, it didn't fit the premise of a TV crime drama. Pilots had no intention to cause their ill-fated outcome. They tried very hard to follow what they thought was the best course of action at the time. Viewed from their in-the-moment perspective, they looked more like victims of a bad chain of events than perpetrators of that bad outcome. More accurately, they appeared to be professional pilots immersed in a confusing and fast-moving situation, making decisions that made sense at each moment, and finding themselves arriving at a bad outcome. It turns out that they are not bad apples after all.

6.2.2 Flaw of Deconstruction Logic

The second logical flaw is the use of deconstruction logic. Like dissecting a frog in biology class, we can't take a complex system, open it up, remove and examine all the parts, remove the "bad" one, put the good parts back together, and expect the frog to jump happily back into the pond. Deconstruction logic assumes the following.[4]

- Separation of the whole into parts is feasible.
- Subsystems operate independently.
- Analysis results are not distorted by taking the whole apart.
- Components are not subject to feedback loops and other non-linear relationships.
- Interactions among components are simple enough that they can be considered separate from the behavior of the whole.

In real-world line operations, none of these are true. Before we completely discard deconstruction logic, we should accept that some aspects are useful. It does help us understand some interrelationships and processes. It reveals *possible* forces and influences that may be useful in further investigation. We just shouldn't use it to declare causation or predictability.

6.2.3 Flaws of Hindsight and Foresight

Consider our early example of Clem's unstabilized approach and runway excursion. Reversing his timeline, we can clearly see where Clem's event trajectory began veering off. He mismanaged his approach energy (skill-based error), chose to continue instead of going around (decision-making error), and failed to safely stop his aircraft. If he hadn't committed either of those "errors", he would have gone around and reattempted his approach. Take a moment and notice how easily we labeled his errors and told his story. Knowing the outcome (sliding off the runway), we inferred that he must have made some bad decisions to cause the mishap. We reversed his timeline and located the point where he first mismanaged his energy – where he should have recognized that he could not achieve stabilized approach criteria. This marked the point where his profile deflected from the desired path. We identify Clem's flawed decision to continue his approach as the proximate cause of the accident. The pieces of our story fall neatly into place. Clem might even have agreed with us as he recalled the event from his hindsight-biased perspective.

Just for comparison, let's flip hindsight around for a moment and imagine magically freezing Clem's aircraft at the point where he should have realized that the approach wasn't working. While frozen in time, we inform him that if he continues his approach, he will slide off the end. What do you think he would do? Since Clem is a well-intentioned, professional pilot, he would go around – absolutely and reliably 100% of the time. This demonstrates a form of foresight bias. With perfect foresight, we would always avoid undesirable outcomes.

Considering hindsight bias and foresight bias, we conclude that both aren't very useful. While we are immersed in the flow of aviation and acting in-the-moment, neither foresight bias nor hindsight bias is immediately available. Intent on promoting our game plan, reinforced by thousands of successful landings in our past, we don't foresee the possibility of sliding off the end of a runway. Instead, we focus intently on flying the aircraft, landing, and stopping successfully – just like we always do. The in-the-moment perspective directs our attention toward what we want to happen, not away from what we don't want to happen.

6.2.4 Flaw of Failure to Follow Rules

Another conclusion is to label Clem as a rule breaker. He should have recognized that his approach was not stabilized, followed the stabilized approach rules, gone around, and reattempted the approach. While true, it doesn't help us solve the problem of pilots flying unstabilized approaches. For comparison, consider driving the speed limit. Let's generalize that most of us exceed the posted speed limit on occasion. Maybe we even do it routinely and harbor no sense of guilt or intention to change our behavior. Does this make us all rule breakers? If so, are any of us going to stop speeding as a result of this revelation? No, probably not. Instead, we might defend our reasons for speeding. We could present a compelling argument that if we followed the speed limit while everyone else around us was speeding, then the other drivers would need to swerve around us to avoid collisions. Following the posted speed in this environment would make us a safety hazard. We could logically conclude that it is safer to match the flow of traffic, even if everyone ends up breaking the rules. This example shows how safe choices and procedural compliance can sometimes conflict with each other.

Another dimension is the difficult topic of *intentional non-compliance* or *selective compliance.* This can lead to morality/loyalty test where those who strive to comply are placed into one category while those who bend the rules are placed into another. A study of military aviators cited two reasons why pilots intentionally violated the rules.[5] First, they thought that some rules were misguided and didn't deserve to be followed. Often, these were rationalized as rules that were outmoded, that were intended for a different subset of pilots, or that were devised using a flawed process. Second, pilots thought that if they broke the rules, they wouldn't be caught or punished. This was rationalized as a discretionary space between "what they say in the manuals" and "what they really expect us to do". Imagine if we were the supervisors reading the results from this survey. We might resolve to write better rules that everyone would support and to find more effective ways to enforce violations. Would this work? Back to our driving example, if they raised the speed limit to something that all of us would consider as reasonable, we might be more inclined to comply. Additionally, if they stepped up enforcement, we would comply with the speed limit to avoid citations. Whenever rules are revised or enforcement is increased, compliance temporarily improves. Over time, however, our behavior tends to drift. Eventually, we might acclimate to the higher speed limit. All of those past reasons why we sped before reemerge. If we are running late, we might be inclined to speed. When no one enforces the rule, we might speed again, even when we aren't in a hurry. Eventually, we might find ourselves routinely speeding. Pushing against limits is a typically human behavior.

In the end, rule compliance works best when it acknowledges the complex interactions between incentives, culture, acceptance, and personal discipline. Ideal rules are the ones that everyone naturally wants to follow, the justification behind them is clearly communicated, and they make sense. They match how we naturally want to fly the aircraft and operate the airline. Our line culture aligns to conform with the rules. We apply peer pressure and align line culture to promote compliance. Finally, we feel empowered to maintain rule compliance as a professional standard.

6.2.5 Flaw of Deficient Rulemaking

An influential component of airline management is its rule making philosophy. Currently, companies maintain safety management systems (SMS) to monitor safety parameters and adjust policies and procedures to mitigate risk and reduce vulnerabilities. One strategy is to use rules to bound behavior. If our SMS program detects an undesirable practice creeping into the line operation, we draft a rule to stop the adverse behavior and arrest the drift. If crews are missing a particular task, then we write a stronger procedure or add a checklist step to ensure that it is accomplished. If organizations only follow this strategy, more and more rules are added over time. While this may solve some problems, it often creates different problems. One airline, for example, slowly added steps to their Before Takeoff Checklist until it reached 24 items.[6] Since many of their operations were conducted from smaller airports with short taxi distances, it unintentionally encouraged rushing behaviors. The airline recorded a steady rise in aircraft configuration errors. Following a procedure redesign, many steps were eliminated or moved resulting in a Before Takeoff Checklist of 7 items.[7] Configuration errors dropped significantly.

New rules can generate unintended consequences. A procedure applied in one area may adversely affect an interrelated process. For example, consider a procedure where crews are directed to depress and release the Master Caution light switch. If it remains illuminated, it indicates a system anomaly. We are expected to stop what we are doing, diagnose it, and correct the problem. This sounds reasonable until we study how this procedure affects daily line flying. Assume that that an anomaly only occurs 5% of the time. That means that 95% of the time, the light extinguishes. Our habitual flow proceeds without interruption. When one of those rare anomalous events occurs, it distracts us. We interrupt our normal operational flow to process the distraction, just as we are trained to do. So far, so good. After resolving the anomaly, we need to rejoin our operational flow. Here is where the unintended problem arises. If we lack a solid anchor point to mark where to rejoin the operational flow, we may select the wrong reentry point and bypass important steps.

6.3 HOW COMPLEX SYSTEMS HIDE THE SOURCES OF ERRORS

If we limit our accident investigation process to linear time analysis and mechanical deconstruction, we miss the underlying forces that contribute to error. The following assumptions frame our understanding of human error within complex systems.[8]

- *Error is a label assigned using hindsight:* While it is true that hindsight distorts the error investigative process, consider the ways that hindsight can improve it. The label of "error" is still useful as long as we can accurately define our perspective. Using linear logic, we can still identify points in a failed scenario where the intended path veered off. This point isolates decisions, actions, or omissions that may prove useful for further examination. We can still call these points "errors" as long as we acknowledge that they probably didn't look or feel like errors to the pilots while they were flying.
- *Actions and labels are not equivalent:* This is a follow-on to the previous point. Error is a label that refers to what happens following decisions, actions, or omissions that caused the path to veer off. The action, inaction, or choice is something that happened that led to something undesirable. What the crew did or failed to do didn't feel like an error to them at the time. It felt like the most appropriate decision or action needed at that moment.
- *The point where error occurs marks where our investigation should begin:* Using linear logic, we identify where the error occurred and who did it. It can feel like we have located the cause. Instead, a better strategy is to use this as a starting point to investigate the underlying contributors. This point marks where the trajectory of the event started veering off. Looking back in time from this point, we can study what factors, forces, and influences caused it to happen.
- *The actions or inactions at this point are symptoms of something deeper:* Pilots don't start their day by intending to bend metal or injure passengers. Look for the deeper forces or processes that explain their actions and choices. What underlying factors, forces, and influences exposed the latent vulnerabilities within the system? We can't undo the past, but we can reduce our mishap potential from those same underlying vulnerabilities. As long as we treat errors as symptoms, we can target our safety efforts effectively. This encourages us to expand the events we investigate to include incidents and close calls. Uncorrected, these forces may generate undesirable events in the future.[9]
- *To isolate the crew component, imagine substituting crews:* A legacy of the bad apple approach biases us to presume that an accident only occurred because particular pilots were involved. A useful exercise is to imagine substituting a different set of pilots into a similar situation and then assessing whether the same event still could have happened. If so, then the crew contribution is less important than systemic vulnerabilities. This keeps us focused on the forces underlying the event.

- *In complex systems, the same conditions can produce different outcomes:* How can nine aircraft can land on a slippery runway and successfully stop, while the tenth doesn't? They all land on the same runway, with similar aircraft, and similarly trained crews. All succeed until one fails. Is everyone making good decisions until the mishap aircraft makes a bad decision? Are they all making bad decisions (meaning they all should have diverted) with only one crew suffering the bad consequence? Are they all making good decisions until some unknown systemic factor triggers the accident by that one crew? All cases are possible. We need to accept that complex systems and situations generate unpredictable outcomes. Results are governed by probability, not certainty.
- *Hindsight biases our conceptions of how processes work:* We want to write a story that makes sense. Using story logic, we may adopt the flawed premise that good outcomes come from good processes and bad outcomes come from bad processes. If a bad outcome happened, we search for the bad process that produced it. If, instead, we accept that a good process can produce both good and bad outcomes, it changes our strategy. How can we solve problems when mishap events still seem to find ways to slip through our safety net? In the end, we must accept that complex processes produce their results based on probability. When we develop new procedures to mitigate underlying vulnerabilities, we should expect our error rate to drop, but not to disappear. After we institute a change, we continue to monitor it, study the results, and make further improvements.
- *There is rarely a single cause of a mishap:* An easy accident investigation would involve only one cause – that one component that is 100% responsible for the mishap. This doesn't happen. Imagine an accident where a nose gear strut mechanically fractures and collapses the nose wheel assembly. On the surface, this would seem to be a fairly open and shut case of metal fatigue. We can hold the broken strut in our hands and examine the fractured surface. When we dig deeper and question why the strut failed, our simple investigation begins to become complicated. What kind of metallurgical break was it? Would a crack have been visible during a preflight inspection? If so, why didn't the pilot detect it? Why didn't the maintainers detect it during their last scheduled inspection? What is the inspection cycle for this part? Was the part poorly designed by the aircraft manufacturer? Was it poorly manufactured by a subcontractor? Did it fail from cumulative stress fatigue? Was the failure due to that airline's landing or taxi culture? Is this broken part an anomaly or is this same component approaching failure on similar aircraft? Did the regulator establish acceptable inspection standards? Did the airline comply with inspection methods and frequency? The tendrils from our simple mechanical failure spread throughout the operation, the maintainers, the manufacturer, its subcontractors, the regulator, and the pilots. When we look below the surface, we expose many underlying forces that affect our analysis of the event. The vulnerabilities that we uncover allow us to apply proactive steps toward preventing a similar accident.

- *Contributing factors are always present in the system:* The greater the complexity, the more interrelated forces are present. It is practically impossible to remove all adverse outcomes. The properly functioning parts of the system can become sources that contribute to its failure. Take "productivity" for example. If we increase pilot productivity, we also increase their flying currency (good), frequency of practice (good), duty day (bad if pilots become fatigued), habitualization of procedures (good at first, then bad as actions become automatic and attention levels drop), and laxity (bad). If we look at how productivity affects the airline, we get increased profits (good), less flexibility (bad), increased delays from disruptions (bad), overstretched resources (bad), and smaller safety margins (bad). Despite all of these disadvantages, market pressures encourage companies to seek more productivity, even when it undermines resilience.
- *Error results from normal work:* It is useful to study crews exhibiting normal behavior while performing normal work.[10] We discover that errors still emerge. This is a worrisome concept – that normal pilots doing normal work can generate events that we will, in hindsight, label as errors. For example, in the early days flightdeck design, similar switches and levers were positioned close to each other. This made sense to engineers. It was cheaper and easier to manufacture. A particular WWII bomber had similar gear and flap levers positioned next to each other. Selecting the wrong lever (raising flaps while intending to raise gear) would cause the aircraft to stall and crash. A trained pilot intending to raise gear after takeoff, reaching for the gear lever, doing what they normally did, but unintentionally raising the flap lever, could cause a crash. As an HF solution, the landing gear lever was replaced by one shaped like a wheel. The flap lever was replaced by one shaped like a flap. This gave the pilot both visual and tactile safeguards against the error. Later models moved the levers further apart to increase the distinction. These were not ideal solutions from an engineering perspective, but were really important from an HF perspective.
- *Systems fail:* Successful organizations manage their operations to minimize bad consequences. Imagine an airline trying to conduct operations into a major hub as a squall line of thunderstorms approaches. At some point, forecasters predict that the airport will close. If the airline stops landing aircraft too early, they risk misconnecting thousands of passengers. If they wait until too late, then they risk diverting dozens of aircraft. If they cease operations and the weather front stalls short of the airport, they will lose the opportunity to keep their entire operation running on time. All three represent undesirable outcomes. Ideally, the airline managers would strive to accurately time their decision, and actively manage the consequences.
- *The system changes and adapts through failures:* Systems are dynamic entities that learn the same way that people learn. Systems constantly strive for success, test boundaries, detect failure, and adapt. Ideally,

failures are detected early enough to recover before adverse outcomes occur. People within this dynamic system also react differently based on how they interpret context and nuance. A borderline approach and landing might encourage an aggressive crew to test limits by managing risk more carefully in the future. A conservative crew might adapt by increasing their safety margin.

- *System failures emerge from multiple sources:* A failure at the "sharp end" of the system involves many influences from the "blunt end".[11] From our example of the fractured nose strut, while the pilots experienced the failure first hand, many other groups far removed from the flightdeck contributed to the failure event. Altitude busts are attributed to pilot actions, but ATC controllers, automation, and outside distractions contribute. Low fuel states are detected by pilots, but dispatchers, ATC, and weather contribute.

6.4 PILOT CONTRIBUTIONS TO ERROR

The previous sections provided context for understanding the events that we typically label as errors. Next, we'll examine the choices that we pilots make that contribute to our errors. Complexity, stress, overload, and bias unavoidably increase our probability of making errors, but we sometimes amplify their effects. Following are some common threads uncovered during mishap investigations.

6.4.1 Reliance on Single Pilot Actions within the Crew Environment

Until the 1990s, the terms used for the pilot roles were pilot flying (PF) and pilot not flying (PNF). This unintentionally created a cultural standard where the PF was responsible for operating the aircraft and the PNF was primarily responsible for answering ATC radio calls. Programming, flying, and managing the aircraft were the PF's responsibility. It was common to witness PNFs gazing out their side window, only returning their attention back inside when ATC assigned a frequency change. A flightpath monitoring role was loosely assigned to the PNF, but was not emphasized outside of the takeoff and landing phases.

Since then, every airline has changed the non-flying pilot's role from PNF to pilot monitoring (PM). The intention was to motivate PMs to actively monitor the flightpath, follow the progress of the game plan, and verify all aircraft system changes. The PF's role was also clarified to encourage actively announcing aircraft system changes, communicating a shared mental model, and responding to deviation callouts. Still, the cultural legacy of PF/PNF roles lingers in line practice.

Each pilot is responsible for handling both shared and individual tasks. While all system changes should be monitored and verified, high workload can encourage independent actions.

BOX 6.1 FO INDEPENDENTLY CHANGES ASSIGNED ALTITUDE WHILE CAPTAIN'S ATTENTION WAS DIVERTED

Captain's report: I was pilot monitoring on this flight. ATC was giving us vectors for a visual approach to RWY XX at ZZZ. We were at 2,000′ on a right base between waypoints ZZZZZ and ZZZZZ1 (1,500′ MSL) when we told Approach we had the airport in sight. We were cleared for the visual and told to contact Tower. I said, "glide slope alive" and the First Officer asked for gear down flaps 15. I looked down to select the Tower frequency (which I didn't have preselected) and told them we were on the visual for RWY XX. We were cleared to land. When I saw we were capturing the localizer I noticed that the glide slope was now full scale deflection (low). *The First Officer had selected a lower altitude (1,100′) on the MCP without telling me, so we were below the final approach fix altitude of 1,500′ at ZZZZZ1.* I called for a go around and we were vectored back around for a successful landing. The cause was a long flight across the country with last minute vectors to final for a visual approach. This led to the First Officer thinking we were high nearing the FAF and incorrectly setting an altitude lower than glideslope intercept altitude without verbally advising the Captain.[12]

In this event, the PF/FO independently performed (and didn't announce) a system change that was intended to be completed and verified as a crew. Additionally, it appears that they did it while the Captain's attention was diverted. Often, these actions are made with good intentions, but anytime one pilot alters the aircraft programming or a system without verification, we short-circuit our error detection and mitigation protections.

Pilot personalities and work dynamics also influence our work processes. During busy flight segments, PFs often become task-overloaded well before their PMs. Verification and verbalization take more time than quickly performing a task. We can imagine this FO wanting to set a lower altitude, seeing that their Captain's attention was diverted elsewhere, and quickly spinning 1,100′ into the altitude window. Perhaps they even announced the change (but the Captain didn't hear them) or they intended to inform the Captain when they returned their attention to the flightpath (but forgot or became distracted themselves). Whatever the case, the required communication and verification didn't happen. The error went undetected until the Captain noticed the full-scale low deflection on their glidepath.

This vulnerability is amplified when the PF's personality is more proactive and aggressive and the PM's is more submissive. We see cases where PFs err while trying to do too much while their PMs remained much too passive, disconnected, or uninvolved. There is also an authority effect where FO/PMs defer to aggressive or high-status Captains. They feel reluctant to intervene when the Captain performs autonomous actions citing that, "they are the Captain" and "it is their aircraft."

Procedures also inherently create brief periods of single-pilot operation. Making cabin PAs, calling the station, coordinating with dispatch, and completing heads-down tasks are examples of short intervals where PFs operate alone.

6.4.2 The Rush Mentality

There is a fine line between working quickly and rushing. Working quickly means accomplishing all of the required steps as expeditiously as we can. Rushing attempts to speed task accomplishment through multitasking, condensing procedures, and shortcutting. Rushing strives to finish "all of the important stuff" as quickly as possible. We may intend to return and clean up the skipped items later, but often forget or remain too busy.

The rushing mindset can spread across the line culture. Rushing pilots take pride in their ability to prepare, execute, and move faster than others. This becomes evident when an airline attempts to procedurally slow the pace of operation. Pilots who are accustomed to rushing express difficulty and frustration with slowing down. The slower pace feels wrong after moving quickly has become so deeply ingrained into their habit patterns.

Instead of responsibly managing the operational pace, we let the flow of the operation push us. This is a subtle, but important, distinction. Rather than slow down or reset a quickening situation, we just work faster to keep up. We saw this with Clem's unstabilized approach. He worked faster and faster trying to salvage his approach. In the end, he accepted his unstabilized approach and landed.

When rushing, we become especially vulnerable to distraction and miscuing. We skip required steps or miss tasks altogether. This becomes worse when the cue to perform a task is floating, meaning that it is not anchored to another reliable, consistent event. The normal environmental cues that remind us to perform a task are masked or overshadowed by the perceived need to hurry.

6.4.3 Task Management at Inappropriate Times

Simply stated, this is doing the right thing at the wrong time. It increases our vulnerability to distraction and error. The guiding standard is *appropriateness* – completing each task at its appropriate time. Problems arise when pilots manage tasks based on when they remember to do them, even if it is not at an appropriate time. Examples include accomplishing discretionary tasks during low altitude or in dynamic flight, setting navaids for the intended arrival while still on departure climbout, and stowing personal items during taxi-in. These actions are motivated by a desire to stay ahead of future workload or to avoid missing tasks. While these are desirable goals, doing them at inappropriate times unnecessarily exposes vulnerabilities.

6.4.4 Lack of Knowledge, Failure to Recall Knowledge, and Poorly Applied Knowledge

These knowledge-based errors all contribute to a failure to recall required knowledge when the situation requires. They include:

- Vaguely or poorly written guidance
- Absence of guidance to address a particular situation
- Guidance that is difficult to locate in the manuals
- Crews failing to recall that guidance exists or not remembering to apply it
- Crews knowing that guidance exists, but erring while performing the steps from memory

These errors are exacerbated by system complexity, especially when operating at stations with specific local procedures that aren't practiced elsewhere in the system.

6.4.5 Flawed Risk Assessment

Pilots may also be aware of the risk, but intentionally choose riskier options that are more prone to error. They accept the risk because they deem that the choice is discretionary and that the risk is manageable. Some complicating factors are that personal risk standards tend to drift with changes of experience, personal priorities, fatigue level, and familiarity. Also, one pilot's risk tolerance may be significantly different from another's. The main purpose of a company's risk management policy is to promote a standardized practice that every pilot is expected to follow. This eliminates personal variability and refocuses decisions toward trained risk management procedures.

6.4.6 Misapplied Personal Priorities

These errors result when we elevate personal priorities over company priorities. "Let's fly faster. I need to make my commuter flight." Pilots may also elevate priorities that justify what they perceive to be informal company policies. "Yes, the book says to fly profile speeds, but they really want us to fly fast so we can get to the gate as early as possible."

6.4.7 Tolerance of Error

Some pilots allow their error tolerance to drift. Like with our example of driving speed limits, our standard of what we judge as an acceptable exceedance can slowly change. Our level of tolerance either aligns with the line culture or slowly drifts with our personal standards. This shift is often reinforced by *consent by silence,* where pilots don't verbalize each other's errors. Even when both pilots know that they are exceeding limits, crew silence is interpreted as tacit approval to continue.

6.4.8 Ineffective Communications Environment

Some pilots promote minimal flightdeck communications. These quiet flightdecks tend to suppress active communication of errors, problems, or concerns. While Captains set the standard for the communications, quiet FOs can also affect this environment. An operationally quiet flightdeck can undermine crew resource management objectives, team cooperation, and the willingness to make deviation callouts.

6.4.9 Intentional and Selective Noncompliance

This final area covers a range of errors from brazen defiance of the rules to acceptable acts of non-compliance. Intentional noncompliance is a professionalism issue that is addressed in later chapters. Culturally acceptable acts of noncompliance include "common sense" deviations. For example, stabilized approach criteria calls for the PF to maintain glidepath on final, but many pilots intentionally fly slightly above the glidepath behind heavy aircraft to avoid their wake turbulence. Pilot discretion and crew communication standards guide how much deviation is acceptable and when PMs should make a callout.

6.5 A Better Way to Evaluate Error – Telling the Second Story

In preface to their book, *Behind Human Error*, Woods and co-authors make the following statement:

> As practitioners confront different evolving situations, they navigate and negotiate the messy details of their practice to bridge gaps and to join multiple conflicting goals and pressures imposed by their organizations. In fact, operators generally do this job so well, that the adaptations and effort glide out of view for outsiders and insiders alike. The only residue left, simmering on the surface, are the "errors" and incidents to be fished out by those who conduct short, shallow encounters in the form of, for example, safety audits or error counts. Shallow encounters miss how learning and adaptation are ongoing – without these, safety cannot even be maintained in a dynamic and changing organizational setting and environment – yet these adaptations lie mostly out of immediate view, *behind* labels like *human error*.[13]

Let's examine some of these ideas. They use the description, "navigate and negotiate the messy details of their practice". When we summarize an event, we often suppress the messy details by viewing it in hindsight. This allows us to form a clean story to describe the accident. Using this story, we wonder, "Why didn't they see that clear warning sign?" The messy reality is that the pilots didn't perceive it because it didn't seem like a clear warning sign while it was happening.

Mishap investigation is difficult because we rely on the pilots' recollection of the event. Recalling what happened changes after they repeatedly relive it in their minds. I recall an investigation of a landing gear system failure. Despite the pilots' best efforts, the nose wheel assembly collapsed during rollout. The manuals covered a similar landing gear malfunction, although only it proved to be marginally useful. The pilots anguished over the possibility that they "must have done something wrong". Over the days following the mishap, they replayed the event over and over in their minds – each time constructing a fresh and slightly altered recollection. Added to this was an interesting twist that the Captain's wife was also a Captain on that same model of aircraft flying for a different airline – so same aircraft, but different procedures and training. By the time we conducted a crew debrief, both pilots struggled to confidently recall what they had actually experienced. What actually happened became mixed with what they thought should have happened, conversations they had, dreams they experienced, and what they reviewed in the manuals. Throughout this process, each of us tried to construct a simple story of the event. Did the pilots err, or not? This proved impossible because, at the time, we all lacked knowledge of a failed computer chip on a controller card tucked in the aircraft's electronics bay. Only after we learned of this "one in a million" anomaly were we able to form an accurate story. In the end, we concluded that the pilots performed wonderfully. Until then, it was difficult to discard the possibility of pilot error.

Back to the earlier quote from Woods and his colleagues, they cite the need to, "bridge gaps and to join multiple conflicting goals and pressures imposed by their organizations." This captures the challenges of market pressures on corporate priorities. Organizations pursue many goals – safety, profit, on-time performance, fuel savings, high productivity, customer service, employee satisfaction, and regulatory compliance. Frequently, these conflict. One often-cited example is NASA's "Better, Faster, Cheaper" philosophy from the late 1990s. This factor was considered to be one of the driving forces behind the shuttle Columbia accident (a foam piece breaking loose during takeoff that pierced the leading edge of the wing and caused the Columbia's destruction during reentry). All three goals were desirable, but together, they generated conflicts. Better is often slower and more expensive. Faster and cheaper are rarely better. Nonetheless, NASA structured its policies around these goals. They unintentionally fostered an environment of shortcutting, suppressed the reporting of errors and defects, and ignored problems that would take too much time or money to solve. In fairness, these conclusions reflect hindsight bias. NASA employs brilliant, well-intentioned professionals. Coming off of the hard lessons of the shuttle Challenger accident and the losses of the Mars Climate Orbiter and Mars Polar Lander, they were under heavy pressure from Congress. They needed to show that they were fulfilling their mission by doing it better, faster, and cheaper.

Woods and his colleagues encourage us to get past the first story and to look for the *second story*. The second story involves "… doing things safely – in the course of meeting other goals – is always part of peoples' operational practice."[14] Telling the second story involves "… people and organizations trying to cope with complexity, continually adapting, evolving, along with the changing nature of risk in their operations."[15] When we view events from their second story perspectives, it changes the discussion. Stripping the pejorative label of "error" from the investigation refocuses us on questions like:

- What were the crew's priorities during the critical moments of their event?
- Setting aside what we think should have been done and understanding what the crew felt was important at the time, what was their mindset?
- What does the crew recall seeing? This focuses on what they recall actually seeing, not what we think they should have seen. This is a challenge because the crew's recall may become distorted by their own hindsight.
- What did [this important indication] mean to them at the time? This provides insight into crew members' contexts and priorities. Why did this particular indication seem more important or memorable than others?
- When did they first notice [this important indication]? This anchors the timeline at the critical point of interest. From here, we can examine backward to understand what other tasks and indications were competing for their attention.
- How did [this important indication] fit in with their priorities and plans at the time? This provides context and meaning that [this important indication] held for them at that moment.
- What did [this important indication] do to their mindset at that moment? This examines whether the event changed their mindset or was accommodated and absorbed into their existing game plan.
- What were the clearest thoughts they recall from this moment? This indicates where their attention was focused.

Questions like these suppress our hindsight bias and work to recreate the busy, messy environment that actually existed at the time when decisions were made and actions were taken. This way, what we label as errors become markers where we can begin our investigation instead of places where we end them and assess blame.

6.6 STUDYING ERRORS TO PREVENT THE NEXT ACCIDENT

An inherent problem of the flawed component or bad apple approaches to error investigation is the assertion that if we can discover and remove the flawed part or person, that the system will return to perfect operation and that future errors will be prevented. In the previous quote, the authors cited that, "Shallow encounters miss how learning and adaptation are ongoing." Errors are not isolated events. They emerge as natural outcomes from the complex interactions between dependent systems, processes, and people – all of which learn and change over time. As James Reason put it, "errors are … the inevitable and usually acceptable price human beings have to pay for their remarkable ability to cope with very difficult informational tasks quickly and … effectively."[16] This opens a compelling avenue for error investigation – that errors are natural byproducts of the adapting and coping that organizations and operators must use to respond to the real-world complexities and changing conditions in aviation.

We use error investigation to uncover the latent flaws and vulnerabilities within our systems, work processes, and interpersonal interactions. Understanding this, we can pursue strategies to eliminate, mitigate, or educate line operators.

- *Eliminate vulnerabilities:* Our first objective is to refine a work process to eliminate the vulnerability. This is performed at a higher, organizational level. Flight operations and training center managers analyze error trends and events, uncover flawed processes, and revise procedures to remove undesirable trajectory paths.
- *Detect and mitigate vulnerabilities:* When elimination is not possible or practical, procedures can be improved or retrained to improve the pilots' ability to detect and mitigate these undesirable trajectories.
- *Educate vulnerabilities:* When neither of these options are practical, we can enhance training and education to give line operators the tools and skills needed to detect and interdict unfavorable trajectories.

Our objective is to focus on *how* and *why* the error happened instead of *who* made the error and *what* they did or didn't do. Tony Kern observes, "… most error-caused accidents or incidents are the endgame of a series of interrelated events, interpretations, decisions, warnings or actions that are allowed to progress without recognition or intervention. The final trigger decision, action (or inaction), may be relatively innocuous, but sufficient in itself to totally remove a margin of safety previously eroded by other events."[17]

Something happened to tip the scales against this mishap crew. Concentrating on how and why depersonalizes the investigation process and builds that second story that examines how *any* well-trained, well-intentioned pilots might find themselves pursuing a similar, failing game plan.

As pilots, we aren't letting ourselves off the hook. We still need to follow procedures, exercise due diligence, and appropriately manage risk. Using this second story perspective, we may still reach the conclusion that the mishap pilots didn't follow procedures, didn't pay appropriate attention, or managed their risk poorly. We may conclude that our procedures remain well-constructed and resilient, but crews failed to apply them.

NOTES

1 Dekker (2015, p. 38).
2 Dekker (2011), Chapters 3 and 4.
3 Dekker (2011, p. 75). – As cited from Dr. Richard Holden (2009).
4 Dekker (2015, p. 61). Reference is cited from: Levenson, N. (2012) *Engineering a Safer World: Systems Thinking Applied to Safety.* Cambridge, MA: MIT Press.
5 This refers to a survey of U.S. Navy aviators conducted probably in the early 1970s.
6 Loukopoulos et al. (2009, p. 111).
7 Loukopoulos et al. (2009, p. 117).
8 Summarized from "Fifteen Premises" (Woods, 2010, pp. 19–31).
9 Woods et al. (2010, p. 23), Woods et al. use the metaphor of "dress rehearsals" for these close call events. While this is a hindsight label, it directs our efforts toward preventing related future mishaps.
10 Woods et al. (2010, p. 25).
11 This is a commonly cited distinction used by many HF scientists. Sharp-end assets are the line operators – pilots and airport workers. Blunt-end assets include managers, trainers, and regulators (Woods et al., 2010, p. 9).

12 Edited for clarity. Italics added. NASA ASRS report #1791268.
13 Woods et al. (2010, p. xix).
14 Woods et al. (2010, p. 1).
15 Woods et al. (2010, p. xix).
16 Reason (1990, p. 172).
17 Kern (2009, p. 6).

BIBLIOGRAPHY

ASRS – Aviation Safety and Reporting System – National Aeronautics and Space Administration (NASA) Online Database. (n.d.). Retrieved from Aviation Safety and Reporting System. https://asrs.arc.nasa.gov/search/database.html.

Dekker, S. (2011). *Drift into Failure: From Hunting Broken Components to Understanding Complex Systems*. Burlington, NJ: Ashgate Publishing Company.

Dekker, S. (2015). *Safety Differently: Human Factors for a New Era*. Boca Raton, FL: CRC Press.

Kern, T. (2009). *Blue Threat: Why to Err Is Inhuman*. Lewiston: Pygmy Books, LLC.

Loukopoulos, L., Dismukes, R. K., & Barshi, I. (2009). *The Multitasking Myth: Handling Complexity in Real-World Operations*. Burlington, NJ: Ashgate Publishing Company.

Reason, J. (1990). *Human Error*. New York, NY: Cambridge University Press.

Woods, J., Dekker, S., Cook, R., Johannesen, L., & Sarter, N. (2010). *Beyond Human Error – 2nd Edition*. Burlington, NJ: Ashgate Publishing Company.

7 Distraction

Distraction refers to events where something diverts our attention away from our operational workflow. With mishap events, a distraction often marks where the intended game plan started to veer off. Our Master Class goal is to develop effective techniques to minimize distractions, detect them as they occur, remedy their causes, and restore our operational flow.

7.1 DISTRACTION AND WORKFLOW

We are easily distracted. Warning lights, bells, chimes, display movement, color changes, and flashing symbols are all designed to attract our attention. In fact, if the engineers discovered that one of their alerting signals failed to sufficiently attract our attention, they would redesign it to make the light bigger and brighter, the alarm louder and scarier, and the colors bolder and flashier. In a classic Gary Larson *The Far Side* comic, the FO's entire side bulkhead is covered by a giant, three-foot diameter flashing red light. The caption reads, "I'm afraid we're going to have to head back, folks… We've got a warning light on up here, and darn if it isn't the big one."

We devote considerable effort to follow our flight's operational flow. We perceive it as a continuous stream of flight with many individual and interdependent tasks sequenced along the way. Through operating procedures, experience, and repetition, we've come to envision an ideal flow of what we plan to happen, how indications should appear as we fly, and what we need to do to guide our aircraft from departure gate to arrival gate. The best choices and actions are the ones that match our planning (past SA), make sense in the current moment (present SA), and guide our flow along the intended path toward our desired outcome (future SA).

7.2 HOW WE RESPOND TO A DISTRACTION

Distractions seem to reach out and grab our attention. This is because our sensitivity to distraction is hard-wired into our brains. From our ancestral beginnings, our foreparents survived because they alerted to the large brown hump in the tall grass (because it may be a lion) or the rustle in the bushes (because it may be a tiger). The survivors developed instinctive reactions that detected and responded to unknown events as possible threats. They always assumed it to be a lion or a tiger until it proved itself to be something non-threatening. As a result of this evolution, distractions compel us to stop what we are doing, analyze them as possible threats, and respond.

One distraction can be important and guide us to make appropriate decisions. Another distraction can prove unnecessary and induce inappropriate decision making. Some distractions are so brief that they barely disrupt our flying. Others deflect our attention away from our operational path.

DOI: 10.1201/9781003344575-8

7.2.1 Handling a Distraction

Responding to distractions and restoring our operational flow are high priorities. Consider examples like a thunderstorm building across our flightpath or an ATC radio call that interrupts us while completing a checklist. We need to:

1. Divert our attention away from what we are doing and toward the distraction
2. Understand what the distraction means
3. Choose how we intend to respond
4. Complete our response
5. Return our attention back to the flight and restore our intended operational flow

From safety analysis, we've learned that as distractions become more powerful and startling, we become more vulnerable to making errors. For example, consider an engine fire during takeoff as we are approaching V_1 speed. Are we expected to *detect* the fire light and warning bell? Absolutely, 100% of the time. We need to shift our attention away from flying long enough to notice that we have an engine fire. Are we expected to *respond* to this distraction? Yes, we need to choose between rejecting or continuing the takeoff. We are also expected to complete a series of remedy steps, both accurately and at the right time. In fact, the airline industry is so committed to ensuring that we choose the right response to an engine fire at V_1, that they make us practice this event during every recurrent simulator session. If we ever encounter one on a real flight, they want our reactions to be deliberate, accurate, and timely.

7.2.2 Recovering from a Distraction

Analyzing pilot errors, we discovered that many errors did not result from the distraction itself, but from the crews' failures to recover to their intended paths after resolving their distractions. Consider this example from a Boeing 737 procedure.

BOX 7.1 THE DISTRACTING MASTER CAUTION LIGHT AND NO-FLAP TAXI EVENTS

Following the engine start, the FO performed a flow to bring aircraft systems on line. The last two steps of the flow were completing a Master Caution light recall (push and release of the annunciator panel to verify the absence of remote caution lights or minor system discrepancies) and then holding the flap lever. The FO would announce, "Standing By Flaps." This established a strong procedural anchor that signaled to the Captain that they were ready to set the flaps and continue with the operational flow. When the Captain completed their flow and heard the FO's callout, they would direct them to set the planned takeoff flap value. Almost every time, this sequence worked as intended. In a small percentage of cases, however, the Master Caution light remained illuminated following the push-to-test step. The task of resolving this Master Caution light ranged from quick switch resets to lengthy MEL deferrals. For most of these events, both pilots investigated the cause of the light, properly identified the problem, and resolved it. During this process, however, FOs invariably removed their hand from the flap lever. This detached their operational flow from the anchor point.

When the problem was finally resolved, the crews surveyed their surroundings for cues of where to rejoin their operational flow. They observed that they had engines running and that the ground crew had departed. Both were typical signs that they were ready to taxi, so they called for taxi with ATC. Their error was not with detecting or responding to the distraction, but with selecting the wrong recovery point to reenter their operational flow (calling for taxi versus completing the after start flow, setting flaps, checking flight control movement, and completing the Before Taxi Checklist).

This error was reinforced by their sense of pacing. We calibrate our felt sense for how long tasks normally take. For example, assume that pushing off the gate, starting the engines, and completing all of our before taxi tasks take about 5 minutes. In that time period, the ground crew disconnects the towbar, announces their departure, and departs. In our distraction example, starting the engines and resolving the Master Caution light took about 5 minutes. In that same time period, the ground crew finished all of their usual tasks and departed. At this point, the flight crew subconsciously expected the flow to rejoin the typical pacing. They looked for familiar cues. They assessed that they had engines running, that the ground crew had departed, and that about 5 minutes had elapsed. They saw what they expected to see and intercepted the operational flow by calling for taxi.

From this example, we see how some segments of the operational flow stop while other segments continue. The flightdeck flow was stopped by the Master Caution light discrepancy, but the ground crew's flow continued at its normal pace. By relying on the ground crew's departure as their cue, the flight crew didn't notice that their flightdeck flow had become interrupted. They perceived indications that they were ready to taxi, which reinforced their feeling that they had accomplished all required tasks.

Safety analysis indicated that this Master Caution light distraction was a common cause of no-flap taxi events. The flow procedure was redesigned to relocate the Master Caution press-to-test task to a later time. That removed the source of distraction and nearly eliminated no-flap taxi events. They weren't completely solved, since there are other types of distraction that also contribute to no-flap taxi events, but it did significantly reduce errors.

Another example is flap overspeeds after takeoff. For aircraft without automated speed warnings, these overspeed events typically result from distraction – typically something like ATC calling out traffic. Both pilots divert their attention, locate the traffic, ensure that the threat is clear, and return to monitoring their climbout. Failing to notice that their flap retraction pacing had been disrupted, crews rejoined their expected timing for normal climbout and missed that they hadn't finished retracting their flaps.

7.3 THE DISTRACTION ENVIRONMENT

Up to now, we have generalized the average human's reaction to distraction. Let's make it personal by examining our own distraction reaction. What conditions best predict our personal vulnerability? For most of us, it is the *level of attention* that we are devoting at any particular moment during the flight. Consider Figure 7.1[1].

The vertical axis indicates how deeply we are paying attention. The horizontal axis reflects the situation's complexity or difficulty. It also correlates with how vulnerable we are to distraction. The sloped line indicates an ideal balance where we match an appropriate level of attention with the existing level of distraction vulnerability. When our attention is appropriately matched to the complexity of our situation, we typically succeed. When our attention level mismatches the situation, we are more likely to suffer unfavorable outcomes. Consider the four points on the graph marked A, B, C, and D.

- Point A (lower left) indicates an easy flight segment where our level of attention is understandably low – like sustained cruise flight in a low-demand ATC environment. Our workload remains low and there is little probability of a significant distraction. We easily handle any minor distractions that occur.
- Point B (upper left) represents a high level of attention during this same low-vulnerability flight segment. While there is nothing procedurally wrong with this, maintaining a high level of attention is fatiguing in the long run. We can't sustain it. Imagine flying the same low-demand flight segment but deeply concentrating on the aircraft and flight conditions. This is not what proficient pilots normally do.
- Point C (upper right) indicates a demanding flight segment where we have raised our attention to an appropriately high level. This satisfies our goal of tracking closely with the sloped line by matching our attention level with the demands of the flight phase. Like flying an instrument approach in

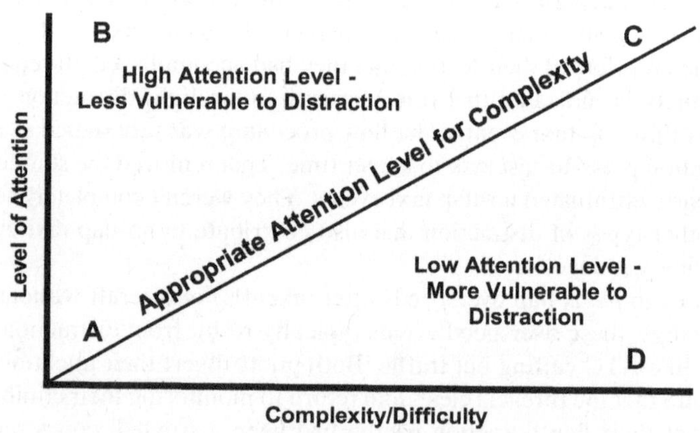

FIGURE 7.1 Appropriate attention level with an increase of complexity.

marginal conditions into a challenging airport, the crew at point C devotes their full attention to closely monitoring the flightpath. They quickly detect and handle distractions.

- Point D (lower right) indicates a situation where we are in a highly demanding situation, but are not paying adequate attention. Perhaps we are fatigued during a late-night arrival into a familiar airport, like the last flight of the night into our home base. Point D represents a highly vulnerable region where the effects of distraction become magnified and outcomes can prove hazardous.

7.4 HOW DISTRACTION AFFECTS US

Point B might reflect novice pilots who quickly gain experience and join the rest of us. We'll limit our study to points A, C, and D.

In Figure 7.2, "Points A and C – Effects" list the advantages of matching our attention level with the complexity and difficulty of the situation. These are all desirable responses. Our concern is with pilots who choose to remain overly relaxed and comfortable despite the rising demands of the flight. "Point D – Effects" reflect the adverse effects that emerge from this mismatch between attention level and complexity. Let's examine each of these effects.

7.4.1 Startle Reaction

A physiological/psychological response to distraction is startle – often called the startle effect. A simplified model of brain function separates the brain into three regions. The lower brain (brain stem and cerebellum) is responsible for basic bodily processes and movement. The middle brain (the limbic system) is responsible for assigning meaning and emotions to sensory inputs. The higher brain (cortex and

Points A and C - Effects	Point D - Effects
Shorter Startle Reaction	Longer Startle Reaction
Quicker Analysis	Slower Reaction
Accurate Plan Recovery	Inaccurate Plan Recovery
Operational Flow Recovery	Operational Flow Disruption
Active Flight Path Monitoring	Flight Path Inattention
PF/PM Role Integrity	PF/PM Role Breakdown
Actions by Choice	Actions by Habit

FIGURE 7.2 Effects at points A, C, and D.

neo-cortex) is responsible for higher thinking like cognition, reasoning, deliberation, and logic.

The middle brain serves as the bridge between our lower brain and the higher brain. During a strong startle reaction, the middle brain becomes highly activated. It initiates an instinctive "fight or flight" response.[2] When this happens, our higher brain (thinking) and lower brain (movement) become temporarily disconnected from each other. The middle brain bridge between the two is temporarily out. Our higher brain struggles to understand what is happening. It becomes overloaded by the emotional and sensory thoughts spilling over from the middle brain. Someone watching us might see that we have frozen our movements and are staring intently.

We have some control over the intensity and duration of our startle effect. Anticipation helps. If we are warned of a distracting event just before it happens, our startle effect is lessened. Imagine flying the simulator and the instructor warns us, "I'm going to fail your #1 engine at 125 knots during your takeoff roll so you can practice a V_1 cut." This would position us at point C on Figure 7.1 – fully attentive, ready, and least vulnerable to distraction. For contrast, if we were overly relaxed while making a normal takeoff on a routine flight (Point D on Figure 7.1), we would probably experience a much stronger startle effect when the engine unexpectedly fails. It would take us much longer to resume constructively thinking and responding.

7.4.2 Analyzing the Source of the Distraction

After the initial startle effect, our minds try to make sense of the distraction. This may be difficult because a strong initial startle reaction freezes the analytical functions of our higher brain centers. Only after the freeze begins to thaw do we begin sorting through the details. For example, with an unanticipated engine failure and continued takeoff, our first priority is to maintain aircraft control and keep the aircraft in a steady, wings-level climb. In these first moments while recovering from the startle effect, that may be all we can handle. As our higher brain functions recover, we scan our instruments to analyze the situation. We see our #1 engine instruments winding down. We notice that we added right rudder to center the yaw from asymmetric thrust.

How quickly we analyze the details to diagnose what has happened depends on our attention level. The higher our level of attention before the distraction, the weaker the startle effect and the more quickly we recover our ability to analyze the cause of the distraction. The lower our level of attention, the stronger the startle effect and more slowly we recover. This is why it is so important to maintain a high level of attention during critical, time-sensitive phases of flight.

7.4.3 Plan Recovery

As we fly, we constantly monitor our operational flow. Our SA tracks where we plan to be, where our current position is at the moment, and where we expect to be in the future – smoothly progressing from one moment to the next. Distractions disrupt this smooth flow. After we process the distraction, our next task is either restoring our original game plan or switching to a backup plan.

Consider an example of a bird strike while on short final. We hear the bang and see the spread of bird remains across our windshield. For a moment, it startles us. Then, we begin processing the event and deciding what to do.

- *Initial reaction:* Continue flying the aircraft down final (maintain aircraft control).
- *Determine what happened:* We had a birdstrike on the radome (assessment of the distraction).
- *Analysis:* Engines look good (make sense of the event and its consequences).
- *Recovery:* Continue with the approach (choose to continue with the original game plan and land).

The more immediate the distraction, the greater the disruption. If an aircraft is slow clearing the runway in front of us causing ATC to send us around, we easily handle the go around. This is because we have time to anticipate the go around. We can mentally rehearse the procedure and coordinate tasks and roles before commencing it. If, however, an aircraft unexpectedly enters our runway while we are in the flare to land, we'll have to initiate an unexpected, immediate go around. This unexpected disruption will feel much more intense. Lacking advanced warning, we have to instantly switch to an unrehearsed, undiscussed go around procedure. Events like this often result in go around errors like over-rotation, forgetting to raise the landing gear, altitude busts, and overspeeding the flaps.

7.4.4 Flightpath Management

The most important task in aviation is controlling the aircraft's path. We can become so focused on the distraction that neither pilot monitors the aircraft's path. Distractions can generate errors like excessive bank events during departure turns (while both pilots are looking for pop-up traffic), incomplete transfer of aircraft control (resulting in no one flying the aircraft during unexpected go arounds), and wingtip damage during taxi (while both pilots are diverting their attention toward the opposite direction).

7.4.5 Time Distortion

Distractions skew our sense of elapsed time. Sensing that we were distracted for only a few seconds, we can discover that a much longer span of time has passed. Consider an example of a Captain taxiing the aircraft. Unsure about an upcoming intersection, they look down to reference the airfield diagram on their electronic flight bag (EFB). They plan to divert their attention for only a moment. While scanning the chart, they have trouble locating the particular taxiway intersection. As they search more intently, they lose their awareness of elapsed time. Suddenly, their FO alerts them that they are drifting dangerously close to the edge of the taxiway. The Captain intended to only look away for a moment, but their self-generated distraction of searching the chart stretched much longer than they had planned.

7.4.6 PF/PM Role Integrity

One of our most resilient distraction countermeasures is the division of PF and PM roles. This division separates how each pilot performs, monitors, and verifies each task. It ensures that at least one pilot is actively monitoring each important parameter at all times. Very few errors slip through undetected.

Consider a situation where the PF becomes focused on handling a distraction. We still want the PM to participate in interpreting and managing the distraction, but only as a secondary role. The PM must remain detached enough to monitor the quality of the PF's decisions and actions. This helps us to avoid situations where both pilots simultaneously divert their attention toward the same distraction.

7.4.7 Choice or Habit

When highly effective crews maintain an appropriate level of attention, they choose actions that match their actual conditions. They detect important nuances and deliberately select the best options that maintain their desired operational flow. Distracted crews tend to revert to habitual, familiar scripts that include a pre-packaged set of decisions. While habitual decisions do free up mental resources to use while processing the distraction, they may be inappropriate for the situation. In extreme cases, they can lead to Recognition Trap errors of trying to force a familiar set of decisions against contradictory conditions. Factors contributing to this habit preference include inattention, fatigue, low motivation, repetition, boredom, and performing discretionary activities during high workload flight phases.

7.5 EXTERNAL CONTRIBUTORS TO DISTRACTION

External distractions divert our attention from monitoring the operational flow. Following are some of the common sources.

7.5.1 ATC Radio Calls

The most common distractions come from ATC radio calls. These are particularly potent because they occur at unpredictable times, we can't ignore them, and they often divert our attention for significant periods of time.

- *Who is ATC calling?* Consider a typical ATC radio call. We need to stop what we are doing, listen to it, and determine if it is intended for us. When the call is directed toward someone else, this kind of distraction is short-lived.
- *Is the ATC call useful to us?* Even if ATC isn't calling us, we don't ignore the message. A call intended for another aircraft may still contain useful information for us. For example, if an aircraft ahead of us receives a change of routing and runway, we can anticipate that we may receive that same reroute. Monitoring transmissions for other aircraft helps us stay ahead of changes and efficiently manage our workflow.

- *What if the ATC call is intended for us?* As we monitor a busy ATC frequency, we screen the first part of the radio call for our callsign. If we recognize that the call is intended for us, we immediately suspend what we are doing and increase our attention level. Our ability to ramp our attention level up-and-down and back-and-forth is an essential aviation skill. We become experts at filtering radio calls while continuing our work. Even so, these up-and-down and back-and-forth transitions are particularly vulnerable to distractions.
- *Attention capacity and distraction:* A problem arises when we are repeatedly interrupted while we are committed to a lengthy task. Consider two examples. In the first case, we are in sequence to land at a busy airport. We have the arrival and approach briefed and loaded in the FMS. Despite the constant stream of distracting ATC calls and replies, we easily manage flying and monitoring. In the second case, everything is the same except that we are given a last-minute change to a different arrival procedure. The complex FMS reprogramming and chart review consume much more of our focused attention. The constant ATC radio traffic now becomes a significant source of distraction. The bottom line is that we have limited ability to switch our attention back-and-forth between what we are doing and the ATC calls. When the demands of the situation overload our available attention or require us to switch too frequently, our vulnerability to distraction error rises.
- *Number confusion:* In aviation, we deal with lots of numbers – callsigns, frequencies, headings, altitudes, sequences, plus all the numbers involved with our aircraft systems. When numbers are repeated and mixed within sentences, we need to channel more of our attention. Consider this readback to an ATC clearance by a Captain, "Cleared to descend to two zero zero, cross two zero miles south of XYZ at two two zero." They acknowledged, "Leaving two two zero for two zero zero." In investigating the crew's resultant altitude bust, ATC claimed that the clearance was only to FL 220.[3] In this case, the confusion was caused by the difference between "to" and "two". Had they included "flight level" in the readback such as "Cleared to descend to flight level two zero zero," ATC would have detected their error and corrected the readback.

  The Flying Tiger 66 CFIT accident (1989) at Kuala Lumpur, Malaysia was largely attributed to number confusion when ATC assigned, "Tiger 66, descend two four zero zero [intended as 2,400′]. Cleared for NDB approach runway three three." The Captain of Flying Tiger 66, who heard, "descend *to* four zero zero" replied with, "Okay, four zero zero" (meaning 400′ above sea level, which was 2,000′ too low for their position on the approach).[4] While the runway is near sea level (about 60′ MSL), significant high terrain surrounds the airport. The crew distraction was intensified by their hurried attempt to locate the approach charts, then prepare and brief for the unexpected and unfamiliar NDB approach. These are examples of confusion between "to" and "two". Other number mistakes occur with:

- *Number confusion:* We have confusion between "five" and "nine". Many compensate by using "niner".
- *Number reversals:* "Heading 120" versus "Heading 210".
- *Callsign/number confusion:* "Trans American 209, climb to flight level 290".[5]
- *Heading/Maneuver confusion:* "…left three-sixty, descend to three thousand, follow traffic…" The crew executed a full-left 360° turn. Controller meant a left turn to heading three-six-zero and to follow traffic.[6]

- *Information overload:* When we see something, our visual sensory memory lasts for only about one second.[7] In that short period of time, we need to decide whether to retain it within our working memory or to let it pass so that it can be replaced by something else. Working memory involves taking what we just saw or heard, comparing it to our knowledge and experience (to apply meaning), and holding onto it until we are finished using it. The consensus in the research community is that the maximum number of distinct pieces of information that we can hold in short-term, working memory averages around seven. The more complex the pieces of information, the fewer we can retain. The more familiar or interconnected, the more we can retain. Most ATC controllers try to limit their transmissions to three or four distinct pieces – like callsign, altitude, and route. For example, "Trans American 12, climb one two thousand, resume SHEAD 1 departure." Most of us would read back this clearance accurately. Consider how this changes as we ratchet up the complexity, "Trans American 5729, climb to cross TARRK at one one thousand, then direct DBIGE, resume SHEAD 1 departure." The first problem starts with the callsign. Our minds might process "Trans American 12" as a single piece of information since "12" is commonly recognized as "twelve". When we complicate the callsign with unrelated digits, "Trans American 5729" in this case, it consumes several distinct working memory information units. At most, it can consume four distinct information pieces (five, seven, two, nine). If we combine them, we can reduce it by two (fifty-seven, twenty-nine). Both numbers remain unfamiliar to most of us, so they are harder to remember. After we stumble through the callsign, now we have a specific crossing restriction at TARRK at "one one thousand". While experienced pilots might simplify this as "eleven thousand", it takes more mental processing to translate "one-one" to "eleven". Finally, we have a change in our route clearance to proceed direct to DBIGE from TARRK. Due diligence compels us to review the SID chart to verify the change and to reprogram the FMS. If we were in the process of doing something else when ATC assigned this clearance, we would have to stop what we were doing, read back the clearance, review the chart, coordinate the changes as a crew, reprogram the FMS, ensure that the aircraft autopilot is responding correctly, and only then, return to what we were previously doing.

  Imagine that we are at high volume airport during the height of a departure push. Controllers might try to save time by combining all required pieces of information into a single transmission, "Trans American 209, turn right on ALPHA to runway one two, hold short taxiway BRAVO, follow the Airbus on BRAVO, taxi onto Cargo pad SIERRA to hold for your departure time." From the controllers' perspective, this might seem like a reasonable clearance. They already have a clear mental picture of what they want Trans American 209 to do. Additionally, they are completely familiar with the taxiway layout and the normal flow of aircraft on the airport. Trans American 209 is just one moving piece within their overall taxi flow SA. This allows them to process this instruction as a single sequencing/aircraft flow instruction, albeit with four or five pieces of information – complex, but workable. The pilots of Trans American 209, however, would need to build their SA of how they fit within the airport taxi flow, while lacking context, and at an unfamiliar airport. Add the challenges of maneuvering the aircraft, clearing for obstacles, referencing the taxi diagram, and coordinating as a crew. There is also a crew dimension. Imagine the FO making the readback correctly followed by the Captain asking, "What do they want us to do?" It would be better to divide this complex clearance into separate transmissions – two pieces of information in the first and two in the second.

- *First ATC call:* "Trans American 209, turn right on ALPHA to runway one two, hold short of taxiway BRAVO."
- *Second ATC call:* "Trans American 209, follow the Airbus on BRAVO, taxi onto Cargo pad SIERRA and hold for your departure time."

7.5.2 Other Distractors from Outside of the Flightdeck

Other sources of outside distractions include aircraft, birds, balloons, wake turbulence, and the whole range of weather-related challenges. As a general observation, the stronger the distraction, the greater the disruption. In the following report, a bird strike during takeoff disrupts the operational flow and the crew fails to notice that they have left their gear down.

BOX 7.2 BIRDSTRIKE ON TAKEOFF DISTRACTS CREW

PF's report: On takeoff at V_1 we struck a bird with the left windscreen. PM didn't call "positive rate," nor did the PF call "gear up." PF called for "After Takeoff check" and PM said, "After Takeoff check complete." The gear remained down without [me] noticing. I was the pilot flying. Upon reaching 15,000′, I realized the gear was down…

Cause of this event was pilot distraction. The bird we struck at V_1 and its remains were distracting. Due to the distraction we didn't make the proper callouts. Then, we didn't execute our After Takeoff Checklist properly. After that, [it] was a matter of misdiagnosing the reason for the ambient noise in the cockpit and not scanning our EICAS system.[8] Had the proper callout been made and the After Takeoff Checklist been executed properly, this wouldn't have occurred. We had two opportunities to grab the gear handle and put it in the up position and did not. Personally, I will no longer assume the gear handle has been put in the up position after I call for the After Takeoff check.[9]

7.6 INTERNAL CONTRIBUTORS TO DISTRACTION

In addition to the outside distractors, we have a number of sources that originate inside of our flightdeck.

7.6.1 Aircraft System Distractors

As introduced earlier in this chapter, many alerting features of the aircraft are designed to be distracting. Warning and alerting tones, bells, lights, and messages are designed to alert us about important aircraft conditions. These generally fall into two categories – alarms and alerts. Alarms accompany the most serious events like engine fires. The alarms continue until the condition resolves or a pilot silences them. For example, we are trained to notice the source of the fire, silence the alarm, and then decide how to handle the emergency.

Alerts are more nuanced. Take the altitude alerter, for example. The designed purpose is both as an alert that sounds as we are approaching the selected altitude and anytime we deviate above or below that selected altitude. The approaching altitude alert is common and expected. It doesn't require any pilot action. Ideally, it should direct us to level the aircraft or monitor the autopilot as it levels the aircraft. The altitude deviation alert is extremely rare, especially when using the autopilot. Hearing that altitude alert tone when we don't expect it should trigger a strong "something is wrong" response from both pilots.

7.6.2 Automation Distractions

Automation distractions have increased as our flightdecks have become more computerized. Aircraft engineers use automation to compensate for known human vulnerabilities. For example, the human mind is poorly suited to continuously monitor a stable condition. Imagine watching an oil pressure gauge for an entire flight. We would soon grow bored with the task. Our minds would wander. Automation, however, is well suited to monitor for out-of-tolerance conditions within aircraft systems. When the engine temperature is exceeded, a warning light illuminates. Displays provide alerts for undesired events (course deviations) and desired events (localizer and glideslope capture), changes in aircraft path, extremes of airspeed, route discontinuities, and many more out-of-tolerance system parameters. All are potential sources of distraction.

One of the perennial causes of automation distraction is FMS confusion. When the aircraft doesn't respond as we expect it to, we often ask, "What's it doing now?" Both pilots then divert their attention inside to "dive into the box". Even when one pilot takes the lead in diagnosing and correcting the problem, the other pilot typically monitors the changes. This can cause aircraft control and SA-building to suffer.

7.6.3 Screen Distractions

As a society, the amount of time we spend looking at screens is steadily increasing. In Adam Alter's book, *Irresistible: The Rise of Addictive Technology and the Business of Keeping Us Hooked,* he describes how computer programmers have tapped into research on human psychology and physiology to make apps and games more engaging. He reports how developers of addictive games study their users' behavior and optimize game quests to promote the highest engagement of time and interest. Many app programmers have tapped into these same techniques to entice us to check our apps more often and to spend more time using them. Alter cites that "41% of the population has suffered from at least one behavioral addiction over the last twelve months."[10] Additionally, "46% of people say they couldn't bear to live without their smartphones and some would rather suffer physical injury than to lose their phones. In 2008, adults spent an average of eighteen minutes on their phones per day; in 2015, they were spending two hours and forty-eight minutes per day."[11] As pilots, we are not immune to the lure of mobile device screens.

When we experience boredom, or any time when our mind senses that it is under-occupied, do we turn on a screen device to check email/texts, open an app, or play a game? Many of us answer, yes. Data from an app that measures smartphone usage shows that the average usage time is just under 3 hours and that users access their phones an average of 39 times a day.[12]

This repetitive habit of checking our screen devices has altered what we consider as normal. While it eases the passage of time, keeps us connected with our friends and family, and keeps us better informed, it also shifts our perception of what a normal environment looks and feels like. We are becoming more dependent on our screens and more accustomed to the high-paced sensory stimulation that our screens deliver.

Consider flightdeck EFBs. Undeniably, they present a great improvement over paper publications. Applications have expanded across all areas of airline life. At first, EFBs were just a convenient source for charts, maps, and flight manuals. We have since added all of the manuals we used to leave at home, training materials, bulletins, read-before-fly notices, safety alerts, station guidance, overnight accommodation guidance, leadership videos, training videos, and much more. We have apps for viewing our schedules, bidding for our trips, trip trading, vacation trading, checking in for flights, downloading dispatch packs, position tracking, accessing current and forecast weather, and viewing real-time weather radar.

In addition to our official EFB, we have our personal screens. What is the first thing most of us do after completing our gate arrival checklist? We turn on our smartphones. If we have a few minutes of free time, how often do we fill them by opening an app? These devices aren't inherently bad, but their constant use carries a cost.

The more we use screens, the more we align our habits around using screens. Consider an example of a typical cruise flight segment. Perhaps we start by opening our weather app. We select the expanded map just to see what is happening weather-wise across the country. Even if the weather along our route is clear, we might look at weather for locations where something interesting is happening. "Look at that line of storms moving through Florida. Glad we aren't going there." Then, we move on to our enroute chart app. We might call up the moving map display to see what cities and features we are flying over. What airport is that? Accessing our chart app, we see that it is KBPG – Big Springs, Texas. How long is that main runway? It's 8802′. Moving on, what's new from the company? Oh, I see Mary Jones just got promoted as a chief pilot in Denver. Now, pause here and notice how far we have strayed from information that directly applies to our current flight. Every single screen app was available and permitted under our procedures, but none were directly relevant to the flight we were flying at the time. Everything we did was just passing time – occupying our mind – discretionary use.

7.6.4 Reliance on Automation to Alert Us

Improvements in automation, combined with the reliability of our aircraft, have decreased our need to monitor aircraft systems. We rely on automation to attract our attention to malfunctioning systems. With automation warnings serving us so reliably, how often do we still need to perform a detailed visual scan of our panels? Consider the following report from a Boeing 757-200 crew:

BOX 7.3 LARGE FUEL IMBALANCE FROM MISINTERPRETATION OF AUTOMATION WARNING

Captain's report: While on base leg to a visual we received a low fuel message. The right main fuel tank had 2.2 remaining. Upon examination we discovered we overlooked the crossfeed switch during preflight and it was open. The entire flight was fed from the right main tank, over 12,000 LBS, leaving a 12,000 LB imbalance. … During the flight we did get the fuel CONFIG warning but the center tank quantity fluctuated around 1.0 – 1.2 and the light continually illuminated allowing us to discount its significance. I do have the totalizer in my scan *but I overlooked the respective amounts in each main tank.*[13]

For perspective, this was an older model of Boeing 757 with digital fuel gauges and rudimentary fuel imbalance warnings. There is no shortage of recent fuel imbalance events, but the numbers rarely reach the 12,000-pound imbalance that this crew experienced. Notice that the Captain (and arguably both pilots) scanned the total fuel without ever incorporating the "respective amounts in each main tank". This may be because the gauges were digital and arranged with the total quantity displayed directly under the right tank quantity while the left tank quantity was paired with the fuel temperature. Lacking a visible split between analog displays needles, the crew would have to notice the difference between the left digital readout and a separate right digital readout. Also, notice how the automated CONFIG warning alerted them to a problem, but they mistook it as a nuisance warning from the residual fuel in the center tank.

7.6.5 Inappropriate Discretionary Choices

One area where we have considerable control is how we choose to handle discretionary tasks. Take a moment to recall how we allocated our time when we were inexperienced compared with how we allocate our time now. Early in our flying careers, our inexperience left us feeling behind much of the time. Many of our decisions were motivated toward catching up or keeping up. As we gained experience, we developed habits, shortcuts, and techniques to work more efficiently. This freed up extra time. At first, we used that extra time to expand our SA. When we felt that our SA was sufficient, we looked for other things to do. The point is that we became conditioned to "doing something" whenever we had extra time. We became attuned to how we felt about our workload – the feeling of being behind, the feeling of being caught up, and the feeling of staying ahead of the operational flow.

Consider a discretionary task like making a required aircraft logbook entry. Further, assume that our procedures direct us to delay this task until reaching a low-workload phase of flight, like stabilized cruise. One day, on a particularly short flight, we felt that itchy feeling to work ahead (and not fall behind). We felt the need to *do something.* We knew that if we waited until the short cruise segment on this quick flight, we might feel rushed to complete it along with our required descent and arrival preparation. To compensate, we decided to complete the logbook entry during climbout. Why not? The PF was flying. The aircraft was coupled to the autopilot. The FMS was controlling the route and level-off. So, we diverted our attention inside and completed the paperwork. Once we leveled off, we had plenty of time to complete our before descent tasks. Completing the discretionary task early worked out well. On the next short leg, encouraged by our previous success, we repeated our new technique. We avoided that itchy "feeling behind" sensation. Nothing bad happened. Success reinforced repetition.

The drift didn't stop there. At first, we adopted the technique of waiting until we were past FL180. Next, we waited until we were above 10,000′. Pause here and notice how far we have drifted from our original practice. Notice how a one-time, well-reasoned deviation became our normal daily habit. The truth is that our practice drifted because of the urge to stay ahead of workload, to avoid the bad feeling of falling behind, and to fill our habit of "doing something". Success encouraged further drift. When we begin to allocate time and workload based on what we feel, the timing of our actions naturally drifts. That drift only stops when we detect it and reset our practice. As Master Class pilots, we periodically and honestly assess whether we are succumbing to feeling-driven drift.

Discretionary tasks divert our attention from monitoring. When our attention is diverted, our vulnerability to distraction and startle effect increases. Consider taxi operations. Everyone accepts that we want pilots to actively monitor the taxi progress whenever the aircraft is moving. Yet, we often witness pilots engaging in discretionary "housekeeping" tasks during taxi-in. Our feeling of wanting to stay ahead of workload, combined with knowing that we will be pressed for time at the gate, creates a strong incentive to complete some clean-up tasks early. Maybe it starts with discarding flightdeck printouts while actively monitoring the taxi movement. Then, it drifts to removing and stowing our EFB – only looking away for a moment or two. Over time, it drifts to full, heads-down clean-up, even approaching the gate area. Complicate this scenario with both pilots trying to do some clean-up tasks at the same time and we have the ingredients for a missed taxi turn or even a taxiway excursion.

7.7 OTHER FACTORS THAT COMPLICATE DISTRACTIONS

In addition to the external and internal sources of distraction, there are a number of additional distraction sources and nuances.

7.7.1 Team Distraction

Flightdeck crew distractions tend to be minor. Pilots work well together because our flightdeck protocols facilitate the quick exchange of information. We can see each other and gauge each other's task loading. If the Captain is busy, the FO can choose to wait for an appropriate opportunity before speaking up. The same applies if the Captain needs to communicate information with the FO. We share a common standard for what needs to be shared and when to do it. We can also assign roles to mitigate distractions. For example, in a holding pattern we can have the FO fly and handle ATC calls while the Captain coordinates with dispatch, station, passengers, and flight attendants.

Our other team members don't share these advantages. Remote from the flightdeck, ground crew, flight attendants, dispatchers, and station personnel don't know our task loading when they interrupt us. Their requests can arrive at unpredictable and inopportune moments. How we handle these interruptions is important for team cohesion and effectiveness. We try to avoid erecting barriers to open communication by appearing abrupt, authoritarian, or unapproachable. Some pilots overcompensate by being especially responsive and approachable. In striving to appear accommodating, they interrupt critical workflow to respond to the outside call, even when it might be better to ask them to stand by or to call back later. Many checklist errors are traced back to interruptions from these outside team members. For example, station personnel, concerned about passenger connections, may initiate long radio exchanges to pass connecting gate information. These calls often arrive while we are in the busy descent and arrival phase. To them, this information is extremely important. To us, it feels like something that we would prefer to handle at the gate.

7.7.2 Distracting Distractions

Sometimes, we are distracted by one event that diverts our attention from a more significant follow-on event. Dr. Barry Turner coined the term "decoy" distractions to describe when one distraction diverts our attention from a more-significant problem.[14] An often-cited example is Eastern Airlines 401 (CFIT accident in 1972). All three members of the flight crew became distracted by a malfunctioning landing gear indication and failed to notice that the altitude hold function of the autopilot had become disengaged. While they were heavily focused on diagnosing the gear light problem, the aircraft slowly descended until it impacted into the Florida Everglades.

7.7.3 The Lingering Effects of Earlier Distractions

Distractions grab our attention. Even after we resolve them, their effects linger. If we encounter a significant distraction, deal with it, and try to move on, our minds sometimes continue to replay the past event. Did I handle that right? What should I have done differently? I personally recall an event where I had to intervene in an operational problem before departure. I was upset by the way the station personnel chose to handle the issue. After we departed the gate, I kept rerunning the scenario through my mind. Did I handle it right? Should I report the event? Reliving it diverted my attention from taxiing the aircraft. I nearly missed an assigned taxi route turn.

Normally, we are excellent compartmentalizers. We set the past aside and focus on the present. Sometimes, however, we prove that we are also humans who remain vulnerable to a wide range of distraction errors. Consider the following account by a Captain who committed several flight errors while distracted from a ground event.

BOX 7.4 CAPTAIN MAKES SEVERAL ERRORS WHILE DISTRACTED FROM A PAST EVENT

Captain's report: … I climbed through 10,000′. Without thinking and forgetting that we are speed restricted at 250 knots until given normal speed in ZZZ, I accelerated to 290 knots. A few minutes later ATC came on and gave us a vector of 20 degrees right for spacing. We scanned the instruments and FO pointed out that he didn't think we were given normal speed. I promptly bugged 250 knots and slowed the aircraft back down. ATC never mentioned our speed. Later in the flight during our arrival into CAK, I also passed through 10,000′ and forgot to slow down until we were around 6,000′ which I caught and slowed to 250 knots. *Both times were due to my distraction from an earlier event in which I experienced a personnel issue with an FO. That took a couple hours to resolve. I realized during the flight that I was mentally distracted and had FO fly the leg back.*[15]

This Captain realized that the previous event was distracting them and causing errors. They let their FO fly the return leg. Like a boxer knocked back on their heels from a stunning blow, we don't immediately recover. We take some time to process the event, set it aside, and refocus on current matters.

7.8 EVOLVING TRENDS IN DISTRACTION VULNERABILITY

Having examined some features and consequences of distraction, let's drill down another level and uncover some vulnerabilities that are emerging within our industry.

7.8.1 Distraction Vulnerability and Screen-Induced Loss of Mental Rest

We engage in discretionary activities to pass the time during the low-workload cruise portion of our flights. As we increase our use of screen devices, we enter a world intentionally designed to keep our minds and senses deeply engaged. Computer app designers carefully select colors, layouts, and pacing to hold our attention. By constantly engaging with our screens, we fatigue the parts of our brains that process visual information. This is important because we lose an opportunity to rest these parts of our minds. Compare it to a gym workout. We wouldn't repeatedly exercise only one set of muscles day after day. Instead, we rotate between exercise regiments – core, legs, arms, and cardio. If we continued to exercise the same way every day, we wouldn't give those muscles time to rest and recover. The same applies to our brains. If we repeatedly engage in the same kind of intense mental activity, our mental performance will become fatigued and suffer.

Consider a typical flight. While on the ground, we have a high workload. We have to remain highly engaged and focused as we complete all of our flight preparation, planning, and briefing tasks. We continue to pay close attention during taxi and takeoff. After takeoff, our workload steadily decreases. We have less to do. Our attention level eases off. At cruise, we settle into our lowest workload phase. Approaching top-of-descent, our workload begins to ramp up again as we prepare for the busy descent, approach, landing, and taxi-in phases. The typical flight sequence is high workload (ground/takeoff/climbout) – low workload (cruise) – high workload (descent/approach/landing).

We can't keep our attention elevated at peak levels. Like muscles, our minds cannot sustain continuous effort. For best performance, we need to alternate between high-workload segments (planning and doing) and low-workload segments (resting). With rest, our brains become revitalized to handle the rising workload of approach and landing.

This resting period during cruise flight becomes quite important. To recharge our mental batteries, we need to relax the parts of our brain that process flying tasks while we are in the cruise phase. Activities like conversation use different areas of our brain than aviation tasks like visually monitoring, planning, interpreting, or physically flying. So, while we can have an engaging discussion with our crewmember at cruise, we are still effectively resting the parts of our brains that we will use most during descent, approach, and landing.

Why is this important? Highly engaging screen apps tend to fatigue many of the same mental processes as flying tasks. These include visually detecting, interpreting, choosing, and acting. Filling our resting time at cruise with highly interactive screen activities continues to fatigue those brain functions. We lose the opportunity to rest. This, in turn, increases our vulnerability to mental errors during the high-demand descent, approach, and landing phases.

7.8.2 Distraction Vulnerability and Multitasking

Another screen-generated vulnerability is multitasking.[16] The truth is that we don't actually multitask. While our senses can experience many stimuli at once, our minds can pay attention to only one thing at a time. What we call multitasking is actually rapidly switching our attention focus between stimuli. Every time we switch, our minds need a moment to reestablish the context of the new stimuli. The faster we switch, either the more shortcuts our minds need to take or the lower our grasp of the context. The faster we think we multitask, the lower the quality of our work.

Our perceived skill to multitask is a myth reinforced by screen culture. With practice, we can become rather skillful with our "multitasking" efforts. I personally witnessed a delivery van driver, smartphone in hand, texting as he drove across a busy intersection in New York City, with cars and people moving all around him. He didn't hit anyone or anything, but it seemed like an extremely risky undertaking. How did he reach the point where driving through the heart of Manhattan while texting became acceptable? Surely, he didn't start by doing this. He probably began with a quick glance at his smartphone at a stop light, then again while waiting for the car in front of him to move, then more often in increasingly risky conditions until he reached the level I witnessed. When we frequently use high speed, high attention-grabbing screen apps, we gain confidence in our "multitasking" skills. Our perception of risk drops. Perhaps we've become so good at it that it no longer feels risky. In the end, feeling safe doesn't make it safe. Risk is measured by conditions and probability, not by our confidence level.

7.8.3 Pushing Safety Limits

Circle back to the topic of inappropriate discretionary choices – which are often attempts at multitasking. Just because we develop skills to do more tasks more quickly, it doesn't alter the underlying accumulated risk. I recall one Captain who routinely pressed their visual approaches. They waited until the last possible moment before reducing thrust and slowing to configure. Admittedly, they were very good at it. They rarely flew an unstabilized approach or go around, but that is not the point. Their personal skill with pushing safety margins didn't change the appropriateness of their choices. Pushing the field wasn't necessary in the first place. It was a discretionary choice based on their desire to challenge their flying skills.

7.8.4 Attention Level and Dynamic Flight

The foundation of flightpath management is awareness of the aircraft's path, especially while in dynamic flight. Whether taxiing or flying, the more dynamically our path is changing, the more vigilance we need to sustain. The more stable and less dynamic it is, the less attention it requires. Consider the differences between steady cruise flight with the autopilot engaged and hand-flying an arrival into a busy airport environment. Each of these examples requires vastly different attention levels.

This is an area where our aviation practice drifts. When we were new, aircraft maneuvering required our full attention. As we gained experience, many of us inappropriately lowered our vigilance because flying felt easier. We lost the connection between matching our attention level with the level of dynamic movement. Instead, we matched our attention level to our perception of difficulty and workload. As long as the task felt easy, we could lower our vigilance.

7.9 SELF-INFLICTED DISTRACTIONS

Engaging in discretionary tasks is an intentional decision. For most of us, the migration of discretionary tasks into dynamic flight happens so slowly that we may not even notice it happening. We start by following the procedural guidance. Then, usually for good reasons, we grant a one-time exception. Over time, the exception becomes a habit. We solidify the habit by making it the new baseline for our next shift. By tiny steps, we move further and further from the ideal standard. Following are some of the forces that fuel this drift.

7.9.1 Normalization of Deviance and Rationalization

One mechanism that drives this slow drift is the normalization of deviance. Consider a task that we completed using a standard, trained procedure. One day, we chose to alter it slightly. Our new modification seemed to work well, so we stuck with it. After a while, we made another incremental change. By tiny steps, our practice strayed further away from the original procedure. As we grew comfortable with each increment, what we considered as "normal" shifted along with it. Eventually, our modifications can unintentionally erode our safety margin and expose a latent vulnerability that may eventually result in error.

Every system has its breaking point. Unfortunately, many systems don't produce warning signs prior to failing. There aren't cracks to warn us of an impending catastrophic break. The system works until it completely fails. Consider the task of checking in for our trip. When we were new, we allowed ourselves plenty of time to drive to the airport, get through security, walk to the pilot lounge, and check in 30 minutes early. As we grew more comfortable, we started shaving minutes. First we cut 5 minutes, then 10, until we settled at 15 minutes early. It felt like a reasonable safety margin. Then, one day, we left home late, traffic was terrible, the shuttle bus from the employee parking lot was delayed, and we checked-in 10 minutes late. If we had maintained our 30-minute pad, we would have accommodated all of these unexpected delays with 5 minutes to spare. Incrementally shaving our safety margin did not affect the tasks we needed to complete, but it did adversely affect our ability to absorb unanticipated delays. Each deviation of our check-in time became normalized with repetition.

Sidney Dekker tracks this same normalization of deviance drift with the aviation accident of Alaska Airlines 261 (January 31, 2000 – MD-80 jackscrew failure). He traces how the jackscrew lubrication service interval drifted from 300 hours in 1985 to 2,550 hours, then switches to every 8 months by 2000. Jackscrew wear tolerance measurement, called an end-play check, started at nothing (no standard set with initial DC-9s) to a tolerance value measured every 2,500 hours in 1985, to every 15 months (as much as 9,550 hours) by 2000.[17]

7.9.2 Experience Leads to Relaxing Our Standards for Completing Discretionary Tasks

When we never experience a bad outcome, we might conclude that we are doing everything safely. Nothing bad happens, so we aren't doing anything "bad". This represents a flawed connection between actions and consequences. Pilots use this faulty logic to justify shifting discretionary tasks toward riskier flight phases. For example, procedures direct both pilots to focus their attention on aircraft movement during taxi-in. Yet, pilots routinely perform discretionary tasks during this phase. This can lead to errors as one or both pilots divert their attention toward discretionary activities while the aircraft is moving.

7.9.3 Sterile Flightdeck Protocols Compared with Discretionary Choices

The underlying motivation for sterile flightdeck protocols is to reduce both the adverse effects and sources of flightdeck distractions during critical flight phases. As we grow more proficient, our perception of risk eases. Adding a few easy discretionary tasks to a dynamic flight phase doesn't feel risky. We become more comfortable with mixing distraction-reducing practices (sterile flightdeck) and distraction-raising practices (discretionary tasks).

Another misguided practice is following individual sterile flightdeck protocols while continuing to engage in self-generated, unnecessary distractions. For example, some pilots diligently prohibit "unnecessary conversation" in sterile flightdeck regimes, but see no problem with diverting their attention inside to perform a discretionary task. As long as they do it silently, they feel like they are complying with the rule. They misinterpret that the underlying goal of sterile flightdeck is to reduce all avoidable distracting practices during critical dynamic flight phases.

BOX 7.5 CAPTAIN DOESN'T FOLLOW STERILE FLIGHTDECK PROTOCOLS

FO report: [The] Captain ignored/disregarded/did not know altitude restriction for the initial turn on the Departure and turned at 400′ AGL instead of 2,000′ MSL per the departure. This turn was therefore initiated approximately 1,600′ early. *As a rule, this Captain does not adhere to sterile cockpit protocol, particularly on taxi out.* The Captain was talking about hotel issues, union issues, not getting a workout that morning, that [they] didn't want to wait in line at a restaurant, and other non-essential topics during the taxi out. … *We all talk about non-pertinent issues during taxi out to some degree, but this Captain's non-pertinent conversation was excessive and distracted from the command and control of the aircraft.* We all want to be friendly, but sterile flightdeck means sterile flightdeck. After three days of this behavior, with similar lapses resulting in larger and smaller deviations I was worn down.[18]

7.9.4 Prospective Memory Challenges Lead to Discretionary Choice Vulnerabilities

Another reason we engage in discretionary tasks is to complete delayed tasks so we don't forget to do them later. These are *prospective memory* challenges. Prospective memory (remembering to remember something later) is a rather tricky undertaking for our minds. Typically, we make mental notes to remember to do something later. Unfortunately, we get busy and forget them. Mental notes are written with disappearing ink. This is especially true when the task isn't anchored to a solid reminder cue.

One solution is to complete the task whenever we think of it, even if it isn't the best time. Imagine that we use a special clip to hold our communications cord against the side rail of the flightdeck. When leaving the aircraft, we have to remember to retrieve it (prospective memory task). During a particularly chaotic aircraft swap, we became distracted and inadvertently left it behind. We really valued that clip, so we made a mental note not to lose one again. The next time while taxing-in, we remembered to retrieve the clip. While steering the aircraft, we reached over, removed the clip, and dropped it into our flight bag. Problem solved. We should have waited until stopped at the gate, but that's what led to us losing the last one. Our new technique seemed to solve the problem nicely. Then one day, something different happened. After we removed the clip, we dropped it under our seat. Rather than wait until parked at the gate, we decided that maybe we could reach it. As our attention was diverted, the aircraft slowly veered away from the taxi line. Unfortunately, our FO had similarly diverted their attention inside completing their own discretionary clean-up tasks. At the last moment, we looked up and saw that we were about to depart the pavement. We slammed on the brakes just in time. For seemingly good reasons, our practice drifted in a way that increased our vulnerability to distraction. On that particular day, with all of the right conditions in place, our discretionary habit almost bit us. The cause was not dropping the clip. The cause was engaging in discretionary tasks during a critical phase of aircraft dynamic movement.

7.9.5 Ill-Timed Diversions of Our Attention

Sometimes we have a good reason for diverting our attention, but the duration and timing of tasks can increase the adverse effects.

BOX 7.6 FO DIVERTS THEIR ATTENTION TOO LONG AND MISSES CAPTAIN'S TAXI ERROR

FO's report: After landing there is a very short taxi from the runway to the terminal, [so I had] minimal opportunity to obtain the 2-minute cool down for the engines to be shut off. The Captain was taxiing slowly to the gate and told me to shut the left engine down as soon as we have 2 minutes. I looked down to look at the times off the FMS [to verify the cooling time] and went straight to shutting down the left engine. At that time, the standard MASTER WARNING and MASTER CAUTION messages were appearing; therefore I immediately checked them and turned off the warnings. [Then,] I looked up and noticed the Captain [had] taxied past our gate and was going into the wrong gate. I told him that we passed [the gate] and he immediately stopped and began turning. The Captain added enough thrust to blow over construction equipment.[19]

The FO had a number of tasks to complete during the short taxi-in. All were justified and reasonable, but the only one that mattered was shutting down the #1 engine after 2 minutes of cooling. We can infer that the FO must have spent a fair amount of inside, heads-down time checking for the cooling time, shutting down the engine, clearing the resultant warning and caution lights, and perhaps some other tasks not detailed in the report. Despite the Captain's slow taxi, enough time had passed for them to overshoot their assigned gate. Granted, the Captain missed it, but that error was something that the FO could have caught by periodically cross-referencing the Captain's taxi progress.

7.9.6 Just One More Thing

Even during taxi, there are times when we need to briefly divert our attention away from monitoring the aircraft's path. We need to choose the best moments to do this. Referring to the previous NASA ASRS report, it appears that the FO performed a long series of inside tasks without glancing outside to crosscheck the Captain's taxi progress.

Striving to become as efficient as possible, we often bundle tasks together into a flow pattern. Flows are beneficial because they join individual tasks together, reduce omission errors, and improve accuracy. Our natural learning process looks for ways to combine interrelated tasks. These groupings start out small – just a couple of quick items. Encouraged by our success, we add one more thing, then another, and then just one more thing. The resulting flow pattern can become rather long. We can imagine that this FO's engine shutdown sequence typically included the items that they cite in their report – check engine cooling time on the FMS to verify that 2 minutes have elapsed, then close the start lever, then check to see the engine rolling back, and then cancel the warning lights triggered by systems dropping off line. These tasks flow naturally from one to the next. Most of the time, the FO probably completed this flow on a long taxiway with little time urgency. This time, however, they were still performing their flow after entering the parking ramp. The critical condition in this incident was the short taxi time. Following their flow technique, they kept their attention inside for too long. In hindsight, they should have included quick crosschecks to monitor the Captain's taxi progress.

- Select the FMS page that displays the landing time – then a quick crosscheck outside to verify taxi progress.
- Verify that 2 minutes cooling is reached – if not yet at 2 minutes, crosscheck the taxi progress.
- When the cooling time is reached, shut down the engine and ensure that it is winding down – then a quick crosscheck outside to monitor the taxi progress.
- When lights appear in their peripheral vision, check inside for warning and caution lights, and clear the lights – crosscheck outside to monitor the taxi progress.

In this way, we exploit natural breaks within the flow pattern to crosscheck important monitoring priorities.

NOTES

1. This chart reflects the concept forwarded by Dr. Mihaly Csikszentmihalyi, in his book, *Flow: The Psychology of Optimal Experience* (1990, Harper & Row). In his chart, the areas reflect anxiety (top section), flow (along the line), and boredom (bottom section). Striving to stay along the central flow channel promotes well-being, creativity, and productivity.
2. This is an activation of the sympathetic nervous system – commonly called "Fight or Flight". Some expanded references list this as "Freeze, Flight, Fight, Fright, Flag, and Faint". These are instinctual brain responses.
3. Cushing (1994, p. 52).
4. Lanfermeijer (2021), https://tailstrike.com/database/19-february-1989-flying-tiger-66/.
5. Callsign is a tribute to the airline cult classic movie, *Airplane!* (1980) by Paramount Pictures.
6. Cushing (1994, p. 57).
7. Flin, O'Connor, and Crichton (2008, p. 20). Sensory hearing lasts for about 2 seconds.
8. Engine Indicating and Crew Alerting System (EICAS).
9. Italics added. Minor revisions for clarity. NASA ASRS report #1375987.
10. Alter (2017, p. 25).
11. Alter (2017, p. 27).
12. Alter (2017, p. 14).
13. Edited for brevity. Italics added. NASA ASRS report #841602.
14. Dekker (2011, pp. 88–89) – citing Turner, Barry (1978).
15. Minor revisions for clarity. Italics added. NASA ASRS report #1698579.
16. The best aviation source is *The Multitasking Myth: Handling Complexity in Real-World Operations*, by Loukopoulos, Dismukes, and Barshi published by Ashgate Press (2009).
17. Dekker (2011, pp. 39–44).
18. Italics added. Minor revisions for clarity. NASA ASRS report #957346. Notice that even the FO acknowledges that, "We all talk about non-pertinent issues during taxi out to some degree."
19. Edited for clarity. NASA ASRS report #1447292.

BIBLIOGRAPHY

Alter, A. (2017). *Irresistible: The Rise of Addictive Technology and the Business of Keeping Us Hooked*. New York, NY: Penguin Press.

ASRS – Aviation Safety and Reporting System – National Aeronautics and Space Administration (NASA) Online Database. (n.d.). Retrieved from Aviation Safety and Reporting System: https://asrs.arc.nasa.gov/search/database.html.

Cushing, S. (1994). *Fatal Words: Communication Clashes and Aircraft Crashes*. Chicago: The University of Chicago Press.

Dekker, S. (2011). *Drift into Failure: From Hunting Broken Components to Understanding Complex Systems*. Burlington, NJ: Ashgate Publishing Company.

Flin, O., O'Connor, P., & Crichton, M. (2008). *Safety at the Sharp End: A Guide to Non-Technical Skills*. Burlington, NJ: Ashgate Publishing Company.

Lanfermeijer, S. (2021). 19 February 1989 – Flying Tiger 66. Retrieved from Tailstrike.com: https://tailstrike.com/database/19-february-1989-flying-tiger–66/.

Loukopoulos, L., Dismukes, R. K., & Barshi, I. (2009). *The Multitasking Myth: Handling Complexity in Real-World Operations*. Burlington, NJ: Ashgate Publishing Company.

Regan, M., Lee, J. D., & Young, K. L. (eds), (2009). *Driver Distraction: Theory, Effects, and Mitigation*. Boca Raton, FL: CRC Press.

8 Safety

Consider how often we use the term *safety* in aviation. We take pride in being a safe industry, providing safe travel for our customers, operating safely, and promoting safety. Safety is such a foundational principle that we assume that it is always present. In a way, it envelops us so completely that it vanishes from our awareness. We take it for granted. Of course we are safe. We are always safe. We wouldn't think of being unsafe. Despite our emphasis on safety, we still have events where we judge that the pilots acted unsafely or committed unsafe acts. Is there a discrepancy between how we judge these events and our understanding of safety? As Master Class pilots, we strive to understand what safety is, how we achieve it, and how our operational environment can threaten it.

8.1 WHAT SAFETY IS NOT

Let's start by dispelling some mischaracterizations that undermine our understanding of safety. Most of these stem from the broad-brush generalizations that we commonly use.

8.1.1 Safety Is Not Completely Present or Absent

We make statements like, "that was unsafe" or "that was safe". Both are inaccurate generalizations. They assume that there is a group of events and behaviors that is inherently safe while another group is inherently unsafe. We use these extremes because they simplify the labeling of events. This, in turn, affects how we classify decisions, motivations, and actions. When we declare that an event was safe, we conveniently ignore unfavorable vulnerabilities or risky aspects. "Nothing bad happened, so all's well that ends well." Conversely, when we judge that an event was unsafe, we ignore features that might have otherwise contributed positively toward a safe operation. "Despite their efforts to ensure adequate separation, they still struck the provisioning truck, so they were operating unsafely."

By using absolute labels, we blur the connections between important features that would otherwise deserve our consideration. Consider an example of a line of aircraft landing at an airport with steady snowfall. Each aircraft lands and stops successfully, but each one slides a bit further down the runway before reaching taxi speed. Were all the outcomes successful? Yes. If we stop our evaluation here, we might overlook the threatening trend that was developing. If the last crew to land did everything exactly the same as all previous crews, but slid off of the end of the runway, was that crew suddenly unsafe? No. By classifying safety by successful outcomes, we miss the critical safety concern – that the runway was becoming dangerously slick.

DOI: 10.1201/9781003344575-9

8.1.2 Safety Is Not a Number or Value

In our metric-driven world, we favor using quantifiable values to measure safety. Consider a factory workspace displaying a sign, "320 Days Without An Accident". While this seems to characterize a safe workplace, we may not have the whole picture. Maybe the workers had experienced a high number of close calls. Maybe the definition of "accident" sets the bar very high. Maybe they were just lucky. Maybe factory floor supervisors were suppressing accident reports to protect monetary bonuses awarded for accident-free operation. While we can gain some insight into an organization through its safety metrics, we need to look beyond their selected numbers.

Conversely, if an organization reports a rise in incidents, does this mean that the operation is becoming unsafe? Not necessarily. The higher numbers might indicate an increase in unsafe events. That would be unfavorable. They may also indicate a rise in the complexity of the operation. While concerning, this may reflect conditions that can't be controlled or removed. The higher numbers may also indicate an improvement in the reporting culture as more workers proactively identify latent vulnerabilities and safety concerns within the system. This would be a favorable trend that would mitigate hazards before they develop into future accidents. So, while metrics are useful, we need to investigate further to understand what they mean.

Some measurements are biased toward profitability goals. Consider the commonly tracked airline metrics of load factor, on-time percentage, misrouted baggage, flight cancellations, and customer complaints. None of these provide useful insight into the company's safety culture. Most safety departments acknowledge this and track a broad range of safety metrics. They attend conferences to share program initiatives that effectively "move the needle" on stubborn safety problems. These efforts breathe meaning into the numbers and promote a proactive, resilient safety culture.

8.1.3 Safety Is Not a Feeling

One of the common statements made by mishap pilots is, "I didn't feel unsafe." Even assuming that our knowledge and experience give us some insight into sensing our level of safety, our feeling of safety, especially while immersed in-the-moment with a stressful situation, remains unreliable. Consider the following report from a Captain who departed with an out-of-limits tailwind component.

BOX 8.1 CONTINUED OUT-OF-LIMITS TAKEOFF WHEN IT DIDN'T FEEL UNSAFE

Captain's report: ...Biggest takeaway for me was that I did not get new takeoff speeds before accepting takeoff clearance. After hearing [that] aircraft taking off in front of us had a tailwind (020/8), we should have told tower we needed new numbers. ...[The] Tower called wind for our takeoff as we were rolling at 020/13. Takeoff data for theses winds [would have] resulted in "T/O NOT POSSIBLE FIELD LENGTH LIMIT". ... Another surprise factor was how

> fast the amount of rain and wind shift changed in only a moment. *I was not thinking about tailwinds because the airport just changed runways to avoid tailwinds. I did not feel unsafe until we were already accelerating and the water was not shearing [from the windscreen] as I have always seen. By that point I felt it would be more dangerous to attempt an abort than continue take-off. Looking back, I also probably felt an urgency to takeoff. We were in a line of at least 10–12 aircraft lined up to depart before the storm hit.*[1]

Notice how events and conditions affected the Captain's mindset. Since ATC had just changed runways to resolve adverse tailwinds, the possibility that the current runway was also out-of-limits seemed unlikely. The wind callout from Tower and the uncharacteristic behavior of rainwater off of the windscreen raised the Captain's level of concern. After reaching cruise altitude, they recomputed the takeoff data and discovered that they had exceeded tailwind limits. The Captain reported that it "did not feel unsafe" and "felt it would be more dangerous to attempt an abort".

We should not completely ignore our feeling of safety. We can recall past events that felt unsafe, so our gut-felt intuition deserves some consideration. Our emotional sensation of "feeling unsafe" arises as we subconsciously perceive subtle details that suggest that something unfamiliar or hazardous may be present. Research by Gary Klein shows how experts use gut feelings and intuition to guide their risk assessment. For example, on-scene fire department Captains reported making fire-fighting decisions based on the *feel* of the fire, or at least by perceiving subtle cues that non-experienced observers typically overlooked.[2]

8.1.4 Safety Is Not Defined by Outcomes

Consider the statement, "An operation performed safely is a safe operation." Clearly not. I recall a film clip of a motorcycle stunt driver performing a jump. While sailing through the air between the ramps, an aircraft flew underneath them while executing a 60° bank turn. The bi-plane cleared the gap and the stunt driver landed safely. While this demonstration was performed without mishap, few could defend it as a safe operation. This demonstrates of fallacy of classifying safety by its outcome. When we assess safety, it is helpful to ignore the outcome of an event. Instead, focus on analyzing the interactions between conditions, the level of risk, risk management, the decision-making process, and crew actions.

8.1.5 Safety Is Not a Reason (or Excuse) for Unwise Choices

Some unstable approaches and landings are discovered during flight data recorder analysis. When the crew is contacted by safety investigators, they acknowledge that while the parameters might have been excessive, that it wasn't unsafe because they successfully stopped. These pilots rationalize that they had a long runway that compensated for their excess energy. They claim that the approach "didn't feel unsafe" or that they felt that the event fell within pilot's discretion. Since their actions didn't actually result in a mishap, it must have been safe.

8.2 WHAT SAFETY IS

Understanding what safety isn't, let's tackle the more difficult (and more useful) question of what safety is.

8.2.1 Safety Is Probabilistic

Safety is governed by probability, not certainty. Some events like takeoffs from long runways in good weather are recognized as highly safe. While there is still a chance for mishap, aircraft performance, pilot experience/training, and good conditions maximize the probability of safe outcomes. Other events, like threading around thunderstorms or landing on slippery runways, are deemed risky. We employ additional safety measures to balance our priorities and choices to reduce the chance of an unfavorable outcome. Consider the decision of whether to turn off the passenger seat belt sign. First, we note that it is a passenger convenience, not a necessity. Under turbulent conditions, we routinely leave the sign illuminated for the entire flight. If we find some smooth air and assess that it is likely to remain smooth, we turn the sign off. On the other hand, if we see convective buildups ahead or if other aircraft are reporting poor rides ahead, we leave the sign on. The discretionary nature of the decision gives us plenty of latitude to balance the conditions and choose the safer option.

Compare the seat belt sign decision with whether or not to land from an approach. Here, we don't have a yes or no choice. We have to land. Our discretion lies with choosing when and where. If the conditions are good, we favor landing at our intended destination. As conditions worsen, we begin weighing other options. Should we continue and land despite the worsening conditions? Should we wait until conditions improve? Should we divert to an alternate?

We operate within an environment of constantly changing conditions, many of which we can't detect or process in real time, while continuously moving forward. Consider the following report of unexpected windshear.

BOX 8.2 HAZARDOUS LANDING CONDITIONS WITHOUT WARNING SIGNS

FO/PM's report: *The ride was smooth and visibility was good as we entered into light rain. As we began the round out and flare portion of our landing, the aircraft began to lose lift and rapidly sink.* The auto thrust was engaged for the entire approach up to the point. The Captain reduced power per the aircraft call outs. *We received no warnings from the aircraft's predictive or reactive windshear alerting systems.* We also didn't receive any excessive pitch or speed warnings as the nose began to rise to maintain our rate of descent. It was at this time the Captain initiated a go around by advancing the throttles

> to TOGA (Takeoff Go-Around). *What seemed like immediately thereafter, the aircraft quickly lost a significant amount of lift and settled onto the runway in a nose high attitude.* The aircraft bounced back into the air as we executed our go around procedure. We then completed our after takeoff checklist and prepared for another approach into ZZZ. We didn't suspect that there was any damage to the aircraft. … Upon performing my post flight inspection, *I discovered that we had impacted the runway with the underside of the rear fuselage and empennage during our go around.*[3]

This is a particularly disturbing event because it lacked all of the typical warning signs of possible windshear. The crew didn't apply countermeasures like adding extra margin to their target speed or hand-flying to quicken response time. Expecting a routine approach, they were surprised by the rapid loss of lift.

8.2.2 Safety Emerges

Emergence is a feature of complex systems. It describes the effect where potential outcomes cannot be accurately predicted from their conditions. Instead, the same set of conditions can produce a wide range of results. Referring back to the previous pilot report, if ten aircraft flew down final, maybe only one might experience the unique combination of conditions that would generate a tailstrike event. All of the other aircraft might experience the same level of rain and turbulence, but maybe only this one would encounter sufficient sink, occurring at exactly the right moment, to cause a mishap.

Emergence also complicates procedure development. Consider an unfavorable event where a crew took the runway with their flaps still up. The investigation revealed that a procedural distraction contributed to their error. The company decided that the procedural distraction was serious enough to revise procedures. The development team tweaked the procedure and removed that source of distraction. Sometime later, another crew entered the runway with no flaps – the same undesirable event. That investigation uncovered poor checklist discipline by the crew. So, leadership added a block to quarterly training to emphasize checklist discipline. This worked for a while until it happened again – this time because of a tripped flap motor circuit breaker and rushing behavior. We see how similar environments can generate the same undesirable outcome from widely different causes. On the flip side, those same conditions were present with an overwhelming percentage of favorable outcomes. A different crew with poor checklist discipline could still set their flaps properly. A distracted crew can still recover and operate successfully. A hurrying crew can still detect a tripped flap extension circuit breaker before entering the runway for takeoff. It is only when conditions interact with existing vulnerabilities in just the right ways to generate a mishap.

This demonstrates why it is so difficult to solve operational problems with procedural changes alone. There are just too many combinations and interactions – too many holes in the cheese for a failure to slip through. Every procedural intervention changes the operational flow, but those changes aren't rigid. Line cultures adapt and drift. Our personal practice changes flight-to-flight. Consider how we might conduct a flight briefing on the first leg of trip to an unfamiliar destination with how we might brief for the last leg home to our domicile.

The same intermixing mass of conditions that produces bad outcomes also produces good outcomes. Many thousands of flight crews fly every day in unpredictable and challenging conditions, do amazing work, and avoid countless opportunities to fail. Aviation is an inherently unstable system. It naturally wants to veer off, accelerate, slow, or drift out of sync. If we focus only on the ways that a scenario can go wrong, each successful flight seems like a succession of miracles. This is clearly not the case.

There must be something that we do right to achieve such a high percentage of successful outcomes – and there is. Pilots are a special group of professionals. We may think we just fly aircraft, but the truth is much larger. We actively manage this unpredictable environment to achieve success by continuously applying tiny nudges and corrections to keep our flights on desirable paths. Consider manually flying an aircraft. We constantly make control and power adjustments to hold path and speed. With practice, we internalize this process and react automatically. If a gust bumps one wing up, we quickly apply flight controls to stop the wing lift, counter it to bring that wing back down, and then remove the control inputs to restore level flight. Safe flying emerges from our continuous stream of timely and accurate actions.

8.3 CREATING SAFETY

Safety isn't something that *is* or something that we *feel*. It is a product we *create* moment-to-moment and flight-to-flight. As safety managers, the Captain is the key. We are not just active contributors toward creating safety. We are ultimately responsible for it. We stand at the pointy tip of the spear. The vast array of aviation and company systems are lined up behind us. Domestic and international aviation systems, ATC, aircraft manufacturers, dispatch, flight operations, maintenance, ground crew, station crew, cabin crew, and flight crew all form our support team. If we manage our players well, they contribute to our success.

Many Captains view our position in the operational hierarchy as falling somewhere near the middle. Around and below us are FOs, cabin crews, and ground support teams. Above us are check pilots, chief pilots, flight operations supervisors, and regulators. On the ground, this perspective may be accurate. When we step onto the aircraft, however, we move to the top. We fully acknowledge that we must follow the policies and procedures from those higher organizations, but in creating safety, we must embrace full responsibility.

Safety results from our direct and active manipulation of current conditions. We are the master crafters that handle the tools that create our outcomes. This is an extremely important perspective to embrace. Mishap crews sometimes fail because they have lost this master crafter perspective. Instead of seeing themselves as the managers of conditions, they view themselves as the victims of circumstances. Effective Captains are able to detect and manage existing conditions to create safe, desirable outcomes. While that sounds easy, it clearly isn't. Some conditions can be missed entirely. Some others can be detected, but misinterpreted. Even if all of the conditions are accurately managed, we can balance conflicting goals inappropriately. We can let quickening situations accelerate until we lose active control and become swept along by a failing situation. On the positive side, the vast majority of pilots navigate their entire careers successfully creating safety with each and every flight.

NOTES

1 Edited for brevity and clarity. Italics added. NASA ASRS report #1662659.
2 Klein (2003, p. 4).
3 Edited for brevity and clarity. Italics added. NASA ASRS report #1673014.

BIBLIOGRAPHY

ASRS – Aviation Safety and Reporting System – National Aeronautics and Space Administration (NASA) Online Database. (n.d.). Retrieved from Aviation Safety and Reporting System: https://asrs.arc.nasa.gov/search/database.html.

Dekker, S. (2011). *Drift into Failure: From Hunting Broken Components to Understanding Complex Systems*. Burlington, NJ: Ashgate Publishing Company.

Dekker, S. (2015). *Safety Differently: Human Factors for a New Era*. Boca Raton, FL: CRC Press.

Flin, O., O'Connor, P., & Crichton, M. (2008). *Safety at the Sharp End: A Guide to Non-Technical Skills*. Burlington, NJ: Ashgate Publishing Company.

Klein, G. (2003). *The Power of Intuition: How to Use Your Gut Feelings to Make Better Decisions at Work*. New York, NY: Currency Books.

9 Time

One final topic in this Core Concepts section involves how we view time. Our pilot perspective of time evolves from the dynamic nature of flight. When our aircraft is moving, we can't stop, analyze our situation, sort things out, and then pick up where we left off. We have to make our decisions literally on the fly.

9.1 HOW WE SEE TIME

How we see time is one of the characteristics that separates pilots from other professionals. While other groups emphasize the present moment, we mark time by how events appear to be progressing, which emphasizes a future perspective.

9.1.1 The Flight Schedule

The schedule drives the air carrier business. Passengers expect to depart and arrive on time. Freight operation managers expect their freight to arrive and depart on time. Airline schedules are optimized to shorten turn times and cycle more flights through each gate. This is where the profit lies. The more productivity a company can gain from workers, aircraft, gates, slots, and support equipment, the better it competes in airline business.

Unfortunately, optimization adversely affects our operational resilience. A fully optimized schedule lacks the slack that we need to dampen the ill effects of disruptions and delays. Without this slack, a delay with the first flight of the day spreads across follow-on flights. Since our operations are so interconnected, a late-departing aircraft from Miami might create delays and passenger misconnections in Chicago, which then affect passenger connections to dozens of other cities.

The stress from systemic delays is not just an airline management problem. It transfers to us. Operations agents feel pressure to make up time during the aircraft's turn. Pilots feel pressure to shave minutes enroute. Misconnected crews hurry to make flights that are holding for them. We feel pressure to work faster. By the end of the day, late flights begin running against airport curfews and crew duty day limits. Flights are cancelled or diverted. Lingering disruptions transfer systemic stress to the following day.

Managers encourage us not to rush, but they value anything we can do to get the schedule back on time. As we internalize our personal responsibility for the success of our company, we increase our pace. How quickly we can accomplish our workload becomes a measurement of our personal skill.

DOI: 10.1201/9781003344575-10

9.1.2 Efficiency

Efficiency is the driving force behind resource optimization, both for the company and for frontline workers. We continuously experiment with techniques and shortcuts to improve our efficiency. Many of these work quite well. Others do not. Misguided efforts may save time, but may inadvertently undermine safety protections embedded within detailed procedures. Misapplying priorities or pushing limits can lead to undesirable practices such as rushing, skipping steps, and excessive speed. We rationalize these unwise choices as well-intentioned efforts to satisfy company efficiency goals. The process where frontline workers rationalize their modification of procedures is termed *local rationality*.[1]

Companies can address local rationality in several ways. Well-managed companies can evaluate locally developed techniques, find them beneficial, and expand their use across the operation. Conversely, they may choose to prohibit a locally developed technique that contains unacceptable risk. Risk-tolerant companies can choose to turn a blind eye to certain local practices as long as they improve efficiency. Detached management may remain unaware of local techniques used by their workers.

9.1.3 Distance

Another way we perceive time is as distance. We value a route shortcut because it saves time. If we can land early, those extra minutes can mitigate future delays. A 30-mile final is worse than a 10-mile final because it takes longer. Flying longer distances is almost always viewed negatively. By contrast, flying shorter distances is typically viewed positively.

9.1.4 Fuel

When dispatchers add fuel to our flight releases, they ensure that we meet weight and performance limits. After ATOG limits and efficiency goals are satisfied, fuel becomes *time*. Many companies even use software that assigns a flight or taxi time value to each fuel load entry on the dispatch release. When we review our dispatch release, we assess the adequacy of our fuel load based on how much extra time it allows for contingencies. A fuel load that forecasts arrival at minimum fuel means that if anything goes wrong, the extra time required for a go around and subsequent approach will dip into safety margins. Planned arrival fuel above this minimum value translates into extra time that we can use to go around, to handle a system malfunction, or to enter holding while they remove a disabled aircraft from the runway.

9.1.5 Sense of Pacing

As we perform tasks, we sense the time it takes to complete each one. Preflight takes about this long. Engine start takes about that long. We use this pacing to gauge whether an event is taking longer or shorter than usual. This sense of pacing becomes deeply integrated into our comfort zone. We really notice this when we fly with an inexperienced pilot by how long they take to complete their work.

When we combine this sense of pacing with our quest for efficiency, we value ways to work faster. Every successful modification encourages another. Unchecked, it can lead to a line culture that values multitasking, shortcutting, combining procedures, and ultimately, rushing – all factors that contribute to error.

BOX 9.1 RESTRUCTURING WORKLOAD FROM PUSHBACK TO TAKEOFF

At one airline, the pursuit of efficiency progressed to a point where pilots began gate pushback immediately after receiving the final paperwork and closing the forward entry door. They started one engine during pushback and began taxi as soon as the tug and ground crew cleared away. Captains taxied and started the second engine while FOs computed performance data and prepared the flightdeck for departure. Checklists were hurriedly completed during taxi-out to be ready for takeoff before reaching the end of the runway. The line culture embraced this process as the most efficient use of time. While it was indeed extremely efficient, it tended to split each pilot's attention toward different sets of tasks. There was little cross-monitoring. Captains were responsible for taxi-out (outside tasks). FOs were responsible for computing the takeoff performance numbers (inside tasks). Checklists were designed to catch errors, but they were completed while moving, so each pilot's focus of attention was compromised. The airline experienced unacceptably high error rates during taxi-out, takeoff, and departure.

The airline restructured their procedures to require engine start and checklist completion before beginning the taxi-out. This moved all of the heads-down work to times when the aircraft was stationary with the brakes set. This dramatically altered the pacing of events from pushback to takeoff. FOs who previously divided their attention between monitoring taxi and pretakeoff preparation were now free to monitor aircraft movement. This significantly altered their felt sense of pacing. Captains who were accustomed to immediately starting their taxi-out following pushback reported frustration with "just sitting there". FOs reported feeling much more comfortable during taxi-out. Pilots initially felt that the new procedures wasted a lot of time, but analysis proved that the effects were negligible. With practice, the pilots readjusted their comfort zones to match the new pacing. Error rates dropped significantly.[2]

9.2 THE SIDE EFFECTS OF FEELING BEHIND

Following are some of the unfavorable concerns that arise when we push efficiency too far.

9.2.1 Impatience

As we gain skill and proficiency, we complete our tasks more quickly. With the extra time that we free up, we look for future tasks that we can complete early. This feels like an efficient use of time. The problem is that many of our tasks cannot, or should not, be completed early. Waiting to perform them can prove frustrating. One urge is pushing us to get ahead of our future workload. Another urge is warning us against waiting and possibly forgetting to complete a required task. Both urges can promote a feeling of impatience. This impatience contributes to a number of undesirable practices like performing discretionary tasks during critical flight phases, preparing the EFB for arrival while still climbing out on departure, and repeatedly pressing the FMS execute button before the prompt light illuminates. Following is a personal story about FO who exhibited impatience with delayed tasks.

BOX 9.2 THE IMPATIENT FO

I flew a 3-day trip with a particularly impatient FO. They were good-natured and competent; however, they exhibited profound impatience with delayed tasks. For example, while in climbout, our procedures required us to announce the altitude departing for the planned altitude with a callout such as, "11,000 for 12,000". The procedure directed us to make this callout as we passed 11,000′. The intention was to focus our attention on the level-off and to detect altitude programming errors or mismatched altimeter settings. As we neared the altitude for these callouts, I noticed that my FO began to get antsy. Finally, they couldn't hold back any longer and announced, "11,000 for 12,000", even though we were still several hundred feet short of 11,000′. After satisfying the callout procedure, they seemed to visibly relax. I attempted to model the required procedure by waiting to make my callout when we actually passed 11,000′. This didn't change their impatient practice. Throughout the pairing, I noticed many other instances where they performed tasks early. Driven by their impatience to get ahead, it affected both their procedural compliance and their enjoyment of the flying.

9.2.2 Combining Tasks

Consider how multitasking-type techniques evolve. We might start by combining two similar tasks. Feeling faster and more efficient, we expand our task combining technique. Next, we add another task. It also works. Over time, the combinations multiply. As we do more, our focus of attention become thinner and more fragmented as it zips back and forth between tasks. We devote more attention toward performing task actions and less attention toward assessing the conditions or system changes. Unbounded, the process can push against the gray zone of increased risk.

9.2.3 Shortcutting

Written procedures follow orderly, sequential formats. Consider a five-item flow pattern. It guides us to start with the first item, then to proceed along a defined path or pattern while checking each gauge or switch along the way. With repetition, we may decide that items #2 and #3 are never in the wrong position, so checking them feels unnecessary. Consciously or subconsciously, we remove them from our flow. Our five-item flow becomes a three-item flow. It works and nothing bad happens. Over time, we form a mental picture for switch positions on the panel. Visually scanning the panel, we expect that we'll detect if a switch or gauge is positioned incorrectly. Our mental snapshot becomes the substitute for the three-item flow. The flow fades in importance. In this way, our practice slowly drifts from the desired five-item flow to a one-item visual snapshot. It *feels* more compact and efficient, so it *feels* like a good thing. Unfortunately, problems arise because this shortcutting undermines built-in protections against latent vulnerabilities.

9.2.4 Rushing

Rushing is our last ditch effort to stay ahead of rising workload. Working quickly assumes that we perform and verify each step as fast as we can. Rushing means that we complete the steps quickly without verifying that they are correct. We rarely intend to rush. We start out intending to work quickly, but conditions push us past our limits. Unfortunately, we are poor judges of when we have exceeded our limits. Our emphasis centers on *doing* everything as fast as we can. Our standard of care becomes "close enough" or "getting it done". We truly believe that we can perform the task, detect any missed steps, recognize problems, and correct them before anything consequential happens.

9.2.5 The Adverse Effects of Time Pressure

We can work quickly when our abilities, currency, and attention level are up to the challenge. When any of these fall short, we experience problems.

- *Increased stress:* Some stress can improve our performance. Especially in situations that we know will be time-pressured, we can rise to the challenge. When our stress rises too much, we can feel overloaded. The quality of our performance drops off. We either miss important steps, skip verifications, or reduce our level of due diligence. We can manage our pacing by slowing down or resetting the situation, but these options feel unavailable when time-pressured. Our egos tell us that we can succeed if we just work a little faster. We rush to get ahead with the hope of restoring normalcy.
- *Abrupt aircraft handling:* Another undesirable side effect of feeling behind is rough aircraft control. There are two aspects of this problem. The first is feeling the need to complete tasks more quickly. We feel like there isn't enough time to perform tasks smoothly, so we speed them up. The second is that our personal standard of flying smoothly doesn't seem as important as it normally would. Gentle control inputs feel like luxuries reserved for times when everything is under control.
- *Frustration:* We pride ourselves in delivering a consistent, high-quality service. We feel frustrated when events start going wrong. "Why did I screw up this approach?" "Why won't they vector us to the airport?" "Why isn't our

gate open?" Even when problems are beyond our control, we feel some ownership. When the flight matches our expectations for the pacing of time, we feel relaxed. When the flight lags behind our time expectation, we feel frustrated. Consider the following report from an FO whose Captain abruptly handled the aircraft when stressed.

BOX 9.3 CAPTAIN'S ABRUPT AIRCRAFT HANDLING WHEN TASK-SATURATED

FO's report: I was the FO and Pilot Monitoring on this flight, and the Captain was the pilot flying. He is highly experienced but had not flown much in the last couple months. We have both flown this route dozens of times or more, and the flight was non-eventful until the approach. We were given a standard descent clearance that left us ample time to descend for the visual to Runway XX. The Captain previously did not brief or plan any specifics for how he was going to descend. He briefed just the approach, the taxi-in, and airport details. After our descent clearance, the Captain selected a very shallow descent than left us 20 miles out at around 15,000′ on a straight in approach – much higher than our typical 3 degrees. The descent rate was then increased, but by then it was too late to salvage the approach. He continued toward the runway. Approximately 7 miles out and on a 30-degree angle to intercept the localizer course we were flaps 20, gear down, and around two thousand feet high by my best guess. *At this point the Captain abruptly turned off the autopilot and quickly banked to the right in what was a poor attempt to lose altitude, which quickly exceeded 30 degrees of bank. I called "watch the bank" just before we exceeded 45 degrees of bank and the aural BANK ANGLE was heard. The Captain corrected back to the left but rapidly over-corrected and I believe we had a bank angle aural that direction too. I believe we experienced a flap overspeed briefly as well. We were rapidly approaching 1,000′ at this point still over 200 knots and I called a go around and notified ATC. On the go around, the Captain was hesitant to employ any automation which led to another flap overspeed as they retracted, despite numerous calls by me to "monitor the speed".*

Fundamentally this was caused by poor planning. No descent plan was briefed or employed, and the initial descent rate was inadequate for us to fly a stable approach. *I believe the Captain had tunnel vision and was hesitant to use me or ATC to help remedy the situation. The bank issues after disconnecting the autopilot were caused by faulty airplane handling skills and made it more difficult for me to assist or provide direction. ... this tunnel vision and lack of automation-use continued into the go around and led to our flap overspeed there. In addition, I believe pride led to improper maintenance procedures being followed after the flight.*[3]

9.3 MASTER CLASS PERSPECTIVE OF TIME

We use pacing to gauge how well the flight is going and how well we are handling the workload. Problems arise when the pacing becomes a goal in itself. We see this when pilots try to force their workload with their expected pacing – to match how long they expect it to take, not what the task actually requires.

9.3.1 Time Pacing as a Tool

Problems emerge when we try to force our pacing beyond what existing conditions allow. This can lead to unwise decision making. Some examples are carrying excessive speed during descent and approach, taxiing fast, and pushing crewmembers to complete their tasks faster than they are able. Often, these choices push others outside of their comfort zones or exceed prudent operating practices. The irony is that many of these techniques don't save much time. We pursue them because they make us *feel* like we are saving time.

We have a number of ways to influence our flight's pacing. We can work faster to speed it up or more deliberately to slow it down. For example, when we are running behind, there are reasonable measures we can take. We can ask for route shortcuts from ATC. We can fly faster or request altitudes with favorable wind components. We can even have station operations call our hotel so they can have the van waiting for us at the curb. The same goes for slowing down. If favorable winds have us arriving to our destination before morning curfew is lifted, we can reduce our cruise speed. After landing, we can taxi to a holding pad and wait until our gate opens.

9.3.2 Acknowledging the Existing Conditions

We need to accept each situation as it is. When we are late, we accept that we are late. Having expertly exploited every opportunity to shave time, we accept the remaining delay. Master Class pilots accurately interpret existing conditions to select choices that exploit available opportunities. Clear weather provides the opportunity to fly a shorter visual approach that saves time, but flying that same visual approach too tightly risks an unstabilized approach and go around.

We look for opportunities to save time without compromising our safety margin. The problem and challenge is that there isn't a clear line between capitalizing on the available margin and forcing past it. Pilots who choose to properly fly a short visual approach and pilots who force past the limits both believe that they are skillfully managing the opportunity. They use the same techniques to achieve their goals. Most times, both groups succeed. Sometimes, the aggressive group fails.

The difference is that Master Class pilots remain keenly aware of their safety margin and select decisions that carefully preserve it. Aggressive pilots push beyond the safety margin boundary to exploit potential savings. If they guess right, they succeed. If they guess wrong, they either land from unstabilized approaches or go around. In the previous NASA ASRS example, it's possible that the Captain was so tunnel-focused that he might have landed from his unstabilized approach if the FO had not directed a go around.

NOTES

1 Woods and Cook (1999, pp. 141–171).
2 From the author's experience and featured in Loukopoulos, Dismukes, and Barshi (2009).
3 Edited for brevity. Italics added. NASA ASRS report #1750649.

BIBLIOGRAPHY

ASRS – Aviation Safety and Reporting System – National Aeronautics and Space Administration (NASA) Online Database. (n.d.). Retrieved from Aviation Safety and Reporting System: https://asrs.arc.nasa.gov/search/database.html.

Dekker, S. (2015). *Safety Differently: Human Factors for a New Era*. Boca Raton, FL: CRC Press.

Kern, T. (2011). *Going Pro: The Deliberate Practice of Professionalism*. Lewiston: Pygmy Books, LLC.

Loukopoulos, L., Dismukes, R. K., & Barshi, I. (2009). *The Multitasking Myth: Handling Complexity in Real-World Operations*. Burlington, NJ: Ashgate Publishing Company.

Woods, D., & Cook, R. (1999). Perspectives on Human Error: Hindsight Biases and Local Rationality. In Durso, F. T. (ed), *Handbook of Applied Cognition* (pp. 141–171). New York, NY: Wiley.

Section II

Introduction to Techniques

DOI: 10.1201/9781003344575-11

This section will examine how we develop and refine our Master Class techniques to improve our effectiveness and resilience. We'll start by distinguishing how techniques differ from procedures.

II.1 PROCEDURES

Procedures are company-directed steps that define a standard sequence for performing operational tasks. Crafted from policy objectives, they form the standardized model for how the company expects us to conduct daily operations.

II.1.1 Types of Procedures

There are several procedure categories. Each relates to a particular set of conditions and assumptions.

- *Daily normal procedures:* These are the repetitive tasks that we accomplish on nearly every flight. They include preflight preparation, flows, normal checklists, and typical aircraft maneuvering. The normal case presented in our manuals assumes an ideal environment without exceptional conditions or complex combinations. This includes situations that differ slightly from the ideal case, but still favor using standardized procedures. In short, normal procedures outline the ideal task accomplishment sequence that all pilots should strive to follow unless specific conditions warrant modification.
- *Rare normal procedures:* Rare normal procedures are additional or modified procedures that apply to specific or exceptional conditions. Examples include engine start using an external start cart, tow-in to the parking gate, deicing procedures, specialized arrival/departure flight profiles, unique procedures for certain stations, departing/landing on contaminated runways, short runway operations, and high-crosswind conditions. We are expected to recognize when these requirements apply and follow the published rare normal procedures. We may need to reference particular sections from our manuals (like for aircraft tow-in to a gate) or recall how to perform the procedure (like landing in high-crosswind conditions).
- *Non-normal and abnormal procedures:* These procedures are required for exceptional indications (caution or warning lights) or aircraft malfunctions. They are guided by flightdeck resources like Quick Reaction Checklist cards (QRC), Quick Reaction Handbooks (QRH), or screen-displayed procedures generated when malfunctions are detected. These procedures are addressed in Section Three of this book.

II.1.2 Policy/Procedure Intent

Since written procedures can't cover every nuance that may emerge during line operations, we are empowered and entrusted to exercise judgment to modify or innovate remedies to deal with real-world situations. Our priorities descend as follows:

1. *Follow the procedure:* Used whenever procedures are available and applicable.
2. *Modify the procedure:* If procedures are unavailable or unsuitable, adapt existing procedures to achieve the same end result.
3. *Innovate the procedure:* If faced with a situation not covered in the manuals, innovate a solution that achieves an acceptable outcome and satisfies the intention of company policy.

With empowerment and trust comes responsibility and discernment. If a procedure exists for a situation, we are expected to follow it. Beyond this, we are empowered to find solutions that achieve a similar result. Consider a modified configuration sequence. Typically, we slow, extend maneuvering flaps, then landing gear, then intermediate flaps, and finally, landing flaps. If we need to slow down more quickly, a modified technique might involve extending the landing gear first, then each flap position as placard limits allow. All of the standardized steps are accomplished, but we alter the order to satisfy a particular operational constraint. If we didn't do this, we might cause a spacing conflict with another aircraft or an unstabilized approach. We modify the normal procedure to achieve the same final outcome – a fully configured aircraft on a stabilized final approach profile.

With novel and unanticipated scenarios, we may lack useful guidance and need to innovate a solution that satisfies policy intentions and safety goals. Consider the following flap malfunction report.

BOX II.1 FLAP MALFUNCTION WITH NO APPLICABLE QRH GUIDANCE

FO/PF report: As the Pilot Flying (PF), I called for flaps 30 and the Before Landing Checklist on the visual approach into ZZZ. The Captain (Pilot Monitoring) ran the checklist and we discovered that the Leading Edge Flap Extend green light was not illuminated. *We executed a go around to give ourselves more time to assess the situation.* At 3,000′ and after ATC gave us vectors, the Captain proceeded to [search for] the appropriate checklist in the QRH. *He could not find any checklist for LE FLAPS EXT green light not illuminated. He then asked if I could look and I transferred aircraft control to the Captain. I looked and didn't find any checklist addressing our exact condition. At that time, I did a push-to-test on the LE FLAPS EXT (Green light), LE FLAPS TRANSIT (Amber light), and LE devices annunciator panel and all tested good. The Captain called for Flaps 1 and we verified visually by looking out the window that the LE Flaps did in fact deploy and the flap position indicator showed flaps 1, but LE FLAPS EXT (Green light), LE FLAPS TRANSIT (Amber light) and LE devices annunciator panel did not illuminate. I looked through the QRH again to see if we had missed anything and the closest checklist we could think [of] was the LE FLAPS TRANSIT (Amber Light) but that light was not illuminated, so we elected to not use that checklist. …*

After discussing together and visually confirming the LE flaps deployed, and no roll or yaw was felt in the controls, and it correctly showed on the flaps position indicator, we elected to land normally with flaps 30. … *In looking back, I believe that we could have used our commuting pilot in the cabin to visually check the inboard LE Flaps and Trailing Edge flaps were deployed. Also [we could have] contacted Dispatch and got Maintenance to get another perspective on the indications we were seeing in the cockpit.*[1]

Detecting a possible flap malfunction, the crew executed a go around to make time to diagnose their problem. Their specific malfunction was not covered in their QRH. They expanded their investigation to confirm that the light bulbs were good. While not stated, we can infer that they also verified that all circuit breakers were intact. Lacking a written procedure, they followed the intention of their company's policy to ensure safe configuration before commencing an approach to a successful landing. In summary, we follow the procedure whenever we can, modify the procedure when we need to, or innovate a solution that satisfies the policy.

II.1.3 Standardization

Procedures form the foundation for standardization. We can pair any Captain with any FO and expect them to perform successfully across the full range of the airline's operation. Standardization provides a common baseline for expected roles, duties, and actions. For example, when our process reaches the point where we complete the Before Start Checklist, we know that the Captain will call for that checklist, that the FO will read the steps in the prescribed order, and that the Captain will verify each system's condition and respond. This predictability generates both advantages and vulnerabilities. An advantage is that repetition builds a familiar sequence that highlights errors and exceptions. If the FO misses reading a checklist step, the Captain will detect the error because it fails to match what they expected to hear. The same goes for incorrect responses. If the Captain replies incorrectly, the FO immediately detects their error. For example, if the current conditions require flaps 10, but the Captain replies "Flaps 5", the FO would detect the Captain's error, identify the incorrect response, and together they would confirm the required flap setting.

Vulnerabilities arise when crews become overly familiarized and habituated to the checklist challenge and response cadence. They can fall into a habit of reciting the steps and stating the responses without verifying the actual switch positions. This evolves from a slow drift in our priorities. Accurately reciting and replying subconsciously become more important than reading, verifying, and responding. An example behavior of this is when the FO reads a checklist step and the Captain responds without either pilot verifying the switch position.

Standardization also prioritizes tasks that should be briefed in greater detail. When threats are low, a normal takeoff conducted under normal procedures doesn't usually warrant a detailed briefing. Everyone knows what to do and what to expect. Exceptional conditions, however, require exceptional briefings. For example, a heavyweight, adverse condition, short-runway takeoff warrants a detailed briefing. This supports our process of building SA. When we build a shared mental model of what we expect to happen, it is easier to detect when something goes wrong (emergence of counterfactuals). Through this process, standardization promotes predictability in both ordinary and exceptional situations.

II.1.4 Procedure Sequences

Another advantage of procedures is that they connect series of tasks or steps into logical, interdependent chains. The first task logically leads to the second task and so on. Linking one task to the next keeps us from missing important steps. When the FO announces, "Standing By Flaps" and holds the flap lever with their left hand, it cues the Captain that they have completed their flow and are ready to move forward to the next task in the chain. The Captain then directs the appropriate flap setting, the FO repeats it, sets the flap lever, and both pilots crosscheck the flap gauge. This focuses each pilot's attention on verifying that the FO has set the flap lever to the correct position and that the flaps are beginning to extend. This chained sequence uses the completion of each step to cue the next.

II.1.5 Initiate – Flow – Verify

Before reciting most routine checklists, we complete individual flows. For example, approaching the end of the runway, the Captain signals their readiness to initiate the before takeoff sequence by calling, "Before Takeoff Checklist". Both pilots then perform their required flows to prepare aircraft systems for flight. An example sequence has the Captain press the cabin crew chime button to alert the cabin crew, scan the overhead panel for warning lights, and then shut down the APU. The FO configures pressurization and then sets aircraft lighting for departure. When ready, the FO holds the checklist card and announces the first checklist step. In this example, Captains initiate the process, then both pilots setup their systems using individual flow patterns, and then both pilots verify accuracy using the checklist.

II.2 TECHNIQUES

Techniques reflect our preferred way to accomplish tasks or task sequences. If procedures describe *what* needs to be done, techniques describe *how* we actually do them. We adopt or develop personal techniques for several reasons.

II.2.1 Techniques Provide Depth to Procedures

Most procedures reflect the basic steps and the end result, but not how to achieve that end result – so the destination, but not the path. For example, if our aircraft has four hydraulic pump switches (common configuration for two-engine aircraft with two electric-driven pumps and two engine-driven pumps) and the procedure states,

HYDRAULIC PUMPS – ON, this gives us discretion in both the sequence and the speed for moving the four switches to ON. Options include:

- Use both hands and select all four switches ON simultaneously.
- Engage them directionally – left-to-right or right-to-left.
- Engage the electric hydraulic pumps first, and then the engine-driven pumps second (or the reverse – engines then electrics).
- Engage one A-system switch, then crosscheck the pressure gauge for the appropriate rise, and then repeat for the B-system.
- Select all switches to ON. Then, check the hydraulic pressure gauges.
- Select them ON, ignore the pressure gauge, and only check that the amber LOW PRESSURE lights extinguish.

All of these options emerge from the simple procedural step, HYDRAULIC PUMPS – ON. The point is that one simple procedural step can spawn a wide range of techniques. Is any option wrong? No, because they all satisfy the directed procedure. Are some of them better than others? Yes, because some reflect more effective ways to verify the correct system response.

II.2.2 Techniques Fill Engineering Gaps

During aircraft development, the engineers and test pilots strive to solve every engineering challenge that they can. Since they don't fly the aircraft operationally, they may not detect certain operational problems. After a new aircraft model joins our fleet, we begin flying thousands of scheduled flights. As issues emerge, we develop techniques to solve them. Following is an example from the Boeing 737.

BOX II.2 UNWRITTEN AIRCRAFT GENERATOR ENGAGEMENT TECHNIQUE

Early models of the Boeing 737 (–100, –200, –300 models) used an electrical system that momentarily disengaged the established power source before engaging the selected power source. For example, switching from the APU generator to the engine-driven generator momentarily dropped APU generator power before quickly engaging the engine-driven generator power. As the generators were engaged, there was an audible clunk as certain relays disengaged and others engaged. Visually, this caused a momentary, but insignificant, flicker from the flightdeck lights. From the pilot perspective, this was inconsequential. Unfortunately, if the pilot engaged the #2 generator, then the #1 generator, it interrupted the flight attendant's PA system for about a second or two – usually while they were in mid-sentence delivering their cabin safety briefing. This was an operational irritant that the Boeing engineers probably never detected. Line pilots discovered that this problem didn't occur if they engaged generators starting with #1, then #2. So, the order of generator engagement became significant. A pilot technique evolved to always engage the #1 engine-driven generator first, then #2 second – simply as a courtesy to the flight attendants so that they could enjoy uninterrupted PA audio during their safety briefing. The manuals were never changed, so this remained as a pilot technique until later models of the B-737 (Next Gen and later) redesigned the electrical system and eliminated this glitch. As the older models were retired from the fleet, this technique became unnecessary and faded into disuse.

II.2.3 Techniques Make Sequences Flow Better

Consider our series of preflight preparation tasks. We need to review the dispatch/ weather packet, acquire our flight clearance, program it into the FMS, set up NAVAIDs, compute weight and balance, and set up our personal items on the flightdeck (called "building our nest", this includes tasks like connecting our headset, mounting our EFB, and filling our coffee cup). When we have a typical amount of time, we probably accomplish these tasks in a standard sequence. Consider how we modify the sequence when the aircraft arrives late to the gate or we have to scramble to a different aircraft. We typically alter our normal sequence to adjust for time-pressured conditions. Pressed for time, FOs might request the ATC clearance first, then unpack their personal items, receive the clearance, set up the FMS, review the weather package, finish setting up their personal items, arrange their charts on their EFB, and so on. Captains pressed for time might start with the dispatch release and planned fuel load since resolving any problems in these areas takes the longest time. Contrast this with arriving early to the aircraft. FOs would definitely not start with requesting the clearance because it probably isn't available from ATC yet. Captains might delay flightdeck setup and spend extra time interacting with their crewmembers to promote an open communications environment.

This demonstrates that while flow techniques have a standard sequence, they also have modified sequences that accommodate time constraints. We may also have different versions for marginal arrival weather, deicing requirements, passenger issues, crew issues, and so on.

II.2.4 Techniques Compensate for Personal Error Vulnerabilities

Many of our techniques compensate for errors we have made in the past. A common source involves tasks that require prospective memory (remembering to do something later). Prospective memory tasks are usually *unanchored,* meaning that they lack a strong, salient cue to remind us to do them. Consider a particular airport station that requires us to call them when we are 20 minutes out from arrival. We brief this requirement before top of descent, but it is too early to make the call at that time. We need to find a way to remember to do it later. So, we build a memory aid – a technique to identify a point that reminds us to make the call. Perhaps we construct a visual cue by programming an arc or a nominal waypoint on our map display. We notice the prominent visual cue whenever we scan our route display. When we reach that point, it reminds us to call the station.

Techniques also work on the back side of a task to remind us that it has already been accomplished. Consider how we can receive our landing clearance from ATC Tower anytime from immediately after contacting them until short final. That is a long time span during a busy flight phase. Also, receiving our landing clearance is a repetitive event that we have experienced thousands of times before. If we are flying multiple patterns, multiple flights, or just flying regularly, we can mix up memories of past landing clearances with our current one. To keep this straight, many pilots adopt a technique to show that they received their landing clearance. At one carrier, many FOs click the wheel well light switch to ON when they received landing clearance. There was no standing procedure to turn on the wheel well light, so the switch was effectively used as a landing clearance verification switch.

We amass a vast repertoire of techniques to incorporate a wide range of tasks. Techniques that may seem unnecessary to one pilot are embraced by another. A technique that I use may prove ineffective for you. On the other hand, your technique might be exactly what I need to solve my reoccurring problem.

Techniques evolve over time. Over my airline career, I found that I would adopt a technique and use it for a while. Sometimes, I continued using it. Other times, for unknown reasons, I stopped using it. Then, for equally unknown reasons, I rediscovered it and started using it again. The point is that our toolbox of techniques is always changing. Some well-worn tools are always close by and used every day. Others settle to the bottom of our toolbox – available whenever we choose to dig them out.

11.2.5 Techniques Keep Us Grounded within Our Comfort Zone

The driving force behind new techniques is our unending quest to enrich our personal comfort zone. Familiarity promotes comfort and techniques promote familiarity. Psychologically, our comfort zone is the lowest stress environment that we can create within the current environment. Since we dislike feeling stressed, we naturally adopt techniques that lower our stress level. We achieve our ideal balance by simultaneously avoiding negatives (conditions that increase our stress) while pursuing positives (techniques that reduce our stress).

- *Avoiding the negatives:* Negatives include errors, conflicts, and unfamiliarity. After making an error, we trace back through our timeline to identify the choice or action that would have prevented that error. "I did this which led to that error." "I missed this indication which would have alerted me to that problem." We then develop or modify techniques to avoid committing that error again. By reinforcing our habits and choices, we enhance operational resilience.

  The second "negative" that techniques avoid is conflicts. Conflicts arise from trying to balance our decisions between opposing priorities. The vast majority of our decisions are repeats of past decisions made in similar situations. However, there are always some scenarios that challenge us. The most challenging decisions are choices that mix positive rewards with negative consequences. "If we land, we'll be on time, but that storm cell is getting close to the field and we risk encountering a windshear. On the other hand, if we divert now, we'll avoid the possible windshear, but we'll be low on fuel and mis-connect all of these passengers." To ease this conflict, we might adopt a technique of adding extra fuel whenever convective weather conditions are forecast near our destination. This reduces the urgency of diverting and eases that conflict. We'll have more time to hold and wait for the storm to move off. If we need to divert, we won't be as concerned about our fuel reserves. Again, following a conflicted event, we debrief ourselves and develop a technique to avoid that conflict in the future.

The third negative that we seek to avoid is unfamiliarity. Familiar decisions and techniques keep us within our comfort zone and out of the gray zone of elevated risk. We know that if we try to force familiar decisions and techniques into unfamiliar situations, we set the stage for Recognition Trap errors. Techniques steer us toward familiar game plans that work.

- *Promoting the positives:* Positives are factors that increase the good feelings we enjoy while settled within our comfort zone. They include aligning with the operational flow and increasing familiarity. We love settling into our groove and riding it from start to finish. When we align where we planned to be with where we currently are and where we want to be (past, present, and future SA), we create a positive feeling of continuity. Everything is going right. This feeling is not limited to simple and familiar situations. It also applies to complex and familiar situations like finding our comfort zone while flying into busy hub airports. It's more work, but our comfort zone is still available. We accept working hard as long as we keep up with the aircraft, maintain our SA, and make effective decisions.

  In promoting their comfort zone, some Captains may impose their techniques on their FOs – forcing the flightdeck climate to align with their preferences. They stay comfortable, often to the detriment of their FOs' comfort. A better choice is to promote an open and balanced flightdeck environment. Following is success story guided by a comfortable flightdeck climate.

BOX II.3 BALANCING PRIORITIES AND LANDING SAFELY IN ADVERSE CONDITIONS

FO's report: Overall, we attempted three approaches into ZZZ. First Approach/RNAV 24: At the time we shot this approach, winds were most favoring Runway 24 out of about 280° gusting to 30–40 knots with some reports of LLWS in the area by a Dash-8 who had recently landed. On the RNAV just inside the final fix, the aircraft was getting kicked around pretty good by gusts of wind, causing speed fluctuations of around 15 knots. After a strong burst, we went from being on-speed, to getting the high speed clacker, to getting the stick shaker in a matter of seconds. *I was a little stunned and slow to react, but the Captain appropriately called for the go around.* When Tower asked of our intentions, we decided to give it another try.

Second approach/RNAV 24: Unfortunately, the second approach's *conditions started to shape up and feel just as the first one did.* After getting moderate turbulence with speed fluctuations outside of our stable approach parameters, *I felt uncomfortable and called for the go around. The Captain promptly initiated the go around profile callouts.* After going around, wind reports and checks indicated that the winds were starting to favor Runway 34. After assessing the landing data, terrain, and communicating with the Dispatcher, we decided we had enough fuel to attempt the ILS to Runway 34 and still be able to divert should we have to go around again.

Third approach/ILS 34: The third approach had significantly less turbulence. *It felt and looked much better.* However, in the last few hundred feet we started to get strong gusts again. We got an amber windshear caution message, *but appeared to be stable* so the Captain said [to] continue. At about 10′–20′ the windshear message went red. *I was slow to react, because my head told me I should call for a go around, but my gut was telling me we are stable enough to proceed 10 more feet down to landing. A low energy go around at a mountainous airport in windshear conditions did not feel like a better option. The Captain must have felt the same, and proceeded to land safely. It all happened very quickly.*

Captain's report: … *I felt that it was safe to continue the approach* because these gusts were very short. Around 20′ above the runway we received a red windshear warning. It took me about a second to think about our state. *At this point, about 10′ off the ground, I felt that it was safer to continue the landing since we were so close to the runway instead of doing a low energy go around with windshear present.* We landed safely with no further incident. … *I felt that we both performed well and worked together very well. My First Officer did an excellent job assisting me. I was very impressed that he called for the go around on the second approach once we were out of his comfort level instead of just allowing the Captain to continue. I understand that I broke operational procedure by not going around with the windshear warning, however, I still feel that this was the safest thing to do.*[2]

While this crew did not follow procedures with their landing decision, they worked well together, balanced priorities, and landed safely. Notice how both their decision to go around and their final decision to land were strongly influenced by how they felt. As you read it, sense how their comfort zone influenced their choices, how they worked to promote each other's comfort zone, and how they worked effectively as a crew in a challenging situation.

II.2.6 How Techniques Make Procedures Work Better

Techniques are ways to accomplish procedures better. They provide the fuel that powers procedural improvement. The evolutionary path of a technique starts with an idea. "I think this would work better for me if I did it this way." Our idea works, so we adopt it as a technique. Other pilots see us using it, like it, and adopt it themselves. Knowledge of our technique eventually travels up to flight operations leadership. They evaluate the technique, validate its benefit, and begin teaching it in classes. It becomes a best practice.[3] Perhaps it ultimately is added to the manuals and becomes a procedure.

BOX II.4 THE EVOLUTION OF THE THRUST LEVER CHECK – FROM TECHNIQUE TO PROCEDURE[4]

In many aircraft, the final protection to ensure correct takeoff configuration is the takeoff warning horn. This alarm sounds when thrust levers are advanced and one or more of the takeoff configuration parameters are out-of-tolerance. If the horn sounds, pilots are directed to reject the takeoff, taxi clear, and diagnose what caused the warning. The cause is typically an incorrectly set takeoff trim value or flaps not properly extended for takeoff. Pilots adopted a technique of quickly advancing, then retarding, one thrust lever while on the taxiway before taking the runway. If the horn didn't sound, then it confirmed that everything was properly set. If it did sound, then it gave them time to diagnose and correct any configuration problems before taking the runway. Flight operations leadership evaluated the technique as a way to mitigate potentially hazardous events such as initiating a takeoff with no flaps extended. Given the severe consequences of this error, flight operations elected to add the *thrust lever check* to the before takeoff flow procedure. The airline enjoyed a significant drop in rejected takeoffs for configuration errors and no-flap events.

II.2.7 The Hidden Vulnerabilities of Techniques

There are hidden risks within some techniques. When a new procedure is evaluated, a development team investigates a range of operational, engineering, and human factors considerations. They start by deeply examining the errors that crews are making. They design a procedure that satisfies all the engineering needs while maximizing error detection and mitigation. Next, they test the new procedure in the simulator with their team members and test crews. They tweak the procedure to get it running as smoothly and accurately as possible. Next, they develop the training product that will accompany the new procedure. They perform additional test crew evaluations using the training product. Finally, the new procedure rolls out. The pilots complete the training and begin using the new procedure. The safety and standards departments monitor line-pilot behavior to ensure compliance and to uncover any missed vulnerabilities.

From our personal perspective, we explore techniques only within our operational context. We experiment with techniques without knowing or remembering the underlying vulnerabilities or errors that the procedure targeted. Conflicts can emerge between our techniques and those underlying procedure development objectives. This can allow latent vulnerabilities to emerge.

BOX II.5 FUEL BALANCING AND THE MASTER CAUTION LIGHT

In most multiengine aircraft, a need may arise to correct for fuel imbalances between the wing tanks. The crossfeed procedure directs the pilots to open the crossfeed valve and then deselect the fuel boost pumps on the lower-quantity fuel tank. This forces fuel to feed from the higher-quantity fuel tank. It takes some time for the fuel to burn off and restore balance.

The problem develops when we get busy and forget that we are still balancing fuel. This is one of those prospective memory problems – remembering to do later what we can't do right now. If the balanced-fuel point is missed, the imbalance flips and the high side becomes the low side. When we detect our oversight, we reverse the fuel boost pumps. Imagine a particularly distracted crew that allows their fuel imbalance to seesaw back and forth several times. They might develop a technique to remind themselves that they are balancing fuel. Ideas might include posting a sign on the forward panel (Fuel Balancing in Progress) or clipping the end of their necktie to the yoke clip.

One technique was to leave the Master Caution light illuminated during balancing. The light illuminated when the pilot selected the low-side boost pumps to OFF. Instead of cancelling the light, pilots purposely left it ON. This did help to solve the problem since the glaring amber light reminded them that they were balancing fuel. Unfortunately, it created an unintended side effect of disabling the alerting function of any other Master Caution system malfunction. For example, if an air conditioning pack tripped off during fuel balancing, it wouldn't generate a Master Caution alert because the light was already illuminated. When the pilots finished fuel balancing, they might clear the Master Caution light, reset the pumps, close the crossfeed valve, and fail to detect the illuminated PACK OFF light. So, the pilot technique of using the Master Caution as a memory aid created an error vulnerability.

II.3 THE TECHNIQUES DEVELOPMENT LAB

Line flying is an innovations laboratory. Pilots, by selection and practice, are skilled problem solvers. Whenever we see something that can be done a little bit smoother, faster, or better, we experiment with it. Some experiments fail while others succeed. As we experiment with new techniques, we should use the following process.

II.3.1 Understand the History Behind the Procedure

We probably weren't on the team that developed a particular procedure, so we don't know all of the considerations they used in building it. We need to understand their underlying assumptions and intentions, the assumed initial conditions, the desired end-state, and how their procedure joined them.

Sometimes, we have procedures that don't quite work. Our technique needs to fix the flaws of the original procedure while satisfying its objectives. A good example might be a procedure designed for conditions that have fundamentally changed. The legacy procedure may not fit the current aviation environment or recently upgraded aircraft technology. Pilots flying fleets with several models may find that some procedures that were designed for the oldest model don't work for the newer models. Until the procedures are revised, pilot techniques may fulfill the operational intentions better than the legacy procedure.

II.3.2 Preserve the Protective Features Built into the Procedure

Procedures satisfy engineering, operational, and human factors needs. Our technique may inadvertently undermine protection requirements in one or more of these areas. The fuel balancing example reflects how a personal technique undermined the engineering function of the Master Caution alerting system. Operational and human factors needs tend to be more nuanced. For example, assume that our procedures still direct us to call the station when we reach a point 20 minutes out from arrival. That made sense years ago. Now, anyone with a computer can track our aircraft and arrival time. Perhaps some station operators have already started using them instead of relying on our in-range call. Still, we don't know the inner workings of each particular station, how they plan their gate manning, the lead times for delivering baggage carts to the gate, and tasks like this. Until the company transitions their procedures to use internet apps instead of radio calls, it is best for us to continue following seemingly antiquated procedures like this.

II.3.3 Investigate the Underlying Conditions

To understand the underlying conditions and interrelationships, human factors use an assessment called a task analysis. This process lists each task, identifies who is responsible for completing it, the prerequisites they need to complete it, the other interrelated tasks that compete for the pilot's attention, the errors made, and each error category (knowledge-based, skill-based, CRM, resource management, training, decision making, risk assessment, distraction, prospective memory cuing, task loading, and climate/culture – to name some). The point is that there are a lot of hidden moving parts that can be adversely affected when we start tinkering with techniques.

II.3.4 Improve the Overall Quality of the Procedure

As new options become available, techniques explore them. The ground taxi procedure may simply direct, "Have the airfield diagram displayed and available." This reflects the legacy procedure using paper JEPP pages. It was written to cue pilots to have the chart open and immediately available for all taxi operations. Since we have moved on to EFB computer "charts", our technique options have expanded to include displaying our own aircraft position over the airfield diagram, expanding the display for a zoomed-in depiction of complex taxi intersections, selection of day or night display mode, and horizontal versus vertical EFB orientation. The main objective remains unchanged – to have the airfield diagram open and available. If our personal technique preserves and enhances the directive, our techniques may prove to be quite useful.

11.3.5 Build Personal Techniques to Improve Personal Performance

The techniques laboratory is our personal workspace. Like our home office or workshop, we have some latitude in arranging our tools in a way that suits our preferences and habits. While remaining grounded in procedural compliance, our techniques fulfill our personal needs and preferences. Many procedures already allow for pilot discretion. For example, they direct us to have our charts ready for our intended arrival, but probably don't dictate how we arrange those pages. The procedure dictates that we have the approach chart displayed as we fly it, but probably not which backup approaches to have ready in case we need them.

11.3.6 If Unsure, Ask the Experts in Standards or Training Departments

If we discover a promising technique, but aren't completely sure whether it generates all the positives and none of the negatives, we should ask. The experts from our training center and standardization department were either involved with the original procedure development or know who was. If they cannot detect any unintended consequences from our technique, it is probably good to go. Ideally, they will favor our technique so much that they will begin the process of modifying our airline's procedures to incorporate our ideas.

Organization of the Techniques Section: This techniques section will expand on the core concepts to explore techniques for risk management, decision making, building situational awareness, time management, workload management, distraction management, automation techniques, communications techniques, and CRM techniques.

NOTES

1 Edited for brevity and clarity. Italics added. NASA ASRS report #1756780.
2 Edited for clarity and brevity. Italics added. NASA ASRS report #1156858.
3 There is some controversy with the term *best practice*. While it is used extensively across aviation, it is not particularly accurate. Better terms are *preferred practice* or *good practice*. Accept *best practice* as a generic designation for a technique that enjoys wide acceptance.
4 Loukopoulos, Dismukes, and Barshi (2009, pp. 110–121).

BIBLIOGRAPHY

ASRS – Aviation Safety and Reporting System – National Aeronautics and Space Administration (NASA) Online Database. (n.d.). Retrieved from Aviation Safety and Reporting System: https://asrs.arc.nasa.gov/search/database.html.

Loukopoulos, L., Dismukes, R. K., & Barshi, I. (2009). *The Multitasking Myth: Handling Complexity in Real-World Operations*. Burlington, NJ: Ashgate Publishing Company.

10 Risk Management Techniques

Most of the time, we manage risk quite well. However, as complexity and time pressure increase, our normal risk management process becomes stressed. We are pushed toward the gray zone of increasingly unmanageable risk. In our desire to simplify our situation and keep up with the pace, we may resort to rushing or shortcutting. Our well-reasoned risk assessment process weakens. We resort to making snap-judgments and choosing biased, rationalized decisions. For some pilots who become especially settled into their comfort zones, this increased risk can lead to task overload, surprise, and tunneled attention.

10.1 THE RISK MANAGEMENT SKILLSET

As we gain experience with daily operations, we become quite skillful at detecting and managing risk. We recognize our situation, apply a successful game plan, use the choices that support that game plan, and make the necessary corrections to maintain our desired course. As conditions become more stressful, we notice that we have to work harder to sustain our game plan. Even if we force it to work, we can reach a point where the game plan still fails. All game plans have their breaking points.

What can we do to avoid this? We start by refining our ability to sense the rise and fall of risk. We may believe that we can accurately rate a situation's risk threshold, but research indicates that the more overloaded we are, the more unreliable our risk assessment becomes. While we can't accurately identify the point where risk reaches a hazardous level, we can reliably sense its trend. As we gauge risk's rate of movement toward the hazardous threshold, we expand our vigilance for warning signs. We become more cautious. When we sense both rising difficulty with sustaining the game plan and the emergence of warning signs, we need to increase our willingness to abandon our original game plan for a backup option. Following are some strategies to guide this process.

10.1.1 Considering the Consequences

One effective strategy is to consider *consequence* as part of our risk assessment process. Mishaps only occur when a flawed game plan is forced to an unfavorable conclusion. For example, we can make a series of unwise choices that cause us to arrive unstabilized on short final (like Clem's mishap of sliding into the overrun following his unstabilized approach), but going around erases that adverse consequence. If we pre-consider which bad outcomes may arise, our decision making tips toward the safer option.

DOI: 10.1201/9781003344575-12

For this to work, we need to consider possible consequences early in the scenario – well before becoming task overloaded. This gives us time to anticipate the early warning signs, consider possible outcomes, and preselect our decision trigger point for aborting our game plan. Absent a clear abort trigger, we often push forward and hope for the best. Once we become stressed and overloaded, we typically lack enough time to deliberately weigh our choices.

Changing our perspective changes our perception. Consider an approach when we started down final knowing very well that we were just too tight, too high, or too fast (maybe all three). We tried to compensate by reducing power, increasing drag, and maybe even performing an S-turn. All of these are perfectly acceptable techniques for salvaging an approach. With most of our past situations, they worked. With this particular situation, however, they weren't enough. The approach remained unstabilized. As we consider this situation, we all agree that we should go around. In the heat of the moment, however, many pilots choose to continue – just like Clem did. They accept the approach instability and land anyway. Master Class pilots don't. We deliberately consider the consequence of overrunning the runway, recognize the threat, and go around. Consider the following strategy.

1. Use *consequence assessment* to anticipate warning signs and preselect game plan abort triggers.
2. Attempt *path deviation management* as long as the game plan appears to be working.
3. Use the *emergence of warning signs* to arm and rehearse our abort option triggers.
4. Use *abort triggers* as solid decision points to switch to backup contingencies.

This strategy requires us to honestly assess the effectiveness of our corrective actions. We can't rationalize flight goals like on-time arrival to justify deviating from the strategy. Clem knew that he had exceeded the company's stabilized approach criteria, but his desire to land eclipsed switching to a contingency option. This allowed him to rationalize that it was better to continue. "I wasn't that fast." "It didn't feel unsafe." "I knew I could stop the aircraft safely."

10.1.2 Modulating Our Vigilance

As we gain experience and settle into our comfort zone, it becomes easier to relax our mental effort. After the 50th or 100th time flying the same city pair, the flight can become repetitive. Consider a hypothetical case of a very comfortable pilot who only flies the same undemanding, familiar flights. With each repetition, they become more relaxed and settled. This pilot might lose their motivation to raise their level of vigilance, even during dynamic flight phases like takeoffs and landings. If they suddenly encounter a challenging event, they might succumb to startle, indecision, and error.

Aviation demands that we concentrate during high workload phases of flight. This is mentally fatiguing, so following these high vigilance periods, we need to rest our attention focus. As Master Class pilots, we deliberately modulate our vigilance to match both the flight phase and the relative difficulty of the flight. Every single takeoff and departure demands our attention. Every single approach and landing demands our attention. Even when the flight is simple, familiar, and monotonous, we pay attention. Alternatively, whenever we enter low-workload flight phases, we deliberately relax and rest our minds.

10.1.3 Searching for Counterfactuals

Counterfactuals are indications that contradict what we expect to see. They appear as mismatches between our present moment SA and predictive future SA. If we expect that our airspeed should be slowing, but it is actually increasing, that is a strong counterfactual. An unknown condition may be adversely affecting our game plan.

The search for counterfactuals requires us to look, see, and process relevant information. This may sound like common sense, but time and again, we witness pilots becoming so task-saturated and tunnel-focused that they don't notice these warning signs. When we interview a pilot who has flown an overspeed approach, they rarely recall their airspeed. Typically, they either have no recollection of their actual speed or they recall a value that was much lower than that which the flight data computer recorded. They often recall the single parameter where their attention was tunnel-focused (like their intended touchdown point), but not other important parameters (like sink rate and airspeed).

As Master Class pilots, we detect rising risk, event quickening, and escalating stress. These cue us to expand our monitoring for counterfactuals. Instead of optimistically assuming that our game plan is succeeding, we skeptically look for signs that it might be failing. If we have time, we can attempt some corrective measures. If not, we switch to our contingency backup game plan.

10.1.4 Preserving an Escape Option

Escape options are almost always available to us. They only vanish when we wait too long to act. To better understand this, consider how our escape options change as we progress along a challenging approach profile. Consider a scenario where the runway landing threshold is clear and dry, but a significant storm looms just beyond the end of the runway.

- *Case one:* Before starting the approach, we detect the threatening storm and ask ATC for holding airspace. Our escape plan is easy to manage. We have plenty of time to evaluate whether to wait or divert.
- *Case two:* Now, let's move in closer, like joining a 7-mile final. Choosing not to land, we request to break off our approach. While Tower and TRACON coordinate airspace, we begin feeling antsy. Approaching a 4-mile final, they finally vector us away from the storm. This escape options is not as preferable as the first one, but it is still safe and manageable.
- *Case three:* Now, let's move our approach to a 1-mile final where we encounter a windshear warning from a microburst. Procedures direct us to go around and to prepare for a windshear recovery maneuver. Our best

option is to immediately turn away from the storm cell. This may require us to exercise emergency authority since ATC cannot react quickly enough to assign vectors. This escape option may present aircraft performance challenges and possible hail damage.

- *Case four:* Next, imagine that we made it to touchdown when a strong microburst crosswind starts forcing off the runway. We could still go around as long as we aren't committed to stopping. Our escape option is a MAX-thrust rejected landing into a strong storm cell. This would be quite challenging and possibly hazardous. It would still be better than being blown off the side of the runway.
- *Case five:* The final case is encountering a severe microburst after landing with our thrust reversers deployed. Most manufacturers designate this as the point where we are fully committed to stopping, so our escape options vanish.

The point of this exercise is to understand how our escape options steadily narrow and become more difficult and consequential the further we proceed into a risky situation. Logically, the sooner we exercise our escape option, the easier it is for us. Monitoring and preserving escape options are important Master Class skills. We anticipate the trajectory of our game plan, monitor for counterfactuals, manage escape options, select trigger points, and protect ourselves from adverse consequences.

10.1.5 Communicating a Backup Plan Using "if-then" Language

Changes in our game plan can require changes in path management, aircraft configuration, radio coordination, expected procedures, and crew roles. The last thing we need during a last-second change is crew confusion. It is much easier when we communicate our plan ahead of time using "if-then" language. I recall an approach where we were tracking down final in a significant crosswind. I asked my FO to request a wind check from Tower. Tower broadcast a wind report that was challenging, but still within our crosswind limit. I shared my contingency plan, "Based on that wind report, I'm continuing this approach. I want you to monitor the crosswind component on the FMC. *If* we don't satisfy crosswind landing limits by 200′, *then* I want you to direct a go around." Our plan recognized the warning sign (strong crosswind approaching limits), set the decision point where I wanted crosswind limit compliance (200′), defined our contingency safe escape plan (go around), and assigned specific roles (monitor crosswind component and direct the go around if winds are out of limits).

10.1.6 Rehearsing Contingencies

When last-second go arounds occur during line flying, they are usually surprising and stressful. We practice them in the simulator, but since they are part of the expected training profile, they don't surprise us as intensely. Consider how much better our real-world go arounds would look if we rehearsed the steps beforehand. Imagine an approach that isn't working out well. It's iffy whether we will achieve stabilized approach criteria. If the PF verbalizes, "I'm going to continue flying this approach. If we have to go around, the steps are…" This brings back-of-the-mind knowledge up to the forefront of our thinking. It gives us an opportunity to mentally rehearse the steps. It prepares the crew for their tasks and primes our decision point for executing the go around. In the case of a go around, it also combines the first few action steps into an organized flow.

10.1.7 Evaluating Options Using a Premortem Exercise

The premortem exercise is an imagination exercise that traces the possible failure paths of a game plan. It assumes that our current game plan fails and asks what would have caused that failure and how the warning signs would have appeared.[1] The premortem serves two purposes. First, it considers the possibility that our game plan might be flawed. It plants a healthy seed of doubt. Second, accepting that it may be flawed, it directs our attention toward indications that would warn us of its possible failure. Armed with this awareness, we can assign specific monitoring duties, consider more resilient game plan options, or rehearse contingency backup plans.

Consider an approach in marginal conditions. The premortem exercise would ask the question, "Let's assume that this approach fails, what would cause it to fail and what indications would we expect to see?" If weather is trending worse, we might anticipate a missed approach for poor visibility, heavy turbulence, or exceeding landing wind limits. Listing the reasons provides us with a range of counterfactuals to monitor. We might ask Tower when landing aircraft are breaking out of the weather on final. We might plan our best flap setting for turbulence. We might brief an approach that accommodates lower ceiling and visibility limits.

The premortem exercise guides us to assign specific monitoring duties or to select flight display options that show critical information in the most accessible format. For example, we could select the FMC page that displays the crosswind component. If we are focused outside flying the aircraft, we could direct our PM to feed us that information. "I've got my hands full with this approach. Please give me regular crosswind component readouts from the FMS. If you see evidence of windshear on short final, direct a go around."

10.1.8 Treating Anything Unique as a Warning Sign

As we gain experience, we encounter fewer events that appear new or unique to us. Settled in our comfort zone, we can develop a mindset where little surprises us anymore. Consider a group of highly proficient pilots that have become so comfortable that they remain unconcerned by events that probably should concern them. They may react with an "isn't that interesting" response instead of a "this is new, unknown, and potentially hazardous" response. Early in our aviation career, we used surprise to prompt ourselves to pay attention. Surprise was a useful warning sign. Highly experienced pilots might never reach this "surprise" threshold anymore. They may only recognize situations as being unique or unpredicted.

As we become proficient, the prudent course of action is to begin treating anything unique or unpredicted as a warning sign. This strategy garners little down-side risk and protects us against failing game plans. The fact that something unexpected happens means that its cause may be unknown. Leaning toward careful, conservative choices protects us against a wide range of unknown conditions and adverse consequences. Master Class pilots treat all unique situations as potentially hazardous because they prompt us to pay attention and choose carefully.

10.2 PROACTIVE RISK MANAGEMENT

Master Class pilots don't rely on momentum or hope to make it through difficult scenarios. We engage risk proactively. Following are some concepts that frame our mindset.

10.2.1 Avoiding the Deteriorating Spiral

Mishaps pilots often waste valuable time trying to diagnose an unexpected event or indication. Consider three scenarios.

- *Scenario one – wasting time trying to diagnose the situation:* A crew tries to make sense of an unexpected event. They sift through the stream of conflicting indications trying to determine which are relevant and which are unimportant. They expend valuable time considering and discarding possible explanations. As time runs short, their SA collapses and their operational pace quickens. They narrow their attention. As they become task-overloaded, they rely on split-second reactions, habits, and instincts. Unfortunately, these tools have been shaped by years of familiar, normally paced, successful scenarios, not unfamiliar, fast-paced, failing scenarios. Their corrections don't work. They continue clinging tenaciously to their failing game plan long past its salvageable point.
- *Scenario two – plenty of time to diagnose the situation:* The crew has enough time to identify the cause of their problem. They rebuild their SA. It guides them to either modify or abandon their game plan. Notice how this guides a successful outcome – no avalanche of problems, overload, quickening, or tunneled attention – just pilots using the available time to successfully handle an unexpected glitch.
- *Scenario three – not enough time to diagnose the situation:* The crew detects the unexpected event, doesn't identify its cause, and decides that they don't have enough time to diagnose it. They switch directly to a contingency backup plan like a pattern breakout, a missed approach, or a go around. This crew also successfully avoids the deteriorating spiral of a failing game plan.

When faced with an unexpected event, we want to follow either the second or third scenarios. That sounds easy, but remember that the unexpected event may not generate recognizable warning signs while we are flying in-the-moment. Instead, all we may notice is that an unexpected event has just happened. Some mishap pilots tried to rely on their gut-felt or emotional assessment. The fallacy here is that our emotions don't process unexpected events reliably or consistently. Our gut-felt sense may fall anywhere between finding the event mildly interesting to seeing it as hazardous. So, while our instincts prove useful with innovating workable game plans, we can't fully rely on them to assess risk. A more reliable starting point is noticing that something unexpected has happened. Consider the following four-step assessment process.

1. *Is the event immediately threatening?* If there is an immediate threat, we need to switch to a safe recovery option. An example would be a wake turbulence encounter while on very short final. There is no time to diagnose the cause or assume that it probably will work out. We need to immediately go around.
2. *Do we have time on our current profile to investigate the cause?* If so, we can use the time to diagnose the cause, cross-reference other indications for verification, and choose whether to continue or abort the game plan. If time is short, we might elect to make extra time by requesting a pattern breakout or going around.
3. *Is the indication something beyond what our current game plan can handle?* If the unexpected indication doesn't adversely threaten our current profile, we can probably continue without consequence. An example might be an air-conditioning pack failure on 3-mile final. We could reasonably assume that our landing will remain unaffected.
4. *Does the indication fall outside our current game plan?* If the indication exceeds something that our current game plan can accommodate, we need to switch to an escape option. Otherwise, we risk spiraling into a deteriorating situation. An example would be flying down final with the aircraft configured, but the airspeed keeps increasing from an unknown cause.

Summarized, these steps guide us to find the cause of the unexpected event only when we have available time. If not, we either ensure that the problem is inconsequential or we abort our game plan for a contingency backup. We don't attempt to force an unknown situation.

10.2.2 Modulating Vigilance to Match Risk

As Master Class pilots, we strive to actively modulate our vigilance to match the needs of the situation. We elevate our vigilance to respond to rising risk. We relax our vigilance to rest and prepare for the next challenge. We focus our attention when the situation becomes more complex/unpredictable or when our response time is short. We relax our attention when the situation becomes normal/predictable and our response time is long. We anticipate future task loading, plan ahead, and conduct effective briefings. In summary, our vigilance matches the risk level, guides our preparation, aids in SA building, governs the parameters that we monitor for counterfactuals, and preserves our escape options.

10.2.3 Managing Priorities and Assumptions

Our everyday goals are to fly safely, efficiently, and on-time. When operations become disrupted, these goals may encourage us to tolerate riskier game plans. While flying in-the-moment and pursing an on-time arrival, 15 knots fast on final may seem "close enough". When we succeed, it encourages repetition and drift. Then, one day, we make a mistake like Clem's – accepting an unstable approach, but missing the latent vulnerability of slippery pavement.

As Master Class pilots, we remain aware of both the friction between conflicting priorities and of the slow drift in how we prioritize our goals and assumptions. We study how our personal biases change with our moods, fatigue levels, and personal priorities. We erect strong countermeasures to guard against our human tendency to rationalize between conflicting goals. This recalibrates our risk management and reverses adverse drifts.

10.2.4 Instilling a Healthy Dose of Caution

Many mishap pilots exhibit the characteristics of overconfidence. They remained strongly optimistic about their game plan in spite of rising counterfactuals and warning signs. It is important to distinguish between self-confidence and the confidence we have with our situational risk assessment. A highly self-confident pilot can still have low confidence with a particular game plan. Mishap pilots, however, seem to link their personal self-confidence with how well they handle a situation. When indications cast doubt on their decisions, they take it personally and react defensively.

As Master Class pilots, we know that we can miss important warning signs, misjudge conditions, and select unworkable game plans. When we discover these flaws, we actively reassess and start over. The key is with how we apply caution. Caution begins with an awareness that no matter how well a scenario is going, it can change. As risk rises, our level of caution rises with it. A game plan that looks perfectly reasonable at one point can fall apart when new information surfaces or conditions change. Our self-confidence lies in our ability to discern these subtle changes, respond resiliently, and guide successful outcomes.

10.2.5 Making Continuous Corrections

Aviation is more like balancing on the peak of a mountain and less like resting in the bottom of a valley. Aviation requires a continuous stream of corrections to maintain a desired path. Even a well-trimmed aircraft will slowly drift off course. There is never a time when we can completely relax while the aircraft remains in dynamic flight. Likewise, there is never a time when the success of our game plan is guaranteed. It constantly needs to be monitored, tweaked, and verified.

10.3 THE RISK AND RESOURCE MANAGEMENT (RRM) MODEL

A useful model for incorporating risk management is the Volant Systems RRM model.[2] The model promotes awareness of rising risk. It guides how we apply

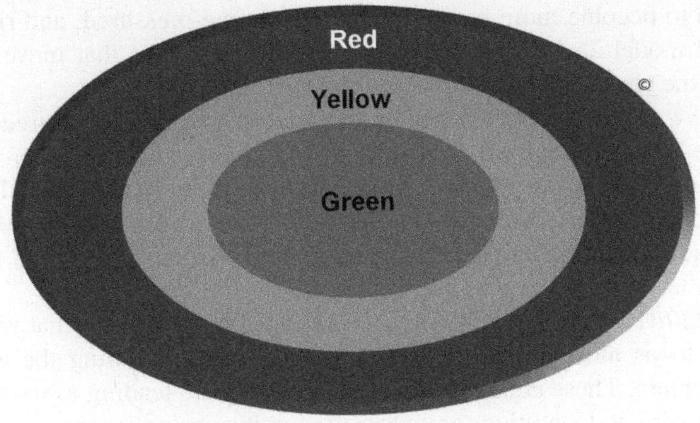

FIGURE 10.1 RRM target with colors depicting perceived risk.

resources to reduce the risk and restore our operational flow. There are three main components of the model – the target, the resource blocks, and the Assess, Balance, Communicate, Do, and Debrief (ABC*D*) process.

10.3.1 The Target and Colors

The RRM target graphic depicts three concentric circles of green, yellow, and red (Figure 10.1). These colors represent a combination of the level of risk and its impact on crew performance. This includes our ability to handle the situation as well as an assessment of our confidence, stress, SA, and comfort level. The green circle in the center (shown as dark gray) reflects a well-prepared, comfortable, confident, and familiar situation. We are rested, alert, and have excellent SA. The next ring is yellow (shown as light gray). It represents increased risk, a felt-sense of unease with the situation, and decreased SA. Something is not going well. Our attention is either focused more intently on the problem or distracted while handling the problem. The outer ring is red (shown as black). Red represents extreme concern or confusion. Our attention becomes highly focused trying to understand or handle a stressful situation. It also represents when we can experience task overload and make undetected and uncorrected errors.

- *In the Green:* When we are "in the Green", we are confident and comfortable. We have a good game plan that is following a predictable path. We are aware of the relevant conditions affecting our flight and are maintaining good SA. We have plenty of extra time to detect and avoid potential threats. It is our intention and desire to remain "in the Green" as much as possible.
- *In the Yellow:* When we are "in the Yellow", it means that we have detected conditions that have increased our risk and stress levels. We aren't comfortable with the progress or trajectory of our game plan. We feel uneasy with changing or challenging conditions. Our SA is becoming unreliable and the probability of making a consequential error is rising. Events and conditions

begin to become more unpredictable, more time-pressured, and risky. The RRM model identifies three categories of risk factors that move us away from the Green toward the Yellow and Red.

- *Task loading:* Our workload is rising. We feel time-pressured because the time available to deal with the situation is shrinking or the time needed to handle the situation is increasing. We feel the need to work more rapidly, make quicker decisions, and respond faster. High task loading increases operational risk. Low task loading can lead to laxity. Either of these can push us into the Yellow or Red.
- *Additive conditions:* These are complicating conditions that require us to focus more attention on understanding and handling the emerging problem. These conditions upset our game plan, leading us to modify or abandon it. Conditions are interacting in increasingly complex ways that create competing priorities or allow unpredictable outcomes to emerge.
- *Crew Factors:* These are individual pilot conditions like boredom, distraction, hunger, drowsiness, stress, illness, attitudes, crew composition, crew interaction, and crew roles. They may either move us deeper into the Yellow or Red or inhibit our ability to move back toward the Green.

External or internal factors may move us out of the Green. External influences (Additive Conditions) include weather, complex airport operations, and aircraft malfunctions. An example would be arriving on final with minimal fuel and experiencing an aircraft configuration issue that requires a go around and QRH procedures. Internal influences (Crew Factors) refer to our personal capabilities and crew actions. Inadequately briefing for a particularly complex airport arrival and failing to arrive to work rested are two examples.

Being in the Yellow means that we have detected risk factors or are being pushed away from our comfort zone. As we narrow our attention toward emerging risk factors, we might miss other risk factors that previously would have been easily detected. Yellow may be unavoidable, but it can be mitigated.

- *In the Red:* Being "in the Red" means that we are reaching the limits of our abilities or are losing control of the situation. Our capabilities are maxed out. The problem demands our full attention. Being in the Red isn't necessarily a sign of personal failure, but it does often mean that we need to focus on CRM or change the game plan. For example, if we are reacting to "WINDSHEAR, WINDSHEAR" and flying a windshear recovery maneuver, we may be appropriately in the Red. It means that, as a crew, we are completely focused on executing the appropriate recovery maneuver and are unable to handle additional task loading. Communications are typically more directive. It is difficult to expand our SA beyond the immediate situation. We may miss some communications, make undetected errors, or miss additional conditions that arise. For example, we experience an engine failure and fly the recovery profile only to discover that we neglected to retract our gear.

RRM colors form a common reference that each of us use to articulate how we perceive our situation and personal mindset. They may not match. PFs may be in the Red while PMs only reach the Yellow. PFs can remain blissfully in the Green because they don't detect the risk factors, while their PMs move into the Yellow or Red because they do. A pilot in a new aircraft or duty position may move into the Yellow simply due to lack of experience or familiarity. CRM breakdowns can also move the entire crew into the Yellow or Red. Poor communication or personality conflicts can place the crew in the Yellow without any situational risk factors present. Whatever the combination, any elevation from Green to Yellow or Red alerts the crew that they need to mobilize resources to manage risk and move back into the Green.

10.3.2 The Five Resource Blocks

After each pilot states their color level, we use the five RRM resource blocks to guide our recovery process back into the Green. Once we reestablish control, they help us to remain in the Green. These resources arrest the deterioration of the situation, mitigate threats, and resolve the disruption (Figure 10.2).

- *Policies, procedures, and flows:* Policies, procedures, and flows create consistency to improve individual crewmember performance, crew coordination, and decision making. Polices provide a foundation for sound decision making and guide us through the approved sequence of resolution steps. As line pilots, we are often unaware of the many nuanced defenses incorporated within particular policies and procedures. Embedded within them are scientifically tested and experience-validated protections against latent

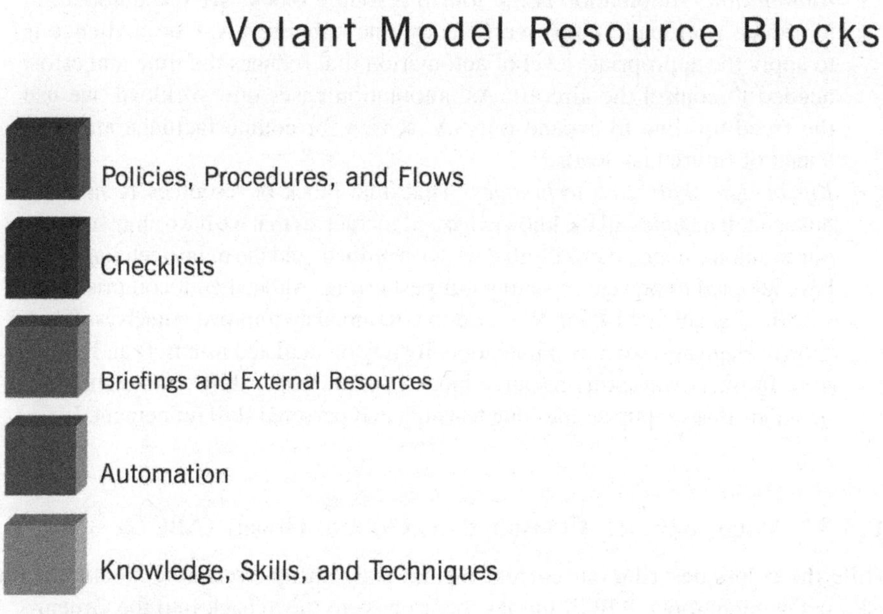

FIGURE 10.2 RRM resource blocks.

vulnerabilities. When we choose to deviate from these procedures, we can undermine these protections. In addition, well-designed policies and procedures help us anticipate what to expect from each other.

- *Checklists:* The second resource block is checklists. Startle, stress, and surprise can cause confusion and CRM breakdowns. Checklists and other job aids decrease reliance on memory and help us identify errors that may have occurred during a procedure or flow. Checklists provide clear and unambiguous sequences that unify our efforts and simplify our action steps.
- *Briefings and external resources:* This third block includes our preparation steps of crew briefing and accessing information from outside resources. Standardized briefings are designed to create a shared mental model of the intended actions, expected risk factors, game plans, and backup contingencies. They align everyone's efforts toward the same game plan and assign monitoring roles to detect deviations and counterfactuals. Errors are recognized and corrected quickly. External resources include dispatchers, ATC controllers, maintenance control, operations control, inflight medical services, and aircraft manufacturers. They are especially useful in rare or complex situations where we may lack knowledge or experience. External resources typically have the same goal as the crewmembers (landing at destination airport or on-time performance) but they often have a different perspective or competing priorities. Because of this, it is important to integrate the information and direction from these external resources into the context of our flight. The linchpin of this resource block is effective communications. By taking the time and effort to promote an open communications environment, we mobilize the highest levels of operational resilience.
- *Automation:* Automation is the fourth resource block. We use automation to reduce workload, improve efficiency, and enhance SA. Our challenge is to apply the appropriate level of automation that reduces the time and effort needed to control the aircraft. As automation eases our workload, we use the freed-up time to expand our SA, search for counterfactuals, and stay ahead of future task loads.
- *Knowledge, skills, and techniques:* This final block of resources is vast and nuanced. It includes all the knowledge and memories that we have amassed over our aviation careers, the skills that we have refined, and the many techniques we have adopted to prevent repeating our past errors. Altogether, it comprises our wealth of aviation wisdom. We need to continuously improve ourselves. If our growth stagnates, we may unintentionally rely on outdated mindsets and strategies. In many ways, this resource block of knowledge, skills, and techniques demands that we pursue life-long learning and personal skill refinement.

10.3.3 Assess, Balance, Communicate, Do and Debrief (ABC*D*)

While the colors describe our current mental state and the resource blocks list our task management tools, ABC*D* guides the process to move back into the Green.

- *Assess:* Assessing is the first step. Since we are in a moving aircraft, we cannot just freeze the situation and sort things out. We have to assess current indications, often while our SA has been degraded, under high levels of stress, with rising workload, and as underlying conditions change – all while flying the aircraft. Startle and surprise have knocked us off balance. To recover, we need to understand what is happening, discover what caused events to change so unexpectedly, and form an appropriate recovery plan. To make it even more difficult, we often need to fix our problem while we are trying to understanding it.
- *Balance:* Balancing is the second step. We compare our original goal with available outcomes. Is continuing the original game plan still a viable option? Do we need to find more time to resolve our problem? Should we switch to a contingency backup plan? For example, if we encounter a flap malfunction while on approach, complex QRH procedures require that we go around to complete the required flap malfunction checklist. The original objective of landing from that approach is no longer appropriate. Balancing requires that we make honest, and often inconvenient, assessments. The complex and difficult option may be the most appropriate. Again, we need to balance options while the aircraft is moving, new indications keep flooding in, SA is still rebuilding, and conditions keep changing.
- *Communicate:* Communicating plans and intentions is the third step. Ideally, assessing and balancing produce a viable game plan. This step directs us to discuss that game plan with the affected team members, invite comments, identify risk factors, and assign roles. Depending on time available, some communication may be directive. For example, given our flap malfunction on final approach, the first communication may direct, "Go Around". In complex situations like this, communication may come in bits and pieces as time and distractions allow.
- *Do…:* "Do" directs us to execute the new game plan. Since we need to continue flying the aircraft, we are probably "doing" some steps while we assess, balance, and communicate a new game plan. It is important to remember that the ABC*D* process isn't necessarily linear. Ideally, the process becomes subconscious. Although we may learn using the steps, it would be rare to be literally thinking "ABC*D*" while dealing with the non-normal. Resolving our flap malfunction, for example, we might use the following steps:

1. *Direct immediate response steps:* Advance thrust, disconnect autopilot, initiate climb, and raise landing gear.
2. *Coordinate with required agencies:* Inform ATC and coordinate climb out instructions.
3. *Move to safe airspace:* Enter airspace that allows sufficient time to manage the task load – establish a stable downwind with autopilot engaged.
4. *Engage the* ABC*D process:* Assess and balance. With the malfunction stabilized, select a game plan for recovery – continue back for landing with new approach speeds and configuration.

5. *Communicate, coordinate, assign duties:* Communicate the game plan with the crew, assign duties, and declare an emergency with ATC.
6. *Perform (Do) required procedures:* Complete non-normal/abnormal checklists.

- *... and Debrief:* The second "*D*" is debrief, which refers to the feedback process. With most situations that drive us into the Yellow or Red, we need to resolve unexpected and unfamiliar situations. The pace of events requires us to deal with imminent risk factors first and to defer other risk factors until later. Mini-debriefs summarize our ABC*D* progress and uncover additional considerations and tasks. They ensure that everyone agrees on the latest assessment and game plan.

Consider an inflight engine failure on final. The ABC*D* process might proceed as follows:

1. Assessing (A) the indications, we conclude that the #1 engine has failed.
2. Balancing (B) conditions and objectives, we decide to go around.
3. We direct a missed approach (Do) and communicate (C) our intentions with ATC.
4. ATC assigns us a new clearance (C).
5. We follow the engine failure checklist and conclude that we shouldn't attempt to restart the engine (A).
6. Balancing risk (B), we decide to divert to a nearby airport with a longer runway and favorable weather.
7. We communicate (C) and execute (Do) this new game plan.
8. While diverting, we expand our communication (C) to include the cabin crew, passengers, and company operations.
9. We solicit (Debrief) inputs from crew and recommendations from company maintenance technicians. Reassessing our past choices (A), we change our plan and perform an engine restart (Do). The successful restart changes the game plan.
10. We conduct further crew briefings (C), assign roles, and prepare for landing (Do).
11. With the time available, we repeat the ABC*D* process to either verify that our game plan is tracking as expected or change it to address emerging counterfactuals.

The assess, balance, and communicate steps of ABC*D* tend to center on the Captain. While all crewmembers can offer their interpretations and suggest viable game plans, the Captain needs to choose the course of action. Given the demands of ABC*D*, Captains can become overloaded. It is useful to let the FO fly the aircraft while the Captain focuses on managing the RRM process, rebuilding SA, monitoring the progress of the game plan, and planning contingencies.

10.3.4 Putting the RRM Process Together

Following is an instructive example from a mishap crew who experienced an engine failure immediately after takeoff.

BOX 10.1 APPLICATION OF THE RRM MODEL DURING AN ENGINE FAILURE EVENT

FO/PF's report: ...Tower cleared us for takeoff ... Within approximately five seconds there was a big boom on the right side of the aircraft and the airframe shuddered slightly as the #2 engine wound down. You could hear grinding noises. *I assessed that the Captain was most likely in the Red, based on his comments at the moment and his breathing hard in the intercom. I assessed that I was personally in the Yellow.* I observed N1 winding down rapidly through 30% and EGT flashed red but then also came down in temperature. ... I had just been through training 2 weeks prior and at that moment I was really appreciating my single engine practice I had on the simulator. I felt like I was in full control of the aircraft and our rate of climb decreased but we were still climbing slowly to 3,000′ MSL. *Once I was satisfied that we still had one good engine, I felt myself coming back into the Green. In reality, I honestly think I vacillated back and forth between Green and Yellow throughout the entire 13 minute divert as we handled threats (some, if not many, self-induced).* The Captain pulled the QRH and said, "I think we have an engine seizure." Between the Captain's calm ATC transmission and his immediate grab of the QRH, *I assessed ... that the Captain was out of the Red and at least at Yellow if not in the Green.*

Captain/PM's report: I cannot speak highly enough of the First Officer! His airmanship and professionalism shined throughout this event. The best thing he did for our Passengers, our Company, and me was fly the aircraft and remain calm and professional. This allowed me to manage the bigger picture of getting this damaged aircraft on the ground safely. The flight lasted 14 minutes. From the time of engine failure to touchdown at ZZZ1 was approximately 13 minutes. *I do not recall many of the details of those 13 minutes as this was the fastest 13 minutes of my life.* ... We did the "big" things right, or at least mostly right. We, as Pilots, want to be robots and do Step 1, Step 2, Step 3, etc. *In real life, the effect of being stunned and the rush of adrenaline causes momentary tunnel vision, fog shrouded initial thinking, and brain hyper drive.*

Then entire airplane shook and shuddered like we had been hit by something large. *I have never experienced an engine failure outside of training and can say with certainty, nothing in training could have prepared me for the loudness, violence, immediate shock and stun of an event like this, or the effects of adrenaline on the brain....* I can recall several things, but I cannot recall the exact order in which they happened. *The tunnel vision effect presented itself. I recall looking at the engine gauges and seeing one of them filled with red. I believe it was the N1 gauge but am not 100% certain.* I recall looking at the flight instruments and these are the basics of what I saw. We were passing 2000′, climbing, flaps 1, airspeed good. ... *My mind was racing and my first thought was bird strike. I remember being stunned with tunnel vision. I looked at the hand microphone and was thinking, the Passengers need to know they will be OK. I regained my senses and the gravity of the situation hit me. This is real! Focus.* We are in a damaged aircraft. This is on me! Another rush of adrenaline, I'm sure. All of this took place in a matter of seconds. Once I "snapped to", the following events occurred (though I am uncertain as to the exact order). I asked the FO if he was OK. I think I asked if he had it under control, but not sure. He responded, yes. I got on the radio, [advised ATC], told ATC we had lost an engine and we would be flying up to ZZZ1. … *I was "in the Red" and was aware of it, still a little stunned and had not completely processed anything more than an engine had failed, we were not on fire, and the aircraft was under control. I was overwhelmed with the number of tasks that had to be accomplished to get this aircraft safely on the ground ASAP. My brain was going in many different directions.* This is where Company Training Department shines! With 100% certainty, one of the best things Company Training has ever done is give us the QRH. The training kicked in. Although I was "in the Red", I did what we do in training. I reached for the QRH. I cannot tell you how valuable this tool was at the time. *It brought me back into the loop and zeroed me in on where I needed to be. With the adrenaline rush/stun factor, my brain was still in hyper drive and I still was not processing information as well as I would have liked.* I recall looking at the QRH and methodically going through the items. … I vaguely recall doing the items from the QRH as a Crew. … At this point, there was short conversation with the FO about what we were dealing with. I do not recall the details. *The only thing I remember is that the conversation gave me conflicting information with my view of the situation and I had to reassess. I stopped the checklist. I did not want to do anything unless I was 100% sure it was correct. … My brain was still in overdrive. I made a decision. Don't spend too much time analyzing and get this aircraft on the ground immediately.* The aircraft is flying just fine. I verified again, we do not have an engine fire. … Back to the basics. What do I know for sure? The FO is flying, the aircraft is under control. We do NOT have an engine fire. We do have an engine failure. Throughout this process, I was also trying to handle the radios so the FO could concentrate on flying. We teamed up though. If I was too busy or missed a radio call, he took it. *We traded off when necessary and worked well together.* I inherently knew he had the aircraft under control because he was calmly answering radio calls when necessary and complying with ATC instructions. I did, however, check back with him at least a few times throughout this event to make sure he was doing OK.

ATC asked which runway we would like. …Still being in the Red and unable to compute numbers, I pulled out the performance computer, and quickly found ZZZ1 and put in the winds with all other fields to the default numbers.

At approximately 700′, the FO said, "You're supposed to land." I said, "No, you're landing." He responded, "Are you sure?" I replied, "Yes, I'm not transferring control of the airplane at this point, you're landing." Touchdown and rollout was uneventful.

> *Overall, I think communication was good given the high stress event and the time crunch. We worked well together and divided our duties. He was responsible for flying; I was responsible for everything else. We interacted as needed, and maintained the big picture with what I thought to be acceptable risk given my decision to get the aircraft on the ground immediately.* …Looking back, I can see where some small picture things could have been done better, but in the end, the big picture items won the day and our training worked. We inherently fell back on the basics. Maintain Aircraft Control, Analyze the Problem, Take Appropriate Action, Maintain Situational Awareness.[3]

The FO's recent practice at the training center prepared him well for this emergency. He instantly recognized what to do and only reached the Yellow. Notice how he maintained effective SA throughout the event. The Captain had a rougher time of it. His reaction to the engine failure clearly placed him in the Red. Notice how long it took him to restore Yellow. Also notice how he used the resource blocks to get out of the Red and back to the Yellow. An interesting twist was that the Captain chose to let the FO land the aircraft single engine (contrary to company policy). Given his stress level, we should consider this as an effective balancing of risk instead of procedural noncompliance. The Captain clearly recognized that the FO was in the Green, while he still remained in the Yellow. Notice the calm clarity of the FO's report versus the elevated emotion of the Captain's report. Everything was working well, so the Captain chose to leave the roles unchanged.

NOTES

1 Klein (2003, pp. 98–101).

2 RRM is a risk management model from Volant Systems, LLC. (http://www.volantrrm.com/risk-and-resource-management.html). Figure 10.1 RRM graphic is a gray-scale depiction of the color-coded Volant version. Figures 10.1 and 10.2 are provided courtesy of Volant Systems, LLC.

3 Edited for brevity and italics added. NASA ASRS report #1456711.

BIBLIOGRAPHY

ASRS – Aviation Safety and Reporting System – National Aeronautics and Space Administration (NASA) Online Database. (n.d.). Retrieved from Aviation Safety and Reporting System: https://asrs.arc.nasa.gov/search/database.html.

Klein, G. (2003). *The Power of Intuition: How to Use Your Gut Feelings to Make Better Decisions at Work.* New York, NY: Currency Books.

11 Decision-Making Techniques

11.1 THE TYPES OF AVIATION DECISIONS

We'll begin by examining the range of aviation decision making from ideal/familiar situations, situations requiring minor adjustments, and novel/unexpected situations.

11.1.1 Familiar Situations Following Familiar Decisions

With most situations, we recognize familiar features that we have seen many times before. We know which common, time-tested game plans to use. These game plans then guide our decision making.[1] We become quite good at this recognize, match, and decide process. Imagine having flown 100 flights between two cities like Seattle (SEA) and Los Angeles (LAX). We would rarely encounter anything new to challenge us. All likely events would be comprehensively indexed in a virtual SEA-LAX database stored in our memory. Like driving down a familiar country road, we'll know every curve and every bump. We just need to stay in the center and stay out of the ditches.

11.1.2 Simple Deviations that We Resolve with Quick Decisions

Rarely does a game plan unfold exactly as predicted. Continuing with our country road metaphor, imagine that we notice that we are drifting away from the center of the road. We can interpret this deviation in any of these three ways.

- *Insignificant deviation:* We accept the deviation as inconsequential and ignore it.
  - Decision: Drifting a bit – no concern – continue monitoring.
- *Minor deviation:* We adjust our path to dampen the deviation, reverse the trend, and restore our desired game plan.
  - Decision: Trending toward the side. Make a correction back toward the center of the road.
- *Significant deviation:* We sense that the drift is becoming consequential. Time to pay attention and restore our path.
  - Decision: Sit up and pay attention. Something important is happening. Make a significant correction back to the center of the road.

How we rate each of these three cases is guided by our SA. It serves as a filter that sorts events between *indications that don't matter* and *indications that do matter.*

DOI: 10.1201/9781003344575-13

The first two categories (insignificant and minor deviations) are relatively trivial. They are so ordinary that we barely notice them. We react quickly and instinctively. We probably don't even notice moving the steering wheel to correct back. Instead, let's focus on the third case – a significant deviation that appears out of the ordinary. Decision making during these events deserves deeper analysis.

- *What does it take to attract our attention?* Remember that our SA maintains a running prediction of what we think should happen next. If everything matches that prediction, we move happily along and enjoy the scenery. If something deviates from that prediction, our minds alert to it. Since our minds are quite good at detecting *difference*, we quickly notice anything that deviates from what we expect.
- *Is it important?* After our mind detects the anomalous event, we need to decide whether it deserves further attention. If we deem that it is important, then we devote attention to it. If not, we ignore it. How do we determine what makes something worthy of our attention? One strategy is to treat any anomalous event as something important. Simply, if it attracts our attention, then we investigate it and decide what to do about it. This strategy takes a significant amount of effort. It is difficult to sustain. The opposite strategy is to treat nothing as important unless it becomes an obvious threat. This also doesn't work because we would miss early warning signs of situations that might become threatening if we don't correct them.

  As with most decision-making processes, we need to find a balance. Ideally, we develop a highly discerning *importance filter* that accurately discards unimportant anomalies and identifies anything else as worthy of our attention. This is a Master Class skill that we need to constantly refine. As we encounter interesting events, we exploit opportunities to debrief them, both individually and as a crew. With every new analysis, our personal importance-determining filter grows more dependable and resilient. This, in turn, achieves the balance that we seek. We learn to relax our attention for the vast majority of ordinary events while appropriately raising it for those cases that truly deserve our attention.
- *Is it a threat?* Imagine that this is our 100th time flying the same city pair. Nestled snugly in our comfort zone, we've easily handled or dismissed dozens of minor deviations without much effort. Suddenly, something novel or unexpected happens. It attracts our attention. How *different* does this event have to be to feel threatening?

  From studying aviation mishaps, we conclude that many pilots accurately detected the early warning signs of their situation, only to dismiss them as trivial. It wasn't that they didn't see what was happening. They just didn't recognize them as signs of an emerging threat. The indications didn't rise above their personal threat-detector threshold. Remember that as we settle into our comfort zone, we tend to bias our decision making toward optimistic choices – choices that support our current game plan. As a result, fewer indications seem threatening. The antidote for this is to maintain a mildly skeptical mindset. When we notice an anomalous event, we investigate further to discover its cause. We don't discard it until we can confidently rule out any adverse consequences.

11.1.3 Novel and Unexpected Events

Novel and unexpected events should attract our attention every single time. We should always take the time to assess their significance, then decide whether to act or not. This becomes especially important as we gain experience. Consider a novice on their first flight. Everything would seem novel and unexpected to them. Their attention would jump at every new experience. They would feel exhausted after that flight. Next, consider the opposite extreme of a highly experienced airline pilot flying a familiar route. Feeling like they had seen it all and done it all, few events would seem novel or unexpected. Even if they detected something unexpected, they might underestimate its importance. Between these extremes lies an ideal balance – an experienced professional pilot applying a keen importance filter that accurately discerns significance every single time. That pilot would be confident and relaxed for all routine flying, but would focus their attention when conditions unexpectedly change.

11.2 HOW WE DETERMINE IMPORTANCE

The flaw that both our jumpy novice pilot and our overly relaxed experienced pilot share is that they filter importance based on familiarity. Familiarity is, in turn, based on a subconscious judgment of how well that situation matches with our past experiences. The novice pilot tends to judge most situations as *unknown* and *important* while the experienced pilot tends to judge most situations as *known* and *unimportant.* In effect, importance is rated in terms of familiarity. This is the source of the fallacy. Both pilots are misguided because importance is not directly related to familiarity. We can have familiar, important events and unfamiliar, unimportant events. In the first moments after we detect an event, even before we judge its importance, we sense its familiarity. If we stop there, we never advance to the crucial step of assessing importance.

Our next step is to classify the event between the two categories, definitely unimportant and potentially important. Unimportant events are quickly ignored or immediately corrected. We need to treat all remaining events as important unless we determine otherwise. Our decision-making process then assesses three specific parameters: our sense of how serious the event is, how quickly the event is happening, and how far it strays from our expectation.

11.2.1 Severity

The first factor that calibrates our importance filter is severity. Our assessment comes from our gut-felt, subconscious appraisal. For example, as we monitor our PF flying down final on a hot bumpy day, we would notice many flightpath deviations caused by the wind gusts, turbulence, updrafts, and downdrafts. We monitor each deviation and judge whether their flight control correction is appropriate or not. If they encounter a strong updraft, we expect them to push forward on the flight controls and reduce thrust. We would expect the opposite for a downdraft. If we could plot our attention level and concern on a graph, we would see an immediate increase when the aircraft enters the updraft and a subsequent decrease as the PF applies an appropriate correction. The PF could be "fighting it" all the way down final, but as long as the flightpath continues correcting back to the desired track and glidepath, we remain alert, but unconcerned.

Let's complicate this scenario. If the PF suddenly made the opposite correction that amplified the deviation, we would immediately assess a rise in severity. For example, if they encountered an updraft at the same moment that they pulled up to avoid a bird, the combined effects might cause the aircraft to exit stabilized approach parameters. We would immediately rate this as a significant problem. Our assessment might shift to determining whether the approach is salvageable, or not. If we decide that the approach is unsalvageable, we would direct a go around.

Continuing with our example, assume that there it still sufficient time to correct the deviation. The PF pulls the thrust levers to idle and pushes the nose over to correct back to glidepath. Just then, the aircraft enters a strong downdraft. Seeing that the downdraft and the path correction are too severe, the PF counters by adding lots of thrust and pulling way back on the flight controls. While all of these are appropriate corrections for the given conditions, we might judge that the magnitude of the corrections has become too severe. Could the PF salvage the approach? Probably yes. Could they also miss-manage their flare and make a hard landing? Yes, again. Lacking time, we might decide that there is too much risk with continuing and direct a go around.

Sometimes the rapidity of the changes overloads our decision-making skills and we effectively freeze up. Mishap PMs sometimes find themselves repeatedly alternating between either calling for a go around and allowing the approach to continue. Often, their bias defaults to allowing their PFs to continue (plan continuation bias). If the approach is teetering on the edge, corrections are becoming extreme, and risk is rising, the prudent course is to call for a go around and try again under better conditions.

11.2.2 Time Available

Time available is a sense of how much time we have to diagnose and correct a deviation. A minor instability on 5-mile final is judged as less important because we have plenty of time to correct it. That same instability on ¼-mile final gives us little time to correct and would be judged as very important. Any time we are in a flight regime where the time available is short, we should be prepared to respond. For example, during takeoff or landing, both pilots should be positioned to immediately assume aircraft control. The appropriate standard is to position our body to apply the necessary response in case something happens. Like a baseball infielder standing in a ready crouch when the pitch is thrown – prepared to immediately catch that 100 MPH line drive off the opponent's bat. Infielders aren't in the ready position because the ball *will* be hit to them. They are in the position *in case* the ball is hit to them. The average infielder assumes that ready crouch over 150 times a game and may not field a single ball. The same goes for us. We may fly our entire career without ever needing to assume the controls, but we should be ready to do it every single time – just in case. Consider the following event of a FO misinterpreting TCAS RA symbology and climbing toward conflicting traffic.

BOX 11.1 CAPTAIN RESPONDS IMMEDIATELY TO FO'S UNEXPECTED ACTION

Captain's report: We were climbing on autopilot to an assigned altitude of 22,000′ and received an advisory from ATC that there was another aircraft descending to 23,000′. As the autopilot was leveling off the aircraft at 22,000′, we got an RA of, "Adjust Vertical Speed". My First Officer (FO)/Pilot Flying (PF) disengaged the autopilot *but instead of leveling the aircraft he climbed, flying into to the RA.* I immediately took control of the aircraft but we ended up climbing approximately 300′ before I was able to return the aircraft to 22,000′. ATC did not mention the deviation. I had to explain to my FO the way a RA works. Apparently, he misidentified the red "do not climb" area of the VSI as the green "safe" arc. *Also, at the time, it took me a second to understand what was happening: having someone fly into an RA was something I absolutely didn't expect and took me completely by surprise.*[2]

Imagine if this Captain had their seat back eating a sandwich when this occurred. They were able to quickly react because they were ready and positioned to take the controls to reverse the FO's error.

11.2.3 Deviation from What We Expect to Happen

The third factor that calibrates our importance filter is deviation from path. Imagine we are flying in the cruise portion of our flight with the aircraft symbol perfectly tracking the magenta line on our moving map display. For some unknown reason, our autopilot navigation becomes uncoupled. We don't notice it. The autopilot is still holding the wings straight and level, but is no longer actively tracking the course. Ever so slowly, our aircraft symbol begins to slide off of the magenta line. After about 10 minutes, we notice it and discover that the autopilot's lateral navigation is disengaged. We correct back, reengage lateral navigation, and proceed as planned. What might cause us to notice this deviation? Outside references offer little help as the aircraft flies straight and level even without lateral navigation engaged. It isn't the rate of change either as the drift is extremely slow. Something else has to trigger in our mind as our eyes scan over our moving map display.

Our minds are skilled detectors of difference. The greater the difference, the more easily we notice it. The rate of deviation is also relevant. We detect faster drift rates easier than slower drift rates. As experienced pilots, we develop a keen sense of the *rightness* or *wrongness* of these differences. If we see a rapid rise in our engine temperature during start, we easily detect it and abort the start. This is because we notice that the rate of the rise is atypical (a sense of wrongness).

One way our minds detect rightness and wrongness is by sensing the difference between where we expect something to be and where it actually is. If we had a slow hydraulic leak, we would probably notice it more quickly by the split between dial-type display gauges (a needle split between A and B hydraulic systems, for example) compared with digital hydraulic gauges (typically displayed as a percentage of hydraulic fluid tank capacity). This is because our perception is more sensitive to aircraft gauges that aren't parallel.

A latent vulnerability arises when both positions are parallel, *but wrong.* Consider a pressurization mishap event where pilots inadvertently leave both pressurization pack switches in the OFF position. Passing through 10,000′, the cabin pressure warning alarm sounds. Their error remained undetected because both pilots had developed a habit of confirming that the switches were parallel (in this case, both OFF) without confirming that they were in the correct position (both should have been in AUTO/ON). This has also been cited in takeoff thrust errors where the crew inadvertently programmed an incorrect assumed temperature. This resulted in a much lower takeoff thrust than desired.[3] The crews only looked for parallel indications instead of an appropriate takeoff thrust.

Rightness or wrongness is also sensed by the rate of change. If we lose thrust during takeoff, a quick glance at the engine instruments would confirm that the failing engine is the one with the needles winding down compared with the engine with the steady needles.

11.3 USING OUR INTUITION FOR DECISION MAKING

We need to understand how intuition works, when to apply it, when to ignore it, and how to improve our decision-making skills when using it.

11.3.1 Intuition and Problem Solving

We are proactive problem solvers. We don't wait for a problem to fully develop. Instead, we use our perception to detect the subtle clues from emerging problems. Our intuition doesn't stop there. Intuitive problem detection and intuitive problem solving work together to detect an emerging problem, generate a solution, guide our decision making, and monitor the outcome to ensure that our actions solve the problem. Our intuition is interwoven through all five of these problem-solving components.

1. Detecting the indications from an emerging problem
2. Investigating what caused those indications
3. Sensemaking to understand the meaning between the indications and their causes
4. Selecting, adapting, or innovating a workable game plan
5. Choosing the actions needed to enact and guide that game plan

Whether the problem is easy or complex, intuition plays a role. For easy, normal operations, it guides a routine matching exercise. We intuitively select a game plan that has worked with similar situations from our past. It just pops into our mind as soon as we detect the problem. The more complex and nuanced that the situation becomes, the more we need to draw on more subtle aspects of our intuition.

11.3.2 Pattern Recognition – The Puzzle Metaphor

An instructive metaphor of this process is assembling a jigsaw puzzle. We'll examine a series of increasingly difficult cases to see how our intuition changes.

- *Case 1 – full puzzle with all the right pieces:* In our first case, we have all of the pieces of the puzzle, but no picture to reference. Instead, we rely on our memory from the last time we assembled the puzzle. Organizing the pieces, we recognize some that some match in color and design, so we connect them together. Some pieces immediately make sense. Other pieces don't. Our guiding assumption is that all of the pieces belong somewhere and that they will eventually assemble into a complete solution. Knowing that a complete solution exists, our strategy unfolds clearly. We begin to recognize prominent features – sky, foreground, tree line, and lake. The puzzle begins to take shape and form an image. This strategy works because we start with a contained problem and the promise of one complete solution. Logical decisions flow easily because we know that there are no exceptions or anomalies – no counterfactuals. This perspective guides our intuition which guides how we form the image. This, in turn, guides our prediction and our prediction guides our decision making.
- *Case 2 – missing pieces:* Now, let's increase the complexity. Imagine that before we started, someone removed a handful of the puzzle pieces. Unaware of their actions, we would still assume that all of the pieces are present and that a complete solution exists. Our strategy begins the same way it did earlier. As we work, inexplicable gaps appear. Searching for a particular piece, we become frustrated that we can't find it. The conflict between our assumptions and the puzzle's reality clashes with our intuitive process. This generates uncertainty. As we struggle to find meaning, our strategy falls apart. At the height of our frustration, someone informs us about the removed pieces. Our perspective, strategy, and the way we apply our intuition immediately change. Our progress resumes because we now have an explanation for the gaps in our picture. Notice how our assumptions affect our strategy, which then affect our intuitive process, which then affect our decision making.
- *Case 3 – missing pieces and extra pieces:* Let's take it one step further. In this last case, someone removes one handful of pieces and adds a handful of similar pieces from an unrelated puzzle. Imagine how difficult this puzzle might be to assemble – especially because we are unaware of the partial removal and mismatched replacements. Our assumptions, strategy, intuitive process, and decision making would all run into trouble. We'd become confused about the missing pieces and extra pieces. Our entire problem-solving process might grind to a halt. At this point, assume that we learn of the removal and replacement. We would realize that we have to radically change our strategy.

We'll need to find patterns between the matching pieces, identify and predict what should fill the gaps left by the missing pieces, and ignore any foreign pieces that don't belong. Our intuition can truly shine in this chaotic and imperfect environment. Once we assemble an accurate set of assumptions, our intuition guides a workable strategy. Notice how important our intuition becomes in this complex and uncertain environment.

11.3.3 Pattern Recognition – Aviation Problems

Decision making in aviation shares many of these same pattern recognition challenges as our puzzle metaphor. Consider how this same set of situations affects our intuition and decision making while flying.

- *Easy aviation problems:* We'll start with a simple aviation situation, like a flight where everything is standard, all the information is available, and our objectives remain clear. The flight is on time and the weather is good. We recognize all of the pieces and know how they fit together. We form an accurate picture for how the flight should progress. We apply a proven game plan as our strategy. Lacking any anomalies, decisions flow seamlessly. Our SA tracks along a familiar path.
- *Complex aviation problems – with gaps and errors:* As complexity rises, interactions between conditions, operations, and processes can form gaps in our information. We might classify these gaps as errors, oversights, and lapses of procedure. For example, assume that our dispatcher got distracted while preparing our release, missed a temporary weather condition forecast for our destination, and failed to assign an alternate. If we assume that the dispatcher sent us an error-free release, we might pursue a strategy that skips making a detailed review. We would miss this error and sign the flight release without discovering their omission. Like our puzzle-maker assembling the complete puzzle, if we assume that everything is complete and familiar, we would apply a strategy that presumes that there are no gaps of information. From this example, we conclude that it is an unreliable strategy to assume that our knowledge is complete and that we know everything about our flight. We should always assume that some missing information or latent errors exist.

11.3.4 Maintaining a Cautious Perspective

Aware of the complexity introduced by potentially adverse weather at our destination, we should adopt a strategy to always check for an alternate. We can't ever know when our dispatcher may have become distracted or task-overloaded, so the only prudent strategy is to *always assume* that there might be omissions or errors. Just like assembling a puzzle with missing pieces, we alter our perspective, assumptions, strategy, and decision making to detect the gaps.

This cautious or skeptical perspective isn't embraced by all pilots. Comfortable pilots sometimes allow their level of diligence to drift. When our dispatchers always get it right, it feels unnecessary to check for errors. We may rationalize that the dispatcher's software will flag and correct any errors. Rechecking their work feels like a waste of time.

The challenge becomes tougher when we consider all of the possible errors that can be made by our dispatcher or by us reviewing the flight release. Some of them include signing a release for the wrong flight, not noticing a wrong aircraft designation, missing a required takeoff alternate, missing a required destination alternate, missing a required second alternate, missing that ATOG was planned based on a closed runway, planning cruise at an unusable altitude, inaccurately planning weights of passengers/freight/fuel, and inaccurately planning destination airport conditions. This is a list of some errors with the dispatch release. Station oversights can include improper fuel uploads, HAZMAT paperwork errors, passenger load errors, jumpseater errors, expired weather packets, and incorrect manifests. Following is a report from a dispatcher documenting multiple workplace challenges contributing to an error on a flight release.

BOX 11.2 TASK OVERLOAD RESULTS IN ISSUING A FLAWED FLIGHT RELEASE

Dispatcher report: *[I] missed applying an MEL with a fuel penalty.* Corrected the mistake and resent the [release] for the Captain to sign. I got to work 10 minutes early for an [early morning] shift. The license had expired for the Jeppesen chart viewer on my computer. I called the help desk. After trouble shooting for 10 minutes, I was told to find another machine. There were no more licenses available. My workload was 20 fights vs. a normal plan of 18. Seventeen of my flights were scheduled out in a 1-hour period. I had a major snow storm out over the Plains and a frequency outage in the vicinity of ZZZ. I had to build around ZZZ and had a couple of transcontinental flights for good measure. After the computer delay of 30 minutes I went to work at XA:30 looking at weather and MELs and contacting ramps with restrictions. I have long advocated for a maximum of 18 flights on [this] desk. … When coupled with the compression (numerous departures in a short time period) it only makes these types of mistakes more likely. *Today, it caused me to rush and miss a MEL item. I feel like I was set up to fail when I walked through the door. An inoperative computer system, excessive number of flights and a tight compression of departures.* I realize that responsibility rests on me for my work, but these issues continue to be ignored time and time again. … This particular flight had two MELs. The first was a broken lock and the other was a fuel penalty. I looked at the lock MEL first, called the ramp with the restrictions and then failed to go back and account for the fuel penalty.[4]

Consider the Captain who received this flawed flight release. It makes sense that a thorough and prudent pilot would *always* assume that there are some gaps and hidden errors waiting to be discovered. Unfortunately, years of successful flying and habituation lead a slow drift in our level of diligence. First, we may start scanning over the NOTAMS more quickly, then we may drop a verification step or two, and finally, we may just sign the release and hand it back to the operations agent without ever looking at it. As Master Class pilots, we maintain an appropriate level of diligence and guard against this kind of drift. We hold a cautious or skeptical perspective because some errors can always slip through.

11.3.5 Assume that There Are Gaps and Counterfactuals

Our third puzzle case involved both missing and extra pieces. In aviation, the extra pieces are counterfactuals that contradict our SA or the trajectory of our game plan. When something happens that we aren't expecting, it is like discovering extra pieces that don't fit in within the image we have in our minds.

When we detect a counterfactual, we have two possible strategies that significantly affect our decision making. First, we can choose to ignore it because it is an anomaly, an effect from another condition that we are already aware of, or that it is inconsequential. Second, we can choose to accept it, alter our game plan, and rebuild our SA to account for the new information.

We should use a strategy that *expects* to find errors and counterfactuals. We can't assume that everything is correct and that all of the right pieces are present. Instead, our strategy begins with the information we know and can confirm – those clear portions of the puzzle. These intact portions form the framework for building our understanding. Next, we form ideas about what the gaps might be hiding. Klein suggests that these ideas arise from our experience and assumptions.[5] Some useful questions are:

- What do we know and how confident are we about it? Use this to construct a framework.
- Where are the gaps in our picture? Can we fill them in by seeking more information? This guides our investigation.
- Do the gaps mean that something important is missing or hidden? If we should be seeing something, why is it missing? This guides our search for meaning.

The same goes for counterfactuals.

- What does this unexpected indication mean?
- What is causing it?
- Is it potentially consequential?
- Does it indicate an influential condition that we didn't anticipate or accommodate within our plan?
- Do we need to alter or abandon our game plan?

- How can we rebuild our SA to integrate this new information?
- Do we need to change our flight trajectory to make extra time to investigate further?

As Master Class pilots, we need to anticipate for the gaps (errors and oversights) and for the extra pieces (counterfactuals).

11.3.6 The Risk of Ignoring Counterfactuals

The worst option is to just ignore the counterfactuals. Some of us make this error, probably because gaps make us feel uncomfortable. To restore our comfort zone, we either discard missing information (conclude that it is not important or assume that some other factor compensates for it) or hope that it will all work out (safety margins or systemic resilience will protect us against failure). Recall the story of the person in their comfortable chair who became frustrated with the annoying fly. Removing the annoyance became more important than investigating the cause or solving the problem. Counterfactuals indicate rising complexity. Complexity indicates rising risk. Both complexity and risk are important, so we pay attention to counterfactuals. Even if we ultimately conclude that they aren't significant, we still need to acknowledge that they increase uncertainty.

11.4 THE DIFFERENCE BETWEEN QUICK/COMMON DECISIONS AND REASONED/UNCOMMON DECISIONS

Figure 11.1 represents the frequency of decisions-making events from quick, familiar decisions to well-reasoned, unfamiliar decisions. Quick, familiar decisions are plentiful and instinctive. We make them all the time, so they are somewhat habitual and automatic. Moving to the right on our graph, events requiring well-reasoned, unfamiliar decisions tend to be fairly rare. Our decision-making process changes significantly between the two extremes. Problems emerge when we try to apply our quick, familiar decision-making process with unfamiliar events. Consider what happens when we don't.

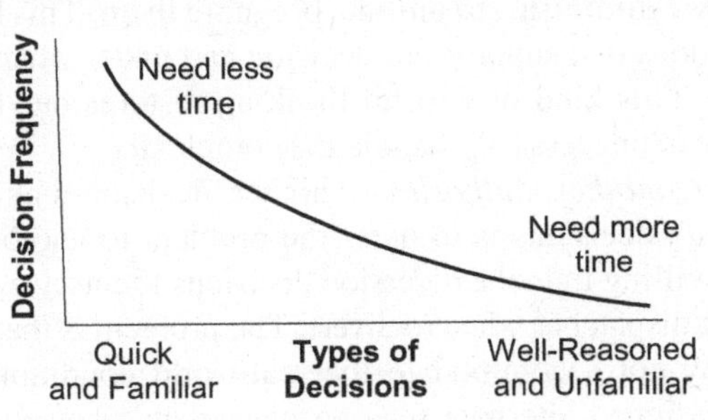

FIGURE 11.1 Decision frequency versus types of decisions.

11.4.1 Problems with Making Inappropriately Quick Decisions

Vulnerabilities surface when we our use quick decision making process for complex situations. Should we continue holding or divert to our alternate? Should we deviate to the left or right around this thunderstorm? Should we continue to wait in the line during a departure delay or return to the gate for more fuel? These are decisions that require a different decision-making process than quick, familiar decisions. This is because quick decisions rely on *matching*, while well-reasoned decisions require *assessing* and *balancing*. If we are short of fuel at the gate, the decision is simple – call the refueler back. If we are stuck in a long line of aircraft for takeoff and it looks like we may fall short of our minimum takeoff fuel, the decision is not simple. Should we shut down an engine? How fast is the line moving? Will we be cleared for takeoff before reaching minimum takeoff fuel? Can we amend our release to reduce our minimum required takeoff fuel? If we continue and then our fuel falls too short, the delay we incur will be much longer than if we go back to the gate early. Decisions become more difficult and consequences become more severe. What should we do? When should we do it? How should we do it? If we choose a particular course of action, how do we recover from the consequences? These are tough questions that balance advantages and disadvantages. Ideally, we take the time to acquire information, consult with others, weigh options, choose a game plan, and coordinate it with our team. For a variety of reasons, this doesn't always happen.

- *Misidentifying threats:* If we see a problem as simple and familiar, it needs to actually be simple and familiar. Ignoring the complexity and applying our matching process, we may end up forcing a familiar game plan into a situation where it doesn't fit. It's like trying to complete a job using the wrong tool. Misidentifying the problem leads us to either stop our investigation before detecting the true underlying conditions or it biases our attention toward favorable indications while ignoring unfavorable ones.
- *Oversimplifying solutions:* Complexity makes us feel uncomfortable. We don't like it. Pilots who rely on their quick match-and-choose process tend to favor shortcuts. By selectively choosing assumptions that match their shortcut, problems become easier to process. Instead of addressing important factors, we diminish, streamline, or ignore them. This biases us to over-weigh conditions that support our decision and under-weigh conditions that contradict it. This kind of wishful thinking restores our familiar comfort zone, but doesn't necessarily handle the complexity.
- *Deferring to another authority:* Another decision-making strategy for avoiding tough decisions is to defer the problem to another authority. For example, unwilling to make diversion decisions themselves, some Captains will ask their dispatcher when to divert. The problem is that the dispatcher's selection may not accommodate the real-world conditions affecting that flight. The assigned alternate may be inaccessible beyond a line of thunderstorms that blocks the route. Also, software-computed diversion fuel assumes that we are the only aircraft diverting, that we can begin our diversion without delay, that there will be no delays climbing to optimal cruise

altitude, that ATC won't descend us early for approach at the diversion airport, or that we won't get additional delays. More likely, we will experience some delays. This is why pilots usually compute higher minimum divert fuels than the software.

Deferring decision making to another authority also transfers responsibility for undesirable consequences. Depending on the company's procedures, this may be appropriate. Many passenger issues are best handled by station personnel who are trained in regulatory requirements. They also know what local resources are available. Knowing when to intervene and when to let others handle a problem is an important Master Class skill. One useful standard is that problems should be handled by people who are *trained and certified* to handle them. If we aren't trained and certified to handle a particular problem, we should refer it to someone who is. If we want to help, we could actively monitor the process while the trained and certified people work. If a process starts bogging down, we can call for additional support from those higher levels rather than directly intervening.

- *Procrastinating:* Pilots unwilling to make tough, reasoned decisions can choose to wait and hope that the situation will become clearer, that it will settle on a particular course, or that it will go away. This follows "if you stall long enough, most problems work themselves out" logic. Procrastinating pilots find themselves torn between selecting a decision and the consequences that might evolve if they choose it. Vacillating between the options can lead to analysis paralysis. To become skill decision makers, we need to learn as much as we can about the underlying processes surrounding our operation. This will help us discern whether the resolution process is working, bogging down, or stalled. Knowing when to leave it alone, when to give it a nudge, and when to intervene sets Master Class pilots apart.

11.4.2 The Precedent Set by False Success

When we select an unwise decision, but the situation still works out, it sets a bad precedent. It creates a false appearance that the decision was right, or at least that it was good enough. Even though that game plan succeeded, risky vulnerabilities remain. If that same decision is used again in the future, it might not go as well.

Most times, mistakes such as misidentification, simplification, deferring to others, and procrastination end up working out. This is mostly due to the resilience of the operation, not because the decision contributed to their success. Falsely attributed success hides the risk and gives that decision a thin veneer of validity. "All's well that ends well." Imagine a pilot who repeatedly follows this flawed decision-making process. Even if they know that it is flawed, they could conclude, "It's easier, it works, and nothing bad ever happens, so I'll keep using it." The increased risk becomes acceptable – even normalized. Every time they reuse the flawed decision, they inch closer toward the line of failure. Eventually, they cross the line and something bad happens. The vast majority of mishaps that I investigated fell into this category – good pilots who slowly drifted toward riskier choices and either remained unaware of the risk or assumed that it was manageable. Every time, luck and resilience worked to produce successful outcomes. Then that one time, it didn't.

11.4.3 Through Experience, the Complex Becomes Easy

As we acquire experience, even complex problems can become familiar match-and-choose situations. The first time we experience a tight fuel situation during a taxi-out delay, we'll find it unfamiliar, stressful, complex, and difficult to resolve. Torn between the options and potential consequences, we struggle. Even after choosing, we may second-guess ourselves. Afterward, we'll talk with more experienced pilots and compare strategies for handling that problem. We'll add their advice and experience to ours. By the tenth time we encounter this same kind of problem, we'll have amassed a full repertoire of options for what worked and what didn't. We'll know which relevant conditions to look for and have a better sense of time and pacing. We'll have useful mental models to evaluate the situation quickly. So, while the inexperienced pilot may see the complex problem as one requiring a well-reasoned solution, we'll recognize it as a familiar match-and-choose situation.

In this way, complex situations can eventually become familiar. We are careful not to confuse the difference between complex problems that we clearly recognize and complex problems that are new to us. The more complex the problem, the more unforeseen outcomes can emerge. So, while an observer might conclude that a Master Class pilot quickly matched the situation to their familiar game plan, they might miss the deeper analysis that the pilot applied. The difference is that while Master Class pilots use familiar solutions from their vast experience, they take the extra steps to communicate their assumptions, assign roles in monitoring for counterfactuals, and set clear trigger points for abandoning the game plan.

11.5 IDENTIFYING COMPLEX PROBLEMS

Now that we have examined some of the cases and strategies where complex problems are inappropriately treated like familiar, simple situations, we will focus on how to identify and solve complex problems. Consider the following list of characteristics that imply complexity.

- Something happens that is new to us.
- It feels important.
- It doesn't match the current game plan.
- There are significant consequences if the wrong choice is made.
- It falls outside of what we expected to happen.
- It involves coordination with crewmembers or outside agencies.
- It presents unclear or uncertain outcomes.
- It adds more complexity to a situation.

11.5.1 Recognizing Wrongness

While this list seems wide-ranging, all of the points share one common characteristic – they describe events or characteristics that fall outside of our future SA prediction. Since they don't match what we expect to see, they feel wrong. Remember that we continuously envision a story of how we expect the flight to unfold. This story is like a movie formed from our experience of how our game plan progressed in past flights and how we adjusted it to fit with conditions. If we were to freeze our aircraft like we can in a simulator, we could describe how we would expect the flight to progress from that point on. When we detect events that fall outside of our story, they feel wrong. We can classify wrongness in three categories.

1. *Small blips:* We expect to see many subtle variations. They contain a small amount of wrongness. We either dismiss or correct them.
2. *Events requiring quick corrections:* As we ratchet up the intensity of these variations, events start to clearly feel wrong. They still remain within the safety margin of our game plan, so they shouldn't generate adverse consequences. We only need to make quick, corrective decisions to preserve our game plan.
3. *Significant events requiring reasoned decision making:* Significant events feel very wrong. These events make us sit up and pay attention. Their complexity requires us to apply well-reasoned decision making. It feels like our current game plan is failing, so we need to modify or replace it.

11.5.2 Recognizing Familiarity

How we use RPDM changes with how well we recognize our situation (Figure 11.2). On the left end of the horizontal axis of Figure 11.2 are familiar situations (recognized patterns) which we match with previously validated game plans (action scripts assessed through mental simulations).[6] These are "Yes, I recognize it" and "Yes, I have a game plan" situations. As we gain experience, these yes-yes situations become most

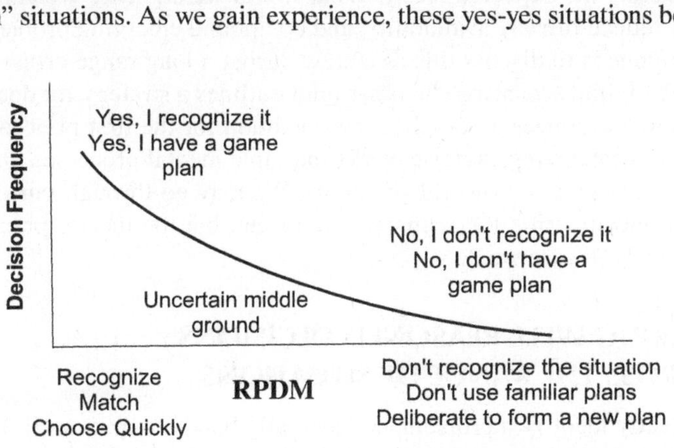

FIGURE 11.2 Decision frequency versus how we apply the RPDM process.

common. They also gain context and depth as we incorporate more encounters. The information playbooks in our memory become thick with ready-to-go game plans.

On the right end are situations that don't match familiar scripts. They will require well-reasoned innovation to form viable game plans. We'll call this the "No, I don't recognize it" and "No, I don't have a game plan" side. These no-no situations are clearly different from past events we have experienced. We realize that there are no game plans available in our memory playbook. Solutions will require modification of existing plans or entirely new game plans. Yes-yes situations comprise the vast majority of events that we encounter. Effectively matching indications with proven game plans works well. Gaining experience steepens the curve as more situations become familiar and quickly handled. The no-no side reflects rare situations that we haven't encountered before. The good news is that we are skillful problem solvers. Once we recognize the situation as something new, we guide our decision making to innovate viable game plans.

11.5.3 The Uncertain Middle Ground

We are skillful at matching our yes-yes situations and skillful at innovating our no-no situations. In the middle, however, our strategy is mixed. Here, we have situations that appear somewhat like familiar ones. Our natural bias is to modify familiar game plans before innovating new ones. Decisions toward the no-no end are inherently more complex and riskier, but forming innovative solutions is also risky. How we handle this uncertain middle ground requires a balanced solution. Again, we are generally quite good at this as long as we take the time to recognize the conditions at play and have enough time to follow an effective risk management process.

11.5.4 Practicing Scenarios in the Uncertain Middle Ground

Practicing middle-ground scenarios help us exercise our decision-making skills. Start with a typical non-normal scenario like one from a QRH manual. Next, complicate it with conditions that create unexpected twists. Imagine that a directed step does not generate the expected result. What would we do next? We can complicate the weather, reduce runway availability, and compound emerging problems.

One technique is to discuss this as a crew during a long-range cruise phase. One pilot poses the initial scenario. The other pilot outlines a strategy for dealing with it. That pilot now introduces a complicating condition for the first pilot to solve. This analyzing and strategizing exercise works the same mental processes that we would need to solve complex, real-world situations. We may go through our entire career without ever encountering the scenarios we invent, but the mental practice ensures that we are ready in case we do.

11.6 Making Well-Reasoned Decisions Across the Range of Situations

When we clearly identify a problem, we generally handle it quite well. It is rare that experienced pilots making well-reasoned decisions still experience undesirable outcomes. Let's expand our understanding of decision making using the continuum we introduced in Chapter 4.

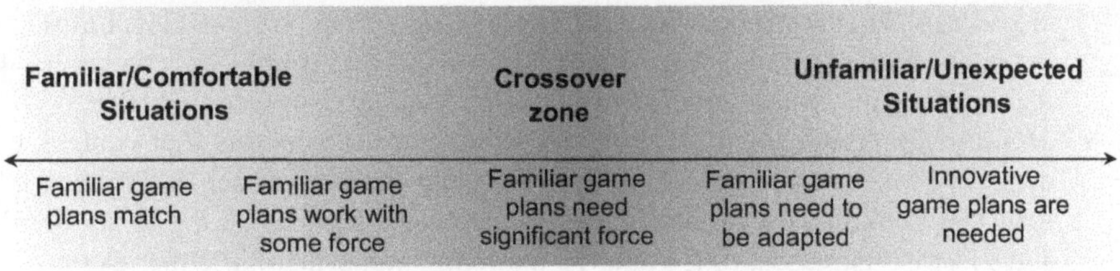

FIGURE 11.3 Game plan continuum.

11.6.1 Situations to the Left Side of the Game Plan Decision Graph

Along the left side of our continuum (Figure 11.3), we have familiar situations that closely match past game plans. The matched game plans guide our decision making. As we move to the right, situations become more complicated and uncertain. Consider a situation that requires some thought, but in many ways, still aligns with familiar situations we have successfully handled in the past. We recognize that the additional complexity increases uncertainty. How should we proceed with modifying our familiar game plan?

1. Identify the complicating factors and determine how they affect our choices.
2. As a starting point, select a familiar game plan that seems viable or workable.
3. Modify that familiar game plan and predict how it will flow.
4. Monitor for counterfactuals.

Let's expand our discussion of these four steps.

- *Identify the complicating factors:* Before launching in, we need to determine what is making this situation different. If we understand the source of complexity, we gain insight into potential conflicts down the road. Traps form when we ignore these factors, force our familiar game plan, and hope for the best. The forced choice assumes that our game plan is resilient enough to handle any adverse consequences. We convince ourselves that we just need to push a little harder and that our game plan is "good enough" to succeed.
- *Select a familiar, workable game plan:* RPDM asserts that we tend to select the first viable or workable option that pops into our head. Trusting our intuition, this game plan is a good starting point.
- *Modify a familiar game plan and predict how it will flow:* We make modifications to deal with specific challenges. Since the modified game plan is somewhat new for us, we monitor it more closely. We pay special attention to the differences between our familiar game plan and the modified one. For example, if we planned to cruise at FL410, but discover that the only smooth rides exist below FL250, we descend. The main complications are fuel burn and flight time. As long as our fuel margin allows us to arrive with an acceptable reserve, this remains a viable choice.

  Once we modify the game plan, we should revisit the decision whenever conditions change. Using our last example (FL410 as the originally planned cruise altitude), many pilots descend for better rides but don't consider climbing back up after passing the turbulent air mass. If fuel is tight, we can modify the game plan a second time by climbing back up to reduce fuel burn.

- *Monitor for counterfactuals:* Counterfactuals are indications that contradict what we expect to see. If we are flying fast on an approach and reduce thrust, we would expect our airspeed to decrease. If, instead, the airspeed stays fast or increases, this would be a strong counterfactual. Other examples of counterfactuals are:

    - Expecting a visual approach, but not seeing the runway due to low visibility
    - Expecting a normal fuel burn, but discovering a significant overburn
    - Expecting a normal arrival flow, but ATC assigns holding with a long EFC delay
    - Expecting vectors to a visual approach, but ATC assigns a complex RNP procedure

    With any modified game plan, we need to be especially wary of signs that it isn't working. We saw this with Clem's unstabilized approach. There were clear counterfactuals that his approach was failing, but he became so focused on forcing it to work that he didn't recognize them. Ideally, our search for counterfactuals increases as we move toward the right on Figure 11.3.

11.6.2 Situations in the Middle of Our Game Plan Continuum

As we move toward the center of Figure 11.3, situations move further away from familiar cases and closer to novel and unique cases. Somewhere in this transition, we lose the option of selecting and modifying familiar game plans. Solutions become more unique and innovative. Our situation becomes increasingly uncertain. Take an example of an airport that closes due to thunderstorm activity while we are still an hour out. Monitoring weather updates and consulting with our dispatcher, we learn that everyone is being assigned holding with lengthy EFC times. We could continue with our current plan, proceed at our planned cruising speed, and join the holding stack. Alternatively, we could innovate a different solution. We ask ATC to slow our cruise speed and remain as high as possible for the remainder of our cruise segment. This saves fuel that might make the difference between landing at our intended destination or diverting. A prudent strategy would be to prepare and brief two separate game plans – one for holding followed by an approach and another for holding followed by a diversion. We plan, build, and brief both options and carefully highlight which counterfactuals are important for each.

11.6.3 Situations to the Far Right of Our Game Plan Continuum

Situations to the far right of our graph comprise a small number of exceptional scenarios. These events may be time-critical or drawn-out.

- *Time-critical situations:* Two time-critical examples are US Airways 1549 (birdstrike and dual engine failure - Hudson River landing)[7] and Swissair 111 (inflight electrical fire accident off the coast of Nova Scotia).[8] In both of these cases, the crews had little time to diagnose their problem and innovate a viable game plan. With US Airways 1549, the crew concluded that they could not reach a landing runway. Instead, they chose to ditch on the Hudson River. With Swissair 111, the crew detected the fire and followed established procedures, but were incapacitated before reaching the airport.

The US Airways 1549 crew followed two game plans simultaneously. The primary game plan was to get an engine restarted and land back at LaGuardia (LGA). The Captain gave the FO the task of restarting the engines while he held best glide speed and coordinated with the cabin crew to prepare the passengers. When it became clear that they couldn't make it to any runway, the Captain prepared a backup game plan of landing on the river. Imagine what the Captain saw through his windscreen – tall buildings everywhere except for a clear expanse down the Hudson River. Holding best glide speed while the FO worked to get an engine started, the river must have felt like the only "runway" available. When it became clear that neither engine would restart, they abandoned the restart option and worked together to ditch the aircraft.

From an HF perspective, their decision making, workload management, timing, and execution were commendable. Following the birdstrikes, they quickly diagnosed the problem and began the trained procedure for restarting the engines while maneuvering back toward the runway. Optimizing workload management, each pilot was fully engaged with the tasks that they were best suited to handle. At some point, the Captain made two critical decisions. First, he decided that he couldn't maneuver back to LGA for landing. Extensive post-accident investigation concluded that they could have made it back to the runway, but it would have taken an immediate turn back and optimal energy management. Anything less would have resulted in failure. Imagine the catastrophic consequences if they had chosen that fragile game plan and crashed into New York City. Choosing loss of airframe over loss of life is a good example of consequence management. The risk of failure was too high compared with the low chance of success. Second, he discerned that they needed a backup game plan in case the FO couldn't get an engine started. Imagine if the Captain refused to consider this option and instead doggedly pursued stretching an unsuccessful glide back toward LGA.

Another HF aspect was how well the crew optimized both game plans simultaneously. The Captain maneuvered down the river to preserve the ditch option while positioning the aircraft for landing at LGA in case an engine restarted. Imagine if he had biased his flightpath back toward LGA, then at the last minute, switched to attempt the Hudson River landing. He might not have had sufficient time and maneuvering space to succeed. Another HF aspect is timing. At some point, the crew made the decision to abandon restart attempts and redirect the FO's workload toward preparing for a water landing. Few airlines taught ditching procedures with this kind of scenario. Most ditching practice assumed an ocean ditching following all-engine failure from high-altitude cruise. In these practice scenarios, crews had a long preparation time while slowly gliding down. This crew had scant minutes to prepare and execute their plan. If the Captain had delayed his decision to make the water landing, the FO and cabin crew wouldn't have had enough time to prepare.

Each pilot's task execution was praiseworthy. Accident investigators commended the speed and accuracy of the FO's restart attempts and his water landing preparation. The pilots' coordination with the cabin crew, coordination with ATC, execution of the water landing, and attention to detail during egress were commendable. While most of society only remembers the water landing, we should learn from the crew's decision making, workload management, timing, and execution.

Comparing the US Airways 1549 and Swissair 111 crews, the US Airways 1549 crew had a clearer choice since they had no thrust and only the Hudson River to land on. The Swissair 111 crew had an unclear situation and conflicting choices. Let's place ourselves in their seats to understand their dilemma. They started with a "smoke and fumes" situation. A search of the NASA ASRS database for the 20-year period between January 2000 and January 2020 yielded over 1,500 reports for "fumes", over 3,200 for "smoke", and over 1,600 for "odor". Considering that many of these reports overlap, we estimate at least ten reported events per month across the U.S. airline industry. Over these 20 years, none resulted in a crash or loss of life. In fact, many resolved themselves before landing. This trend creates a collective mindset that "smoke and fume" events are both fairly common and rarely serious. If faced with one of these events, we estimate that we'll have plenty of time to run the checklist, resolve the problem, and land safely. Assuming that the Swissair pilots shared this mindset, it is understandable that they followed checklist procedures, turned toward a landing airport, and started dumping fuel. In fact, this is exactly what was taught across the industry at the time. Not knowing the location of the fire or how quickly it was spreading, they did exactly what they were supposed to do – exactly what any of us would have done. In their case, the fire spread very rapidly. Only 3 minutes elapsed between their "PAN, PAN, PAN" call and catastrophic systems failures. This situation went from concerning, to serious, to dire in minutes. Because of the lessons-learned from this accident, we now teach to immediately land any aircraft with an unresolved smoke and fumes situation.

Time-sensitive, critical situations give us little time to gather information and weigh options. In both of the previous examples, we see the benefit of preparing both a primary option that we intend to use and a backup option to use in case the first one deteriorates. In both cases, the end goal was clear, to land the aircraft safely at the nearest airport. In both, that goal wasn't achievable. Consider these decision-making considerations for time-critical situations.

- o Maneuver the aircraft toward the main goal.
- o Preserve a backup option if the main goal becomes unworkable.
- o If the situation deteriorates, assume the worst, and choose the safest option.
- o Set a decision point for abandoning the primary option and devoting full effort toward the backup.
- o Divide workload appropriately.

- *Possibly critical – non-time sensitive situations:* In both of the previous examples, the severity of the problem was immediately apparent. Most situations, however, are more nuanced and confusing. Consider a subtle smoke and fumes scenario. The only fumes case I personally encountered involved what was probably a discharge of a passenger's personal pepper-spray device into an overhead bin. The cabin crew reported localized odor and a burning sensation with no smoke. The odor quickly dissipated, so we continued to destination.

Smoke and fume events may begin with the cabin crew reporting, "something smells funny". Our mindset biases us to suspect galley sources – coffee makers, food heaters, smoking in the lavatory, and such. We start by determining the seriousness of the situation. Is it just an odor or something more? What does it smell like? Where is it coming from? Does it smell like something burning? Is there smoke? If there are visible flames or sparking, the decision is clear. If not, we continue investigating while preparing our two game plans. One game plan is to continue monitoring the situation with the expectation that it will dissipate. The second game plan is to begin preparing for an expeditious descent and emergency landing with possible evacuation. Again, dividing the workload proves useful. Since the main decision rests with Captain, they should focus on communicating with the cabin crew while the FO flies the aircraft and begins surveying landing options.

- *Novel situation with plenty of time:* These are cases at the far right end of our graph. Consider the JetBlue Airways 292 (LAX landing with the nosewheel stuck 90° off) mishap.[9] After takeoff from Burbank (BUR), the crew received a nose gear shock absorber fault code on their monitoring display. Lacking any additional indications, the crew continued on course while the Captain consulted the Flight Crew Operating Manual (FCOM) and with company maintenance technicians. There was a note that this fault may indicate that the nose gear "may be caught at 90 degrees". Here, the Captain's choices were to continue to destination (JFK) or divert. The Captain elected to divert to Long Beach (LGB). They flew down the runway while ATC visually checked the nose wheel. The tower controllers confirmed that the nose wheel was, indeed, stuck 90 degrees off. It was clear at this point that they would have to land with this misaligned nose wheel assembly. All of the crew's choices from this point on were focused on limiting the undesirable consequences following a collapse of the nose wheel assembly during landing roll-out.

Given ample time with a situation like this, our priority is to minimize undesirable outcomes. The crew's first choice was to select a landing runway. They elected to divert to LAX for the additional runway length and superior emergency response capability. Next, they determined the most favorable landing configuration. They chose to fly for several hours to burn fuel and land light-weight. They ensured that the cabin crew and passengers were well informed of the situation and game plan. They thoroughly discussed their landing technique and crew roles for various contingencies. Additional mitigation steps included using normal braking, holding the nose wheel off for as long as possible, shutting down the engines during roll-out, and disabling ground spoilers, reversers, and autobrakes. After touchdown, the crew could smell burning rubber from the scrubbing tires. Tower assisted by confirming no fire, which the Captain relayed to the cabin crew. They stopped on the runway and the passengers deplaned safely using air stairs. Several features of novel situations that allow plenty of time include:

- Ample time to investigate the cause of the malfunction
- Time to coordinate with experts
- Time to consider which option maximizes success and minimizes potentially adverse consequences
- Extensive briefing of roles and responsibilities with all involved agencies
- Rehearsal of exceptional procedures and techniques to maximize success

- *Trained situations with plenty of time:* Another category involves situations we train for, but rarely see in line flying. Simulator training offers excellent practice at handling emergencies, but not necessarily the associated of real-world details. For example, for an engine failure during takeoff, the procedure directs us to initially climb straight out (unless a turn is required for terrain). The intention is to optimize climb performance until reaching a safe altitude. We then accelerate and raise flaps. Our simulator instructor often prompts our timing for a turn to downwind. With an actual engine failure in the aircraft, ATC might not be as helpful. They might assume that we wish to continue straight out until we ask for a turn. In one event, a crew experienced an engine failure departing to the west from the California coast. The crew remained on runway heading while diagnosing their problem and running all of their checklists. They didn't request a turn back toward the airport until they were many miles out over the Pacific Ocean. It took them a long time to return to the airport.

  Another interesting example is where we choose to hold when an airport suddenly closes due to thunderstorm activity. Often, ATC asks close-in aircraft where they wish to hold. Most pilots choose holding patterns on an extended final. This places them in the best position for landing assuming that the airport reopens using the same runway. Very often, however, wind shifts following a storm favor the opposite direction runway. So, their holding pattern proves unhelpful. Another option is to hold in airspace between our destination and our planned alternate. This way, we balance both the option of landing at our intended destination and diverting to our alternate. Another useful option is holding upwind from the movement of the thunderstorm. This allows us the quickest access to all runways as the storm moves off.

  The bottom line is that simulator scenarios often disguise the complexities of real-world flying. Master Class pilots mentally rehearse simulator-type scenarios within line-flying conditions. Consider the example of the thunderstorm on the airport. As we are preparing for arrival, we could anticipate the adverse effects of isolated storms in the area, note the speed and direction of their movement, anticipate the need to enter close-in holding, and have a plan ready to go if the airport unexpectedly closes.

11.7 MASTER CLASS DECISION-MAKING PRACTICES

Following are some ideas to guide our decision-making process.

11.7.1 Choosing Instead of Reacting

Our instinctive reactions to unfamiliar situations can be strongly influenced by our mental biases. Often, we gravitate toward familiar and recent game plans that may not be appropriate for the conditions. When those game plans meet resistance, we tend to force them to get them to work. Ideally, we should interrupt our biased response, pause long enough to accurately interpret the situation, select a viable game plan, execute it, and then continually assess it to ensure that it remains viable.

- *Recognizing our biases:* We need to accurately recognize two important aspects about our situation. First, we need to dispassionately and honestly assess our status. If we mismanaged the approach, we must honestly accept that we have mismanaged the approach. This helps us switch to a safer, but undesirable, alternate game plan – in this case, going around and trying again. Second, should guard against our personal biases, whether optimistic or pessimistic. We learn from past events that didn't work out as we expected to improve our discernment. For example, consider an event when we stayed too long in a holding pattern hoping that an airport would reopen. Unfortunately, it remained closed and we ended up diverting with a low fuel reserve. As we reflect back on the event, we realize that all of the negative warning signs were present, but that we biased our expectations toward the positive signs. From this, we might conclude that we have an optimism bias – that we tend to favor an optimistic mindset over a pessimistic one. Master Class pilots aren't free from bias, but we strive to become aware of our personal tendencies.
- *Interrupting our natural biases:* If we don't recognize our biases, we revert to our habits. This is the path of least resistance that our minds naturally want to follow. To break this tendency, we need to interrupt our biased mental processes and construct tripwires to flag when these kinds of events reemerge. For example, aware that the first options that usually pop into our heads are overly optimistic ones, we resolve that we need to pause, mentally step back, and scan all of the information more skeptically. The reverse holds for pilots with a skeptical bias. This encourages us to build backup option trigger points. "We can continue in holding until our fuel reaches 10,000, then we will divert." This way, if events begin to quicken and stress rises, the trigger point will interrupt our biased mindset and redirect us toward a safer game plan.

11.7.2 The Dos and Don'ts of Using Our Intuition

As our need for intuition rises, the gaps and counterfactuals increase. Gary Klein, in his book, *The Power of Intuition*, offers a number of useful suggestions.[10]

- *Start with intuition, not with analysis:* Use intuition as the starting point. With the experience we have amassed over many years in aviation, we each have a finely tuned sense of intuition to guide our snap decisions. As those first intuitive ideas pop quickly into our minds, they will form the foundation for a workable game plan. Start with that intuitive idea, form an initial game plan, and see if it fits. We can still use our analytical process to evaluate its workability and guide our decision making.
- *Accept the zone of indifference:* Klein advises us to guard against obsessing over picking the best choice. We rarely know which is the best choice while we are immersed in a scenario. If we have two good solutions, it really

doesn't matter which one we pick. Klein calls this the zone of indifference. Instead, focus on the choice that seems easier to execute. A "pretty good" solution that is easier to implement is generally better than a potentially better one that might become mired in complications and details. If two options seem about the same, pick the one that our gut feeling tells us to use.

- *Use mental simulation to evaluate options and guide monitoring:* This is a natural process for pilots – mentally flying our game plan profiles. When we use mental simulation to perform a fly-off comparison between two competing ideas, we should conduct it honestly. Guard against attaching too much importance to the original goal. This can lead to plan continuation bias. When we apply too much weight to a particular outcome, we bias our detection of counterfactuals.
- *Monitor our simulation for rising complexity:* If a game plan is flawed, we may notice that our mental simulation becomes increasingly problematic. The more fragile the simulation, the more resistance we'll encounter when actually flying it. Notice the rising trend of complexity. Instead of forcing the solution against the headwind of complexity, look for other options or leverage points to uncover an easier path.
- *Simplify the analysis:* Guard against paralysis by analysis. If options seem essentially equal, choose the one that appears easier to implement. If they seem about the same to implement, choose the one that our intuition prefers. Don't simplify the conditions or challenges. Accept them as they are. Instead, find the natural path of least resistance toward a desirable outcome.
- *Know when to apply procedures and when to apply intuition:* This topic was initially covered under the Introduction to Techniques section, but is worth revisiting. Remember that procedures are written for the normative case – for the conditions that are most common and expected. As we stray further toward the unexpected and novel, procedures become less applicable. At first, we may need to adjust them to fit the unique combinations of existing conditions. At some point, however, the situation will become so unique that no procedures exist. Approaching this extreme, we need to innovate new procedures to satisfy the underlying policy intentions. Klein cautions us to listen to our intuition as it may guide us toward a much better conclusion than scripted procedures that no longer match the scenario.

11.8 USEFUL DECISION-MAKING TECHNIQUES

Following are some techniques to improve our decision-making skills.

11.8.1 Basic Steps for Deliberative Decision Making

As we move to the right along our decision-making continuum, we transition from making quick decisions to making deliberate decisions. Following are the basic strategies that guide unexpected event decision making.

1. Ensure the safe path of the aircraft. Maintain aircraft control.
2. Determine if we have the time available to make deliberate decisions.
3. If little time is available, create time by aborting the game plan and resetting the problem.
4. When plenty of time is available, follow the RRM process to assess, balance, communicate, do, and debrief. Repeat the process until an effective game plan returns us "in the Green".

11.8.2 The In All Situations Checklist

If we launch into solving a problem before we have investigated its cause, we may unknowingly commit to a flawed game plan. We need a powerful reminder to pause and assess. The In All Situations Checklist interrupts our reflex to immediately act. It is often published on the cover of our non-normal or abnormal checklists manuals. The following is known as MATM for the beginning initials.

- **M**aintain aircraft control.
- **A**nalyze the problem.
- **T**ake appropriate action.
- **M**aintain situational awareness.

None of these steps are actually procedural or specific. They guide a general strategy that applies to all unexpected events. They interrupt our automatic *reflex reaction* and direct us to follow a *reasoned response.* They remind us to stop (interrupt the reflex reaction), take care of the big priorities (control the aircraft), understand the situation (analyze it), build a viable game plan (take appropriate actions for the conditions), and maintain situational awareness (continue to monitor our game plan to make sure it remains effective and appropriate).

11.8.3 Examining Our Mindset

Our mental mindset is an instantaneous snapshot of our "attitudes, motivations, expectations, knowledge, feelings, plans, goals, and self-image" as they exist within the flow of events.[11] This list is both wide-ranging and incomplete, but it reflects the dynamic and complex nature of our mental mindset.

From mishap debriefs, we notice two different mindset perspectives. The first involves pilots who become consumed by the urgency of the moment and *react* to each newly discovered condition during a rapidly changing scenario. They react at an ever-increasing pace (event quickening). Afterward, these pilots generally can't recall their mindset. The fast pace of events pulls them swiftly along from one moment to the next like riding a kayak down challenging rapids. Much of what they recall is distorted by hindsight. Their hindsight perspective doesn't help us understand why they continued their ill-fated event. We hear comments like, "I don't know what I was thinking."

The second perspective of mindset comes from pilots who are able to maintain a mental detachment from rapidly changing events. Rather than being consumed *within* their event, they maintain a perspective *apart* from it. They recall more details. Their detachment allows them to manage their *response* to rapidly changing conditions. Unlike the first group of reacting pilots, these responding pilots recall what they were thinking at each critical decision point. They also recall where their attention was focused throughout the event. When pilots lose this detached perspective and dive into the unfolding scenario, their attention begins to tunnel in on singular parameters. They lose the big picture.

Master Class pilots cultivate a keen awareness of our own mindset, or more accurately, our ever-changing mental perspective. We incorporate it within our decision-making process. For example, when we feel tired, we recognize that we tend to exhibit plan continuation bias. Because of this, we know that we need to actively avoid risk. We slow the pace of events to accommodate our fatigue level. Conversely, when we are on top of our game, we can handle more risk by expanding our SA and planning. We sense when conditions change and take active steps to monitor for warning signs.

11.8.4 Add a Qualifier to the Game Plan

We know that any plan can fail unless we actively ensure its viability. One way to do this is by adding qualifiers to our decisions. These qualifiers (if, as long as, after, unless, and until) acknowledge that there are always interrelated and conditional events that must happen between the present moment and our goal. RPDM guides us with conducting a mental simulation of the expected flow of events. While that simulation guides our expectation, these qualifiers keep us grounded in reality and alert for warning signs.

- *We can do it IF:* The first qualifier is "if". This acknowledges conditions that we need to have before continuing the game plan. We can do this IF that happens. "We can safely accept this flight plan IF we add 1,500 pounds of fuel." "We can make a stabilized approach, IF we fly an S-turn to the North." This qualifier identifies the critical step that must occur to keep our plan viable. If the critical condition fails to occur, then the game plan fails and we switch to our backup plan.
- *We can do it AS LONG AS:* This qualifier acknowledges conditions that need to remain valid or constant throughout the scenario. A change in those conditions warrants a reevaluation or rejection of the game plan. "We can accept a higher altitude AS LONG AS the ride remains smooth." "We can accept that reroute AS LONG AS our headwind doesn't increase." This qualifier guides us to develop and update a backup plan. It creates a trigger criterion for switching to that backup game plan and keeps all crewmembers alert for counterfactuals.
- *We can do it AFTER:* This qualifier acknowledges preconditions that must be satisfied *before* we will attempt the game plan. It identifies what we (or some outside agency) must do before we can proceed. "We can use this flight release AFTER we amend it with a closer alternate." "We can address

that problem AFTER we complete this checklist." This qualifier establishes a STOP sign within our chain of events. It forces us to complete a critical task before continuing on to something else.

- *We can do it UNLESS:* This qualifier acknowledges warning flags that might invalidate a game plan. These are conditions that exclude or prevent our game plan from working. They force us to identify tripwire events that make the plan unusable. "We can continue to destination UNLESS they assign us holding, then we'll have to divert." "We can continue UNLESS that storm cell moves toward the airport." Once the tripwire springs, we immediately discard the original game plan and execute the contingency backup.
- *We can do it UNTIL:* This qualifier identifies the endpoint when the present course of action must change. Some common examples are weather holding and crew rest. "We can continue in holding UNTIL we have 7,500 pounds of fuel remaining, then we must divert." "We can monitor this delay UNTIL 2300Z, then our duty day will time-out." This qualifier protects us from moving the goal posts. It establishes a clear point in time where we will go no further.

NOTES

1 Klein (2003, p. 28). Applying RPDM terminology, these steps would be: cues lead to recognized patterns, patterns activate action scripts, action scripts are assessed through mental simulation, and mental simulation is driven by mental models.
2 Edited for clarity and brevity. Italics added. NASA ASRS report #1592804.
3 Most modern jet engines use takeoff thrust settings below maximum thrust to reduce engine wear and tear. This is accomplished by programming the FMS with an assumed temperature that is higher than the actual outside air temperature. The FMS displays a lower N1 thrust value. Pilots routinely accept the displayed reduction (computers are always right, aren't they?) and just check that all of the engine displays are parallel, regardless of whether the thrust values are appropriate.
4 Edited for brevity and clarity. Italics added. NASA ASRS report #1779705.
5 Klein (2003, pp. 52 and 124).
6 Klein (2003, p. 28). Words within the parentheses reflect Klein's terms.
7 (NTSB, NTSB/AAR-10/03 , 2010) – US Airways 1549 NTSB final report –January 15, 2009.
8 TSB (1999) – Swissair 111 final report –September 2, 1998.
9 (NTSB, LAX05IA312, 2006). JetBlue Airways 292 final report – September 21, 2005.
10 Klein (2003, pp. 80–84). Although not written exclusively about the aviation profession, this book offers insight into the use of intuition with decision making in complex, highly conditional, and dynamic occupations.
11 Wischmeyer (2005).

BIBLIOGRAPHY

ASRS – Aviation Safety and Reporting System – National Aeronautics and Space Administration (NASA) Online Database. (n.d.). Retrieved from Aviation Safety and Reporting System: https://asrs.arc.nasa.gov/search/database.html.

Klein, G. (2003). *The Power of Intuition: How to Use Your Gut Feelings to Make Better Decisions at Work.* New York, NY: Currency Books.

NTSB. (2006). *LAX05IA312: Jet Blue Airways flight 292 – Airbus A320, N536JB – nose wheels cocked 90 degrees –21 September, 2005*. Washington D.C.: National Transportation Safety Board.

NTSB. (2010). *NTSB/AAR-10/03: Loss of Thrust in Both Engines After Encountering a Flock of Birds and Subsequent Ditching on the Hudson River – US Airways Flight 1549 – Airbus A320-214, N106US – Weehawken, New Jersey – January 15, 2009*. Washington D.C.: National Transportation Safety Board.

TSB. (1999). *A98H0003: In-Flight Fire Leading to Collision with Water – Swissair Transport Limited – McDonnell Douglas MD-11 HB-IWF – Peggy's Cove, Nova Scotia 5 nm SW – 2 September 1998*. Gatineau: Transportation Safety Board of Canada.

Wischmeyer, E. (2005). Why'd They Do That? Analyzing Pilot Mindset in Accidents and Incidents. *Proceedings of the International Symposium on Aviation Psychology* (pp. 639–644). Dayton, OH: ISAP.

12 Techniques for Building Situational Awareness

Ideally, SA tracks smoothly from what we planned to happen (past SA), what is happening right now (present SA), and what we expect for the future (future SA). As we build and apply our SA, we placed high priorities on:

- *Knowing* what is currently happening and what is likely to happen
- *Monitoring* indications to quickly detect important events and changing conditions
- *Anticipating* future threats and preparing contingency options
- *Maintaining* resilience against disruptions
- *Recovering* from deviations from our game plan

Let's examine these five dimensions of SA by considering factors that degrade them and techniques that enhance them.

12.1 BUILDING THE SA GOAL OF *KNOWING*

Good SA confirms that we have built a good game plan and that we are guiding it toward a desirable outcome. The sequence of events flows smoothly and predictably. When disruptions occur, good SA helps us to quickly detect them, understand them, and resolve them.

12.1.1 Planning As If We Know What Is Going to Happen Ahead of Time

Consider a simulator LOFT scenario. We typically don't know what will happen. Imagine instead, that we are given the entire scenario ahead of time. We'll know which challenges and emergencies will happen, when they will occur, how they will unfold, and what past crews did well or poorly. With this perfect foresight, we can preselect our game plan, identify important parameters to monitor, prepare contingency options, and assign tasks, roles, and responsibilities. Knowing everything ahead of time would make building and maintaining our SA easy. Granted, this is an unrealistic scenario. When we fly, we don't know what will actually happen. Uncertainty is unavoidable. Despite this, nearly every airline flight is conducted successfully every day. We succeed because we employ many of the same *knowing* techniques in our daily flight preparation.

Our experience gives us a highly reliable expectation of what is likely to happen. We then plan and brief for those likely events and conditions. For unanticipated events, we prepare contingencies *just in case*. We review the RTO procedure as we take the runway, *just in case*. We position ourselves to respond to an engine failure during takeoff, *just in case*. Even if we believe that such events are highly unlikely,

DOI: 10.1201/9781003344575-14

our preparation reduces the startle effect, orients our visual scan to quickly make sense of the situation, and maintains our SA.

12.1.2 Planning for What Is Likely to Happen

While we cannot know what will happen before a flight, we can apply our past experience to infer what is likely to happen. Imagine flying between two particular cities a dozen times. Those 12 flights would give us experience across a range of typical profiles and events. After the second dozen flights, we may experience a few more unique events. By our third dozen flights, we may encounter only one additional unique event. In time, we will have gained such extensive experience that we become experts at flying that city pair. If someone asks us to teach them about flying that city pair, we'll have plenty of knowledge to share. We would *know* what typically happens.

The same goes for conditions. Our first snowy weather landing into Detroit Metro (DTW) would present a rich learning experience with lots of new details to learn. By our tenth encounter, we would feel well-experienced. We would know which events are likely to happen. Moreover, we could generalize that knowledge. What we learn during our snowy flights into DTW becomes highly useful for flights into any airport subject to similar lake-effect snow and visibility conditions. In time, we will have amassed a vast repertoire of experiences to draw upon. While we can never know exactly what will happen, we can effectively predict what is likely to happen. Experience creates knowing – not complete knowing, but close enough.

12.1.3 Reaching a Sufficient Level of Confidence with Our Level of Knowing

When faced with an unexpected event that we don't fully understand, we have two choices. We can either accept it and press forward or we can study it until we reach a sufficient level of confidence with our understanding. The first option is biased toward continuing the current game plan. The second option may lead us to change the game plan to make more time to diagnose the problem. Since we can't freeze our situation to study our problem to reach a perfect understanding, we need to balance between studying the problem and remaining on profile.

The desire to stay on profile is strong. Accident analysis reveals that crews sometimes aggravated their problem by trying to do both – staying on profile while analyzing it. While struggling to make sense of what was happening to them, they ran out of time, became frustrated, and abandoned their efforts at SA-building. While in-the-moment, moving forward in spite of their uncertainty felt like a better decision. As their SA shrank, their decision making became rushed and reactive.

Do we break off our profile to make more time to diagnose the situation, or do we accept our current level of knowing and press forward? The choice can be difficult. If the situation compels us to keep advancing or if we believe that we are sufficiently confident with our understanding, then we typically press forward. Our intuitive judgment helps us to determine how much is enough. Every time we encounter an unexpected situation, we study it to determine how to improve this discernment. Consider the following report of a crew that experienced a smoke and fumes situation.

BOX 12.1 CREW BUILDS THEIR SA DURING A SMOKE AND FUMES EVENT

FO's report: …The takeoff and departure were uneventful. Weather was clear with unrestricted visibility. The climb was continued until we leveled off at FL340. At approximately XA:13Z, I started to smell something. *It smelled like burning plastic. At the same time I noticed the smell the Captain asked, "Do you smell that?" Before replying, we both saw smoke coming from the "FL" button on the transponder located on the center console. At this time we both executed the items for "Smoke, Fire, or Fumes", donned our oxygen masks, and established crew communication. At that point the Captain gave me both aircraft control and the radios so he could focus on the [problem]. While I monitored the autopilot the Captain continued with the "Smoke, Fire, or Fumes" Checklist. I confirmed with him that the proper switches were being turned off in accordance with the checklist.* We noticed that the smoke and fumes were decreasing and did not need to run the "Smoke or Fumes Removal" Checklist. *While the Captain attempted to locate the transponder circuit breaker to isolate the problem, he noticed that the "ELEC PANEL LIGHTS AFT/CENTER" circuit breaker located at the base of the front of the center pedestal was popped. We were confident the problem had been isolated. The Captain informed me that he was about 90% sure we were going to divert but he wanted to check with company to see if we were missing anything. We quickly realized that a step in the checklist had us turn off the internet and he would be unable to use the crew phone app to communicate with the Dispatcher. The Captain then took the radios back and attempted to get phone patch with the Center Controller. The controller was unable to give us the phone patch but was able to relay information.* After communicating through the controller and using ACARS, it was decided that we would divert to ZZZ2. The controller asked if we want ARFF (aircraft rescue firefighting) and the Captain said "yes". The Captain then gave me back the radios so that he could run more checklists. The Captain continued with the checklists the controller asked what runway we wanted to land on. The Captain and I agreed that we wanted the longest runway due to the fact we would be landing overweight. I then requested Runway XXR. I got the airplane set up for an ILS approach to XXR. I started the descent into ZZZ2 while the Captain finished checklists and briefed the passengers and Flight Attendants. Then the Captain told me that he ran the "Non-Routine Landing Considerations" Checklist and that the passengers and flight attendants had been briefed. He talked about the heavyweight landing and did a landing assessment. He asked if I had anything to add or could think of anything else we might need to do before landing. I told him that I think we had covered everything and I had nothing else to add. He then briefed the approach and took aircraft control. We completed both the descent and before landing checklists. After landing we cleared the runway. I relayed to ARFF that unless they see anything abnormal from the outside, our plan was to continue taxiing to gate. They informed us that they did not see any unusual readings on the thermal scan. ARFF followed us to the gate. The Captain had me make a PA to the passengers that we would be taxiing to gate and they would be accommodated after deplaning. The taxi and engine shutdown at the gate was uneventful.[1]

Notice the process that they followed. Initially, they used very little time trying to understand or locate the source of the smoke. Instead, they performed their trained smoke and fumes response (boldface memory steps). With smoke dissipating, they had time to diagnose the problem. While performing the remaining checklist steps, they discovered the tripped circuit breaker. This increased their confidence level that the malfunctioning system was safely isolated. At this point, they had plenty of time to coordinate their diversion game plan. Notice that the Captain transferred all flying and ATC duties to the FO so they could coordinate with their cabin crew and company. This crew demonstrated excellent CRM.

12.2 BUILDING THE SA GOAL OF *MONITORING*

Monitoring techniques strongly influence how we build and expand our SA.

12.2.1 Active and Subconscious Monitoring

Effective monitoring relies on a range of techniques. Consider the task of monitoring engine parameters. We closely monitor them during engine start, but rarely at other times. Instead, we rely on automation to maintain steady parameters and alert us of exceedances. The more reliably that automation alerts us to out-of-tolerance conditions, the more likely we are to remove those tasks from our active monitoring. We still devote a small amount of attention toward monitoring for anomalies. Consider a situation where we are cruising along at FL 350 and our #2 engine begins to fluctuate. Automation wouldn't alert us, but we would still notice the anomaly. We might hear an unexpected sound or feel a subtle vibration. We might detect movement in the engine readouts in our peripheral vision. The point is that we maintain SA by applying both conscious and subconscious scan patterns to monitor for indications that fall outside of our normal expectations.

12.2.2 Expectation, Comfort, and Drift

As we gain experience, we become skillful at knowing when to look for an indication, what to expect to see, and what it means. If we become overly comfortable and confident, however, we might reduce the scope and quality of our monitoring. In effect, our monitoring skills drift toward confirming expectations and away from detecting anomalies.

BOX 12.2 NO "B"S, NO BRACKETS

I was on the jumpseat of an airline that used an onboard performance computer to determine landing data. It displayed approach reference speed, target speed, and computed rollout distance. Two particular exceptions were when the computed landing distance exceeded available distance for a planned autobrake setting (the distance block was bracketed) or if the planned landing exceeded brake cooling limits (a "B" was displayed). Both of these occurrences were somewhat rare. "Brackets" only appeared for short or contaminated runways (snow and ice) and "Bs" only appeared for high-altitude airports on warm days. On this particular day, the PM/Captain computed the information for the landing runway, showed it to the FO, and confidently declared, "No Bs, No Brackets". The FO looked at the Captain quizzically. It was clear that he didn't quite know what to say. Finally, he declared, "The runway has a 'B' for brake cooling."

So, what happened here? Perhaps stating, "No Bs, No Brackets" had become so commonplace that it became habitual with this Captain. Not expecting to see a "B", it became invisible. The Captain's expectation (future SA) overshadowed the quality of their monitoring.

12.2.3 Task Overload and Single-Parameter Fixation

The more overloaded we become, the more our SA and monitoring narrows. This is especially prominent with landing mishaps. Focused intently on the landing zone, mishap PFs can fail to notice important counterfactuals. They recall details about the landing zone and their efforts to align the aircraft, but completely miss major problems with sink rate (like a "PULL UP" warning), airspeed (that they were 30 knots fast), and configuration (that landing flaps weren't set).

BOX 12.3 NARROWED SCOPE OF MONITORING WHILE LEARNING TO USE THE HUD

Events during one airline's HUD training demonstrated the hazard of single-parameter focus. During initial HUD orientation in the simulator, Captains were taught to align the HUD's flightpath symbol with the touchdown zone until it was time to transition to flare. Many pilots became so intently focused on "flying" the symbol that they lost awareness of other hazards. During one practice approach, the instructor introduced a hazard of an unannounced aircraft taxiing onto the runway. In many cases, Captains did not see the intruding aircraft since their attention was fixated on the task of aligning the flightpath symbol. Instead, it was their FOs, who didn't have a HUD, that detected the intruder and directed a go around. This highlights the importance of PMs expanding their monitoring whenever their PFs become intently focused on their flying.

12.2.4 Developing High-Quality Monitoring

As we gain proficiency and experience, we gradually transition from monitoring the quantitative aspects of parameters to monitoring their qualitative aspects. Consider a typical final approach. After arriving at the gate, few pilots would be able to report their quantitative parameters like N1 thrust setting or sink rate. They typically could relate qualitative memories like the level of turbulence, the frequency of thrust corrections, or how well they held their target speed. With experience, we change the way we monitor and manage parameters like pitch and thrust. Holding our glidepath angle takes over for sink rate and maintaining target airspeed takes over for thrust. We instinctively adjust pitch to track our aimpoint and thrust to maintain on-speed. If we encounter a wind shift that makes us 10 knots fast, we reduce thrust a little bit to compensate. Our hand automatically knows how much to pull the thrust levers back and how much to move them forward to hold target speed.

- *PF monitoring:* With experience, we adopt intuitive measures for deviations. If we detect a wind gust that bumps us 5 knots fast, we rate it as a minor deviation. If it jumps 10 knots, we rate it as a moderate deviation that deserves increased attention. If it jumps 30 knots, we rate it as a serious deviation that might be evidence of windshear. None of these assessments are typically attached to actual number values. We assess them intuitively.

  While monitoring the quality of our approach, maintaining good SA demands that we verify a range of important details. Is the aircraft fully configured? Have we completed the Landing Checklist? Do we have landing clearance? Is the runway clear? Many pilots incorporate these assorted details into mini-flows or scan patterns. It is useful to attach these flows to solid anchor points that remind us when to complete them. For example, many airlines require or program a callout at 500′ above the touchdown zone. This serves as a good anchor point to perform our personal mini-flow.
- *PM monitoring:* As PMs, we qualitatively monitor for *appropriateness*. Is the PF's correction appropriate for a 5-knot gust, a 10-knot gust, or a 30-knot gust? In addition to this, we expand our perception beyond the PF's immediate focus of aimpoint and airspeed. If the PF is working hard on a bumpy day to maintain final approach path and speed, we expand our monitoring to ensure aircraft configuration, checklist completion, landing clearance, and that the runway is clear of obstructions. As increased workload causes our PFs to tunnel their attention inward, we expand our perspective outward. We strive to stay out of sync with each other.

12.3 BUILDING THE SA GOAL OF *ANTICIPATING* FUTURE CHALLENGES

As we monitor our flight parameters, the indications either support our expected SA flow (present SA agrees with future SA – indications verify the smooth flow of our game plan), appear neutral (present SA neither supports nor contradicts future

SA – continue to monitor indications until a clear trend emerges), or contradict our expected flow (present SA doesn't agree with future SA – treat contradictory indications as counterfactuals and investigate). Let's examine some issues and techniques that affect this process.

12.3.1 Detecting Task Saturation, Stress, Bias, or Rising Complexity

Our first line of defense is to avoid becoming task-saturated, stressed, biased, or confused in the first place. When we detect the early signs of rising task load and complexity, we direct more attention to understanding the cause and to resolving problems early.

- What is causing these unexpected indications?
- Are the indications temporary or part of an adverse trend?
- How do we resolve these adverse conditions and rebuild our SA?

In most cases, we deal with the situation and restore our SA. Later, we circle back and debrief the event to see what we can learn from it.

- Were there counterfactual indications that we could have detected earlier?
- Did we misinterpret early warning signs?
- What made this event unanticipated or unpredictable?
- How can we improve our Master Class monitoring skills to handle events like this better?

Like our metaphor of that annoying house fly, the important issue is how the fly got inside (the cause), not how we can remove the irritant (the symptom). We view disruptions as possible warning signs emerging from underlying problems – not nuisances. If we view them as nuisances, they become irritants to suppress or remove. When we successfully swat them away, we may rationalize that the underlying problems are resolved and that they won't reemerge. Viewing counterfactuals as nuisances inappropriately confirms flaws in our SA. It does not refine or rebuild accurate SA.

When stressed, we become subconsciously attracted toward indications that support our future SA and blinded to factors that contradict it. In extreme cases, this can lead crews to miss glaring warning signs. Mishap crews have missed warning horns and directive "GO AROUND" callouts as their minds tunneled in on limited parameters. Listening to CVR recordings later, they reported that they "never heard" the loud gear horns, automated voice warnings, or callouts from the other pilot.

When we consciously accept that our SA degrades under stress, we place less trust in it. This helps, but is only half of the solution. We need to share our SA confidence level and stress level with our crewmembers. "I'm in the Red. Do you understand why this is happening?" This alerts them that they need to increase their vigilance because we may be too overloaded to accurately detect problems. It empowers them to share their ideas and prepare for backup contingencies. For example, if we are visually following traffic to the runway and lose sight, it is useful to admit, "I've lost visual with the traffic to follow." The PM can either help us to re-acquire the traffic or support the need to go around.

12.3.2 When We Recognize Task Saturation/Stress/Bias/Complexity, Search for Counterfactuals

When one pilot becomes overloaded, the other pilot needs to increase their vigilance for counterfactuals that precede emerging problems.

- *PF is overloaded, PM is not:* This is the most common case, especially when PFs become overloaded flying a challenging approach and landing. They may not have time for short-term tasks, planning, or SA-building. PMs need to widen their future-SA perspective to ensure the quality of flightpath management and associated landing details (like landing clearance and wind limits). If the flightpath trend becomes unacceptable, they need to make additional callouts, prompt necessary actions, or direct a go around.
- *PM is overloaded, PF is not:* This situation is not as common. These are situations where the PM is either new (and easily overloaded) or is engaged with a time-consuming task like reprogramming a complex reroute into the FMC. Here, a useful PF technique is to engage the autopilot to reduce workload. Then, they can assist the PM. Anytime the PM is overloaded or their attention is diverted, PFs need to assume responsibility for the quality of path management.
- *Both pilots are overloaded:* When both pilots become overloaded, roles can overlap, tasks can be missed, or work processes can breakdown. A common situation occurs when PMs abandon their duties and essentially become non-flying PFs. Both pilots align their attention on the same flying problem – one focused on flying the aircraft and the other focused on what they would do if they were flying the aircraft. Take the example of a visual approach where the aircraft is too high and too fast. In an effort to manage the excess energy, PFs focus their attention outside looking at the runway. Needing additional drag to deplete excess energy, they rapidly call for landing gear and flaps. Because PMs have their attention diverted outside also, they may not check for compliance with flap placard speed limits or verify that the flaps are actually tracking to the intended setting. This situation can lead to flap overspeeds, mis-setting of the requested flap positions, and undetected flap malfunctions. PFs know that they shouldn't ask for flap settings when they are over placard speed and PMs know they should check for speed compliance before moving the flap lever, but in the heat of the moment, both pilots have their attention focused outside.

  The underlying cause of this problem is that we are most comfortable when we are flying as PFs. We are in this business to be pilots, after all. We didn't choose this profession to become flightpath monitors. The PM role is intentionally unaligned with the PF role, but under stress, PMs tend to revert to a pilot-like, PF-like mindset. Ironically, it is under these failing scenarios, when crews desperately need effective PMs, that they lose PM safeguards.

For this reason, it is essential that PMs maintain their separate duties, especially when their PFs become overloaded. We never want two pilots looking at the same place, seeing the same things, missing the same things, making the same unwise choices, and committing the same mistakes.

12.3.3 How Perspective Affects Our Detection of Counterfactuals

As we search for counterfactuals, we need to maintain the most advantageous perspective.

- *The optimistic perspective:* If we are optimistic about our game plan, we tend to look for and see indications that confirm that it is succeeding. These *profactuals* support continuing with our game plan. The optimistic perspective feels good, so we tend to favor it. The problem is that an optimistic perspective can also lead us to miss or discount counterfactuals.
- *The skeptical perspective:* If we view our game plan with some skepticism, we'll have an easier time detecting counterfactuals. We'll assess the *probable* chain of events, not our *desired* chain of events. As our eyes open to the full range of indications, we'll see both the positive signs that support our SA and the warning signs foreshadowing collapsing SA. As a general rule of thumb, the more complex the situation, the more skeptical we should be. This encourages us to carefully monitor for counterfactuals and set trigger points for aborting our game plan.

12.4 BUILDING THE SA GOAL OF *RESILIENCE*

Resilience is demonstrated by how well our SA guides our handling of disruptions.

12.4.1 The Range of Expected Events

Consider Figure 12.1 depicting a distribution of events as difficulty and complexity increase.

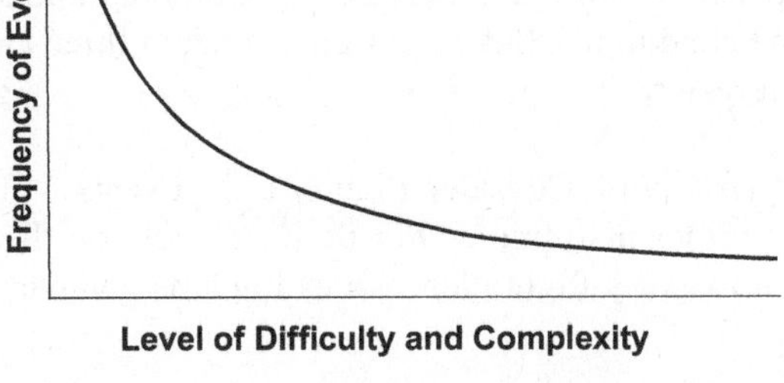

FIGURE 12.1 Distribution of event frequency versus difficulty and complexity.

The left side of the graph represents a high frequency of simple and easy events encountered during everyday flying. The right side of the graph shows how we experience very few events that are challenging and complex.

A perfectly resilient game plan is one that can encompass the full range of events, is easy to implement, and easy to maintain. The reality is that no game plan is completely resilient. As we move to the right on our graph, we reach a point where our everyday game plans can no longer accommodate disruptions. Consider the following list of increasingly challenging events.

1. Unable to climb to our planned cruising altitude due to traffic or turbulence
2. ATC reroutes us for enroute weather or traffic sequencing
3. Flight is running late and we have many connecting passengers onboard
4. Significant traffic congestion resulting in an arrival delay
5. Assigned two turns in holding before being cleared for an approach
6. ATC changes the runway to land in the opposite direction
7. ATC directs us to go around for an aircraft that is slow to clear the runway
8. Disabled aircraft on the runway requiring a go around and diversion to a planned alternate
9. Disabled aircraft on the runway requiring a go around and diversion to an unplanned alternate
10. Double diversion – divert to a first alternate, can't land, then divert to a second alternate
11. Flap malfunction requiring us to complete a complex non-normal checklist while at minimum fuel

All of these scenarios represent unplanned events that disrupt our original game plan. The first few are common and are easily accommodated within our original game plan. It is resilient enough to accommodate them. The cases in the middle of this list would require significant modification of our original game plan. The last cases in the list would require us to abandon our original game plan and switch to a contingency backup.

12.4.2 SA and Safety Margins

To improve resilience, we maintain safety margins that protect us against performance limits (like crosswind landing limits), time delays (like FAR and aircraft fuel minimums), and extreme conditions (like adding additional knots to our target airspeed for gusty wind conditions). Safety margins come in three categories: Built-in, Procedural, and Emergency.

- *Built-in safety margins:* Consider Figure 12.2. Events 1 through 3 from our list depict minor disruptions. For these, we rely on the safety margin that is build-in to every flight plan. An example is company-standard fuel

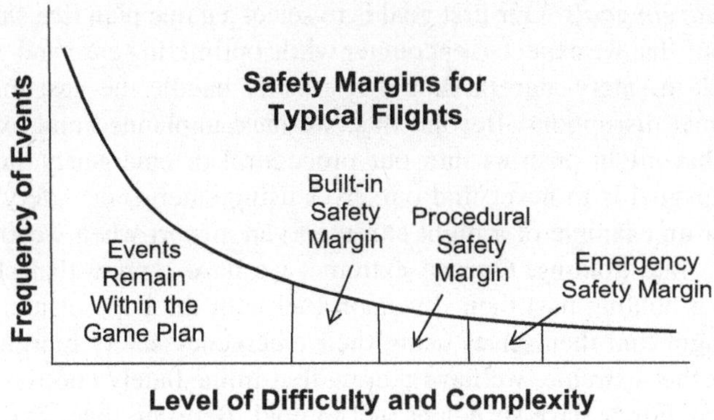

FIGURE 12.2 Distribution of event frequency and safety margins for typical flights.

minimums programmed in every flight plan. This planned arrival fuel gives us enough time and options to handle minor disruptions.

- *Procedural safety margins:* The next events from the list (events 4 through 8), represent events that require an additional *procedural safety margin* for significant disruptions. These events definitely attract our attention, both from their rarity and because they threaten operational resilience. Consider event number 7. On short final, the preceding aircraft fails to clear the runway, so ATC sends us around. We circle back for a visual approach and landing. We might land below contingency and extra fuel add-ons, but still above the company-standard minimum fuel. The more-consequential example number 8 describes an event where the aircraft preceding us lands, blows a tire, and closes the only available runway. Lacking fuel to wait for the runway to reopen, we divert to our pre-planned alternate airport. Neither of these cases are contingencies that we would pre-brief, since both are unpredictable events that rarely occur. Procedural safety margins typically provide sufficient margins to handle events like these.
- *Emergency safety margins:* The last three examples from our event list (numbers 9 through 11) reflect extremely rare emergency situations. We may need to exceed our built-in and procedural safety margins, but remain within emergency safety margins like minimum or emergency fuel. While built-in and procedural safety margins reflect those provided through company policy, emergency margins are set by the aircraft manufacturer or mandated by the regulator. Another example of emergency margins is modified approach speeds directed by non-normal checklists. For example, event number 11 would mandate an increased target speed based on our most restrictive flap condition.

- *Safety margin goals:* Our first goal is to select a game plan that satisfies the conditions that we expect to encounter while optimizing ease and efficiency. Our built-in safety margin should resiliently handle the vast majority of operational disruptions. Beyond this, we have unplanned and exceptional events that might push us into our procedural or emergency safety margins. Our goal is to never find ourselves using emergency safety margins. Consider an example of a flight enroute to an airport when weather forces everyone into holding. On one extreme, we have a crew that chooses to remain in holding past their diversion fuel with the hope of making it in. They might find themselves using their emergency safety margin to land. On the other extreme, we have a crew that immediately chooses to divert even though they have sufficient fuel to hold. Between these extremes, we have a range of options that crews can use to either complete their flight at the scheduled destination or comfortably divert while preserving their built-in and procedural safety margins.

  Adverse event mitigation begins at the gate. As we review our dispatch release, we assess whether our dispatcher has planned for holding and diverting. If their holding fuel and assigned alternate are unsuitable, we can amend our release to add fuel or switch to a more favorable alternate.

  Enroute to our destination, we can update our information of weather conditions to expand our future SA. We can change our altitude or speed to increase our arrival fuel. We can also reassess alternate airports to verify their usability. Entering holding, we can refine our divert decision fuel based on our holding altitude, the number of aircraft likely to divert to the same alternate, and our EFC time. If we near our divert decision fuel, we can coordinate a revised game plan to improve the resilience of our diversion option. In all cases, we select decisions to keep our situation within the built-in and procedural safety margins.
- *Safety margins as complexity increases:* Let's increase the complexity and see how our safety margins change (Figure 12.3).

  Consider a flight to a complex, high-volume airport. The gray curve represents the previous distribution for typical flights. The black curve depicts how complexity increases the number and severity of potential events within complex, high-volume operations. Examples 4 and 5 might be fairly common for a high volume airport. To compensate, we would add more fuel to increase our safety margins.

  We would also enhance our planning and briefing by forming strategies for commonly occurring threats and disruptions. For example, imagine that we are dispatched to an airport with forecast marginal weather. There is a good chance that we could attempt a CAT I ILS approach, fail to see the runway, and go missed approach. Our preflight planning might include a detailed review of a lower-minimum approach like a CAT II or III ILS, how many approaches we might attempt, our decision fuel for diverting from the missed approach point, and a deeper review of our alternate airport's

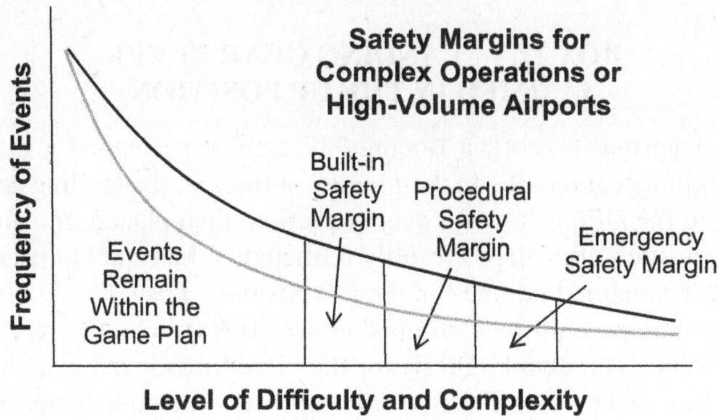

FIGURE 12.3 Distribution of event frequency and safety margins for complex operations.

suitability. Again, our goal is to stay within the built-in and procedural safety margins.

12.4.3 Increasing Our Resilience for Novel or Surprising Events

Complexity causes unforeseeable events to emerge. Many exceed our imagination, let alone the inherent resilience of our game plan. For these, we rely on general strategies for exiting the game plan, maneuvering the aircraft to uncongested airspace, managing time-available, diagnosing problems, innovating new game plans, and communicating revised shared mental models. A useful procedure for unplanned and complex events is MATM.

- Maintain aircraft control
- Analyze the problem
- Take appropriate action
- Maintain situational awareness

MATM is a strategy for managing available time while applying the ABC*D* process to rebuild our SA. Of course, serious emergencies (like an onboard fire), overrule going around and making extra time for analysis and preparation. In most novel cases, however, we have enough safety margin to exit the current game plan, evaluate the situation, and innovate a new game plan.

12.4.4 Building Resilience during Unique Events

Not all situations are addressed in our manuals. When we lack specific policy guidance, we craft a solution that satisfies the underlying intentions of company policies. The foremost intention is to land the aircraft safely. The following event demonstrates this process during a highly complex aircraft malfunction.

BOX 12.4 LANDING GEAR LEVER JAMMED IN THE UP POSITION

Following a normal takeoff, a Boeing 737 crew experienced a malfunction with the landing gear handle. In their model of the 737, the landing gear handle was raised to the UP position for gear retraction, then placed into an intermediate OFF position after flaps are fully retracted. When the FO tried to move the handle, it remained jammed in the UP position. The Captain directed the FO to leave it alone until they climbed above 10,000′. The FO referenced the Quick Reference Handbook (QRH) for the "Landing Gear Lever Jammed in the Up Position" Checklist. The first few steps were intended to address a solenoid malfunction and allow the crew to move the handle to OFF. This didn't work. The remaining steps were intended for a more complex malfunction that seemed to be more appropriate for their situation. Since the flight was scheduled for a destination airport with unfavorable weather, the Captain elected to divert to a closer airport with better weather and maintenance resources. Completing the remaining steps of the QRH procedure should have completely depowered hydraulic pressure to the landing gear and allowed them to gravity-fall and mechanically lock down. It didn't work. Unbeknownst to them, their aircraft had a defect in the gear hydraulics controller resulting in a malfunction that was never expected to occur. No procedures were ever written or trained for their unique problem. At this point, they called for outside assistance from company Maintenance Control. They suggested various options which delivered only partial success. They performed several flybys down the runway and confirmed that the main landing gear was only partially extended. Continuing to work the problem, they eventually achieved three gear-down indications. The Captain performed the landing. Unfortunately, the compression of the main gear struts re-pressurized the hydraulic system enough to cause the nose gear to partially retract. The Captain detected this from the change in wind noise as he began lowering the nose. The aircraft settled on its nose and slowed safely to a stop.

This event is noteworthy because this freshman Captain and their new-hire FO expertly handled a situation that wasn't covered in the QRH, was never trained, and shouldn't have ever occurred. They managed their workload, maximized their landing runway advantages, performed all available procedures, innovated a workable solution, and safely landed. In practice, they expanded the MATM checklist with some additional steps.

- Maintain aircraft control
- Analyze the problem
- Take appropriate action

  - Maneuver toward the most favorable runway
  - Run appropriate checklists and determine that a new solution is needed
  - Consult outside expertise to increase their SA
  - Innovate a workable solution
  - Optimize all factors to achieve a favorable outcome
  - Achieve an acceptable landing configuration
- Maintain situational awareness

In this expanded MATM strategy, we discover that we may select a possible game plan only to discover that it won't work. This prompts us to revisit our analysis and investigate additional options. In this event, the crew had plenty of time while they burned fuel to reduce landing weight. Their continued efforts finally achieved safe gear-down indications. When the unanticipated nosewheel retraction happened, the Captain elected to continue the landing by gently lowering the aircraft down on its forward fuselage.

12.5 BUILDING THE SA GOAL OF *RECOVERING* FROM DEVIATIONS IN OUR GAME PLAN

After a disruption has upset our game plan, SA helps us recover.

12.5.1 Deciding Whether to Recover or Change Our Game Plan

Consider how our decision making changes across a range of approach airspeed deviations. A mild deviation, like a wind gust that bumps our airspeed five knots, wouldn't warrant a go around. We would classify it as a momentary deviation, apply a correction, and continue down final. On the opposite extreme, visual evidence of a microburst along with a windshear warning ("GO AROUND, WINDSHEAR AHEAD") would compel us to go around. Logically, there must be a point between these two extremes where we stop trying to continue the approach and consider switching to a go around. Unfortunately, we can't consistently locate that point while we are flying. We rely on our judgment to gauge when conditions have become unmanageable. Let's look at some influential factors that guide this judgment call.

- *Breakdown in our SA:* Our confidence in our SA depends on how smoothly it transitions between past, present, and future. The approach can be challenging, but as long as the flight matches our SA, we confidently continue. When we detect a significant disagreement between what we expected to happen and what is actually happening, we become less confident with our SA. The prudent course of action is to investigate (when we have enough time) or abort the game plan (when we don't). Consider the following aircraft accident report.

BOX 12.5 CONTINENTAL 1943 – GEAR-UP LANDING IN HOUSTON (IAH) –FEBRUARY 19, 1996

Due to a procedural omission, the crew failed to move their DC-9's hydraulic switches to the HIGH position. This resulted in insufficient hydraulic pressure to achieve full extension of landing gear and flaps. The FO detected that the aircraft wasn't slowing as expected. He made two statements to alert his Captain: "I can't slow it down here now," and "We're just smokin' in here." The Captain encouraged him to continue, "You're alright." About 10 seconds later, the FO offered the landing to the Captain who assumed control and landed with the gear and flaps only partially deployed.[2]

On a normal approach, we need flaps and gear to provide a source of drag to slow the aircraft back to approach speed. Both pilots detected the excessive speed, but not what was causing it. Their SA between what they were seeing (present SA) and what they expected to see (future SA) didn't match. Procedurally, they should have gone around. Instead, they chose to press ahead.

While this outcome was exceptional, this "failure to slow" type of scenario is not unique. In similar cases, crews achieved successful gear extension, but for various reasons, didn't get full landing flaps, usually due to tripped circuit breakers or split flaps.[3] In some cases, mishap crews followed the same flawed decision making as this accident crew. What causes this kind of SA breakdown? We discovered that mishap crews tended to rely on flap lever position instead of the actual flap extension as indicated on the gauge. Similar to the Continental 1943 crew, these mishap crews detected their inability to slow down, but misattributed the cause. Everyone attempted to understand what was happening, but they only detected indications that supported their flawed SA (flap handle was in the intended position), and not indications that contradicted their SA (flap gauge showed no flaps extended). Continuing with this flawed SA, it is likely that they blamed the failure to slow on an external or environmental condition, like a steadily dropping headwind or a steadily increasing tailwind. Real-time readouts of wind speed and direction were available to confirm a relative wind shift. However, lacking time and drawn in by the imminent landing, they continued down final expecting their airspeed to ultimately slow. It didn't.

We rely on PMs to mitigate this type of error. PFs can become overloaded trying to fly the aircraft and manage their parameters – in this case, the need to reduce their airspeed. Their ability to rebuild their SA vanishes. When PFs are overloaded, PMs need to be the ones that guide rebuilding SA or direct the contingency backup.

Analyzed logically, the cause of the unexpected condition is either outside, inside, or unknown. The outside cause can only be a significant wind shift or a combination of wind shift and high-density altitude. A quick glance at the relative wind readout should resolve this. The inside cause can only be from configuration or thrust. A quick glance at the thrust and configuration gauges should resolve this. If after these two steps, the cause remains unknown, then the choice is guided by company policy. If we cannot achieve approach speed limits, then we must direct a go around. This process highlights why PMs need to remain detached from the PFs' perspective and mindset.

The bottom line is that whenever the loss of SA cannot be resolved within the time available, we should abandon the game plan. Unstabilized approaches represent instructive cases because the time to act steadily diminishes as the landing threshold nears. As our stress rises, our attention tunnels in, and we succumb to plan continuation bias. In effect, the runway presents a familiar, desirable lure that draws us in while going around feels like an unfamiliar, undesirable alternative that repels us away.

- *Negative trends or stagnated progress:* Deteriorating game plans exhibit negative trends or stagnated progress. After we have exhausted all of our corrective measures and we feel committed to continuing, all we have left is hope. For example, when we are fast on final and have reduced the thrust levers to the idle stops, we can rationalize that our corrections will soon start to work if we only give them a little more time. We trade the need to reach target speed for the hope of reaching target speed. Waiting for the airspeed to slow, the runway threshold looms closer and closer. Far too often, hope becomes resignation. With the touchdown zone seconds away, we rationalize that we have done the best we can. We resign ourselves to the excessive airspeed and land the aircraft.

  Master Class pilots remain focused on parameters and trends. We don't rely on hope. If trends aren't improving, then we go around and try again under better conditions. This is extremely difficult when the goal seems so close. "Just another 10 miles and we will clear this line of weather." "Just land fast and deplete the excess speed on the runway." "Just push harder on the brakes and we'll slow down." All of these rationalizations accept adverse trends or stagnated progress.
- *Event quickening:* Deteriorating game plans exhibit a quickening effect. We detect a deviation, apply a correction, see that it isn't working, quickly apply another correction, and so on. In failing situations, we soon run out of corrections. Adverse parameters seem to stack up against us at an increasing rate. Workload and stress rise. In the heat of the moment, it becomes difficult to discern whether to continue the corrections or abort the game plan.

  The Master Class skill is accurately recognizing the quickening effect. When we detect it, we know to abandon attempts to make sense of it. Instead, we just assume that our game plan is failing and switch to a safer contingency. We need to use the quickening trend as a warning sign. We train ourselves to become aware of how a quickening situation feels. When we sense that it is beginning to quicken, we switch to our contingency game plan to slow the pace and reset the profile.
- *Unclear trigger for switching to a backup game plan:* Recall the earlier example with the Alpha and Omega crews and how each handled convective windshear conditions. The Alpha crew anticipated the volatile conditions and briefed a diversion plan. They reviewed the specific criteria that would

trigger switching to their backup game plan. On final, they were primed to monitor for both the effectiveness of their approach and the presence of counterfactuals that would trigger their go around. When they detected those counterfactuals, they easily switched to their backup plan. Their decision was predetermined and rehearsed.

The Omega crew didn't brief or plan for contingencies, so when conditions deteriorated, they became indecisive. As stress mounted, biased habits replaced deliberate responses. Their lack of a backup plan depleted their SA. Landing felt like the only viable option.

- *Goal switching:* With everything we do in aviation, we have an objective in mind. Our objectives are reinforced by the hundreds, even thousands, of flights that we successfully completed in the past. Abandoning our familiar objective can feel like failure. Many of us have pressed through questionable weather, continued unstabilized approaches, and landed in unsuitable conditions because we were overly committed to our familiar objective. Within each objective are a series of subgoals. For a normal flight, these may include on-time departure, smooth flight, timely arrival, stabilized approach, and on-time gate arrival. Goal switching is when we skip or swap a goal to achieve our familiar objective.

  On final with a mildly unstabilized approach, we mentally weigh the difference. If we are one knot fast of our stabilized approach criteria, does it warrant a go around? Of course not. We would land. How about 5 knots fast? How about 20 knots fast? At some point, we need to decide how much is too much. Experience seems to guide which personal and operational priorities we choose to follow. What if this is the last flight of our pairing and we have a tight connection for our commute flight home? What if we are running late and the station is holding departures for our transfer passengers? Factors like this tend to encourage us to minimize intermediate goals to achieve a larger objective. In this way, goals, priorities, and objectives often find themselves in conflict. We judge which ones outweigh others. We only have to elevate one goal to justify abandoning another. Consider the Captain who lands 20 knots fast, makes the last turnoff, and gets to the gate in time for paramedics to save a heart attack victim. The company would shower them with accolades. Consider the Captain who lands 20 knots fast, makes the last turnoff, and makes their commute flight home in time to attend their child's birthday party. Their family would shower them with accolades.

  It takes planning, commitment, and professional maturity to switch from our original objective to a backup plan like diversion or go around. Clearly, anytime we apply personal goals to justify goal switching, we compromise our professionalism. Applying operational goals may feel more justified, but the distinction is small. Applying safety goals does seem justified, as long as they can be applied within the guidelines of Captains' emergency authority and conducted safely. In the end, we need to maintain a level of detachment with specific outcomes. Sure, we should make every reasonable effort to fulfill the flight's objective, but when conditions warrant, we need to choose the safer alternative.

12.5.2 Power over Force

Power is achieved by applying the right amount of effort, at the right time, along the path of least resistance to achieve a desired goal. Force only prevails when we apply excessive effort to overcome resistance. Power flows easily and reduces workload and stress. Force pushes against resistance causing additional workload and rising stress.

It is important to discern the difference. Difficult conditions require effort, but yield tangible progress. Parameters improve. Force also requires effort, but parameters either do not improve or they worsen. We interpret the rise of workload and stress as warning signs that we are forcing a situation. The right tools help us to preempt or resolve problems. They include:

- Assessment of probable conditions during preflight planning
- Planning and briefing of the game plan and possible contingencies
- Managing deviations to maintain the desired profile
- Monitoring for counterfactuals
- Detecting and correcting mismatches between our SA and apparent trends
- Setting trigger points for abandoning the game plan
- Empowering PMs to identify adverse trends and direct game plan changes
- Assigning clear roles during high-workload phases

The right tools make the job easier. The wrong tools might still work, but will require excessive force. Power resolves the problem. Force often doesn't. The Alpha crew applied power when they smoothly executed their go around and diverted to their alternate. The Omega crew forced their landing under increasingly hazardous conditions.

12.6 BUILDING TEAM SA

The process of building team SA begins with preflight planning and briefing. The crew shares a mental model of how they expect the flight to progress. This forms the framework of past SA which influences their mindset moving forward. Ideally, team SA produces an additive advantage where the combination yields more than the sum of its parts (1 + 1 = something greater than 2). Everyone works together to highlight problems, resolve confusion, clarify options, and raise each other's personal SA.

12.6.1 Factors that Degrade Team SA

A number of factors can negatively affect the process of building, maintaining, or rebuilding team SA.

- *Inadequate planning and briefing:* The Omega crew underestimated the adverse arrival weather conditions and failed to prepare a backup plan. Their SA was based on their preflight assessment of conditions. Their mindset compelled them to force their failing game plan against a steadily worsening situation.

- *Ineffective communications:* Often, one mishap crewmember accurately diagnoses their situation and succeeds in rebuilding their SA, but fails to alter the other pilot's SA. We see this most often when a PF/Captain fixates on their flawed SA. Either from indecision, lack of assertiveness, or ineffective communication, the accurate-SA crewmember can't change the PF's mindset.
- *Complexity and stress magnify our biases:* Rising complexity and stress tend to stimulate our human biases. Our biases then oversimplify our situation. We end up sticking with our original game plan because it feels safer. Competing versions of SA are discarded or minimized.
- *Roles breakdown:* Team SA relies on different pilots approaching problems from slightly different perspectives. PFs naturally see problems from a PF perspective. PMs need to maintain their qualitative monitoring perspective which may conflict with the PF's SA.

12.6.2 Factors that Improve Team SA

As airline pilots, we receive extensive training in crew resource management (CRM).[4] We review case studies and learn many techniques. The challenge is to take these classroom techniques and apply them to the flightdeck. Consider the following ideas.

- *Empowering pilots to speak up:* The most prevalent mishap crew combination involves a Captain/PF and an FO/PM. Mishap Captains become task-saturated and choose to continue their failing game plans. FO/PMs typically detect the unfavorable trends and make informative statements or callouts. When their concerns are ignored or rebuffed, they back down, resign themselves to the situation, and hope for the best. When we debrief these mishap FOs, they report:
  - "I made the required callouts, but they wouldn't respond."
  - "I told them, but they said it was okay to continue."
  - "They already knew what was happening. I had nothing to add."

  From the Captains we hear:
  - "I never heard them say that."
  - "I thought our parameters were close enough."
  - "As the Captain, I made the decision to continue."
  - "I had no idea that our parameters were so extreme."

  These comments reflect CRM vulnerabilities. When highly stressed, we may not hear or process callouts because our minds become overtaxed with managing the situation. FOs assumed that their callouts were received and rationalized the reasons why they didn't get a response. Even when the Captains heard the callouts, some believed that they had the discretion whether to respond, comply, or override.

  Many Captains assume that their FOs will feel empowered to speak up, so they don't see the need to expressly encourage it during their Captain's briefing. Especially strong-willed Captains may unintentionally erect barriers against contrarian messages. The vulnerability is that FOs are biased to accommodate their Captain's comfort zone far more than they are motivated to confront them. This forms a barrier that many FOs are reluctant to overcome. Inexperience and mismatched personalities can make it even harder. An inexperienced FO can rationalize, "This Captain has been flying here for 20 years. Who am I to tell them what to do?" As Captains, we need to make dedicated efforts to open the communications environment and to empower FOs to assertively make callouts and intervene.

Another barrier is our reluctance to speak certain words – as if saying them makes them more real and consequential. We also resist speaking up if we think we may be wrong. This leads some pilots to hint at problems instead of making clear callouts – "We're still a bit fast" instead of, "Airspeed plus 30, Go Around."

There is also a fear of saying something that might be recorded on the CVR. Procedurally, we counteract this by making callouts factual and mechanical. "Airspeed plus 15" feels like an automated callout. It doesn't come across as personal or judgmental. In the end, we don't want pilots weighing their decision whether to or how to verbalize deviations. We want them to be made consistently and reliably.

- *Rebuilding SA using pockets of expertise:* In especially novel and complex situations, we may find it difficult to acquire the right information to overcome our flawed mindset. Lacking this knowledge ourselves, we can use others to fill the gaps. As an illustrative example, consider a fuel starvation event experienced by the crew of Air Canada 143.[5] This crew was forced to quickly rebuild their SA and innovate a solution to save their passengers, crew, and aircraft.

BOX 12.6 AIR CANADA 143 – FUEL STARVATION EVENT – JULY 23, 1983

The crew's brand-new Boeing 767 had an MEL requiring deactivation of its fuel quantity gauges. This deferral required them to verify their fuel load by comparing drip sticks measurements to MEL charts. Unfortunately, the charts were published in pounds while Air Canada used kilos. Both the ground and flight crews misapplied a conversion value in their computations. As a result, they unknowingly uploaded an insufficient fuel load for their flight from Montreal to Edmonton.

Their process illustrates a vulnerability with problem solving. Every person involved detected a problem, discussed it, and agreed that they had converted the numbers correctly. This is exactly the process that we encourage crews to use to mitigate errors. In the end, their consensus gave their math error a sense of "rightness". Solving "a problem" made them feel confident that they had solved "the problem". This probably contributed to their future reluctance to discard their flawed SA.

They took off and were comfortably established in cruise when a fuel boost pump light illuminated. Their SA and mindset led them to categorize this as a unique mechanical problem unrelated to the fueling issue they solved at the gate. The crew referenced a checklist procedure and secured the "malfunctioning" pump. The pilots expressed frustration with encountering multiple problems (MELed fuel gauges and now a boost pump) on a brand-new aircraft. While they discussed the situation, a boost pump light on the opposite tank illuminated. Something was definitely amiss. In hindsight, this is a classic indication of fuel starvation since the boost pumps are intentionally mounted at different locations and levels within each tank. This is done to ensure that the engine continues to receive fuel regardless of the pitch attitude. In straight and level flight, the "higher" boost pump would become uncovered, cavitate, and cause the LOW PRESSURE light to illuminate.

At this point, the Captain recalled that a senior mechanic was riding in the passenger cabin and called him forward. This was an effective use of a pocket of expertise. The mechanic entered the flightdeck, saw two boost pump lights, blank fuel gauges, and the MEL stickers. Recall that this mechanic was not involved with resolving the fuel upload issue, so he did not share their flawed SA and mindset. He diagnosed the situation from scratch and concluded that they might be running out of fuel. The mechanic correctly diagnosed the fuel starvation problem, but he had to overcome resistance from the Captain's past SA that asserted that it *couldn't possibly* be a fuel quantity problem.

The FO began to question their flawed past SA. While on the ground, he had suggested that the fuel load might be incorrect. He ultimately accepted the Captain's assurance and didn't press any further. When the second boost light came on and the mechanic expressed an alternative explanation, his earlier suspicions were confirmed. The Captain remained committed to his flawed SA. Looking for a plausible explanation, he began investigating the possibility of fuel contamination. Unable to sway the Captain's mindset, the FO started searching for nearby divert fields. Shortly after this, their first engine flamed out from fuel starvation.

At this point, the Captain may still have believed in his fuel contamination theory, but losing the engine led him to the same remedy – the need to land immediately at the closest available airport. The FO immediately suggested diverting to Winnipeg. The Captain agreed. The crew coordinated with ATC who gave them timely vector and distance information (another pocket of expertise). Next, their remaining engine flamed out. The Captain realized that Winnipeg was beyond gliding distance. The FO rechecked his map and located a closed military airfield (Gimli) that he had used during his previous military career (his pocket of expertise). The Captain turned the aircraft toward Gimli. The field had no NAVAIDs and the crew had to penetrate a cloud layer. ATC provided excellent vectors and positional awareness. The FO related the runway features that he could recall while the mechanic continued to monitor their actions. When he needed help the most, several pockets of expertise were feeding the Captain with timely, accurate, useful information. When they broke out of the clouds, they realized that they were too high. Luckily, the Captain had extensive glider-flying experience (his pocket of expertise). Recognizing a familiar solution from glider aviation, he applied full sideslip controls to increase drag and decrease lift. They landed the aircraft with only minor damage.

To solve problems that we may encounter, we have a number of available pockets of expertise.

- *Real-time EFB apps:* Most EFBs provide access to real-time weather information. For example, we can track the movement and severity of convective buildups. This is useful for selecting routes and estimating the arrival of convective weather at our airport.
- *Weather packet:* This resource can inform decisions about our fuel load and MEL restrictions before leaving the gate. We can mentally rehearse the flight and identify areas where we might require more information.
- *Dispatch:* How appropriate is the fuel load? Do we need extra fuel for thunderstorm activity or weather deviations? When we share our concerns, we can work together to improve safety margins against specific threats to our flight.
- *FO:* Is this our first time into a new city? Maybe our FO has already been there and knows the local procedures and threats. If we are becoming task-saturated, our FO may have better SA and detect an impending problem before we do.
- *Captain:* We can draw on the Captain's years of knowledge, wisdom, and experience. Remember, it is better to learn from another's mistakes, than from our own.
- *FMS:* Use the FMS map display or the moving map on our EFB to monitor the relative location and distance to possible diversion fields.
- *Our cabin crew and deadheading pilots:* Sometimes, someone with a different perspective can identify a problem that we can't. I recall a story of a crew that was experiencing electrical problems. They had exhausted all of their own ideas to diagnose the problem. They consulted a deadheading pilot who came to the flightdeck, looked at the lights, and recognized that they had a Ground Service Bus failure. He quickly located a tripped circuit breaker on the back of the center console that both pilots had repeatedly overlooked.
- *ATC:* ATC has a larger perspective of the airspace and aircraft movement. Unfortunately, they don't always share it with us. When we need their help to understand our place within the flow of surrounding aircraft, we need to ask them.
- *Pilot reports:* The most timely PIREPs we need are the flight conditions on final and the condition of the runway surface. Unfortunately, most crews wait to report this information after switching from Tower to Ground Control. To speed the sharing of information, we can make a quick PIREP on the Tower frequency before we switch to Ground Control. We can also ask for conditions from the crew that just landed.
- *Station operations:* If conditions are questionable, ask ATC or company operations about conditions like snowfall, rainfall, winds, ramp closures, aircraft go arounds, and snow removal.

- *Central operations experts:* For unique problems we can call Maintenance Control or the Chief Pilots at central operations. They may have already encountered and solved our exact problem.

12.7 MASTER CLASS SA-BUILDING TECHNIQUES

Following are a collection of ideas to improve our SA-building. Some may not align with our company procedures, so apply them appropriately.

12.7.1 Building Better SA

A particularly effective model of briefing uses a threats, plans, and considerations format.[6] This Briefing Better concept empowers PMs to assume a stronger role with identifying threats and monitoring for counterfactuals. The four goals are:

1. *Threat forward:* The law of primacy asserts that we retain and recall the first information that we view. The PM begins the briefing by highlighting the main threats for the flight. The team discusses which indications might arise from those threats, the monitoring priorities to detect those indications, and the specific backup plans that may be necessary. The second benefit of the threat forward format is that it *informs the plan* by building a shared mental model. This improves the smooth transition of our SA between flight phases.
2. *Interactive:* Instead of the one-sided *my leg* or *your leg* briefing, this format encourages open discussion. Both pilots are empowered to own both the game plan and the outcome of the flight.
3. *Scalable:* One of the main problems with traditional briefings is that they can devolve into monotonous recitation as pilots step down through a list of required briefing topics. The Briefing Better format is tailored to the crew's shared experience. If the crew is flying a repetitious route between the same destinations, common knowledge is understood. Instead, each briefing focuses on exceptional conditions or threats that apply to that particular flight. This keeps the briefings relevant and authentic.
4. *Cognitive:* Just as our minds remember the first item reviewed (law of primacy), we tend to retain the last item reviewed (law of recency). For this reason, the briefing concludes with a recap of the main threats, indications of those threats, monitoring for counterfactuals, backup plans, and specific roles and duties.

12.7.2 The Briefing Better Process

Putting it all together, the briefing process follows this general flow.

1. The PM's assessment of threats
2. The PF's game plan for the flight and how they plan to manage those threats

3. Open discussion of roles, counterfactuals that might indicate that the plan needs to be changed, and backup options
4. The discussion continues until both pilots are satisfied with the game plan (shared mental model)
5. A recap of the main points to highlight priorities

Current briefing lists still serve to guide discussion of important items, but the focus on relevance improves information retention. For example, "Rejected Takeoff" is an important briefing item, but the conditions affect its relative importance. Compare a reduced-thrust takeoff from a long, dry runway with a maximum-thrust takeoff from a short, wet runway. Under Briefing Better, the second situation would warrant far more attention and discussion. Additionally, applying the law of recency, we would review the rejected takeoff steps just before taking the runway.

Whatever format we use, our Master Class challenge is to conduct an effective and relevant preflight briefing. Due to repetition, our natural tendency is to allow our diligence to drift. Consider how our briefings have changed over the years. Have they become better or worse? Are they procedural squares that we fill or do they actively contribute toward building useful SA? Master Class briefings should:

- Identify the threats the flight is likely to encounter.
- Construct a viable game plan.
- Communicate a shared mental model.
- Identify counterfactuals that may indicate that the game plan may be failing.
- Empower both pilots to share their detection of deviations and counterfactuals.
- Open lines of communication to encourage crewmembers to share their concerns.

12.7.3 Improving Present Moment SA

Effective present moment SA is guided by where we focus our attention, our monitoring priorities, and how we manage our workload. The goal of present SA is to guide the smooth flow between what we planned and briefed (past SA) and our prediction of the flight's likely progression (future SA).

- *Disciplined attention focus:* Present SA depends on how we focus our attention. Vulnerable flight phases like taxi movement, takeoff, and landing demand higher levels of attention. Lower vulnerability phases like stationary ground operations with the brakes set and sustained cruise flight on autopilot allow lower levels of attention. Our challenge is to maintain an appropriate level of attention for each flight phase. Our natural tendency is to become more relaxed as we gain experience. Sometimes, we allow ourselves to become too relaxed. This leads us to miss important conditions and events that we would have noticed if we had paid adequate attention. Debriefing events helps us to refine our habits and techniques. Through steady refinement, we learn to maintain an appropriate level of attention for each phase of flight.

The second aspect of attention is where we focus our attention. Using our experience, we know which indications are most relevant, how they should appear, and when to look for them. Refining our scanning techniques allows us to monitor the most relevant parameters during each flight phase.

- *Mindfulness:* Mindfulness is the practice of intentionally holding our mental focus on whatever we choose. It does not keep us from rapidly switching between important parameters, when necessary. Instead, it makes our switching intentional and deliberate. We can choose to practice mindfulness whenever we wish. If someone offered us a monetary bonus if we maintained our airspeed within five knots of target speed on final, every one of us could succeed. We would sit up, pay attention, and make a conscious, disciplined effort to focus on parameters. Do we routinely apply this level of conscious, disciplined mindfulness to all of our flying? Probably not. More likely, our discipline wanes as we slowly drift toward subconscious habits. We learn to pay just enough attention to keep the flight moving along our intended path.

  The practice of mindfulness is a learned skill. At first, it's difficult because our minds like to flit back and forth between objects of interest. Staying focused fights against the innate processes of our busy minds. To overcome this tendency, we need to practice applying mindfulness. With repetition, we will steadily improve the quality of our present moment SA.
- *Learning from others:* As we fly with many different pilots, we encounter situations that some pilots handle particularly well. Use these opportunities to learn from them. Delve into the processes they used to solve problems. Study how they build their SA. When a debrief opportunity arises, ask questions:
  - When did you first notice our problem developing?
  - What were the first indications or cues that you detected?
  - Was this event similar to a past experience?
  - What was different about this event compared with your past experience?
  - Why did you choose your course of action?
  - What concerns or factors influenced the choices you considered?

We incorporate their experience into our wisdom. What happened to another pilot becomes useful and relevant to us after we weave it into our own story.

12.7.4 Keeping Workload Manageable

We know that once we become task-overloaded, our SA suffers. Unprepared pilots can become overloaded with unforeseen challenges. Prepared pilots rarely become overloaded because they anticipate the challenges, allocate resources, and have a viable backup plan ready to go. We saw the advantages with the Alpha crew/Omega crew comparison. Workload management creates time and attention that we can use to fill our SA balloon.

12.7.5 Building Future SA

When most pilots think of SA, they envision how well they can predict the future position of the aircraft. A useful term is *projected SA*.[7] Viewed along a continuum, on one end we have close-in projected SA. This reflects our awareness of the immediate flightdeck environment. This includes the status of aircraft systems, the immediate flightpath movement, and what is likely to happen in the next few moments. It represents the minimal inflation of our future SA balloon. In the middle of our continuum, we have normal projected SA. In addition to all of the components of close-in projected SA, this would include the expected flightpath and which events are likely to occur in the near term. For example, normal projected SA would predict that ATC will assign us a base turn within the next 3 miles. To the other extreme of our continuum, we have expanded, projected SA. Expanded SA describes our larger understanding of external factors that may affect our flight. For example, if we are aware of the aircraft preceding us toward the same runway, we could anticipate when ATC might assign our base turn, our relative spacing on preceding aircraft, and whether they might pose a wake turbulence threat. Expanded SA can also project behind us. For example, we notice that the aircraft following us has minimal spacing. We would anticipate that ATC will ask us to expedite clearing the runway to allow their landing. Building projected SA is a learned skill. The more we practice expanding our awareness, the more subtle details we discover and the more skillful we become. Consider the following examples that highlight the advantage of good projected SA.

BOX 12.7 THE SURPRISE BASE TURN

Crew Omega is being vectored for a visual approach to a familiar runway. The weather is good and they expect a typical arrival. Unexpectedly, ATC delays their descent due to a VFR aircraft transitioning underneath them. Both pilots divert their attention to locate the VFR aircraft. ATC then assigns them the tight left base turn and clearance for a visual approach. The FO/PF doesn't acquire the runway until rolling out on base. The Captain/PM, who sees the runway the entire time, doesn't provide any help. The FO discovers that they are too high to achieve a stabilized approach. They end up going around. They blame ATC for providing poor service.

Crew Alpha flies the same profile. This FO has built excellent projected SA and knows that aircraft in front and behind are equally spaced. They predict that ATC will try to maintain that spacing due to the high frequency of arriving aircraft. When ATC delays their descent, they anticipate that they will be high on profile. The FO starts slowing and calls for maneuvering flaps. After clearing the VFR aircraft traffic, ATC assigns a tight base turn and clears them for a visual approach. The FO calls for landing gear down and continues slowing. Unable to see the runway, they ask the Captain for energy assessments. The Captain/PM advises that they are high and that more drag is needed. The FO ensures that they are below flap placard speed and calls for intermediate flaps. As they roll out on base, the FO acquires the runway, sees that they are still high, and calls for landing flaps. The FO successfully depletes their excess energy and intercepts final on speed and stabilized for a normal landing.

Mismanaging the transition from ATC-controlled vectors to a pilot-controlled visual approach is a common problem area. ATC controllers have limited perspectives since they are sitting in a dark room managing images representing dissimilar aircraft using flat screen displays. Unintentionally, they can place us in a position that looks acceptable on their display, but places us too high and too fast. We compensate by expanding our SA to manage the transition from ATC vectors to the visual approach. Crew Omega failed to anticipate their energy problem and didn't work as a crew to share useful information. Their only option was to go around.

- *Extending our time awareness of future weather conditions:* Many of our EFB weather apps can project arcs that predict the movement of convective storm cells. They give us useful predictions of weather that may block our route or destination. For example, when it is clear that a significant storm cell will arrive over our destination near our landing time, we may wish to slow our airspeed to delay our arrival time or remain at a higher cruise altitude to conserve fuel.

  We can also apply our experience to project worsening weather trends. Examples include a narrowing temperature/dewpoint spread that precedes ground fog, dropping temperatures that threaten to freeze water on runways or taxiways, steadily dropping ceilings that compromise approach minima, and low sun angles that reduce approach visibility. ATC controllers, in their dark control rooms, often can't detect these emerging trends. Until someone tells them otherwise, they only know that every aircraft is making it in. Take the example of a low sun angle landing to the west into a setting sun. The ATIS information remains unchanged and reflects perfectly acceptable weather for visual approaches. Even on downwind, we can easily see the airport. Once we roll out on final with the setting sun in our eyes, haze obscures the runway environment. We suddenly need instrument approaches even though the airport is still reporting VFR. Controllers scramble to increase spacing to satisfy IFR requirements. Shortly after sundown, we return to VFR conditions and resume visual approach spacing.
- *Building our future SA skillset:* As we become experienced, fewer flights challenge our daily SA-building skillset. All of our flights go smoothly and predictably. In time, we can drift into a relaxed and trusting mindset. Even when events challenge us, our game plans contain enough resilience to absorb the challenges. Latent hazards hide beneath daily normalcy. Everything works well until it doesn't.

  As Master Class pilots, we know that our skill level affects our success with adverse events. With regular practice, events rarely surprise or overload us. For example, if we practiced a unique, challenging LOFT scenario in the simulator every week, we would become experts at handling unique, challenging situations. We don't normally get this kind of practice. Most of us only see the simulator once or twice a year. That is not often enough to hone these skills. As an alternative, we can imagine adverse scenarios and exercise the skills required to build expanded SA. Following are some examples.

- Established in cruise flight, consider what we would do if we suddenly experienced a rapid decompression. We mentally review the boldface steps to don oxygen masks and establish crew communications, the initial steps of an emergency descent, how we would communicate our problem with ATC, and cabin crew coordination.
- Established in cruise flight, consider what we would do for an engine fire that requires an immediate landing. We mentally review QRH procedures, ATC communications, crew roles, the best landing runway options, and cabin crew coordination.
- Enroute to a single-runway airport, consider what we would do if the preceding aircraft lands, blows a tire, and closes that runway. We mentally review how much time we have to loiter, where we can divert, and selecting our diversion decision fuel.

By engaging in these thought exercises using hypothetical scenarios, we exercise the mental processes that we would need to engage in a real emergency. Like rehearsing the steps of a rejected takeoff before we take the runway, we reduce the startle factor, prearrange contingency options, and organize our recovery from disruptions. This practice sharpens our future SA-building skills.

- *Improving future SA through story-building:* Consider the difference between a person watching a movie and the person who wrote the screenplay. The movie-viewer could be surprised or confused by events as they unfold. The writer, however, would already know all of the actions and motivations of each character, the flow of the plot, and the outcome. The writer would not be surprised or confused by any events.

The same experience holds for us as we fly. Consider when we are on vectors for a busy airport. We could shrink our SA, only monitor our aircraft's path, and respond to each vector that ATC assigns (watching the movie). When something unexpected happens, we might experience surprise and scramble to salvage our approach. Now, imagine that we expand our SA to anticipate the story that our flight will follow (knowing the screenplay). We could project our SA to monitor the ATC calls for the preceding aircraft. This helps predict when we will get those same vectors. We might hear ATC giving vectors to a transiting VFR aircraft. We could crosscheck our TCAS display, locate that aircraft, and predict if they will impede our normal altitude stepdown. When ATC holds us high to allow the transiting aircraft to pass, we aren't surprised by it. We have already modified our plan to deplete the excess energy.

NOTES

1 Edited for brevity and clarity. NASA ASRS report #1807534.
2 NTSB (1997, pp. 4–5).
3 In most airline category aircraft, automatic flap asymmetry protection freezes the flaps in transit whenever an excessive flap split occurs. This typically happens early in the flap extension process. This prevents a flap asymmetry from adversely affecting aircraft handling.

4 Some airlines use different terms but follow similar training strategies.
5 Hoffer and Hoffer (1989). Also, this event was depicted in the film, *Falling from the Sky: Flight 174,* Pacific Motion Pictures. 1995, and in other TV reenactments.
6 Loudon and Moriarty (2017).
7 Endsley (2022). Dr. Endsley writes about how we *project SA* into the future. While I have been using the term *projected SA* in my writings for several decades, I acknowledge her work as the inspiration.

BIBLIOGRAPHY

ASRS – Aviation Safety and Reporting System – National Aeronautics and Space Administration (NASA) Online Database. (n.d.). Retrieved from Aviation Safety and Reporting System: https://asrs.arc.nasa.gov/search/database.html.

Endsley, M. (2022). *SA Technologies – Publications*. Retrieved from SA Technologies: https://satechnologies.com/publications/.

Hoffer, W., & Hoffer, M. M. (1989). *Freefall: A True Story*. New York, NY: St. Martin's Press.

Loudon, R., & Moriarty, D. (2017, August 1). *Better Briefing*. Retrieved from Royal Aeronautical Society: https://www.aerosociety.com/news/briefing-better/.

NTSB. (1997). *NTSB/AAR-97/01 and PB97–910401: Wheels-up Landing – Continental Airlines Flight 1943 – Douglas DC-9 N10556 – Houston, TX, February 19, 1997.* Washington D.C.: National Transportation Safety Board.

13 Time Management Techniques

We perceive time by our perception of pacing, the flight schedule, and how time equates to distance or fuel. Let's examine some techniques to manage delays across the range of flight operations.

13.1 WHEN DELAYS AFFECT OUR SENSE OF PACING

Passenger delays, crew misconnects, bag/freight issues, mechanicals, gate holds, and weather delays are the most common time-related challenges that we encounter.

13.1.1 Understanding How Timelines Interact

Before we can address how to handle delays, we need to understand the nature and interaction of the various timelines spanning across our operational process. This helps us determine where to apply our leverage. For example, if we have a problem with the fuel load on our flight release, we identify three separate timelines involving our dispatcher, our operations agent, and our refueling agent. The most time-sensitive action involves the refueling agent because their timeline is more restrictive than the other two. So, our first step is to contact the refueling agent to have them hold off refueling until the discrepancy is resolved. If the refueling agent completes the assigned upload and departs, it lengthens our delay to call them back to resolve the fuel load. The next step is to contact our dispatcher to revise our flight release. This acknowledges that the resolution process runs from the dispatcher (who authorizes the amended fuel load), to the operations agent (who revises the fuel slip), to the refueling agent (who uploads the fuel indicated on the revised fuel slip). Each task is dependent on the previous one. While refueling is the final step in the process, it is the most-restrictive timeline that we need to manage first to minimize our delay.

To manage the timelines skillfully, we study each operational process to understand the conditions influencing how they interact. Early in my career, I flew with a Captain who could accurately predict our turn time as we pulled into the gate. It seemed like magic until he explained that he assessed the number of baggage carts, how full they were, the number of ramp agents waiting to offload/onload bags, the presence/absence of a fuel truck, the number of passengers to deplane, and the number of people gathered in the gate area waiting to board the flight. Knowing how each process interacted and interdepended with each other, he mentally connected them to predict the turn time.

Understanding each process gives us the insight into where to apply our leverage. For example, with high passenger loads, the most limiting timeline is the process of passenger deplaning, cabin preparation, and passenger boarding. Within that

DOI: 10.1201/9781003344575-15

timeline, the most time-consuming subtask is transporting wheelchair passengers. After approving the follow-on flight paperwork, I would enter the jetway and see if we had enough wheelchairs and attendants standing by. If not, I would acquire more or push some chairs myself. By applying my leverage to the most restrictive timeline, I reduced potential delays.

Having flown for an airline with tight turn-times and no crew meals, another time-consuming factor was crew food runs. If a flight attendant left the aircraft to get food, we had to delay boarding until they returned. One technique I used was to offer to make the food run for the crew. During cruise, I asked everyone to write down their orders. After arriving at the gate and ensuring that my preparation items were completed and that the turn was running smoothly, I entered the terminal and filled their food orders. Everyone stayed on task and we kept the timeline moving.

13.1.2 Unavoidable Delays

Sometimes, we just have to wait out delays imposed on us. For example, if we are assigned a metered departure slot time, we typically can't alter it. While we are waiting, we can monitor related operational timelines that are affected by the delay. These include passenger connections, scheduled crew changes, duty day limits, gate availability, and airport curfews.

Two common examples are metered gate holds and adverse weather delays. During these delays, we need to decide whether to allow passengers to deplane or to keep them onboard to remain ready in case the gate hold is suddenly lifted. We have a number of resources to help us with this decision. We can monitor real-time weather to track convective storm cell movement. By comparing the lightning strike display with local ramp closure procedures, we can estimate when the airport might resume ground operations. We can see how much more work the ramp crew has left. We can then estimate how long it should take them to finish their work after the ramp closure is lifted. We can call company operations to learn their plans for accommodating disconnected passengers. If it looks like our flight might be cancelled, we can ensure that the station has extra agents ready to handle passenger accommodations. As we expand our awareness of the various timelines, we learn to identify the chokepoints where we can skillfully apply leverage to minimize delays.

During long delays, we should make ourselves available to passengers, crew, and other employees. Sometimes, passengers just need to hear a pilot explain what is happening. This is especially relevant during mechanical delays. To us, a mechanical delay is a minor inconvenience. Whenever passengers hear that there is a mechanical issue, their anxiety level rises. They may only retain that they are about to fly on a "broken" aircraft.

Gate delays are common and boring to us, but rare and scary to passengers. Passengers deserve to know what is happening, but we should carefully choose how we communicate the details. If we transmit boredom through our tone, they may interpret that we don't particularly care. A much better option is to make periodic PAs using a calm and confident tone. Include enough detail so they can understand the cause, but avoid technical details or jargon. Passengers need to feel confident that we have the situation under control. Periodically, update them with the progress of the repair. "The maintenance technician is awaiting a replacement part that is on its way here." "The repair is complete and the technician is completing the paperwork." If the process is taking longer than expected, it might be useful to assure them, "We only fly safe aircraft."

13.2 MAKING UP TIME FOLLOWING A GROUND DELAY

Once we finally get moving, we should guard against frustration and rushing.

13.2.1 Succumbing to Frustration

Imagine that we've been delayed at the gate for an hour and are finally released. Ideally, everyone will work efficiently and purposefully to get the aircraft moving. To succeed, all of the teams need to cooperate and complete their tasks. For example, we can do everything right to prepare the aircraft, but then encounter prolonged delays because the ramp crew has left to work another flight.

Operational delays following a long gate hold can feel frustrating. Frustration encourages shortcutting and rushing. Consider the following report from a freight carrier crew who were informed just before pushback that they had over 1,000 KG of unreported dry ice aboard.

BOX 13.1 SHORTCUTTING PROCEDURES FROM LAST-MINUTE CHANGES FOLLOWING A LONG DELAY

Captain's report: …the *flight was not loaded and ready to push until 2 hours and 30 minutes after our scheduled departure time.* We would later recall the ramp agent remarking as he closed us out that he had been called out to work this flight, and that it was not his normal flight. As we received pushback clearance, we also received an ACARS message indicating "FOM 10.37 Dry Ice Supplemental Procedures in Effect." I was asked to acknowledge the message via ACARS, which I did (as I understand it) to modify the flight release and make the flight legal to dispatch. I held the push as the First Officer got out of his seat, pulled out the [CO_2] monitors, and powered them up. I verbalized that the APU was already running with the packs on, that we would perform a 2-engine taxi, and we would need to start the APU after landing at ZZZ1. *I assumed (wrongly) that I had missed the FOM 10.37 reference in the DG (Dangerous Goods) paperwork I had signed, AND that the FO had already provided a copy to the ramp agent while closing, and had RETAINED a folded copy of in his back pocket.* We continued the push, now 2 hours, 34 minutes late for an uneventful departure. When we leveled off, we reviewed the DG paperwork again and pulled out the 10.37 procedures in the FOM. *It was then I discovered that the paperwork did not have more than 1,000 KG of dry ice, that it did not have a reference to FOM 10.37 and that the FOM procedure also prohibited lives onboard, which we had (chicks). I felt relieved for our safety that we had air on the aircraft while conducting our preflight and waiting on the load for well over 90 minutes, although we had no knowledge of the dry ice during that timeframe. The hot weather conditions were the only thing driving us to maintain airflow on the aircraft.* While I am confident that we did not willingly break any FARs, I am not confident that we received dry ice updates in a timely manner, that we were dispatched with the right paperwork, nor that we met company policy prohibiting live animals while traveling with more than 1000 KG of dry ice per FOM 10.37. *Excessive delays and ground crew changes likely drove some of these issues, and we should have pulled the paperwork and recalled the "lives prohibited" part of the exemption. However I do believe receiving a 10.37 exemption while conducting a push was excessively late in the game and was not a recipe for successful compliance for this flight.*

FO's report: … *The Captain acknowledged the message and said he was familiar with the procedure and we would continue as he discussed items regarding the procedure.* I've only dispatched with the exemption a few other times, however, I felt comfortable with what was required. I further stated that we were already planning a two-engine taxi and that we would need to run the APU upon arrival in ZZZ1, leave the APU running at the gate and notify the ramp personnel in ZZZ. I also retrieved the Drager CO_2 which the Captain turned on for us both and we wore per the procedure. *The Captain made the comment that we could review the rest of the procedure while in cruise.* The Taxi and Departure was routine. Once we got to cruise, the Captain and I reviewed the Supplemental Procedures and our DG paperwork. We realized that the paperwork was incorrect and showed live animals onboard … which is not allowed per the FOM.

Cause: I think multiple things contributed to the event. First, the hub has been severely delayed since the COVID crisis started. I think everyone is a bit tired of this continued "Peak-like Operation." I know the ramp agent who closed out our flight prior to push made a comment that he was unfamiliar with our flight and was working another flight when told to close our flight out so we could push. *Further, it was a long duty day for me and the Captain was eager to get to the layover. It was our first leg together, and while the Captain seemed very capable, we weren't used to flying together yet. I also believe the long and unfortunate cultural norm of "rushing" started to creep in given that we were so late.*

Suggestions: *My biggest frustration is that normally I am very good at slowing down a situation that is rushed and take pride in my attention to detail assuring that things are accounted for and speak up when I think we need to take more time. Unfortunately, given the late nature of the GOC message regarding the Supplemental Procedures coming during the pushback, and assuming the Captain was more familiar with the procedure, I did not suggest to the Captain that we return to the gate and or sit off the gate once pushed back to review the procedure and paperwork to assure we complied with all company procedures. At the time I felt comfortable enough with the Captain's brief of the procedure and reviewing the procedure while enroute given that we verbally talked about the procedures required. That was a mistake and something I normally don't do. I never assume in the airplane and always prefer to verify/review information on my own. I will be more diligent on not allowing such complacency. I should have insisted that we stop and assure we were in full compliance considering that I hadn't had the Supplemental Dry Ice Procedure often.* Further, what is disturbing, is that the ramp agent that closed out our flight obviously was not aware of what was onboard the airplane. He was unaware, and it had not been commented to him that we had the excessive amount of Dry Ice and Live Animals on board. Further, a checklist needs to be better developed for the ramp to use, especially with notifying dispatch and the flight crew in a timely manner of the dry ice supplemental procedure need. *We had been on the airplane with doors closed and engines off/packs off in various stages of configuration with no idea that we were carrying the excessive amount of Dry Ice. …* Further, I need to do better as a First Officer to question the Captain of the need to properly review the procedures to assure compliance as soon as something comes up like this comes up. Again, I'm very frustrated with my performance and certainly will be more diligent.[1]

Notice that the Captain immediately assumed that he knew all of the precautions and procedures for carrying the dry ice. While they were careful to ensure that their own CO_2 monitoring equipment was used, they didn't account for the restriction against live animals on board. Solving one problem doesn't mean that we have solved every problem. The FO accepted the Captain's assurance and they departed. It was only when they were established in cruise that discovered their oversight. Notice the FO's statement, "My biggest frustration is that normally I am very good at slowing down a situation that is rushed and take pride in my attention to detail assuring that things are accounted for and speak up when I think we need to take more time." Despite this resilient mitigation technique for avoiding errors like this, they allowed themselves to succumb to frustration and rushing. This response can also be related to the reported factors of operations fatigue, paperwork errors, long duty day, desire to get to the overnight, and rushing culture.

13.2.2 Feeling Like We Need to Hurry

Many of the elements in this NASA ASRS report are common with frustration-driven events. Crews hurry their procedures, taxi faster than normal, and fly faster than flight-planned. In truth, many of these catch-up techniques don't really make up much time. We do them because they make us *feel* like we are making up for lost time. The crew in the last report departed over 2½ hours late. There was nothing they could do to make a significant dent in that delay.

13.2.3 Start-Up Lag

When the delay finally ends, every group needs some time to reassemble their teams, rebuild their SA, and get their timeline moving. Teams may need to begin anew or with different crewmembers. Like starting a car engine on a cold day, we'll need to let it warm up a bit before we start moving. Even if we are caught up with our flight-deck workload, we need to ensure that the other team members are fully up to speed before resuming our operational pace. Otherwise, we may inadvertently encourage them to rush to keep up with us. As we assess the needs of each timeline, we identify chokepoints and constraining processes. We begin with the most time-consuming processes. Once they are underway, we follow with the remaining processes that can restart more quickly.

13.2.4 Priorities Following Delays

When we are finally released from the delay, we'll need to rebuild our SA (past, present, and future), resume our timeline pacing, and assess our crew's readiness. Operational forces and frustration encourage us to move quickly. We need to overcome "pushes" and move deliberately at our selected pace. It may be useful to back up and repeat previously accomplished preparation tasks and briefings. We may discover that we need a fresh weather packet or a revised fuel load, for example. Ideally, we will have some advanced notice of the restart. Our dispatcher may learn of a weather release before local ATC. We can begin our preparation and be ready to move when the word finally reaches ATC.

13.3 MAKING UP TIME INFLIGHT

For a 2-hour flight, a 5% speed increase saves only 6 minutes.[2] Is it worth the fuel burn, airframe buffet, stress, and additional risk to save 6 minutes? Probably not. Still, pilots do this every day because flying faster makes them *feel* like they are actively doing something about the delay. Undesirable "catch-up" techniques include flying against cruise Mach limits, flying at turbulent altitudes, shaving our clearance distance around convective build-ups, and planning faster descent speeds. None of these measures make up much time. Also, the fuel burn penalty may limit our contingency options during arrival. As Master Class pilots, we resolve to use logic and facts to assess potential time-saving techniques instead of how they make us feel. There are exceptions such as making a tight curfew or carrying passengers to a hub that his holding numerous aircraft for our transfer passengers. So, balance the risk and make deliberate choices.

13.4 MAKING UP TIME DURING ARRIVAL AND TAXI-IN

High-volume operations force ATC to assign matching speeds to all arriving aircraft. We may plan on increasing our descent speed to make up for lost time, but ATC typically slows everyone down to standardize aircraft spacing. Additionally, higher descent speeds may increase the risk of turbulence injuries to flight attendants.

When we are running late, the arrival station also needs to compensate. Many stations have tight gate availability and limited manning. No gates may be available if we arrive after our scheduled arrival time. Also, they may not have enough station workers to operate the jetways or process baggage.

Rushing during taxi-in contributes to excessive taxi speed events, damaging ground equipment from our excessive thrust, and wingtip damage. For example, instead of waiting for wing-walkers, they might try to visually estimate their wingtip clearance. Unfortunately, it is very difficult to judge clearance while sighting down a wing line at a lateral obstruction.

13.5 TECHNIQUES FOR HANDLING DELAYED OPERATIONS

When there is time-slack in the operation, it dampens the consequences from delays. As airlines squeeze more efficiency from the operations, they remove this slack. This forces every flight to remain within its narrow time window. When delays push us out of our window, we no longer fit into the overall operation. We are like an unexpected guest at the dinner table. Station managers have to find room to accommodate us. After we become delayed, resource limitations conspire to keep us late or make us even later. We might succeed in overcoming one delay only to find another rising across our path.

13.5.1 Use Warning Flags to Slow Down and Investigate Further

In the previous NASA ASRS report with the mishandled dry ice procedure, it seems that the crew was relying on the Captain's memory instead of consulting the manual. They successfully recalled the requirement for personal CO_2 monitoring equipment, but not the documentation requirements or live animal restriction. Referencing our procedure categories, this appeared to be a rare normal procedure – one requiring us to consult the manual to review procedural steps that we don't handle every day. Notice how they perceived the need to *do something* about the problem. It might have felt like they were *doing enough* to address the problem. This is a common feature of rushing-type errors.

What if the crew had interpreted the ramp agent's report that he wasn't familiar with their flight due to a last minute personnel swap and the dispatcher's late ACARS message about the dry ice as warning signs? Both appeared to be the kinds of events that happen when an operation is under stress. They might have concluded that procedural, error-capturing safeguards had failed. What if they stopped and asked themselves, "What else have they missed?" Acknowledging that we are the last line of defense to detect and mitigate errors, the prudent course of action is to take some extra time to investigate warning signs. Consider how the event would have evolved if the Captain had stopped the pushback and called the dispatcher while the FO reviewed dry ice procedures in the manual. The error would have been caught and corrected. In retrospect, both pilots clearly acknowledged that they should have slowed down, consulted the manual, identified the error, and resolved it before departing.

13.5.2 Search for Lesser-Included Consequences

Hanging over this event is the hazard that they were unknowingly sitting in an aircraft with over a ton of dry ice venting CO_2 for over 90 minutes. Luckily, they did have the APU and packs running (only because it happened to be a warm day). They probably felt that they had dodged the most serious hazard of crew CO_2 poisoning.

It may have felt like a win – that they had successfully caught the worst consequence of the problem. Unfortunately, solving this big problem overshadowed the lesser-included problems – dangerous goods documentation and the prohibition against carrying live animals with dry ice in the enclosed freight cabin.

A useful strategy starts with detection of the error. In this case, it was the discovery of the undocumented dry ice. Next, we would examine the consequences emanating from that error. The most significant was CO_2 poisoning of the flight crew. Resolving that, the next step is ensuring that protective equipment was activated and operating. The crew satisfied both of these procedures. The next step is asking, "What else was missed?", or "What else do we need to check for dry ice procedures?" This would lead them to consult the manual, discover the prohibition against live freight (baby chicks, in this case), and lead to fully resolving this oversight before departure. A final step is documenting the event through a hazard report to the company safety department. They could then investigate how the problem was allowed to progress to the pilot level without earlier safeguards detecting or mitigating it first.

13.5.3 There Is Always Time to Do the Right Thing

A useful adage is that there is always time to do the right thing. Bad things happen when we rush. Whenever we are running late, we are already in the Yellow. Pushing ourselves faster feels like we are helping our situation, but we are probably only piling on more risk. Safe operation is the gold standard that should guide our choices. In a strategic sense, we are all committed to operating safely. In a tactical sense, however, we sometimes lose our safety focus. How does this happen? We lose our focus because safety feels constantly present. Anything that is constantly present in our lives can eventually be taken for granted. Like the air we breathe, the pursuit of safety becomes obvious and unnoticeable. As we lose awareness, our attitude toward safety becomes dismissive. "Of course we are safe. We are always safe." The Master Class technique is to stop and ask ourselves, "What is the right thing to do with this situation?" Especially when forces are pushing us to move faster, slowing down is the inconvenient, right thing to do.

BOX 13.2 THE BROKEN LAVATORY DUMP HANDLE

I experienced an event where the ramp agent unintentionally broke the lav dump handle while servicing our aft lav. He came to the flightdeck and informed us. This was the last leg of our pairing. Our flight would terminate at our crew base, which was also a company maintenance station. My FO suggested that we should just close off the lav and write up the discrepancy after we landed at our home base. I disagreed and called for contract maintenance to defer the handle and secure the service door. We experienced a significant delay. My FO became increasingly frustrated. He used the length of the delay to justify his recommendation that we should have carried the malfunction and written it up after landing.

Doing the right thing is often inconvenient and time consuming. We need to accept this. For a decision-making standard, apply due diligence. When we are late, we can still make reasonable efforts to speed things up and reduce delays. At the same time, we need to increase our vigilance for counterfactuals – signs that important details may have been missed and might allow errors to surface. Recognizing that our pacing will be thrown off, we apply reasonable due diligence to our monitoring and verification. This moves us back into the Green.

13.6 FAST-PACED AND EMERGENCY OPERATIONS

A related challenge is slowing our pace when events start moving too quickly. Quickening is a warning sign of a failing scenario. The ideal solution to a quickening scenario is to slow the pace. Aviation operations often resist our efforts to slow down.

13.6.1 Managing Time during Emergency Situations

Simulator training unintentionally simplifies managing time during emergency situations. Instructors place the simulator on freeze or adjust headwind/tailwind components to give us just enough time to resolve our problem. Additionally, simulator training tends to reduce distractions like calls from ATC, dispatchers, station operations, and the cabin crew. They create a protective time bubble for us to diagnose the problem and complete the required checklists before taking us off-freeze to fly the approach and landing.

Time management during actual emergency events can be much more difficult. Ideally, we would like to arrive at a convenient position on downwind or base just as we finish our preparation and briefings. We don't want rush this process, but we also don't want to finish our preparation while still flying away from our intended landing runway. This means that we need to simultaneously manage two timelines. The first is the time needed to diagnose the aircraft problem, run required QRH remedy checklists, and coordinate with affected agencies and crewmembers. The second is the time needed to manage our flight pattern to arrive at a convenient position to land. Ideally, we manage these two timelines to match. We estimate how long it will take us to complete our preparation. We manage our pattern to fly half of that time away from the runway or in delay vectors and the other half of the time returning. If we measure it accurately, we will arrive at a convenient downwind position when we are ready to fly the approach. If we need more time, we can extend the downwind and accept a longer final.

13.6.2 Managing Pattern Distance and Location

Mishaps analysis reveals that many flight crews don't consider how to effectively manage their time during their emergencies. For example, crews experiencing engine failures prefer flying straight out on runway heading while they climb, accelerate, and deal with the problem. This makes sense early in the scenario. Unfortunately, many crews continue flying away from their intended runway while securing the failed engine, attempting restarts, and preparing for an approach. As they diagnose and manage their problem, their distance away from the airport grows. Only after they complete their preparation do they realize how far they will have to fly to return for a landing.

Another consideration is possible diversion. If our original airport is experiencing marginal weather, we might consider using maneuvering airspace that lies between that airport and our planned diversion airport. If the original airport proves unfavorable for landing, we would be better positioned to divert to our alternate.

The opposite consideration applies when we have a lengthy fuel burn-off requirement as with many landing gear, flap malfunction, and controllability problems. In these cases, we should consider finding some free airspace that doesn't interfere with airport arrivals and departures, but keeps us close enough to land in case additional problems develop.

Our objective is to form a *time management game plan* that runs concurrently with our *problem management game plan*. The three indented steps within this MATM Checklist reflect our time management game plan.[3]

- Maintain aircraft control.
- Analyze the problem and perform any time-sensitive remedy steps.
    - Decide on the best landing option.
    - Estimate how long it will take to complete all checklists and preparation.
    - Coordinate maneuvering airspace or pattern size that fits our time estimate.
- Take appropriate action.
- Maintain SA.

13.6.3 Refining Our Emergency Mindset

We engage simulator training events with an emergency mindset. We know that the instructor will generate systemic and scenario challenges. Our SA aligns with our emergency mindset. They work together to accurately monitor, detect, and handle any emergencies and operational challenges.

With line flying, however, everything usually goes well. Our prevailing mindset is that the flight will progress normally and that no emergencies will occur. We may go years between experiencing emergencies.[4] The differences between the simulator world and the real world can lead us to subconsciously adopt different mindsets. When something goes wrong in the aircraft, our startle lasts longer. Our recovery from surprise is slower. Meanwhile, the aircraft continues to move forward, workload builds, distractions increase, and complexity mounts. Training programs that use the Advanced Qualification Program (AQP) format present realistic scenarios that require effective timeline management within emergency scenarios. While they still lack much of the startle factor than we experience with line-flying emergencies, AQP training provides useful experiences for handling emergencies within a line-flying context.

As Master Class pilots, we practice switching between our daily flying mindset and our emergency mindset. By mentally imagining and rehearsing possible emergency situations, we practice the mental process of quickly switching to an emergency mindset. Practices include mentally rehearsing engine failure procedures before taking the runway for departure, windshear recovery steps before commencing our approach near convective weather, and missed approach steps before commencing an approach to weather minimums.

13.6.4 Responding to Quickening

Quickening is an effect where problems and tasks seem to multiply at an ever-increasing rate. This happens when we try to force a failing game plan. The trap is that the techniques we use to handle a difficult, salvageable situation are the same techniques that we use with a quickening, failing situation. Ideally, some warning sign triggers our decision to switch to a contingency backup plan.

- *Corrective measures aren't solving our problem:* For an unstabilized approach, we configure full flaps to maximize drag and pull power to idle, but discover that the airspeed still isn't dropping.
- *We are down to our last corrective measure:* Still unstabilized and out of ideas, our airspeed remains above stabilized approach limits.
- *Our attention is tunneling toward a single parameter:* We focus on our touchdown point to the exclusion of all other parameters.
- *Conditions are marginal and trending worse:* Weather at the airport is already marginal and the weather is trending worse.
- *Complexity is generating unexpected indications and warnings:* Airspeed is high. It doesn't seem to be dropping. Automation begins announcing, "Too Low Flaps", then "PULL UP".
- *Parameters are not corrected as we reach a limit criterion:* We are using all available corrective measures, but the airspeed is still too high as we reach our company's unstabilized approach limits.
- *Insufficient time is available to diagnose and correct a problem:* Turning base, we extend flaps and discover a flap malfunction.

13.6.5 Respecting Limits

Notice that all of these triggers share a *running out* aspect. We run out of corrective measures, time, ideas, or safety margin. Consider the concept that every problem has a window of time available for corrections to take effect. While we are within that window, we are free to use every reasonable technique to solve the problem. At some point, however, that window closes and time runs out. This decision point is typically defined by procedural or personal limits. Once we reach that limit, the window for diagnosing closes, the window for trying closes, and the window for hoping closes.

As conditions become particularly volatile, we may set personal limits that are higher, further, and earlier than procedural limits. For example, if our company's stabilized approach limit is set at 500′, but the approach is clearly unsalvageable at

1,200′, we can go around early. We don't have to wait until 500′. Industry data supports that most go arounds do begin well above procedural limit altitudes, so many of us are already following this technique. Ironically, when pilots continue unstabilized approaches down to the procedural limit, they usually choose to continue and land.

NOTES

1 Edited for brevity. Italics added. NASA ASRS report #1753421.
2 Assuming that the planned cruise at FL390 is .78 Mach (447 TAS), we would need to cruise at .82 Mach (470 TAS) for 2 hours of cruise flight to achieve a 5% reduction. Depending on our aircraft, this may press against the Mach limit or cause airframe buffet.
3 The MATM Checklist is (1) Maintain aircraft control, (2) Analyze the problem, (3) Take appropriate action, and (4) Maintain SA.
4 My personal experience from over 10,000 airline flights is that I had two landings requiring declared emergencies. Both were relatively simple problems requiring very few modifications from normal landings.

BIBLIOGRAPHY

ASRS – Aviation Safety and Reporting System – National Aeronautics and Space Administration (NASA) Online Database. (n.d.). Retrieved from Aviation Safety and Reporting System: https://asrs.arc.nasa.gov/search/database.html.

14 Workload Management Techniques

The two main issues with workload are distribution and planning. Distribution refers to how we allocate workload within each flight phase. Planning concentrates on gauging our future workload, how it might rise or fall, and how we intend to manage it.

14.1 WORKLOAD DISTRIBUTION

To understand how we allocate work, let's examine it from both a short-term snapshot (micro perspective) and from how it transitions across each flight phase (macro perspective).

14.1.1 Workload in-the-Moment – The Micro Perspective

Within one small segment of time, there are four workload categories (Figure 14.1). Our most pressing duties are the *immediate* tasks of flying the aircraft, managing aircraft systems, and making essential crew communications. These are all tasks that we need to accomplish right away. If we are engaged with a less-pressing task, we would interrupt it in favor of the immediate task. The next level is *short-term* tasks such as answering ATC communications, near-future planning, configuring the aircraft, completing normal checklists, and time-sensitive decision making. We normally address these as soon as they arise, but we may defer them for a few moments to handle immediate tasks. Next, we have *long-term* tasks such as building future SA, planning for future flight phases, deliberative decision making, modifying our current game plan, preparing contingency plans, and detailed discussions such as coordinating our shared mental model. Any time left over is discretionary time that we can use for activities like conversation or just watching the clouds float by. Figure 14.1 displays these priorities vertically starting with immediate tasks. If immediate tasks are satisfied, we attend to short-term tasks and then long-term tasks.

Our workload potential is capped by an upper limit that represents 100% effort. The ceiling height rises or falls depending on our state of alertness (lack of fatigue), motivation (how hard we feel like working), and capability (our cumulative ability gained through experience, currency, skills, knowledge, and professional maturity[1]). Our ceiling peaks when we are fully rested, engaged, and experienced. It falls when we become fatigued, lack motivation, are inexperienced, or have allowed our skills to atrophy.

14.1.2 Workload across the Flight Phases – The Macro Perspective

Expanding our viewpoint outward to a macro perspective, we see how workload changes across each flight phase.

DOI: 10.1201/9781003344575-16

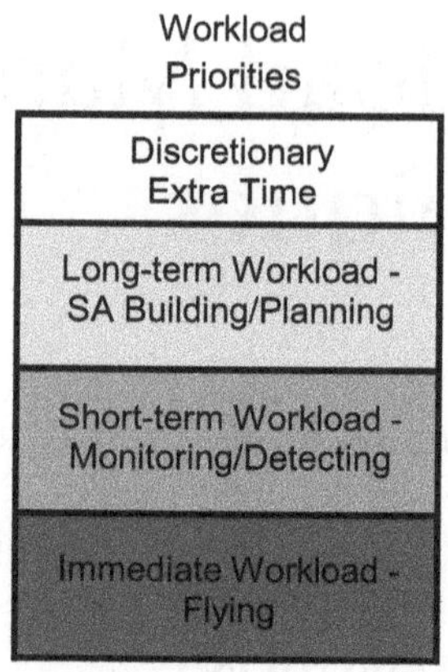

FIGURE 14.1 Workload priorities at any moment.

In Figure 14.2, notice how the four categories rise or fall depending on the workload demands within each flight phase. Notice how some phases are particularly busy, while others are relatively relaxed. Typically, high-workload phases include pushback, engine start, taxi-out, takeoff, and departure on the front-end of the flight, and descent, approach, landing, taxi-in, and gate arrival on the back end of the flight. Between these, we have the cruise phase which tends to feature a low workload.

Immediate and short-term tasks absorb almost all of our capability within high-demand flight segments. Long-term planning and discretionary extra time are effectively squeezed out. Conversely, preflight, cruise, and postflight present very few immediate and short-term tasks, so they offer opportunities for long-term planning and discretionary pursuits.

- *Preflight:* Starting at the left side of Figure 14.2, the early preflight phase contains very few immediate and short-term tasks. Most of our workload centers around long-term planning as we research conditions affecting the

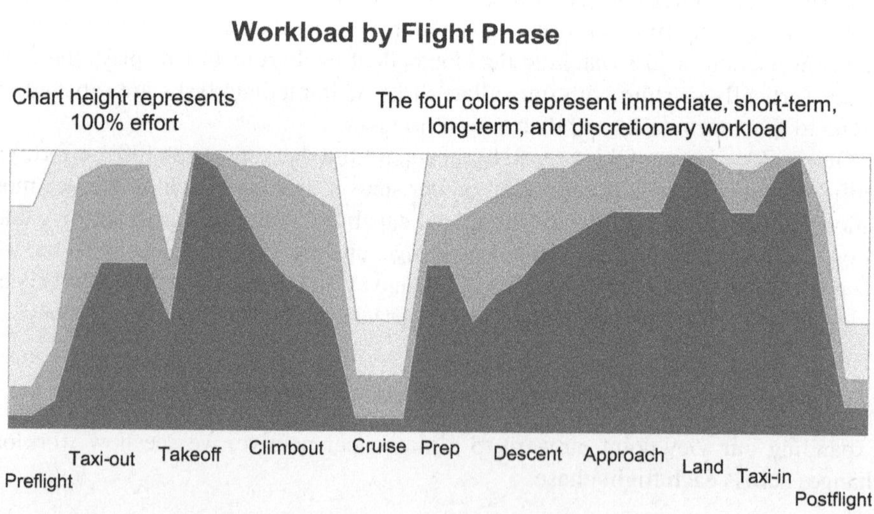

FIGURE 14.2 Workload priorities across all flight phases.

upcoming flight. Time permitting, we can attend to personal discretionary priorities. After we complete our personal preflight preparation, we transition to crew coordination, briefings, and required checklists. As we prepare for pushback, with all long-term planning tasks completed, immediate and short-term tasks begin to consume most of our attention.

- *Taxi-out:* During taxi-out, most of our workload is engaged by immediate and short-term tasks. The graph depicts a small dip between taxi-out and takeoff to reflect an operational delay such as joining a line of aircraft waiting for takeoff or awaiting a departure slot time. If we need to handle a disruption like a passenger cabin issue or to process a deferrable aircraft discrepancy, we would stop and set the parking brakes.
- *Takeoff:* Immediate tasks consume 100% of our attention.
- *Climbout:* Climbout shows a steady decline in immediate tasks as we gain altitude. Opportunities for short-term tasks increase. As the climb reaches the high-altitude structure, we open some opportunities for long-term tasks.
- *Cruise:* Our workload drops off after we transition to cruise flight. There are very few immediate and short-term tasks that require our attention. We have plenty of time for long-term planning and discretionary activities.
- *Before top of descent:* Approaching top of descent, we prepare and brief for the descent, approach, landing, and taxi-in phases. We complete landing performance computations, brief arrival and approaches procedures, and complete required checklists. Starting at this point, we suspend all discretionary activities until parked.
- *Descent:* The descent phase begins with a fairly evenly distributed workload between immediate, short-term, and long-term priorities. The lower we descend, the more our immediate and short-term task loadings rise.
- *Approach:* Immediate tasks continue to rise as we advance toward the terminal portion of the approach. There is no time allocated for long-term planning. If a significant change requires detailed planning and briefing, we may need to break off the approach pattern to make extra time.
- *Landing:* Immediate tasks demand 100% of our attention.
- *Taxi-in:* After clearing the runway, task loading drops. It remains fairly steady except during runway crossings, approaching the gate area, and parking where immediate tasks again demand our full attention.
- *Postflight:* Task loading drops off significantly after we park, shut down the engines, and complete our checklists. We regain opportunities to engage in discretionary activities.

14.2 MANAGING WORKLOAD USING PLANNING AND BRIEFING

The more we can plan ahead and *backload our workload*, the easier it is to manage the flight.

14.2.1 Preflight Planning and Briefing

We start by reviewing flight planning materials including our dispatch release, weather packet, NOTAMS, fuel load, passenger/freight load, and company information about our departure and destination airports. Using our past experiences, we assess the threats and challenges that the current conditions might create.

Next, we select our game plan. Recognition Primed Decision Making (RPDM) steers us toward game plans that are familiar and reliable. An added benefit is that these plans are probably the same ones that the other pilot also views as familiar and reliable. Line culture aligns our common perspectives. They tend to be consistent within each airline. Next, we complete our briefings and checklists. They solidify our shared mental model, ensure that FMS computers are accurately programmed, and guide the configuration of aircraft systems.

14.2.2 Handling Distractions during Preflight Preparation

The preflight planning process is particularly vulnerable to distraction. This is because the planning steps are interrelated with each subsequent step building upon the previous one. If we are interrupted, our train of thought can become disrupted. We either need to back up and rebuild our thought process or accept the disruption as an acceptable risk and keep moving forward. If we choose to move forward, we may rush the process and degrade the quality of our planning. Surrendering to operational time pressures, we can find ourselves hurrying through the planning steps to "check the boxes". This is another reason why we tend to favor our familiar and reliable game plans over innovative or complex ones. The following narrative demonstrates how distraction can breakdown preflight planning.

BOX 14.1 MISSED REQUIRED DOCUMENTS DURING A DISTRACTION

Captain's report: … [had] PDGS (Preliminary Dangerous Goods Summary) printed with the [Maintenance Release]. I separated the FDGS (Final Dangerous Goods Summary) from the [Maintenance Release] noting mentally that it said, "aircraft batteries" and placed it on the console while I reviewed the [Maintenance Release]. *Unfortunately, I subsequently got distracted by other preflight duties and personnel interactions and forgot about the PDGS (Preliminary Dangerous Goods Summary) which got covered up with all the other numerous documents that are printed during preflight. Upon reviewing my own personal departure checklist while pushing back, I realized that we had pushed without having received a Final DG Summary.* This was likely due to the fact that we had an extremely light load *and the agent approached the cockpit several minutes before scheduled departure time saying that everyone was on board and they were ready if we were ready. About the same time the push crew called on the flight intercom saying they were ready to push. I somewhat hastily called for the Before Pushback Checklist and unfortunately failed to consult my own personal checklist that I normally refer to in order to ensure that I have all documents for pushback*, resulting in our pushing before having received the FDGS which had most likely not come yet simply due to the fact that we were several minutes before departure time.[2]

Note how this Captain had developed a personal checklist to ensure that all requirements were completed before departing the gate. We presume that this personal technique evolved from errors made during previous distraction events like this one. This Captain's technique would have compensated for both the fluid nature of preflight preparation and to capture errors like this one. The carriage of aircraft batteries as freight requires Hazardous Materials (HAZMAT) procedures and additional documentation. This rarely occurs during most passenger flights. Since processing HAZMAT paperwork is a unique Captain duty, it is not surprising that the FO didn't detect or mitigate the error. We can imagine the Captain's inner dialog after the door was closed and the pushback started (and the distractions ended), "Okay, now what did I forget?" When running this personal checklist, the Captain discovered the error.

14.2.3 The Planning and Briefing Process

The preparation process starts with an unstructured, individualized portion. This is followed by a structured, crew-oriented portion. We start by getting ourselves ready, then getting the flightdeck crew ready, then getting the aircraft ready, and finally getting the support crews ready. This process highlights four distinct portions of the preflight planning and briefing process.

- *Individual preparation:* This is the most individually tailored and informal portion of our preflight planning. Each of us needs to fulfill our own preparation needs. If we have been away on vacation, we will need to spend more time preparing than the other pilot who flew this same flight yesterday. While intended as an individual task, it doesn't preclude us from discussing information and sharing knowledge.
- *Captain/PF game plan and shared mental model:* The second planning portion begins with a coordinated crew dialog. Depending on the operation, either the Captain or the PF briefs the game plan. Many airlines publish lists of briefing topics and considerations. These may be labeled as "Captain Briefing" or "Pilot Flying Briefing" on flightdeck reference cards. The intention of these guides is to facilitate communicating a shared mental model. This step typically includes a review of the flight clearance and a verification of the waypoints in the FMS to ensure that they match the clearance. The objective is to finish this preparation portion with a shared mental model of the game plan.
- *Challenge-and-response checklists:* The third portion of planning and preparation is completing the normal challenge-and-response checklists. These may be called the Preflight Checklist, Before Start Checklist (required before starting the engines), or the Before Pushback Checklist (required after receiving the final weight and balance numbers).
- *Coordination with the support crews:* The final portion of this planning and preparation phase starts by coordinating for pushback with the outside support crews and finishes with verifying the readiness of the flightdeck crew. The operations agent informs us that everything is ready. We confirm this

by hearing or seeing that the cabin entry door is closed and noting that the all aircraft door indication lights are extinguished. The cabin crew confirms their readiness by telling us or by closing the flightdeck door. Next, the operations agent backs the jetway away and the pushback crew communicates that they are ready. Before proceeding, a useful Captain technique is to perform a before pushback flow to ensure that all items are satisfied. Starting left, we check that the jetway is pulled clear of the aircraft. Then we scan to the right to see that the ground crew marshaller is in place. Then we transition inside to scan the flightdeck panels (ground power is disconnected and the aircraft is on APU power) and that all door and panel warning lights are extinguished. Finally, we confirm with the FO that the pushback clearance has been received and that they are ready for pushback.

14.2.4 Preflight Planning from the Gate through Departure

A typical flight offers two time windows for planning and briefing – one on the ground before departure and the other before top of descent prior to starting down. At the gate, our planning fulfills three main objectives. First is that we are legal to dispatch. This includes flight identification information, routing, alternates, fuel load, weight and balance, and additional planning considerations like crew duty day and curfews. We do this first because these discrepancies take the longest to resolve. Next, we ensure that we have acceptable routing. Depending on the weather and planned cruising altitudes, we need to satisfy questions such as:

- Do the field conditions and weather allow us to operate?
- What contingency plans should we consider?
- Do we have adequate extra fuel for deice procedures, long taxi delays, take-off alternate, enroute restrictions on routing and cruise altitudes, enroute winds, destination alternate, and holding?
- Does the game plan support the turbulence predictions for the planned route and cruise altitude?

Before departure, we concentrate on the earlier flight phases – pushback, taxi-out, takeoff, and departure. Since we probably won't know the actual arrival conditions until we get closer to our destination, we plan those using forecasts. Some useful planning considerations are:

- What weather conditions can we expect during arrival and approach?
- Do we have the aircraft equipment needed to land under these conditions?
- Will we require approaches lower than CAT I ILS? Are those systems fully available?
- Do all pilots have appropriate certification and currency for contingencies?
- For planned instrument approaches, is the required ground equipment operational?

One exception is short flights where the cruise phase is too brief to complete the before top of descent planning process. In these cases, we'll need to perform our detailed descent, arrival, approach, landing, and taxi-in planning before departing the gate.

- *Time-pressured gate departure:* Everything we have considered so far assumes that we'll have ample time at the gate. In operational flying, this doesn't always work out. Aircraft swaps, crew swaps, and flight schedule changes conspire to shrink our planning time. Policy guidance directs us to ignore operational delays and complete our preflight planning and briefing without rushing. Realistically, with slot departure times approaching, passengers waiting, operations agents nervously pacing in the doorway, ground crew informing us that they are ready to push, and the incoming aircraft holding out for our gate, most of us feel significant operational pressure to hurry. We don't just feel this pressure externally, we feel it personally. It is one thing to blame flight delays on outside causes (ATC, passenger/freight loading, and late aircraft arrivals, for example) and quite another when we are the cause of the delay.

  If we arrive late to the aircraft, perhaps our crewmate has tried to help us out by completing some of our preparation tasks. In some areas, this may prove useful. For example, Captains can perform the exterior preflight (walkaround) for their FOs. If we are the ones trying to help out a late crewmember, we should limit our assistance to areas that are truly useful. Every pilot typically rechecks all of their assigned responsibilities, so our assistance may be better spent reducing their perceived pressure to hurry. As Captains, we can project a calm and relaxed demeanor toward our time-pressured FOs. Encourage them to take their time. This reduces their stress level and communicates that thorough preparation is more desirable than speed.
- *Finding our time-pressured comfort zone:* How we respond to operational pressure is up to each individual. On one end of the spectrum are pilots who choose to follow their normal, methodical preparation process regardless of the delay. In the middle are pilots who work quickly and use personal error-capturing techniques to make sure they haven't missed anything. On the other end of the spectrum are pilots who view the delay as a professional challenge to see just how quickly they can complete their preparation and get the flight moving. Like a sports team with the clock running out, they see how much they can accomplish in the shortest amount of time. The right answer here is the one that best suits each individual and the crew as a team. We each need to assess ourselves and select the preparation plan that works best for the given conditions.

  On our road to mastery, we have studied other pilots, adopted techniques that work for us, discarded techniques that didn't, and modified the rest to match our personal style. We have refined our process to form the most efficient and complete planning process. Perhaps we have discovered that we are most comfortable following our standardized, meticulous preparation process. Perhaps we work quite well under pressure and have developed an effective time-limited preparation process. Maybe we fall somewhere in between.

Our choice also depends on our capability in the moment. One day, we may operate at peak efficiency and can speed things up. Another day, we may feel that our abilities are degraded, so we slow down. Perhaps we are morning people flying late evening schedules or vice versa. We work toward our strengths and guard against our vulnerabilities. The bottom line is that when we arrive late to the aircraft, for whatever reason, everyone will have to wait until we are ready to go.

- *Compensating for time-pressured departures – using shortcuts:* We can use effective techniques to complete our preparation more quickly. Many of our techniques involve shortcuts. Shortcuts save time, but they tend to compromise some of the safeguards that are built into procedures. We need to balance the scales to compensate for exposing those vulnerabilities. If we add risk to one side of the scale, we need to add monitoring and contingency planning to compensate on the other side.

  Typically, we apply some time-saving assumptions that accelerate and abbreviate some of our more time-consuming tasks. For example, if the arrival weather is good, we may minimize detailed planning for later phases of flight. We assume that we can easily handle anything that emerges because our game plan is resilient enough to compensate for minor disruptions.
- *Compensating for time-pressured departures – deferring tasks:* The more efficiency that we extract from a system, the more risk we add. As we accept this additional risk, we need to compensate in some way. Perhaps we are leaving from a busy hub airport during the departure surge and expect to spend some time waiting in a long line of aircraft for takeoff. We can defer some planning tasks until joining that line (ideally, while stopped with brakes set). Perhaps we will have a long cruise flight phase. We can allocate some extra time after leveling off to make a thorough review of arrival NOTAMs.

  We can also mobilize our CRM protections. We can highlight our planning shortfalls and ask our crewmembers to be extra diligent for missed planning items. This cues them to raise their diligence while we catch up. Again, when we engage shortcuts, we need to proactively balance the scales.
- *Discovering what we might have missed – the detective perspective:* Like the Captain from the previous NASA ASRS report, we ask ourselves, "What did I miss?" We use an investigation perspective that starts with the assumption that we probably missed something. This motivates us to search for missed items. Compare this against a "I hope I didn't miss anything" perspective. Hoping that we didn't miss something biases us toward accepting our hurried preparation and assuming that we successfully accomplished everything. The *investigation perspective* biases us toward finding anything that we might have missed while the *hope perspective* promotes plan continuation bias.

14.2.5 Before Top of Descent Planning and Briefing

The second planning and briefing opportunity occurs before the top of descent. This planning and briefing window begins well before our FMS-computed top of descent point. For uncomplicated arrivals, somewhere about 50 NM prior works well. For complex arrivals, we'll need to start further back. The distance is not as important as the time we allocate. Since our minds do not recognize time markers as easily as visual markers, we might create mile references on our moving map displays. 50 NM prior gives us at least 5 minutes before starting down. On arrivals with ATC-directed intermediate step-downs, we'll identify an equivalent point that ensures that we'll have enough time to plan and brief.

- *Planning blocks before top of descent:* We'll need a sufficiently long, low-interruption block of planning time before entering the busy arrival environment. Just like with the preflight planning, there is an individual planning portion, a crew coordination portion, and a checklist/briefing portion. The 50 NM point corresponds with the middle, crew coordination portion. This means that we will need to complete our personal preparation and review prior to that point. Consider flying to an airport that we haven't visited in a while. We would give ourselves more time – maybe 80 NM prior to top of descent. Conversely, if we flew the same arrival yesterday under similar conditions, a shorter review block of time should be sufficient. For a particularly complex airport or a large hub operation, we'll need more time. For a small, quiet airport, we'll need less. The point is that our personal planning block varies.
- *Personal preparation before descent:* Our personal preparation should include a review of the current and trending weather conditions, a prediction of which arrival STAR will be assigned, the expected landing runway, available runway exits, standard taxi routes, taxi hotspots, complex taxiway intersections, gate considerations, company planning materials that outline gate and ramp operations, unique airport procedures, and a review of common pitfalls/errors. Notice that the majority of these planning considerations occur after landing. A common vulnerability among experienced airline pilots that we plan adequately for the STAR, approach, and landing, but not as well for ground operations. This seems to embrace a flawed mindset that the flight is effectively ended following our landing. Inadequately prepared for ground operations, we make taxi errors, commit runway incursions, and encounter ramp confusion events.

  Before top of descent provides an excellent opportunity to use CRM. Consider sharing your recent experience with the other pilot. "I haven't flown into this airport in about 6 months. How about you?" Perhaps the other pilot has flown there recently and can update us with the latest information. We can turn our personal review into a crew discussion that constructs an accurate prediction of typical ATC handling practices, airfield construction hazards, and gate procedures.

- *Game plan and contingency planning before top of descent:* At least 5 minutes before top of descent, communicate the game plan, and build a shared mental model for the arrival. We can't postpone arrival preparation items. Unlike departure where we might have time to finish preparing while waiting for takeoff, arrival workload steadily rises. Takeoff and departure start with a high workload and then ease off. Our workload during approach and landing steadily increases. If we haven't planned for possible contingencies early, we probably won't have time later. This is significant because once we become task-saturated, our viable options narrow to either breaking off the approach or pressing forward with the planned arrival. If we continue while task-saturated, it sets the hook for plan continuation bias and Recognition Trap errors.

  We need to build and coordinate our main game plan and contingency plans early. Many mishap crews planned an uncomplicated visual approach only to discover that the field had unexpectedly gone IFR. Hurriedly preparing and briefing an instrument approach while being vectored at low altitude is risky. If we aren't prepared for the instrument approach, the only way to create sufficient planning time is to break off the approach and vector back around. We hate doing this because it creates additional problems and delays. Getting vectored back around at low altitude is typically a fairly busy, noisy, and distracting environment. The quality of our planning and briefing will be degraded compared with calmly preparing while in cruise.

  Many approach and landing mishaps share a common feature – the crew felt compelled, empowered, or limited to continuing with their original plan even though adverse conditions were causing it to fail. As the saying goes, "The body cannot go where the mind has not gone first."[3] Mental rehearsal makes the switch easier. Consider examples from the Alpha and Omega Crews (where Crew Alpha diverted and Crew Omega landed under windshear conditions). Crew Alpha planned for the approach with game plan A, but due to complexity and unstable weather conditions, they fully briefed a backup contingency plan B. During their before top of descent review, the Captain reviewed some of the typical indicators that might trigger a change to plan B. Both pilots watched for those warning signs. Crew Omega briefed game plan A, but they didn't prepare for contingency plan B because they felt familiar with the airport and comfortable with the conditions. They weren't mentally prepared to actively search for warning signs or to switch to a contingency plan. As soon as they could see the runway and had clearance to land, their decision to continue became unshakable. This comparison demonstrates the resilience that we gain through preparation and briefing. Mental preparation supported Crew Alpha's shift to their contingency plan. Lack of mental preparation by Crew Omega narrowed their perceived options. Through their preparation, Crew Alpha actively monitored for the signs of windshear and microburst. It is much like a simulator profile when we are forewarned that we are about to get a windshear training event. Mental rehearsal eliminates the surprise factor so we easily switch to the required windshear escape procedure.

- *Contingency planning for the next worse condition:* When conditions are complex or uncertain, a useful technique is to plan for the *next worst* condition. If VFR, plan for the next worse condition by reviewing the main features of the instrument approach, having the chart displayed, and programming the FMC. If we expect to descend through a cloud layer to VMC conditions on final, the next worse condition warrants a full instrument approach briefing. If the weather minimums are approaching CAT I ILS minimums, then plan the next lowest minimum approach, like a CAT II or CAT III.

  The same goes for runway braking action reports. If the ATIS is reporting RCC 5 (GOOD), we should check our performance data for RCC 4 (GOOD TO MEDIUM) or RCC 3 (MEDIUM). If it hasn't rained in a while, surfaces can become extremely slippery as dust, oils, and rubber deposits mix with rainwater. If we rely only on the initial light rain (-RA) report or RCC 5, we might be surprised when the aircraft in front of us reports RCC 3 (MEDIUM). If this happens when we are on short final, we'll need to know whether we are legal to land. If we haven't checked that next worse condition and we aren't sure about landing performance, our only option is to go around. If, however, we had planned ahead and ensured that we were legal to land with RCC 3 (MEDIUM), then we could confidently continue. The following report shows how RCC 5 (GOOD) may not accurately reflect airport conditions.

BOX 14.2 FIELD CONDITIONS MUCH WORSE THAN REPORTED ON THE ATIS

Captain's report: Landed ZZZ Runway XXL. Weather was near mins. *ATIS was reporting RCC 5/5/5, with taxiways and aprons covered with snow covered icy patches.* Prior aircraft reported braking action MEDIUM-POOR [RCC 2]. The FO and I ran the numbers, after considering elevation, gross weight, and reported braking, the [data] showed plenty of room to brake with a 1X,000′ long runway. After landing, we also reported MEDIUM-POOR braking action to Tower with the midfield having the poorest braking action. We taxied clear and had to advise Tower of our exit and position as visibility continued to drop. …Taxi was slow as the taxiways had icy patches. We were also advised that the prior aircraft had taken 30 minutes to taxi to their gate. On making left turn from [Taxiway] 2 to 3, it was readily apparent that the taxiways were indeed slippery as even a turn at five knots still resulted in some traction issues. Upon nearing XXR westbound, we noticed XXR and [Taxiway 1] appeared snowed over (not cleared). *After advising tower that we were clear XXR, and upon applying brakes for the turn northbound onto [Taxiway 1], it became obvious the braking action was worse on the west side of the airport … with the taxiways in worse condition than I've ever experienced, ….*

Cause: Bad weather, mainly icing under snow covered runways and taxiways. Allow PIREPs to more directly affect the RCC reported on ATIS. *Even after we reported MEDIUM-POOR braking action after the previous landing aircraft had reported the same, the ATIS was still indicating RCC of 5/5/5.* This in my opinion is misleading and contrary to the whole purpose of reporting RCCs.[4]

This report reflects how ATIS and field condition RCC reports often lag behind pilot reports. The pilots expected worse conditions than the RCC 5 report on ATIS. By planning ahead for the next worse condition, they ensured that they had adequate stopping performance.

- *Checklists before top of descent:* The last preparation step before top of descent is completing required checklists. These are referenced under titles such as Approach Checklist or Before Descent Checklist. Their purpose is to ensure that aircraft systems are configured and that FMS waypoints are verified. If we allow latent errors to slip through, they may not surface until reaching the busy approach phase.

  An often-cited example is Continental Airlines 1943 (gear-up landing at Houston Intercontinental [IAH], February 19, 1996).[5] In this accident, the crew unintentionally missed moving their DC-9 hydraulics switches from LOW (used for the climb, cruise, and descent flight phases) to HI (HIGH - used for takeoff/departure and approach/landing). The HI setting is needed to provide sufficient hydraulic pressure to raise or lower the landing gear and flaps. The Captain (flying as PM) read the items from the In Range Checklist, but inadvertently skipped "Hydraulics". The FO did not detect his error. With switches in the LOW position, the aircraft did not produce sufficient hydraulic pressure to fully extend the landing gear and flaps. The consequences of this latent error did not surface until much later in the flight. They might have caught the error during their Approach Checklist, but they were interrupted twice. Lowering the gear and flap levers didn't fully extend either the gear or the flaps. Lacking the necessary drag, they couldn't slow down. They became fixated on their *airspeed symptom* and never detected their *configuration cause.* With events quickening, they missed completing the Landing Checklist. The aircraft landed on the airframe causing damage and some minor passenger injuries.

  Consider the significance of two effects – the snowballing of task overload that a single missed checklist step created and the sensory blinding that task saturation had on future error detection. As we read the CVR transcript from accident report, we envision the timeline for how events unfolded. We perceive two pilots fully settled into their comfort zone, while flying an uncomplicated approach, into their home domicile, and on the last leg of their pairing. Everything was going wonderfully until events began to quickly deteriorate. They became too busy dealing with the airspeed symptom to diagnose the configuration cause. They probably thought the energy problem was due to a wind shift, like an unexpectedly strong tailwind on final. They made normal corrections to handle this kind of energy problem, but the lack of drag made it impossible to slow the aircraft. As time ran short, we see their attention tunneling more tightly on their speed/energy problem. Their normal problem-solving habits were abandoned because 100% of their effort was focused on slowing the aircraft and holding glidepath. Their focus was so overpowering that neither pilot recalled hearing the continuous gear warning horn or "WHOOP, WHOOP, PULL UP". This is a testament to the blinding effect that task overload can cause during Recognition Trap errors. The FO reported, "I don't have any flaps … want to take it around?" The Captain replies, "no that's alright." In the end, two highly experienced, fully trained, well-intentioned pilots scraped an unconfigured aircraft onto a runway after missing the hydraulic switches from a single checklist step. Consider the following analysis questions.

- Do we have personal habits that may contribute toward us missing a checklist step like they did?
- Was the Captain reciting the checklist from memory or reading/verifying each step? If he was reading it, how could his habits contribute to missing the checklist step?
- Was there a laxity about checklist discipline? If so, how did it evolve? Are we also becoming lax with how we complete our checklists?
- How did they drift from the disciplined checklist usage that they demonstrated in training to what they demonstrated that day?
- Should the FO have detected the "wrongness" when the Captain missed the "Hydraulics" checklist step? Would we detect an omission like this?
- Were there problems with their flow patterns that contributed to making this error?
- Assuming that we missed this step like this mishap crew, do we have personal mini-flows or qualitative scans that would have captured an error like this?
- Do we have personal techniques that would detect the quickening pace of Recognition Trap errors? Would we go around or tunnel our attention and continue?
- Do we have a verification practice that checks for gear and flap positions following configuration changes, or do we move the lever and assume that they work as expected?
- Do we have a strong anchor to ensure that we complete the Landing Checklist?
- Do we have a short final mini-flow that checks aircraft configuration and landing clearance?

14.3 CHALLENGES OF WORKLOAD MANAGEMENT

This accident provides a useful opportunity to examine how we gauge and manage our future workload.

14.3.1 Effects of Stress on Workload Management

This accident report shows the detrimental effects of stress on workload management. The greater the stress, the more we succumb to our biases, tunnel our attention, and weaken our error mitigation. It affects us most when we are deeply settled into our comfort zone. Flying the same city pairings day after day can feel like repeatedly watching the same TV show. We easily follow the plot line because we already know what will happen next. We can allow our attention to wander because we feel like we are in full control of the situation. We relax because we know what has happened, what is happening, and what will happen next. Now, imagine that someone at the TV station suddenly switches the episode to a similar spot of an entirely different episode from the same series – an episode that we haven't seen before. The actors and settings are the same, but the plotline suddenly doesn't make sense. We might wonder, "What's going on?" Our attention level jumps from overly relaxed to fully engaged. We sit up, lean in, and search for meaning. As the disconnected plotline continues, we scan for context and connections. Some details are a little different from before while others are radically different. We can't follow it because the new story doesn't match our SA. We begin to question ourselves.

Compare this metaphor with what the crew of Continental Airlines 1943 must have experienced during their event. Everything was going well – just like they expected that it would. They were flying the final leg of their pairing to land at their home domicile. Just one more landing, an easy taxi to a familiar gate, and they would be done for the week. Unknowingly, they had missed switching their hydraulic pumps to HI – a simple slip – a mistake any of us could have made. Normally, the mistake would have been detected during the checklist. Unfortunately, they both missed the checklist step that would have captured the error. Was there a distraction? Were they just too relaxed to notice their oversight? Had their checklist discipline weakened? They continued their approach completely unaware of their error. There wouldn't have been any alert at the time. No effects or warnings of their latent error would surface until they started to configure for landing. Nearing the airport, the Captain lamented not being able to play tennis due to the rainy weather. Then, while they were comfortable in their environment and confident with their SA, the entire script suddenly changed – someone switched the TV show. The jet wouldn't slow down despite the flap lever being properly set. Indications didn't match the script that their SA had built. They became completely focused on the mismatch between their expected profile and what was actually happening. They missed or skipped all of the highly trained and practiced procedures that would have detected their error. While the gear handle was clearly down, they didn't verify that the gear was lowered. They didn't verify that the flaps matched the flap lever. They skipped the Landing Checklist. As stress and overload grew, they didn't even hear the continuous gear warning horn or the "PULL UP" warnings of the GPWS. Every available brain cell on the flightdeck was firing at 100% effort trying to make sense between what was happening and what should have been happening. The pace quickened as they fell into the Recognition Trap.

14.3.2 How We Allocate Time

We might conclude from the Captain's comments about not being able to play tennis that he was undisciplined or unprofessional. This applies a hindsight label that attempts to write his story. It allows us to dismiss the accident as an example of exceptionally undisciplined pilots making exceptionally bad choices. This is both untrue and unhelpful. It keeps us from looking into the mirror. Our efforts are better spent looking for similarities with our own habits and techniques.

Earlier in this chapter, we noted four categories of how we prioritize workload – immediate tasks, short-term tasks/planning, long-term planning, and extra/

discretionary time. Since the FO was flying, the Captain had no immediate tasks to manage. He was easily handling the short-term tasks/planning of aircraft configuration and responding to ATC radio calls. That left him plenty of what he apparently perceived as discretionary time. The fact that he was contemplating playing tennis later implied that his mind was filling this undemanding extra time with discretionary, non-flight-related thoughts. It implied that his mind was in a comfortable state – that everything was proceeding exactly as he expected it should. He had seen this TV episode many times before and knew exactly what would happen next.

As we gain experience, we become highly practiced at allocating our time. We know that immediate tasks demand our full attention. As soon as an immediate task pops up, we attend to it. When all immediate tasks are satisfied, we attend to short-term tasks. These short-term tasks are also handled quickly. Unlike continuous immediate tasks like hand-flying the aircraft, short-term tasks are distinct events that appear, are quickly resolved, and disappear. When the PF calls, "Landing Gear Down", we acknowledge the command, check that parameters are satisfied (speed within placard limits), grip the gear handle, and lower it (and hopefully monitor green gear-down lights). The same goes for answering a radio call. Until ATC calls, there is nothing to do. When ATC calls, we reply, comply with the instruction, and that task ends. In between these short-term tasks, the remaining time should be devoted to monitoring and long-term planning.

Long-term planning requires us to build a sequential chain of thoughts – a step-by-step progression of future tasks. We can imagine how the Captain might have become preoccupied with the long-term planning steps related to playing (or not playing) tennis later. Given this, his mind may have wandered to contemplate the details for how he would spend his afternoon. Again, it's not important for us to know what he was actually thinking. It is better to turn the mirror toward ourselves to evaluate whether we have unknowingly adopted some of these same habits. While serving as PM on the last flight of our pairing, do we sometimes contemplate how we will spend our day after we finish the flight?

14.3.3 Task Management and Appropriate Behaviors

The Master Class path challenges us to study how we allocate our time and attention. Ideally, we should always focus our attention on the aircraft flightpath during highly task-loaded phases like takeoff and landing. Starting from this ideal, let's examine some real-world variations.

While on final into our home domicile on the last leg of our pairing, imagine that we quickly glance out the side window to check for traffic backups on the highway that we use to drive home. Are we being undisciplined, lax, or just using a moment of extra time to survey a relevant piece of information that affects our future? We may conclude that our glance is essentially harmless as long as this is done quickly, that it doesn't distract from our monitoring duties, and that we don't violate sterile flightdeck protocols.

Let's dig deeper. Is anything lost by our momentary glance? Perhaps that moment glancing away was our first chance to detect a bird appearing in our flightpath. We would have lost the few seconds that might have made the difference between dodging the bird or hitting it. So, that glance did add some small amount of otherwise avoidable risk.

Let's ratchet up our momentary glance and see what other problems might emerge. What if our momentary glance becomes a long stare at the road traffic? This would significantly degrade our flightpath monitoring duties. Adding on, what if we make a verbal complaint about how the traffic backup will delay our drive home? Now, we would be breaking sterile flightdeck protocols. Adding more, what if we ask our FO if they use that same highway for their drive home? By this point, we would have clearly crossed the line as our actions become crew distractions that can adversely affect the performance of both pilots. By examining this range of scenarios and the ways they increasingly degrade our attention, we develop a deeper understanding of how seemingly innocent actions can affect the quality of our workload management.

Moreover, our behaviors tend to drift over time. Maybe the first dozen times, we silently make the quick glance. Subconsciously encouraged by the absence of adverse consequences, we stare a bit longer the next dozen times. With the third dozen times, we verbally comment on the traffic. Still, nothing bad happens. Supported by the absence of undesirable outcomes, we unwittingly disrupt crew roles by asking our FO distracting questions. Notice how our behavior drifts in small, seemingly insignificant increments that eventually reach an undeniably, unacceptable end state. This reflects normalization of deviance – the slow drift of behaviors that become subconsciously normalized because of the absence of undesirable outcomes.

It isn't as simple as labeling one act as harmless and another as hazardous. Most acts scatter between these extremes. Assuming that the ideal behavior represents the lowest risk, any deviation from it represents some increase in risk. Encouraged by the absence of bad outcomes and buoyed by confidence in our personal proficiency, we allow our habits to drift. Daydreaming, checking our phones, and digging through our flight bags are all behaviors that slowly become acceptable unless we consciously choose to arrest our drift.

We need to become students of pilot behavior – both in ourselves and in others. Perhaps one day, comfortably monitoring the other pilot's flawless approach down final, we will feel comfortable enough to gaze out the side windshield. Returning our attention to the flightpath, we recognize that we have spent too much time looking for our house, our bike path, or our tennis courts. We notice how long it takes us to rebuild our SA by visually clearing the flightpath for birds, assessing the PF's flightpath management, appraising the aircraft parameters, scanning for visual indications of winds and weather, clearing for surrounding aircraft landing on parallel runways, acquiring aircraft preceding us to the same runway, and verifying that our runway is clear. As Master Class pilots, we detect our early drift behaviors while they remain small and inconsequential. We measure our behaviors against the scales of appropriateness and make frequent and proactive corrections back toward the ideal.

14.3.4 Managing Tasks and Our Attention Level

For pilots who are especially comfortable and experienced, maintaining their comfort zone becomes a desirable goal. When workload rises, they pursue opportunities to work ahead. They may even shift the times when they normally accomplish certain tasks. Consider a short flight with a minimal cruise segment – say only about 5 minutes of steady cruise before starting down. Assume that we didn't prepare our EFB for arrival while we were back at the departure gate. With the aircraft steadily climbing and on autopilot, we choose to take a few minutes to set up for arrival. Since the cruise portion will be so brief, we balance priorities and work ahead – both desirable workload management practices. We go heads-down and start arranging our approach charts. We recognize that this elevates risk, so we proactively resolve to stop our preparation as we near our assigned altitude to monitor the autopilot level-off and autothrottle thrust reduction. While we are focused on reviewing our approach charts, the altitude alerter sounds. We stop what we are doing, monitor the level off, then return to complete our preparation with plenty of time for briefings and checklists.

Let's evaluate our choices to see what was gained and what was lost. The gains include efficiently backloading our workload – something we favor. We maintained sufficient SA and avoided rushing our arrival preparation. What was lost was active flightpath monitoring. Our attention became distracted by our preparation. We would have missed the level-off were it not for the aircraft's altitude alerter tone and autopilot. In effect, our good intentions successfully backloaded our workload, but unintentionally created a vulnerability for missing the level-off. Still, the aircraft alerted us and the autopilot handled the level-off, so no harm.

A significant portion of NASA ASRS reports document altitude busts. No matter how automated our aircraft and how robust our procedures, we still seem to struggle with consistently making altitude restrictions. One reason is that our practice continuously drifts. Consider this evolution of level-off procedures.

BOX 14.3 THE EVOLUTION OF LEVEL-OFF CALLOUTS

Earlier models of airliners had autopilots with limited capabilities. Pilots had to hand-fly the aircraft to the intended level-off altitude, manually level the aircraft, and then engage the autopilot in Altitude Hold. After that, the autopilot would maintain level flight. To ensure that both pilots focused their attention on the level-off, procedures directed them to announce "One To Go" at 1,000′ before the assigned altitude. If their attention was diverted, the aircraft's altitude alerter tone (which sounded at approximately 600′ prior to the set altitude) reminded them to redirect their attention.

Eventually, improved autopilots incorporated automatic level-off capabilities. Pilots didn't need to manually level the aircraft or engage Altitude Hold. In effect, responsibility for the level-off was transferred to the automation. This generated an unintentional drift in pilot monitoring behavior. Since nothing had to be done to level the aircraft, the "One To Go" callout became less relevant. Their compliance with the procedure became reactive. Pilots heard the altitude alerter and reflexively announced, "One To Go" (even though they were actually passing 600′ to go). A pilot rummaging through their flight bag might hear the tone, announce "One To Go", and continue digging without ever looking up to verify the altitude or monitor the level-off. The improved autopilot induced a slow drift both in pilot priorities and in their monitoring behavior. What was lost was pilot awareness of their level-off altitude and detection of autopilot level-off failures – of which there were many. Another effect was the blending of the climb and cruise phases. The late climb phase became essentially the same as the cruise phase. More discretionary tasks migrated from the cruise phase into the climb phase.

Concerned by the steady rise in climb and level-off errors, some airlines imposed new procedures to improve altitude awareness and refocus crew attention on level-off. One procedure was to call the current altitude for the level-off altitude – for example, "15,000 for 16,000". If pilots missed the callout until the altitude alerter sounded, they were directed to call the actual altitude – "15,400 for 16,000". While this may seem like a subtle difference, it highlighted to the pilots how their attention had drifted. Calling "15,400" was perceived as a sign of inattention or distraction – a verbal admission that they weren't actively monitoring their flightpath during the climb. It also highlighted when they were engaged in discretionary tasks that should have been delayed until reaching cruise.

This procedure corrected line-flying behavior back toward the intended priorities. Unfortunately, procedural drift did not stop. Tasked to make a callout 1,000′ prior, some pilots tried to "work ahead" by making their callout early. Some pilots began calling "15,000 for 16,000" while actually passing 14,900′, or 14,800′, or even 14,700′. This drift shows how we can unintentionally lose sight of the main purpose of a procedure. The procedure was not intended to make a callout as an end in itself. The intention of the callout was to remind us to do something important – to focus our attention and actively monitor the level-off.

14.3.5 Balancing the Scales between Efficiency, Workload, and Risk

As we manipulate efficiency, workload, and risk, we discover that these factors sometimes conflict with each other. Much like when we squeeze one side of a filled balloon (modify our workload to exploit efficiency), we cause the opposite side to stretch out (increasing risk somewhere else). Sometimes, operational situations require us to squeeze the balloon and shift risk. To do it skillfully, we need to guard against emerging vulnerabilities. As we increase risk by exploiting efficiency in one area, we need to balance it by increasing our vigilance against the emerging vulnerabilities.

Imagine that we have a short flight and choose to begin our arrival preparation early, say with 10,000′ left until cruise level-off. Assume we are climbing at about 2,000 feet per minute. This translates to roughly 5 minutes until level-off. We mentally apportion about 4 minutes of preparation work before we intend to shift our attention back to actively monitoring the level-off. If we manage this successfully, it proves to be an effective technique for managing workload on an exceptionally short flight. The bottom line is that if we modify a procedure, even for a good reason, we need to build a compensating plan to mitigate the additional risk and error vulnerability.

14.3.6 Quickening and Rejection Triggers

Part of our work is understanding how quickening develops and progresses. This knowledge may not prove useful while we are immersed in-the-moment with a worsening problem. We just need to recognize how quickening feels. We don't even need to know what is causing it. We just need to recognize that it is happening so we can abandon our game plan, reset the situation, and try again.

Consider a situation where we detect a deviation and apply a normal correction. When that correction doesn't work, we add another one. Anytime we find ourselves applying multiple corrections against a growing problem, we should experience a gut-felt "something is wrong" feeling. We may not have time to diagnose the problem. If time is short, we shouldn't even try. Instead, we need to set a rejection trigger that is linked to specific point. "If I can't get my speed back within limits by 500′, I'm going around."

Imagine flying an unstabilized approach. For an unknown reason, our speed is fast and trending faster. We start with a correction. It doesn't work, so we apply a second correction. Convincing ourselves that our corrections are just about to yield results, we give it a few more seconds. We see the runway getting closer while our speed remains too fast. Our attention is tunneling and our stress level is rising. Something definitely *feels wrong*, so we set an abort trigger point. When we reach it, we switch to our contingency plan and go around. We still don't know why the approach failed. We just recognize the quickening and the wrongness. As we go around, we feel our stress level subside. The bottom line is that we need to recognize the quickening, set a trigger point, and automatically switch to the contingency backup.

We build this Master Class skill by learning from our experiences. Maybe we had an approach that deteriorated, but we salvaged it and landed successfully. If we only accept the successful conclusion, we lose the lesson and set the stage for a future quickening failure. Instead, take a few moments to evaluate our experience.

- How close were we to going around?
- Was the profile quickening? How did it feel?
- What thoughts went through our minds as the approach progressed?
- Did we notice a strong psychological pull to accept the instability and land?

Through introspection, we hone our recognition skills and calibrate our game plan rejection triggers. Knowing when a situation has deteriorated, what that point looks like, and how it feels is an important Master Class skill.

14.4 ACTIVE FLIGHTPATH MONITORING, DYNAMIC MOVEMENT, AND SAMPLE RATE

At the heart of workload management is active flightpath monitoring. Active flightpath monitoring relies on two main concepts: dynamic movement and sample rate. Dynamic movement is measured by how much the aircraft path is changing or is likely to change in the near future. It is lowest when we are stationary (stopped with the parking brakes set) or in steady, unaccelerated cruise while flying. It is highest when path, speed, and thrust adjustments are continuously changing.

Examples include making taxi turns or hand-flying the aircraft through turns, climbs, or level-offs. We classify dynamic movement into three levels (low, medium, and high). It begins when we push back from our departure gate and continues until we shutdown at our destination gate.

Sample rate is a relative measure of how often we return our attention toward monitoring aircraft path.[6] As a general rule, the more complex the dynamic movement, the faster our sample rate. Sample rate is highly correlated with *time to fail.* Making taxi turns and hand-flying down short final are examples of short time to fail environments. We need to continuously monitor and correct our path. Taxiing slowly straight ahead or flying at high altitude on autopilot are examples of long time to fail environments. They offer safe opportunities to direct our attention away from monitoring flightpath.

14.4.1 Dynamic Movement – Low

Low dynamic movement applies when the aircraft is stationary or in steady cruise flight with autopilot and autothrottles engaged.

- *Low dynamic movement – during taxi:* This applies when we are stopped with the parking brakes set. Immediate workload tasks are suspended, so we can devote our full attention toward short-term tasks and long-term planning/briefing. Consider a situation where we are taxiing to the departure runway and ATC assigns a runway change that requires us to reprogram the FMS, validate performance data, and complete checklists. As reported in the following NASA ASRS report, trying to do this while taxiing could lead to task overload, rushing, or errors.

BOX 14. RUNWAY INCURSION WHILE RUSHED TO MAKE HOLDOVER TIME

FO's report: … After making the left hand turn from Quebec to Alpha taxiway, Ground Control informed us of a wind shift from 300° to 320°, a 27-knot crosswind for planned Runway 28 departure. They offered us Runway 33 instead. We accepted the Runway change given that the nearly 40 knots gusts aligned with Runway 33. *The Captain elected to continue taxiing on Alpha given the concern of additional snow, while I completed the runway change in the FMS and briefed the updated departure heading of 330°.* Approaching the intersection of Alpha and Mike taxiways Ground Control directed, "Taxi Mike, cross Runway 28." *Remembering the initial taxi instruction of Alpha cross Runway 33, the Captain mistakenly continued straight instead of turning right onto Mike, leading to a Runway incursion on Runway 33. I was distracted with verifying that all runway change steps had been completed and did not recognize the error until holding short of Runway 28 on Alpha.* I called Tower and reported our position, at which point we reoriented, finished the runway change discussion, and departed Runway 33 without issue. *The root cause was misapplication of previous taxi clearance by the Captain, combined with insufficient First Officer backup due to the Runway change tasks.*

Captain's report: ... *There were several things going on that contributed to the error. De-icing and holdover considerations, wind shear advisories in effect forcing non-standard takeoff procedures and a late runway change with taxi instructions similar to the original. Also, failure of both pilots to follow along on the 10–9 chart at night with blowing snow.* We reviewed the expected taxi route from the ramp to the expected departure Runway 28. There were no Hot Spots to brief and getting there via Alpha is very simple. I think we could benefit from a Hot Spots circle and reference on the 10–9 chart here in SYR. *We may have felt a little rushed as we blocked out late and were trying to get airborne before the snow started again.*[7]

The perceived urgency to get airborne before the snow resumed encouraged this crew to divide tasks during dynamic, night, snowy weather taxi operations. Ideally, the crew could have created a low dynamic movement opportunity by stopping and setting the parking brakes. Then, they could have safely performed individual flows, programmed computers, configured the aircraft, rebuilt an appropriate game plan, communicated a revised shared mental model, and completed required checklists. These tasks require undivided attention and crew coordination. In this case, concern for the deicing holdover time accelerated their process and pushed the crew to attempt some shortcuts.

- *Low dynamic movement – while flying:* In the air, cruise flight segments with the autopilot and autothrottles engaged are our only steady and unaccelerated environments. As with stationary on the ground, immediate tasks are suspended (or transferred to aircraft automation). We can divide our time between short-term tasks, long-term planning, and discretionary pursuits. Our workload remains low until we begin planning and briefing before top of descent.
- *Sample rate during low dynamic movement:* The sample rate under low dynamic movement is infrequent. This is because there is a low probability of encountering time-pressured tasks. We'll have plenty of time to diagnose and recover from them. Low dynamic movement environments allow us to engage in discretionary activities. An occasional scan is sufficient to monitor aircraft systems and flightpath. Our autopilot accurately holds altitude and dutifully follows the programmed magenta line. Workload remains low unless we encounter an exceptional event.

14.4.2 Dynamic Movement – Medium

Medium dynamic movement applies when our trajectory is changing, but is autopilot-assisted, remains in a stable trajectory, or is easy to monitor.

- *Dynamic movement medium – during taxi:* Examples during ground movement are taxiing on straight taxiways, making gentle turns on a curved taxiway, and slowly moving in a line of aircraft. Captains directly control the taxi path while FOs monitor. FOs may safely divert their attention away for short periods. For longer diversions, they should coordinate their intention to look away so that Captains know to assume full responsibility for path monitoring.

  If there is a minor change in the departure routing, FOs can make the appropriate programming changes while Captains continue taxiing. When appropriate, the Captains can (as directed by company policy) stop the aircraft, set the parking brake, review the changes, and initiate any required briefings and checklists.

  Many airlines allow crews to complete their Before Takeoff Checklists during taxi movement. Ideally, Captains will select a straight taxiway segment so that they can easily glance inside to verify systems before replying to checklist challenges. During reduced visibility taxi operations, both pilots should remain focused outside while the aircraft is moving. Most airlines direct crews to stop the aircraft before completing detailed FMC reprogramming and checklists.
- *Medium – while flying:* Inflight medium dynamic movement includes flight that changes in a single dimension, is autopilot-assisted, and is above the low-altitude environment. Some airlines use third segment climb (approximately 3,000′ AGL) for the low-altitude environment boundary. Other carriers use 10,000′ as the division between low and high altitude. Under this standard, medium dynamic movement is defined as a steady climb between 10,000′ and planned cruise altitude while under autopilot and autothrottle control. PFs can divert their attention inside for short periods, but they still remain responsible for flightpath monitoring. PMs may divert their attention away from flightpath monitoring for longer periods, as duties require. In any case, discretionary tasks are deferred until the cruise phase.
- *Sample rate during medium dynamic movement:* Sample rate during medium movement is frequent. For example, while accomplishing the Before Takeoff Checklist during taxi, Captains can glimpse in, check a switch position, and then return their attention outside while responding to the checklist challenge item. Normal instrument and aircraft system scans are appropriate while ensuring that the autopilot maintains the desired trajectory.

14.4.3 Dynamic Movement – High

High dynamic movement requires that both pilots actively monitor the aircraft's trajectory. This applies anytime when multiple flight parameters are changing (pitch, bank, and thrust), the aircraft is not actively controlled by flight-directed autopilot, or within the low-altitude environment where outside threats (other aircraft, birds, or terrain) are prevalent. High dynamic environments require both pilots to suspend all deferrable operational tasks that may distract their attention away from flightpath monitoring.

- *High – during taxi:* On the ground, this includes gate departure/arrival, congested ramp areas, tight turns, crossing runways, taking the runway for takeoff, and exiting the runway after landing.
- *High – during flight:* While flying, this includes anytime the PF is hand-flying the aircraft, during all autopilot-controlled path changes (entering or rolling out from turns and starting or completing climbs/descents), or in low-altitude environment where outside threats are prevalent.
- *Sample rate during high dynamic movement:* Sample rate during dynamic flight is continuous for both pilots. At low altitudes, even with the autopilot engaged, both pilots should use a high sample rate to clear the aircraft path for possible threats such as birds, balloons, drones, conflicting aircraft, and terrain.

14.5 AREAS OF VULNERABILITY (AOVs)

A useful convention for managing monitoring priorities during aircraft movement is AOVs. The industry publication, *A Practical Guide for Improving Flightpath Monitoring*, defines the monitoring priorities for low, medium, and high AOVs.[8]

14.5.1 AOVs during Taxi Operations

Figure 14.3 depicts how the AOVs change between low (green – black in the graphic), medium (yellow – light gray in the graphic), and high (red – dark gray in the graphic) from departing the JFK gate area through takeoff on runway 22R. It begins with a high/red AOV due to the close proximity of other aircraft and ground vehicles (dark-gray taxi segment near the terminal). In this example, the crew is assigned a new taxi clearance to a different departure runway 22R. To process the change, they stop on

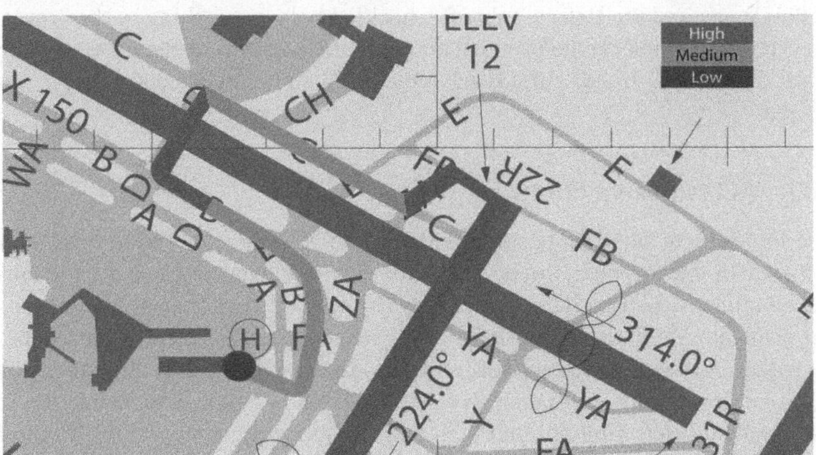

FIGURE 14.3 Taxi areas of vulnerability.[11]

the edge of the ramp and set their parking brakes to create a low/green AOV opportunity (black circle). They both divert their attention inside to reprogram the FMS, recompute aircraft takeoff performance, conduct briefings, and complete checklists. When finished, they call for taxi, depart the ramp, and turn left onto taxiway Bravo. This portion reflects a medium/yellow AOV (light gray taxi segment). While the *Practical Guide* designates this portion as a medium AOV, we might consider treating it as a high/red AOV until established on Bravo. This is because a failure to make the left turn onto Bravo might lead to a runway incursion onto 22R. Additionally, since Bravo is a curved taxiway at a busy airport, the crew's attention to path may be more appropriately treated as a high/red AOV. Taxiing on Bravo and approaching taxiway Delta, ATC directs the flight to cross runway 13L/31R at taxiway Delta. The crew uses high/red AOV monitoring priorities for the runway crossing and for their right turn onto taxiway Charlie. The next portion on Charlie is a straight, parallel taxiway, so the AOV drops back to medium/yellow. The crew could use this opportunity to complete their Before Takeoff Checklist and to prepare for departure. Approaching the active runway, ATC Tower clears the flight for takeoff. They observe high/red AOV priorities for the left turn onto taxiway Zulu Alpha, the right turn onto Foxtrot Bravo, and the right turn onto runway 22R.

14.5.2 Workload during Taxi

Notice how the priorities for path and threat monitoring change with each AOV transition. The runway change causes a significant spike in planning/briefing workload. The crew creates a low AOV for their intensive inside work by stopping the aircraft and setting the parking brakes. When completed, they redirect their attention back outside to monitor aircraft movement. The medium AOV opens an opportunity for the FO to complete some operational tasks on taxiway Bravo, but that window closes with the runway crossing of 13L/31R at Delta. The runway crossing requires both pilots to direct their attention outside. Established on Charlie, each pilot completes their flow items and the Before Takeoff Checklist before switching to the Tower frequency. If, for example, some problem emerges, the appropriate decision would be to stop, set the parking brakes, and resolve the problem before accepting their takeoff clearance.

14.5.3 AOVs during Flight

Figure 14.4 depicts the flight portions and how the AOVs shift between low, medium, and high.[9] In the graphic, high AOVs (dark gray) happen close to the ground and at all circles depicting lateral trajectory changes, vertical trajectory changes, and speed changes. Medium AOVs (light gray) reflect sustained climbs, descents, and intermediate ATC-assigned level-off altitudes. Low AOVs (black) are only available at sustained-level cruise segments above 10,000′. The graphic depicts one low-AOV cruise segment after climbout and then another following a step-climb to the final cruising altitude.

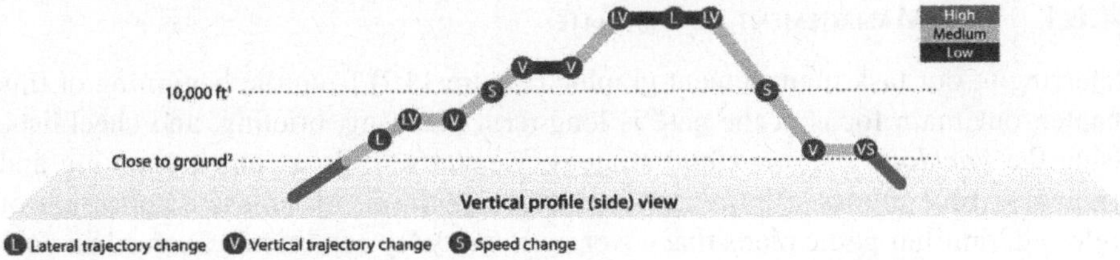

FIGURE 14.4 Flight profile areas of vulnerability.

14.5.4 Workload during Flight

Notice how the priorities for path and threat monitoring transition across each AOV. From takeoff until clean-up altitude (or above the company policy designating "close to the ground"), the crew follows high-AOV priorities. After clearing low-altitude airspace, they shift to medium AOV priorities. With the autopilot engaged and maintaining programmed path, they can perform certain operational tasks. Any time the path requires changes in pitch, roll, or speed, they revert to high-AOV priorities to monitor the transition. Upon reaching sustained cruise flight, they enter a low AOV. This continues until step-climb to the final cruise altitude. Low AOVs allow them to engage in discretionary activities. Nearing their descent to the arrival airport, the AOVs again follow medium and high-AOV priorities.

14.5.5 Summary of AOV Guidelines

Following are some useful rules-of-thumb for managing appropriate AOV priorities.

- Inside tasks are appropriate within medium and low-AOV segments during the taxi operations. Review the taxi route before departing the gate to preplan where to complete required inside tasks and checklists.
- We can create a low AOV by selecting an appropriate temporary parking spot, stopping the aircraft, and setting the parking brakes. The only time during ground operations for discretionary tasks is when stopped. Normally, these only occur during prolonged delays awaiting departure times, during runway closures/ATC route closures, and waiting for our gate to open.
- Necessary operational tasks during aircraft movement should only be accomplished in low and medium AOVs with both pilots fulfilling their monitoring roles.
- All deferrable tasks should be suspended in high-AOV phases so both pilots can devote their attention to monitoring aircraft movement.

14.6 MASTER CLASS WORKLOAD MANAGEMENT

Following are a number of line-flying situations that challenge our workload management skills.

14.6.1 Task Management at the Gate

Referring to our task management graphic (Figure 14.2) from the beginning of this chapter, our main focus at the gate is long-term planning, briefing, and checklists. Using the *backload our workload* strategy, we strive to do as much planning and briefing as time allows. With experience and repetition, we amass a collection of preferred, familiar game plans that cover most everyday conditions. Even when conditions don't exactly match, these plans are resilient enough to handle minor contingencies. The more adverse that our conditions become, the more contingency options we consider.

- *Drift:* As a general tendency, the more familiar that situations become for us, the more our attention to detail wanes. After many successful flights without mishap, we gain confidence in our abilities, game plans, and workload management. Detailed preparation can feel excessive and redundant. Taken to the extreme, an apathetic pilot may become rather lax toward their flight planning. That is why we, as Master Class pilots, conduct periodic resets of our practice. We review training materials and videos that model the ideal process and pacing. As we compare these ideal models with our current practice, we might identify areas where our practice has drifted. We then make corrections to restore the quality of our work.
- *Running late:* When we are late, everyone strives to work faster to get back on time. Especially when passenger loads are light, the station team may be able to complete their work before us. This increases pressure on us to work faster. Assess the preflight planning needs of the flight and identify the most time-consuming problems to resolve first. Perhaps our fuel load is short. It takes time to get the refueling agent back to our aircraft. If we only discover this problem at scheduled push time, we will incur a lengthy delay. After we get the longest timelines moving, we can engage the shorter timelines in the most efficient order. Remain aware of the time pressure, but don't sacrifice quality for speed.
- *Distractions:* Distractions are common during preflight preparation. Unfortunately, many distractions arise from important details that we need to address. We can't just ignore them. One option is to assign the least-burdened pilot to handle the distractions. If the Captain is new or is handling a detailed procedure such as deferring a maintenance discrepancy, then the FO should intercept and handle the distractions. If the FO is new or is busy with preflight preparation, then the Captain should intercept the distractions. At some point, however, both pilots need to create a distraction-free window of time to complete their briefings and checklists. Some airlines position the lead flight attendant at the flightdeck door to intercept jumpseaters or operations agents until the pilots complete these tasks. Lacking a mitigation like this, we need to become attuned to when distractions occur and where we were in our work process before the interruption. If there is any question, we back up to the beginning of the briefing or checklist and start over.

- *Personal factors:* Personal factors play a significant role in workload management errors. Morning pilots flying at night, night pilots flying in the morning, or any pilots flying during their circadian lows are all examples when our personal vulnerabilities increase. We perform self-assessments to identify times when we are not performing at our best. We slow down, take extra time, and become more meticulous. There is an old adage that they will never remember how fast we did it, only how well we did it.

  It helps to inform the other pilot of our vulnerabilities. "I'm a night person who just got rerouted to this morning flight. I feel good, but I'd appreciate you watching for anything I might miss." "We have a new baby at home and I've not been getting the best sleep lately. I'd appreciate you backing me up today." Follow the IMSAFE self-assessment tool every flight.[10] Medication and alcohol are clear and unambiguous. Illness, stress, fatigue, and eating are not. If I have a slight headache, is that enough? What if I missed a meal? What if I'm feeling a bit tired? Each of these assessments are variable. Each pilot needs to make informed and honest self-assessments. Experience has shown that most pilots are overly optimistic with their self-assessments. I don't think I've ever heard a pilot report, "I shouldn't have called in sick. It turns out I was healthy enough to fly." Many times, however, I have heard pilots report, "I shouldn't have flown that flight. I was more tired than I thought I was." Bottom line, if we aren't ready to fly, we need to call for a replacement.

14.6.2 Workload Management in Low AOVs

Low AOVs occur on the ground with the brakes set and in sustained cruise flight.

- *Handling delays and working quickly:* Imagine that we are taxiing off the gate and receive a runway change that requires FMS programming changes, performance computations, briefings, and checklists. Our first consideration is finding a good place to stop. Pick a spot where we won't be time-limited or forced to move while we are in the middle of a procedure. One useful location is in a designated holding pad. Ensure that we are clear of moving traffic and won't be distracted. Take care of any coordination that may distract us during our process. For example, inform the flight attendants of the delay and make a passenger PA before beginning briefings and checklists.

  Divide the workload, then compare results. For example, one pilot reprograms the departure while the other pilot checks performance data. As with normal preflight preparation, each pilot completes their personal preparation first. We rearrange our EFB pages with the new departure SID and review the new taxi route. Next, we complete any FMS waypoint verifications, NAVAID changes, and performance computations. Following that, we reaccomplish the PF briefing and Captain briefing items. Finally, we complete any other required checklists.

The key lies in how well we manage our workload pacing during delays. We'll feel pressure to hurry. Recognize the difference between working quickly and rushing. Rushing occurs when we sacrifice quality for speed. The priority of rushing is speed. Conversely, working quickly preserves the accuracy and quality. The priority of working quickly is quality.

Accelerating the operational pace is especially challenging when experienced Captains are paired with inexperienced FOs. This is a good time to take a deep breath and remember what it felt like when we were new. We can't let our frustration with the delay add to their stress. Communicate an easy calmness. Develop an awareness of the other pilot's comfort level. Take steps to either help them out or remove distractors so they can focus on their work.

- *Ground delays followed by a mass launch:* A common challenge is the mass launch following a delay or departure runway change. When ATC gets the green light, they announce to all aircraft simultaneously, "Call when ready to taxi." The quickest crews to declare that they are ready get to taxi first. This affects where we end up in the departure queue. If there is already a long, slow line for takeoff, it may be appropriate to taxi, take our place in line, and finish all of our preparation steps while waiting in line. Especially when ATC is managing a significant "miles in trail" for aircraft on the same departure SID, there will be plenty of time to set the brakes and complete any remaining details. Of course, the opposite is also true. We don't want to immediately call for taxi and zip to the head of the line without finishing our required work. We may find ourselves with a surprisingly quick clearance for takeoff before we are ready.

  Another aspect is managing passengers and cabin crew mobility. During a long delay, we can turn off the seat belt sign and allow the passengers to stand or use the lavatories. For especially long delays, the cabin crew may choose to provide a drink service. When ATC releases us to depart, it will take time for them to secure the cabin. A technique is to work with our dispatcher to anticipate the release time. Often, local ATC has to wait for higher authorities to authorize aircraft release. Our dispatcher may provide quicker notification. This will give us a head start moving passengers back to their seats.
- *Taxi delays:* Another category is ground delays after we depart the gate. These are caused by runway closures, runway changes, ATC route closures, and assigned departure slot times. We may be directed to push from the gate, taxi to a holding area, and wait out the delay. In these cases, our main concern is complying with our required minimum takeoff fuel. Ideally, we could clear the gate, set the brakes, transfer aircraft power and air conditioning to the APU, and shut down the engines. Even then, the APU continues to burn fuel, so we can still find ourselves running short. Most releases are planned for an average taxi time. Anytime we exceed this, we'll need to closely monitor our fuel burn. If we have contingency fuel elsewhere in our flight release, we may have the option of coordinating an amendment to reduce our minimum takeoff fuel to remain within limits. Anytime we incur a ground delay in excess of what is planned in our dispatch release, we should develop a contingency game plan to manage fuel burn. If our taxi delay still exceeds our best efforts, we'll need to return to the gate for a top off.

- *Keeping our minds ready to fly:* Another consideration is keeping our mindset ready to fly. Our internal clock tracks the expected pace of a normal flight. When we have a lengthy delay, either at the gate or in the holding pad, we suspend that pacing. The longer the delay, the more we can lose our mental focus on flying. Restarting our flying mindset can feel like trying to start a cold engine on a frosty morning. If passengers are up and about, we'll need to get them moving back to their seats. We'll need to restart engines, transfer aircraft electrical power, reconfigure air condition and pressurization, and complete our Taxi Delay Checklists – all while ATC is pushing us to accept a takeoff clearance.

14.6.3 Workload Management in Medium AOVs

Medium AOVs reflect uncomplicated taxi movement on the ground and during steady climbs and descents while flying.

- *Medium AOVs while taxiing:* Since Captains are focused on taxiing the aircraft, managing other tasks is primarily the FO's responsibility. FOs may need to make passenger PAs, change pressurization panel settings, and configure aircraft systems for takeoff. Some of these tasks will divert their attention away from monitoring the taxi path. FOs need to periodically interrupt their inside tasks to crosscheck taxi movement, detect deviations, and prevent taxi errors.

  As FOs, we should preplan what tasks we need to do and when to do them. Start by forming a mental picture of the expected taxi route. Note the taxiway turns, hot spots, and potential conflict points. When we reach a medium-AOV segment, we can initiate our required inside tasks. If they require significant heads-down time, coordinate with the Captain first. Determine an end point for the latest time to complete required checklists.
- *Discretionary tasks during taxi-in:* Some pilots compromise medium-AOV protocols during taxi-in. There is a strong motivation to begin preparing for the next flight or for an aircraft swap. Pilots may put away their EFBs, clean up the flightdeck, and even turn on their overhead speaker and stow their headsets. All of these are discretionary tasks that should be delayed until parked at the gate. While there are no AOV differences between taxi-out and taxi-in, many pilots treat them differently. Probably because the flight feels "finished" during taxi-in, many pilots lower their guard.

  The erosion of taxi-in protocols is a common example of drift. While all discretionary tasks should be delayed until parked at the gate, it is well within the ability of experienced pilots to complete some quick discretionary tasks. As we gain proficiency, we choose to perform a discretionary task. Nothing bad happens and it feels good to get that task out of the way. Since it feels harmless, we do it again. Over time, our practice drifts further from the ideal practice. As Master Class pilots, we avoid this temptation and maintain AOV discipline.

- *Medium AOVs in flight:* Medium AOVs occur during sustained climbs and descents above the low-altitude environment. The PF's priority is to monitor the aircraft's flightpath. Even when we transfer direct aircraft control to the autopilot, we still need to monitor its performance to ensure that it continues to follow the intended profile. Flight directors are notorious for shifting between modes at inopportune times.

  During medium AOVs, PFs can perform brief operational tasks. If a prolonged task is needed, consider switching roles with the PM. Another consideration is hand-flying. For proficiency, PFs may wish to hand-fly portions of the departure or arrival. Ideally, engage and follow the flight director to ensure path accuracy. If something arises that requires us to divert our attention, reengage the autopilot.

  PMs should still monitor the aircraft path during medium AOVs. Since PFs are assuming primary responsibility, PMs serve as backups. Considering sample rate, the PF's sample rate is greater than the PM's. PMs can divert their attention inside for longer duration tasks like passenger PAs, intercom communications with the cabin crew, and radio communications. There is a strong incentive to accomplish discretionary tasks here, but they should be delayed until cruise or after shutdown.

14.6.4 Workload Management in High AOVs

High AOVs happen during dynamic taxi operations and during all pitch, roll, and thrust changes in flight. All high-AOV movements require that both pilots monitor the flightpath.

- *High AOVs while taxiing:* Taxi hazards are defined by the characteristics of the movement and location, not by the prevailing conditions. We all accept that a taxi hotspot at a busy airport under marginal weather conditions demands our full attention, but that same hotspot also demands our full attention on clear, calm days with no other aircraft nearby.

  For comparison, consider three sets of conditions at the same runway crossing. In the most challenging case, it is night with ½ mile visibility and fog. When assigned the runway crossing, both pilots remain fully vigilant for aircraft landing lights, errant ground vehicles, and other threats. For the second case, imagine a clear morning with many aircraft moving around the airport. The good visibility makes it easier, but the busyness of the airport encourages us to remain attentive. For the third case, consider a clear, calm day where we are the only aircraft moving on the airport. It is very easy to monitor for threats and cross the runway. Despite the different conditions between these three cases, they all require us to follow high-AOV priorities, attention focus, and sample rates.

- *High AOVs and radio calls:* A nuanced issue involves high AOVs and radio calls. If ATC calls us during a high-AOV movement, should we respond to their call, respond "Standby", or say nothing? The most common practice is for the FO to respond to the call while the Captain focuses on the taxi path. If we do this, what do we lose?

  For example, while maneuvering out of a busy ramp area, ATC assigns a detailed change to our taxi routing, "[Callsign] turn left onto Bravo, right on Zulu Mike, and left on Charlie, hold short of Runway 13." Assuming that we aren't familiar with this airport layout, we'll need to translate this four-part verbal instruction into a mental picture of where to taxi. That picture requires us to understand the layout, distance, and timing of the taxi route. Typically, we consult our airport diagram chart, reference where we currently are and where we make turns. We build a mental understanding of how the taxi route will progress. This involves a number of sequential steps that take time to connect. It is not a task we want either pilot doing while moving in a high-AOV environment. Let's examine our options for handling this four-part taxi instruction.

  - *Case 1:* The first option is for both pilots to maintain their attention on the taxi path and allow the FO to respond to ATC. Both pilots would rely on their memories of the instruction until established on taxiway Bravo. Then, each pilot references their airport diagram to form their mental picture. This option has both pilots glancing back and forth between their chart and outside during taxi movement.
  - *Case 2:* A second option is for the Captain to focus on turning the aircraft onto Bravo while the FO responds to the call. The FO quickly references their airport diagram to form a mental picture, then talks the Captain through the next turn onto Zulu Mike. "Turn left onto Bravo. About 1,000′ down is your right turn onto Zulu Mike followed immediately by a left turn onto Charlie. Then hold short of runway 13." This also works, but during that quick reference of their chart, the FO may be heads-down during the turn onto Bravo.
  - *Case 3:* A third option is to make the turn onto Bravo before referencing the chart. Now in a medium AOV, the Captain directs the FO to reference their airport diagram to form a mental picture of the taxi routing. The FO describes the taxi route to the Captain. Again, this works, but the Captain relies heavily on the FO to remember the taxi instruction accurately.
  - *Case 4:* The fourth option is to turn onto Bravo, then direct the FO to reference the taxi diagram. When the FO is finished, the Captain directs the FO to monitor outside while they glance down at their airfield diagram to form their own mental picture.
  - *Case 5:* A fifth option is to stop the aircraft and sort out the instruction before continuing.

  All five options are viable and each is commonly used. Each also depends on the conditions at the time – visibility, day/night, wet/dry, the length of the taxi operation, familiarity, crew experience, fatigue level, and Captain's direction. Whichever one we choose, we need to ensure that someone remains fully focused on monitoring the taxi path. Our learning point is acknowledging that there is a wide range of available options that each requires active CRM communications and workload management.

- *High AOVs with both pilots distracted:* What we don't want is Case 6 where both pilots divert their attention to their airport diagram while still maneuvering in the high-AOV environment. Unfortunately, this does happen. Several factors contribute to this practice. First, it is doable. We hear the new taxi route, respond to ATC, and make few quick glances down at our airport diagram while turning the aircraft onto Bravo. We talk back and forth until we coordinate a shared mental model. Most of the time, it works out and we keep the aircraft moving. Like other examples of drift, it works until it doesn't. Over time, our drifting practices can undermine AOV priorities. The Master Class practice is to follow AOV priorities every time, even when our skills and abilities allow for multitasking-type behaviors.

BOX 14.5 A DRIFT STORY TOLD BY TIRE MARKS ON A TAXIWAY

An interesting exercise is studying aircraft tire marks on taxiways. Ideally, they should track equidistant from the painted centerline. That is normally true. There are, however, some significant deviations at minor turns and curves, especially where a long straight segment meets a dogleg turn to another long straight segment. At one particular airport, aircraft exiting the runway join a taxiway for a straight segment. About 500′ down, there is a significant 30° left turn that continues onto the next straight segment. At that turn, there are a number of tire marks straying perilously close to the outer edge of the pavement.

Imagine how this event unfolds. The crew lands and exits the runway. They feel a significant reduction in workload and stress as they begin their taxi-in. This "let down" subconsciously leads them to perform some discretionary tasks. Anticipating a high workload at the gate, they begin to divert their attention inside while on this first, straight, medium-AOV taxiway segment. One or both of them have their attention diverted inside while approaching the 30° left turn. Cued by some sense of wrongness, they look up and suddenly realize that they are about to depart the pavement. The Captain jerks the tiller left leaving a pronounced line of tire rubber perilously close to the pavement edge. It is difficult to imagine a new Captain doing this or a new FO allowing it to happen. More likely, it happens with experienced pilots who have allowed their AOV priorities to drift.

- *High AOVs in flight:* High AOVs in flight occur in the low-altitude environment and during all transitions in aircraft path (starting and stopping turns, initiating and leveling from climbs and descents, and changing speeds).

Examples of drift are pilots moving short-term tasks and discretionary tasks into these high-AOV phases. Threats feel negligible, so we sense little harm with diverting our attention inside, especially since the autopilot will complete the maneuver and autothrottles will stabilize the thrust.

- *High AOVs during landing rollout:* A useful high-AOV exercise involves ATC transmissions during landing rollout. To set the stage, we have landed and are rolling out. It is noisy because our thrust reversers are still roaring. Our risk level is steadily dropping, but we still have much to do and remain in a high AOV. We need to reduce reverse thrust, monitor the disengagement of autobrakes, transition to wheel braking, confirm the stowing of the thrust reversers, and often transfer aircraft control from the FO to the Captain for taxi. During this very busy time, well-intentioned ATC controllers may transmit instructions like, "[Callsign] expedite your exit, aircraft on a one-mile final. Contact Ground." How should we handle this? Our three options are to ignore the radio call and comply, reply "Copy" and comply, or make a full reply, "[Callsign] expedite exit and contact Ground." Each option generates increasing levels of risk.

  Now, change the ATC call and add some procedural drift from the Tower controller. Perhaps our airline occupies different terminals and our taxi route depends on our assigned gate. They might helpfully add, "Say Gate." Where their procedural drift combines with our procedural drift is when the PM responds, "[Callsign], expedite exit, contact Ground, gate B-19", all while completing intensive rollout tasks. Again, this is workable, but increases risk and compromises high-AOV priorities. Since this callout is neither directive, nor timely, it might be best to wait until clear of the runway to respond.

NOTES

1 Professional maturity in this reference is intended as a subjective measure of our personal mastery of aviation.
2 Edited for clarity and brevity. Italics added. NASA ASRS report #1737966.
3 The listed quote is from Brian Stuart Germain, but may originate from a widely used quote from physical training, "The body cannot go where the mind doesn't push it." In any case, mentally rehearse to prepare for future actions.
4 Edited for brevity and consistent references. Italics added. NASA ASRS report #1780155.
5 NTSB (1997).
6 This use of "sample rate" was coined by Captain Steve Guillian of the Flightpath Monitoring Working Group. The descriptions of dynamic flight, sampling rate, and Areas of Vulnerability (AOV) are from the Working Group's report as published by the Flight Safety Foundation (Active Pilot Monitoring Working Group, 2014, pp. 18–22).
7 Edited for brevity and clarity. Italics added. NASA ASRS report #1791286.
8 Active Pilot Monitoring Working Group (2014, p. 19). The Guide depicts the AOVs as green, yellow, and red for low, medium, and high, respectively.
9 Active Pilot Monitoring Working Group (2014, p. 19). This graphic was provided by the Flight Safety Foundation.

10 Active Pilot Monitoring Working Group (2014, p. 19). This graphic is provided by the Flight Safety Foundation.
11 IMSAFE is an acronym for I – Illness, M – Medication, S – Stress, A – Alcohol, F – Fatigue, and E – Eating.

BIBLIOGRAPHY

Active Pilot Monitoring Working Group. (2014). *A Practical Guide for Improving Flightpath Monitoring*. Alexandria, VA: Flight Safety Foundation.

ASRS – Aviation Safety and Reporting System – National Aeronautics and Space Administration (NASA) Online Database. (n.d.). Retrieved from Aviation Safety and Reporting System: https://asrs.arc.nasa.gov/search/database.html.

NTSB. (1997). *NTSB/AAR-97/01 and PB97–910401: Wheels-up Landing – Continental Airlines Flight 1943 – Douglas DC-9 N10556 – Houston, TX, February 19, 1997.* Washington D.C.: National Transportation Safety Board.

15 Distraction Management Techniques

Since distractions are inevitable, we need to learn how to manage them effectively by honing our skills of recognizing when we have become distracted, mitigating any adverse effects, and recovering from game plan disruptions.

15.1 ISOLATING THE SOURCES OF DISTRACTION

One technique is to isolate ourselves from distraction sources. For example, some airlines either close the flightdeck door or position a flight attendant across the doorway during crew preflight briefings and checklists. This prevents other team members and jumpseaters from inadvertently interrupting the flight crew.

15.1.1 Creating Distraction-Free Bubbles around FOs during Preflight Planning

FOs are typically the pilots who receive the flight clearance, compute performance data, and program the FMS. Since each of these tasks affects the next, it is important for FOs to maintain an unbroken train of thought. A distraction early in this process can allow an error to spread. To create a low-distraction bubble around the FO, some operations designate the Captain as the distraction manager. If anyone enters the flightdeck or calls on the intercom, the Captain deals with them. This is particularly important when ATC assigns an unexpected departure transition or an amended clearance. FOs need to detect the change, understand the new routing, change the previously entered FMS routing, and verify an accurate waypoint sequence. Errors made early in this sequence can be covered up by the normal tasks and habits that follow. By creating a distraction-free bubble around the FO, we help them to focus their attention and ensure accuracy.

15.1.2 Creating Distraction-Free Bubbles around Captains during Diversion Planning

An inflight example is entering a holding pattern with a possibility of diverting to an alternate. Here, the distraction bubble needs to surround the Captain. The FO handles distractions, ATC radio calls, and manages the aircraft. The Captain focuses on contingency planning with the dispatcher, station, and cabin crew. After plans are established, each pilot resumes their normal crew roles.

DOI: 10.1201/9781003344575-17

15.1.3 Creating Low Distraction Time Windows

Distraction-resistant time windows form the foundation under sterile flightdeck protocols. The objective is to reduce or delay interruptions until reaching lower-workload phases of flight. Both the flight and cabin crew follow procedural guidance outlining what they can do and what they shouldn't do within "sterile" windows.

Creating low distraction windows is particularly important when we are handling unexpected and complex situations. Urgency demands that we coordinate our work while completing specific procedural steps – some of which are unfamiliar, unpracticed, and cannot be reversed. Additionally, these events tend to be fairly rare, so crews may not have recent experience resolving their problem.

Non-critical ATC radio calls are particularly problematic. A commonly reported example is repeated requests for "fuel and souls on board" by ATC controllers. Even after we provide this information, it may not be passed forward to the next ATC agency. The subsequent controller then asks for the same information that we just provided. To create a low-distraction window, we can request a discrete frequency or ask ATC to give us 5 minutes to attend to a particularly difficult procedure before calling us.

This same technique can be used with the cabin crew. When something happens to the aircraft that may cause concern in the back, flight attendants typically call forward to ask what is happening. Take a complex electrical problem, for example. The flight attendants may notice the cabin lights flickering or that they have lost galley power. Before we dive into the checklist, a useful technique is to call back, tell them that we are working an electrical problem, and that we will call them back in 5 minutes.

15.2 MITIGATING THE EFFECTS OF DISTRACTIONS

Most distractions are unavoidable. It follows that if we can't avoid distractions, we should develop robust strategies to mitigate their adverse effects.

15.2.1 Anticipating Probable Distractions

Recall the example we used earlier where we were given forewarning of a V_1 cut event in the simulator. Since we knew exactly when the event would occur, our vulnerability to startle, surprise, and distraction was dramatically reduced. We conclude that anticipation is a strong countermeasure against the effects of distraction. Anticipation prepares us in two ways. First, it significantly reduces the startle and surprise effects. Second, it kick-starts our recovery by rehearsing our initial response steps. We can recreate these advantages by anticipating common distractions associated with particular conditions. Some examples are:

- When flying an approach in turbulent convective weather conditions, review the caution and warning alerts for predictive windshear. Review the steps for a go around, missed approach, and the windshear recovery maneuver.

- When flying an approach with minimal spacing behind a preceding aircraft, review the go around steps in anticipation of being sent around by ATC.
- When the flight attendants are assisting an ailing passenger, review the steps needed to initiate an emergency descent and medical diversion.
- When operating near significant migratory bird activity, review the initial birdstrike indications (bangs, smells, windscreen obscuration), priorities (engines, visibility), and initial game plan (continue the approach or go around).
- When landing under marginal weather conditions, review the criteria for landing versus executing a missed approach.
- When landing on a slippery runway, review directional control techniques, stopping priorities, anti-skid cycling, and taxi considerations.
- When the landing may require an emergency evacuation, review the criteria for when to evacuate versus when to direct everyone to remain seated. Review the process and the main action steps.

As we analyze past experiences of distraction, include specific details of how the distraction appeared. Consider the sounds generated from outside sources (bangs and bumps) and from inside sources (automated aircraft alerts and alarms). Prepare for how the startling event may unfold, the first steps for overcoming startle, how we will initiate our recovery process, and specific role assignments for each crewmember.

15.2.2 Recognizing that a Distraction Has Occurred

There is an important distinction between recognizing distraction events from our past and recognizing that something is currently distracting us. The first is a product of hindsight. For example, if we made an error because we were distracted responding to a traffic alert, we can look back and recall how that disruption led to our error. While this is helpful, a more powerful skill is recognizing how we feel while being distracted in-the-moment.

While we are immersed in the continuous flow of flying, we sense each distraction as it happens. We choose either to ignore it or to deal with it. After we recover from startle, we need to decide whether that event is a trivial blip within the operational flow or whether it is something that is distracting us away from the operational flow. There isn't a clear line between these two decisions. Consider four bird encounters.

1. *Case 1:* We see the bird and watch it as it zips by. Since we don't hear an impact, we conclude that we successfully missed it. Most of us would conclude that this is not a distraction, or at most, a very minor one. If we don't hit the bird and maintain our attention flying down final, we would treat it as an inconsequential event. We saw it, we missed it, and we moved on.
2. *Case 2:* We only detect the bird at the last instant. It startles us and we flinch. Not hearing an impact, we conclude that we probably missed it. We return our attention to our flightpath and continue down final, albeit a bit shaken. This would be a strong, momentary distraction. It would divert our attention from flying for a few moments, but we would return our attention to

flying down final without further disruption. We might retain some adverse residual effects from the distraction.

3. *Case 3:* We see the bird at the last moment. We hear it impact somewhere behind us on the fuselage. This is a much stronger distraction event. We would be startled. Recovering from the startle, we check our engine instruments to ensure that they are running normally. We smell the air for the odor of burnt bird. For a longer time, our attention might be diverted from flying. Our monitoring of the operational flow is probably disrupted. After processing the distraction, we would look for cues that could provide context for where to rejoin the operational flow. We would also weigh whether to continue the approach or to go around and assess the effects of the birdstrike.
4. *Case 4:* The bird impacts on the radome with a loud bang. Bird parts spread across our windscreen. One of the wings lodges under our wiper blade and continues flapping wildly. This would undoubtedly qualify as a significantly distracting event. Unless we recovered and assessed quickly, we would favor going around to regain our composure, analyze the damage, rebuild our SA, and then return for an approach.

These four cases span the range from inconsequential to very distracting. Consider the operational task of completing Before Landing Checklist. In Cases 3 and 4, if we recognize in-the-moment that the birdstrike has probably distracted us and that we might have missed something, we would look for cues about possibly missed tasks. We would either clearly recall completing the checklist or be unsure whether we had completed it. If we have any doubt, the prudent choice is to run the checklist. The worst consequence is that we would run it twice – far preferable than missing it entirely.

If we don't have an in-the-moment awareness that the birdstrike has distracted us, we might just try to rejoin our operational flow based on our current position on final approach. If that position doesn't match our usual checklist completion point, we might assume that we had already completed it. The critical difference is recognizing that we were distracted and deliberately investigating what we might have missed during the lost time.

Considering this range of cases, we identify three questions to determine whether something rises to a concerning distraction-level or not.

- *Intensity or severity:* How much of our attention was diverted by the distraction?
- *Duration:* How long did the distraction divert our attention from our operational flow?
- *Operational flow disruption:* Following the distraction, how different is our position from where it was before the distraction?

We may need to assess these questions very quickly. If we decide that the event is fully understood, processed, and recovered, we can safely continue. If not, we need to assess how much time we have available. If time is short, we should consider making extra time to process the event, settle down, and try again.

15.2.3 Understanding the Effects of Distraction on Our Operational Flow

There isn't a clear distinction between distractions that prove consequential and those that don't. They feel similar as we experience them in-the-moment. They both feel like flying the aircraft, dealing with events as they come up, and keeping our game plan on track. We only recognize their distinctions in hindsight. Our challenge is to translate this hindsight clarity into present-moment awareness. We develop this skill by analyzing our encounters with distracting events and learning to recognize our personal biases, strengths, and weaknesses.

We begin by constructing two stories comparing how we expected the flight to flow (future SA) with what actually happened. Holding these two stories side-by-side, we run the timeline back and note the onset of the distracting event and our reactions to it. Next, we identify indicators that were present. They may not have registered as important at the time. Maybe they felt unimportant or they felt like minor anomalies mixed in with the background noise of everything else that happens while flying. Maybe we were startled or the distraction drew our attention away. We also assess our personal and crew factors. Were we too relaxed or tired? Were we engaging in discretionary activities? Were we inappropriately diverting our attention?

By comparing the two stories, we realign our practice to respond more skillfully in the future. We identify monitoring techniques that would have accurately detected the adverse effects. We recognize decisions that would have aided our recovery. Each time we perform this kind of debrief analysis, we refine our response to distracting events. Revisiting our three criteria for determining whether something rises to a concerning distraction-level or not, we evaluate the effects on our operational flow.

- *Intensity and severity:* Compare the difference between three bird encounters – missing the bird, maneuvering to avoid it, and hitting it. Imagine how each one would feel, where we focus our attention, and how we manage our flying. If we see and miss the bird, we register the event subtly while easily maintaining the aircraft's path. We don't lose focus. Our operational flow remains intact. For the second bird encounter where we maneuver to avoid it, we need to divert a significant portion of our attention to maneuvering the aircraft. So, the bird distracts us, but we still remain sufficiently connected to the operational flow. For the third encounter, the bird splatting against our windscreen completely grabs our attention. The distraction diverts our attention away from the operational flow of flying the aircraft. The greater the severity of the distracting event, the more our attention is diverted away from holding a stabilized final approach. As we improve our awareness of how we personally respond to each of these cases, we refine our awareness management skills. When a similar event happens to us in the future, we won't experience the same startle and distraction. We'll engage it more mindfully.
- *Duration:* Duration is a measure of how long the distraction diverts our attention from the operational flow. Imagine a scenario where we pass one bird while on final approach, followed immediately by another, and then

another. Each time, we successfully avoid each bird, but the succession of distractions keeps our attention diverted from the actively managing the operational flow. With each encounter, we start recovering, but another distraction arises. We can't fully return our attention back to flying the approach. Next, imagine that during this series of distractions, an unrelated distraction emerges, like an aircraft incurring onto our runway. At what point would our SA become so deflated that we miss that new threat? The duration of a distraction or a chain of distractions might affect our ability to return our attention back to flying. When successive distractions prevent us from fully returning our attention to the operational flow, we need to consider exiting the game plan (like going around) and resetting the operational flow.

- *Learning from our personal distraction encounters:* As we study our personal experiences with distractions, we may notice trends. Perhaps we discover that we tend to respond differently when we are tired or relaxed. Perhaps we lower our vigilance during the last leg of our pairing back to our home base. Perhaps we discover that events from our home life adversely affect our ability to remain focused. Perhaps we notice that we respond differently when paired with particular types of pilots. For example, I discovered that when I was flying with a particularly competent FO, I would subconsciously lower my level of vigilance. Perhaps because they were so good at detecting and mitigating any errors, I allow myself to relax into my comfort zone. I subconsciously lowered my level of vigilance because they raised theirs. Looking deeper, I recognized that the opposite also applied. I would raise my level of vigilance when flying with an inexperienced or especially lax FO. Discovering my personal bias, I learned to control my attention level more appropriately.

  The more we learn about ourselves, the more accurately we can optimize our resilience against distraction. Sensing that we are tired signals us to increase our vigilance. Noticing that we perform better during daytime encourages us to bid for early schedules. Noticing that we are less alert during early morning flights encourages us to bid for later schedules. We should also monitor how we change as we age. We are constantly changing individuals, so we need to continuously reassess ourselves.

15.2.4 Following AOV (Area of Vulnerability) Protocols to Appropriately Manage Distractions

AOV protocols align our attention level to match the aircraft's level of dynamic movement. To understand why this is important, consider how our perception of risk changes as we gain experience. When we were new, everything seemed difficult and felt risky. Our concerns eased as we gained experience. Those previously difficult situations now seem easy and familiar. As our sense of perceived risk falls away, our risk assessment process can subconsciously shift from *managing risk by situation* toward *managing risk by feeling*. Managing risk by situation appropriately uses the severity, complexity, and unpredictability to assess risk. For example, we rate landing on an icy runway as riskier than landing on that same runway while it is clear and dry. While this is true, this perspective can unintentionally minimize the risk inherent in any landing, or how we measure *risk by phase of flight*. Relying on how risk feels as we gain experience, familiarity, and proficiency, we might lower our assessment of risk for busy flight phases like takeoff and landing. In truth, risk by feeling, risk by situation, and risk by phase of flight are all important ways of assessing risk. We just can't let one inappropriately influence either of the other two.

AOV protocols restore *risk by phase of flight* to our risk assessment equation. By following low (green), medium (yellow), and high (red) AOV protocols, we improve our distraction resilience. In high AOVs, we elevate our level of attention and readiness, regardless of how easy the situations feel. This improves how well we respond to distraction events. Consider flying down final with nearby convective weather. While the PF focuses on maintaining approach parameters in challenging turbulence, the PM scans for warning signs of windshear. The PM detects adverse trends in their flight parameters and sees a microburst dust plume on the airfield. They warn the PF to anticipate a windshear encounter. The PF remains focused on flying the approach, but also mentally prepares for a go around and windshear recovery maneuver. Suddenly, the windshear detection system broadcasts, "GO AROUND, WINDSHEAR AHEAD". While the event would still be startling, the crew is ready to handle it. They quickly recover from the startle effect and execute the appropriate recovery procedure.

Considering the opposite case of a benign approach under ideal conditions, our roles remain the same. The PM would still expand their monitoring for anything that might affect landing. Perhaps they notice a lone dust devil (thermally induced whirlwind typically generated by summertime conditions) approaching the runway threshold or a vehicle crossing halfway down the runway. They would alert the PF so they can anticipate switching from flying the approach to going around.

In green AOVs, our readiness to respond to distraction is lower, but the time to recover and successfully respond is longer. Consider an extreme example of a rapid decompression while in the low-AOV environment of steady cruise. It will be extremely startling, but as long as we follow our well-trained procedures, we'll have ample time to successfully execute the emergency descent procedure. Even if we take a few extra moments to regain our wits, as long as we get our masks on and oxygen flowing, we'll have time to recover from startle effect, diagnose the problem, refer to the checklist, and initiate the emergency descent.

15.2.5 Avoiding Internal Distractions

We should guard against creating our own distractions. Common examples include engaging in intense conversations while in dynamic flight, rehashing past events while in high-workload flight phases, and performing heads-down tasks during inappropriate flight phases. Consider the following extreme example.

BOX 15.1 CAPTAIN CREATES FLIGHTDECK DISTRACTIONS

FO's report: While holding short between the active runways in SEA, I observed my Captain using his cell phone to send text messages. I also observed him listening to music throughout all four flights we had together. I should have said something during the trip.[1]

This report reveals two points about intentional internal distractions. First, this Captain's comfort zone had drifted very far away from company standards. Their discretionary choices had evolved into openly noncompliant behaviors. We can surmise that they reached a comfort zone where they saw nothing wrong with their actions. The FO doesn't report any errors arising from these distractive habits, but the potential for error is undeniable. Second, it reflects a line culture where the FO did not feel comfortable voicing their displeasure directly with their Captain. Instead, they elected to use a NASA ASRS report to document their discomfort.

15.2.6 Avoiding Habits that Intensify the Adverse Effects of Distraction

Following an event where we mismanage or become surprised by a distraction, we ask ourselves some questions:

- What was I doing before the distraction occurred?
- Did my mindset at the time intensify my adverse reaction?
- Were there warning signs that I missed or ignored?
- Was I doing something discretionary?
- Did my actions or attention-focus contribute to missing these warning signs?
- Was I following an appropriate level of attention focus for the AOV?

We may discover that certain discretionary choices or habits, which are only appropriate for low AOVs, had migrated into medium and high AOVs. Remember that our personal practice drifts so slowly that we only recognize it when we compare our current practice with ideal practices. It is through our personal introspection and event debrief that we discover the gap between where we are and where we should be.

We form habit patterns to successfully manage chains of individual tasks and to capture errors that we may have committed in the past. Every time we develop a habit pattern, we alter the way that conditions interact. Some possible effects include:

- A seemingly small distraction generates a stronger disruption than expected.
- We've come to rely on a particular habit pattern to catch errors only to discover that it allows other errors to slip through.
- Shortcutting within our habit patterns undermines some verification steps that are designed to capture specific errors.

Ideally, our habit patterns work well, but since aviation complexity allows unexpected vulnerabilities to emerge, we need to periodically review the quality of our techniques. Whenever we make an error, we should evaluate whether our habit patterns contributed to the problem. Consider the following report of aircraft movement after shutdown.

BOX 15.2 PROCEDURAL DRIFT RESULTS IN AIRCRAFT MOVEMENT AT THE GATE

Captain's report: ... aircraft moved forward after a nose chock was inserted. Both engines were shut down at the time of aircraft movement. Movement was immediately after shutdown of #2 engine. ... *When I got home, I reread our shutdown procedures. I noticed that I did not follow the cautions of checking the accumulator pressure being in the green band. The truth be told, I have not been checking this for years. I have been checking for aircraft movement only after releasing the parking brake. A huge mistake that thankfully did not cause any injuries or damage. Most embarrassing for a seasoned Captain. I guess I need to go through all my normal procedures again highlighting all cautions and warnings. Surprising that in so many years of line checks and training events that this bad habit never caught up with me.*[2]

It was only through their hindsight evaluation that this Captain discovered how their habit pattern had drifted. As Master Class pilots, we periodically perform comprehensive reviews to see how well our techniques match standard practices. This recalibration arrests unintended drift and shrinks latent vulnerabilities.

15.2.7 Mindfully Managing Distractions

Distractions divert our attention from what is currently happening toward something else that is also currently happening. Even as we deal with the distraction, the aircraft continues to move forward. Consider two pilots taxiing-in on their last flight after a long duty day. Operations reassigns them to a problematic gate – one that requires stopping in the alley, shutting down the engines, and being towed-in. They talk about the rare-normal procedures required for the tow-in. While discussing relevant issues and proactively planning, they inadvertently overshoot their assigned taxi transition onto a parallel taxiway. This error results in a conflict with an opposite-direction aircraft. How might this error evolve? If we consider their tasks during this event, they had three priorities.

1. Continue tracking the taxiway centerline toward the gate area.
2. Make the assigned transition onto the parallel taxiway.
3. Plan for the rare-normal, tow-in procedure required by the unexpected gate change.

They handled priority #1 well. They got a head start on #3. They completely missed #2. When we evaluate where these three tasks were situated, we see that they all resided in different locations. Taxiing straight ahead was immediately in front of them – visible, familiar, and easy. The tow-in procedure was a half mile away – not visible, not familiar, and complex. Additionally, since the gate change was a distraction to their expected operational flow, it absorbed much of their attention and translated their focus of attention toward the ramp area. We can imagine both pilots mentally visualizing the ramp layout, where they would stop the aircraft, what they would say in their PA to the passengers, how they would manage the shutdown procedure, how they would establish communications with the ground crew, and how the tow-in would proceed.

The missed taxi route transition was initially located a quarter-mile in front of them – visible, but not immediately relevant. It became a prospective memory task – something they needed to remember to do later. They weren't completely distracted, but too much of their mental focus was projected a half-mile away in the gate area. Neither pilot remained mindful of the upcoming turn in their taxi route assignment.

This kind of distraction/prospective memory error is common in aviation. We constantly need to shift the focus of our attention between competing tasks. Mindfulness guides us to stay focused on what is happening, while still planning for upcoming tasks. We start by realizing that the late assignment of an unplanned gate introduces an unexpected change to our expectations. We recognize this as a distraction. While continuing to move safely forward (#1), we protect any upcoming relevant tasks (#2), and begin resolving the distraction (#3).

Consider how the event would have transpired if the Captain had guided crew priorities. "I'll continue monitoring the taxi and our upcoming transition onto Taxiway Alpha while you begin reviewing the tow-in procedure that we'll need for this gate change." This satisfies all three priorities, assigns roles, and reinforces the importance of complying with the taxi route transition. While still in a medium AOV, the FO could reference the manual. They would set the manual aside during the high-AOV transition onto Taxiway Alpha, then return back to their preparation after established on the next straight taxiway segment. Another option would be to delay preparing for the tow-in procedure until stopped in the gate area. They could then work together to review the tow-in procedure.

15.2.8 Avoiding Mutual Distractions

Another common problem is when both pilots become distracted by the same event, detect the same indications, miss the same warning signs, and make the same errors. One of the objectives of PF/PM role assignment is to maintain separate perspectives. We want each pilot to remain enough out-of-sync with each other so that at least one pilot detects each relevant counterfactual. When one pilot becomes captivated by the distraction (usually the PF), we want the other pilot (usually the PM) to remain detached enough to monitor the big picture. Consider the following report of mutual distraction resulting in a slow-speed stick shaker.

BOX 15.3 BOTH PILOTS MUTUALLY DISTRACTED RESULTING IN A SLOW-AIRSPEED EVENT

FO/PF's report: Approximately 2 miles prior to ZZZZZ Intersection on the RNAV Arrival, and due to weather being below CAT I minimums at ZZZ, we were instructed to hold at ZZZZZ Intersection as published. By the time the controller had finished her transmission, we had passed ZZZZZ Intersection. I was the Pilot Flying (PF), selected HDG SEL and began a right-hand turn to enter the hold directly while the Pilot Monitoring (PM) worked the FMC. *Of note, the autothrottles were deferred on this flight.* I slowed the aircraft to 230 knots (226 was our clean maneuvering airspeed) and ensured we were complying with the holding instructions while monitoring the PM's activities with entering the hold into the FMC. *At some point, I became distracted with the FMC and tried to help the PM, completely forgetting that I had pulled back the throttles to slow down. We received stick shaker and I noted approximately 195 knots.*

… There are well established procedures in place that ensure the PF is focused on flying the airplane, while the PM manipulates the FMC as needed. I inadvertently disregarded these procedures in an effort to help with an unusual situation regarding holding at a point that had just sequenced in the FMC. Add to that that it was the end of a 7+ hour leg, with the autothrottles deferred, and I should've have been more aware of the possibility of this kind of error on my part. *This is the kind of mistake that I always thought would never happen to me, and yet here it did, completely of my own doing. I will refocus my efforts on ensuring I'm focused on my primary duties as the PF and ensuring if help is needed on the FMC, to transfer control of the aircraft before diving into those issues.*

Captain/PM's report report: … The PF started the holding manually with "heading select". I confirmed we were turning to the right heading and started to set up the holding page. *It took about four or five times of selecting the fix and "next hold" on the FMC before the next hold page would populate* [Note: The pilots were attempting to program an FMC feature that would direct the autopilot to automatically maintain the assigned holding pattern. When properly programmed, the FMC would "populate" the necessary fixes and turns on the display screen]. During this time the PF looked down at the FMC in an attempt to see if he could help figure out why the next hold wasn't populating or offer assistance. *At that point we were both heads down for a few seconds.* With the power setting that was manually applied to slow to holding speed at 10,000′, the aircraft's speed kept slowly rolling back. The stick shaker activated, we both looked up and the recovery maneuver was performed. *The main issue is to remember the separation of duties. As the Captain, I should have made sure that the PF was staying with the aircraft while I was loading the FMC.*[3]

Both pilots fulfilled their roles until a prolonged distraction occurred. For a few critical moments, they aligned their attention focus. No one was actively monitoring the aircraft. A latent vulnerability caused by inoperative autothrottles emerged as the airspeed decayed until the stick shaker activated.

We refine our instincts to understand how distractions affect us. When our attention is diverted for too long a period, we want our internal warning timer to alert us. When it activates, we assess whether we have become too fixated on the distraction. While the PF focuses their attention on handling the distraction, the PM should broaden their perspective to assess the operational flow. How is the flightpath? Are we following the intended game plan? Did we interrupt another task to handle this distraction? What do we need to do to resume the intended operational flow? On the other hand, if PMs are handling a lengthy distraction, PFs must ensure the quality of the operational flow. They can still help their PMs, but they need to maintain an appropriate sample rate of aircraft flightpath.

15.2.9 Developing Distraction-Resistant Habit Patterns

Good habit patterns are strong countermeasures against distraction-caused errors. Take the example of scheduling the Climb Checklist passing through 10,000′. This coincides with a speed transition from 250-knot climb to an FMS-computed climb speed. It also marks the end of sterile flightdeck protocols and often includes a chime notification to the cabin crew. Together, they serve as strong anchors that remind us to complete the checklist. On the other hand, 10,000′ often coincides with an ATC frequency change. Since PFs are actively engaged in a high-AOV flightpath transition, PMs own the responsibility for remembering to complete the checklist. One technique is to pull the checklist card out and hold it. If a distraction occurs, they retain a strong reminder that the checklist hasn't been completed. Consider what might happen if the PM chooses not to employ a reminder technique. At 10,000′, the Captain chimes the cabin crew. The flight attendants know that they are no longer bound by sterile flightdeck protocols and immediately call to inform the pilots about a passenger cabin issue. The PM takes the call and diverts their attention toward resolving the problem. When they eventually return from the distraction and lacking a solid reminder, the checklist can be missed. Consider the following report from a Captain who didn't follow their distraction-resistant habits resulting in a no-flap condition taking the runway.

BOX 15.4 CAPTAIN MISSES THEIR ERROR-MITIGATING TECHNIQUES AND TAKES RUNWAY NO-FLAP

Captain's report: Day two, leg four of eight in two days. Pushed onto taxiway, started, after start procedures all done, called for taxi, and I didn't call for Flaps 5. My normal plan, on almost every other After Start flow is to state, "Clear left, clear right, Flaps X", but for some reason in this situation, I didn't. No excuse, no reason, I just didn't. Taxi commenced, FO made the PA for seating the FAs (Flight Attendants). *My attention was possibly momentarily switched to the [aircraft] that had just landed and had turned onto our taxiway in front of us, but just for a second or two. The only other distraction was when Ground called shortly thereafter and told us about a flow time back to ZZZ1.* He mentioned we might get off earlier than the estimate which would only amount to a three-minute wait at the end of the runway, still not anything big, but somewhere in there, where I would've normally called for the "Before Takeoff check", I didn't.

When we got down nearer the hold short area, I called back to Ground to ask if they wanted us out of the way for departures, short discussion, said we were fine there for the three-minutes wait time. *Minor distraction, we were still feeling all done, ready to go and once cleared for takeoff, yes, of course the next noise I hear is the takeoff warning horn for the flaps still being up.* The only good thing I did from the after start to the five-knot abort was to abort the takeoff, clear the runway, do the circle of shame back to the takeoff position, set the flaps, and actually perform and complete the Before Takeoff check. My other normal procedure is to do the configuration check as I am calling for the before takeoff check. So, yes, I didn't get the configuration check either, not because I blew it off, just because I never called for the Before Takeoff check.

In debriefing the incident with the FO, who, by the way was rock solid the whole last 2 days, mentioned to me that when he heard the horn, he couldn't believe, in his mind, that the flaps were still up, and in hindsight neither could I. BOTH of us thought we were all good, everything was normal, all systems go.

Uneventful departure thereafter, a good debrief on how dumb the Captain was, humility check complete. *I think this is, once again, an example of sequential thought. When my mind was "slightly" distracted with the flow time, that's when we would normally be accomplishing the Before Takeoff check. Once we were past that point where I would normally do that, there's nothing to remind me to say, "Hey, you're forgetting something" until the horn chimes in. Lesson learned is habit patterns and checklists are good, not so good if I don't perform them, in order, when the flight calls for them. I'm personally going back to my old habit of doing a configuration check whenever I enter the runway.* Yes, it might be redundant, but if I miss the primary, maybe my space-limited brain will pick it up on the backup.[4]

This Captain cites "sequential thought" to describe how habit techniques link tasks together into a chain. In their case, breaking one link caused failures down the chain. There are also interesting references to their habit pattern drift and "feeling" ready to go as they took the runway.

Reflect on the habit patterns that we have adopted over the years – some to simplify tasks, some to catch errors, some to remind us to do something, and some just because they seemed to make our work flow more smoothly. Consider these questions for evaluating each technique.

- Is our technique still working as intended?
- Is it still current given procedural changes or aircraft system changes?
- Has it drifted in ways that adversely affect its effectiveness?
- Are there any error vulnerabilities generated by it?
- Does it improve the quality of our work or do we just do it because it is convenient?
- Did it contribute to any errors that we have made?
- Did it assist us to detect and mitigate potential errors?
- When we hear of another crew's error, would our technique have improved our ability to detect and mitigate their error?

15.2.10 Studying Our Distraction Encounters

We encounter plenty of distraction events. Instead of just handling them and moving on, we should study them. After we reach a low-workload phase of flight or following the flight, evaluate:

- How did the distraction emerge?
- What were the indications? Which indications did we detect? Which did we miss?
- Did we detect the distraction quickly enough to prevent a disruption?
- Did we detect the distraction accurately?
- Did we effectively return to our previous task or was our operational flow disrupted?
- Did we effectively manage the distraction as a crew?
- Did the PF and PM roles remain intact and separate?
- How could we have managed it better?
- Does the event expose a safety vulnerability that should be reported?

If we simply resolved our distractions and otherwise ignored them, we would advance our skills, but only in a slow and haphazard way. It would be like learning to fly by trial and error without the guidance of an instructor. If, instead, we evaluate our handling of each distraction using questions like these, we proactively refine our recovery process and supercharge our learning progress.

15.3 RECOVERING FROM DISTRACTIONS

The goal following any distraction event is to recover our intended operational flow. The following steps guide our process.

1. Perform any immediate action steps.
2. Evaluate the time available to resolve the disruption.
3. Mentally note what was happening as the distraction occurred.
4. As a crew, determine the cause and significance of the distraction.

5. Determine the tasks required to resolve the distraction.
6. Assign or maintain roles.
7. Resolve the problem.
8. Restore the operational flow.
9. Assess any residual effects.

15.3.1 Perform Any Immediate Action Steps

This is a highly familiar process for us. Even if the PM was heads-down working on something and the PF announces, "Birds" and pulls back on the controls, the PM would understand both the cause and the immediate action. In truth, there are relatively few aviation problems that require immediate actions. Even most "Immediate Action" steps can be delayed until we recover from startle and assess the problem. In fact, our recovery may be hindered by taking immediate actions that don't match the problem. For the vast majority of our distraction events, we are advised to take a few moments to assess the situation before taking any action steps. Consider a warning message in the FMC, for example. If one pilot detects the message and silently clears it before the other pilot has a chance to read it, a useful error mitigation feature is lost. The better process would be to announce the warning message, agree on its significance, and then clear it.

15.3.2 Evaluate the Time Available to Resolve the Disruption

Our attempts to resolve disruptions sometimes go astray because we dive into handling them before we evaluate how long they will take to resolve. A quick maneuver to avoid birds resolves the problem almost immediately. Conversely, if we experience an electrical problem, it may take quite a while to resolve. If we are taxiing for takeoff, we could ask for permission to hold our position, set the parking brakes, and begin a deliberate resolution process. Airborne, this problem might require maneuvering airspace. In any case, we need to estimate the time needed to resolve it before proceeding on to the next step.

15.3.3 Mentally Note What Was Happening as the Distraction Occurred

Our working memory tends to be quite short. We are accustomed to handling events as they arise and then quickly discarding them to be ready for the next event. Past events quickly fade from our awareness. The longer we become engrossed with handling a distraction, the more likely we are to forget what we were doing before the distraction occurred.

The distraction introduces a new timeline into the operational flow. We still have the tasks and procedures required by the phase of flight, but we also have the tasks and procedures required to resolve the distraction. To effectively recover, the distraction timeline must be integrated into the existing operational timeline.

15.3.4 As a Crew, Determine the Cause and Significance of the Distraction

Ideally, all crewmembers need to agree on the meaning of the distracting event. This should include a verbal announcement clearly identifying the problem. For a fire alarm and a red light on the #2 engine handle, someone (typically the PM) should announce something like, "Engine fire, #2 engine." For ambiguous problems, pilots should identify relevant indications, "The flaps aren't extending." Other distractions may be more subtle and require discussion. An example would be, "Do you smell something burning?" In any case, we need to reach a clear consensus about the problem before proceeding.

15.3.5 Determine Tasks Required to Resolve the Distraction

This step acknowledges that the steps to resolve the distraction need to be scheduled within the existing operational flow. With our gate-change example, the Captain recognized that they would need to prepare for the unplanned gate tow-in procedure while still complying with their ongoing taxi route assignment. The Captain took full responsibility for the upcoming taxi route transition while the FO began preparing for the tow-in. Their solution integrated upcoming operational tasks with distraction-resolving tasks. There are two advantages with this strategy. First, it preserves the salience of operational tasks so they don't become overshadowed by the distraction. Second, it smoothly integrates both timelines within the operational flow.

15.3.6 Assign or Maintain Roles

For routine distractions, standardized procedures guide our process. For example, most airlines assign the task of responding to cabin crew inquiries to the PM. This keeps the PF fully engaged with controlling and monitoring aircraft movement. PFs should guard against becoming mutually distracted by a passenger cabin problem. If the Captain is PF, but needs to coordinate with the cabin crew, consider changing roles to have the FO fly.

A core objective of maintaining assigned roles is PF and PM role separation. We never want both pilots becoming equally distracted by the same event. The most common occurrence is when PMs divert their attention toward whatever is challenging the PF. Take the example of a PF pushing stabilized approach parameter limits. PMs have the responsibility to direct a go around if the approach cannot be stabilized, so it *feels* like something that demands their attention. Is this approach acceptable? Should I make a callout? Is the PF going to stabilize it in time? When both pilots become fully aligned with the same problem, they can tunnel their attention on the same parameters, miss the same warning signs, and ultimately, make the same mistakes. Instead, we want PMs to maintain their monitoring role and remain out-of-sync with their PFs. This means that the PF stays focused on managing the aircraft to satisfy parameters. The PM stays focused on the limit point – the altitude where stabilized parameters must be met. This keeps the PM centered on the larger perspective of approach quality so they can detect warning signs and errors.

As events become more complex and task loading rises, it may become useful to completely divide duties. An example is when weather unexpectedly closes our destination airport. We need to enter holding and prepare for the possibility of diverting to an alternate. In this example, the Captain would transfer aircraft control and ATC responsibilities to the FO while assuming responsibility for coordinating with dispatch, calling station operations, and informing the passengers and cabin crew. The FO remains fully engaged with flying the aircraft, managing ATC instructions, and handling any "inside the flightdeck distractions". If the FO becomes overloaded, they can always alert the Captain for assistance. Meanwhile, the Captain remains focused on coordinating for possible diversion and handling any "outside the flightdeck distractions". While this division temporarily removes CRM verification and redundancy, it is a good way to accomplish the most work in the shortest amount of time. Notice that despite the level of complexity, one pilot remains fully engaged with monitoring the operational flow. When the excess workload is completed, a vital task is to reform the crew and update each other on what transpired while their workflows were separated. This gives the Captain the opportunity to verify and validate any clearance changes. It gives the FO the opportunity to update their game plan and shared mental model.

15.3.7 Resolve the Distraction

Notice that we don't resolve the distraction until well into the process – the seventh step, in fact. When crews mismanage their distraction events, we often trace the problem back to omitting or rushing some of the previous six steps. Some resolutions are immediate. When we see the birds, we pull up to avoid them, they zip past, and we restore our flightpath. This fully resolves the distraction. Other distractions take much more time and effort.

15.3.8 Restore the Operational Flow

Now that the distraction is resolved, we need to restore our operational flow. In most cases, we return to the original flow that we were following before the distraction. In others, we switch to a new game plan (like with a go around or a diversion) and join that new operational flow. Since the distraction creates a unique *distraction* timeline, we need to fully resolve it first. The previous seven steps guide this process. After restoring the operational flow, we rebuild our SA by restoring the smooth transition from past SA, to current SA, to future SA.

Problems emerge when we only assess the current conditions for cues for rejoining our operational flow. Any tasks that were interrupted when the distraction occurred can be lost since those cues are now past. Consider an ATC call that distracts us just as we were about to call for the Landing Checklist. After resolving the ATC instruction, our attention could transition to landing without completing the checklist. While this is not necessarily consequential, imagine that it occurred just before setting final flaps. We can see how missing final flaps can cause us to subsequently miss the linked, interdependent Landing Checklist. Ironically, the checklist is the procedural safeguard that would have caught and resolved the flap-setting error. Missing the one causes us to miss the other.

Sometimes, events happen so quickly that we don't have the time to step through all seven of the preceding resolution steps before we need to act. In these cases, we should do our best to resolve the distraction, then revisit any past steps just to make sure we didn't miss anything.

15.3.9 Assess the Residual Effects

The final step is to assess whether past distractions are still adversely affecting us. In one simulator training event, the FAA directed airlines to include a minor distraction during takeoff. Typically, crews were given a nuisance Master Caution light. They were expected to determine that the event did not warrant an RTO and continue their takeoff. An unintended side effect was a significant increase of crews neglecting to raise their landing gear. The residual effects of the distracting Master Caution light affected subsequent operational tasks. While not a nuisance light, this crew's bird-strike caused them to miss raising the landing gear.

BOX 15.5 STARTLING TAKEOFF FROM BIRDSTRIKE CAUSES CREW TO MISS RAISING LANDING GEAR

FO/PF's report: I was the pilot flying when our Boeing 737-700 ingested a goose that was on the runway with a flock of geese – into the left (#1) Engine at/approaching V_1. We continued the takeoff roll and rotation with nearly immediate compressor stalls/surges coming from the engine. Tower reported smoke and flames coming from the left engine within the first several hundred feet of flight. I swapped controls with the Captain when safely airborne. At approximately 1,000′ AGL, we cleaned up the aircraft, and executed the Engine Fire/Severe Damage/Separation/Seizure Checklist immediate action items. At no time did we have any associated warning/caution lights or warning horn. The engine kept running with nearly symmetrical power until we shut the engine down in accordance with the checklist. After accomplishing the engine shutdown, I declared an emergency with Tower…

Captain/PM's report: … The geese remained sitting on the runway until we were almost on top of them. Two or three flew up and toward the Captain's side of the runway. This distraction caused me to not make the rotate call. At very close to rotate speed, I felt multiple impacts with the plane's left side, followed immediately by loud, continuous compressor stalls, and accompanied by erratic engine indications on the number one engine. The First Officer continued the takeoff and began a climb. I assumed control of the aircraft at approximately 100′ AGL, by saying, "I have the aircraft." A few seconds later the Tower Controller advised us that we had smoke and flames coming from the left engine. There were no fire or overheat indications in the flightdeck. At approximately 400′ above the ground, I reverting back to 20 years of PTs and PCs. I attempted to call for the exact right checklist name, but stumbled on the exact verbiage when calling for the memory items associated with the "Engine Fire, Severe Damage, Separation". I eventually said something like "Engine Fire Memory Items" and the FO responded appropriately. We completed the appropriate memory items and checklist, and the number one engine was successfully shut down.

While guarding the #2 throttle for the shutdown procedure, I realized that we had accomplished a reduced thrust takeoff, and had more power available. I pushed the throttle almost all the way up and noticed a slight increase in our climb rate, but we were still climbing at only about 300 feet per minute, which is less than a normal simulator profile. *I then noticed the gear was still down, so I called for "Landing Gear Up" right after we completed the memory items a few seconds later. I am fairly certain that we did not raise the landing gear due to the distraction caused by the quick transfer of aircraft control just after takeoff.*[5]

Crews are typically trained to retain roles until safely airborne. By taking control at 100′, this Captain altered their trained habit patterns and reversed PF/PM duties at a time that the FO didn't expect. We need time to switch our mindset between PF and PM roles. This may have set the stage for missing the landing gear. The Captain was clearly "assessing residual effects" when they noticed that they had extra thrust available and that the gear was still down.

NOTES

1 Edited for brevity. NASA ASRS report #1670741.
2 Edited for brevity. Italics added. NASA ASRS report #1716069.
3 Edited for consistent PF/PM terminology and brevity. Italics added. NASA ASRS report #1517510.
4 Edited for clarity. Italics added. NASA ASRS report #1584436.
5 Edited for brevity. Italics added. NASA ASRS report #1113186.

BIBLIOGRAPHY

Active Pilot Monitoring Working Group. (2014). *A Practical Guide for Improving Flightpath Monitoring*. Alexandria, VA: Flight Safety Foundation.

ASRS – Aviation Safety and Reporting System – National Aeronautics and Space Administration (NASA) Online Database. (n.d.). Retrieved from Aviation Safety and Reporting System: https://asrs.arc.nasa.gov/search/database.html.

Dismukes, R. K., Berman, B. A., & Loukopoulos, L. D. (2007). *The Limits of Expertise: Rethinking Pilot Error and the Causes of Airline Accidents*. Burlington, NJ: Ashgate Publishing Company.

16 Automation Management Techniques

Few areas stretch HF science more than pilot interactions with aviation automation. It surrounds us on the flightdeck and across the aviation environment. It is a constant topic of technique innovation and procedure development. It is also a persistent contributor to confusion, distraction, and error.

16.1 AUTOMATION CONCEPTS

Following are some basic concepts that describe our interactions with automation.

16.1.1 Direct Control and Supervisory Control

When we started learning to fly in general aviation aircraft, we exercised *direct control* over the aircraft. Our flight controls were directly connected to the aircraft's control surfaces. As we moved the yoke/stick or depressed the rudder pedal, cables moved the flight control surfaces. A basic set of flight instruments were our only sources of flightdeck automation.

As we progressed to more advanced aircraft, we added higher levels of automation. Perhaps we flew an aircraft with a partial autopilot that could hold our heading or maintain a cruise altitude. At this intermediate level of automation, we continued to exercise direct control from takeoff through level-off, but engaged the autopilot through much of the cruise phase. At this level, we exercised *supervisory control* over the autopilot.

Eventually, we advanced to aircraft with fully integrated flight directors, FMS computers, and autopilots. Once our route was programmed and engaged, the aircraft could control the entire climbout, departure SID, cruise, arrival STAR, approach, and even the landing. At this level of automation, we primarily exercised supervisory control.[1]

16.1.2 Operating and Managing

From our pilot perspective, automation doesn't feel very different across the aircraft we've flown. We still control the flightpath. On balance, we make fewer *operating* inputs and more *managing* inputs through computers and other systems that manipulate the flight control surfaces for us. Still, whether we are operating or managing, we are still the pilots controlling the aircraft.

DOI: 10.1201/9781003344575-18

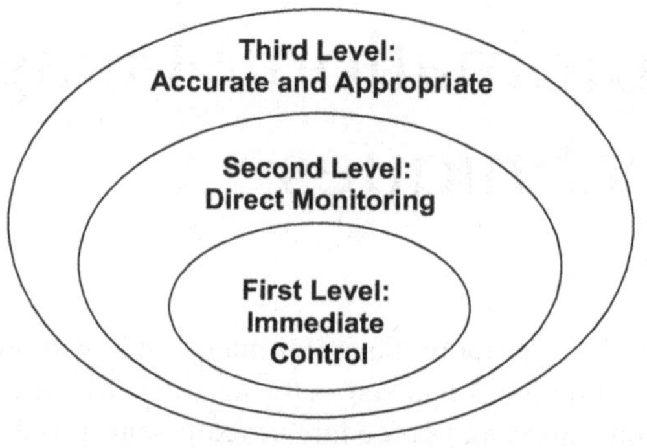

FIGURE 16.1 Levels of qualitative monitoring.

16.1.3 The Levels of Monitoring

Our monitoring skills become extremely important as we manage the many computers and systems that affect flight. Monitoring has both quantitative and qualitative dimensions. The quantitative dimension simply means that we use a lower sample rate during low AOV and low dynamic movement environments (such as steady cruise) and a higher sample rate during high AOV and high dynamic movement environments (such as hand-flying a CAT III ILS approach). The more compelling aspect of monitoring is its qualitative dimension. Consider Figure 16.1 depicting the three levels of qualitative monitoring.

- *First level – immediate control:* When we are hand-flying, we simultaneously control and monitor the path of the aircraft (direct control). The connection between controlling and monitoring is so strong that we rarely consider them as separate tasks. We sense a need for a flightpath correction, make the correction, immediately assess the result, make the next correction, and so on.
- *Second level – direct monitoring:* The next level of monitoring applies to the pilot next in line from whoever or whatever is manipulating the flight controls. When the autopilot is engaged, the PF is the direct monitor. When the PF is hand-flying, the PM assumes the role. The direct monitor ensures that the aircraft is correctly following the path that we expect it to follow within the game plan. We use short-term future SA to rate whether our immediate path remains appropriate and that the *next* change in path is understood, programmed, or displayed.
- *Third level – accurate and appropriate:* The third level of qualitative monitoring is maintained by both pilots, but is the PM's responsibility. The PM not only monitors that the PF is accurately following the game plan, but that the game plan itself remains *appropriate* and *accurate* for the current conditions. This level requires both pilots to build their SA, verify their

short-term flightpath, and then ensure the quality of their projected, long-term flightpath. Third-level qualitative monitoring encompasses the whole picture. While short-term priorities direct the PF's attention inward, the whole picture perspective expands the PM's perspective outward.

16.2 AUTOMATION POLICY

Automation policy guides the application of procedures and operational strategies. Some guiding policy principles are:

- Maintain proficiency with all levels of automation.
- If a level of automation results in task saturation or loss of SA, skillfully shift to a less demanding level or disconnect the automation and hand-fly the aircraft.
- When task saturation subsides and SA is rebuilt, reengage the automation at the level that is appropriate for the phase of flight.

16.2.1 Maintain Automation Proficiency

The first policy statement encourages pilots to maintain proficiency with all available automation capabilities. Unfortunately, many pilots resist this. Accustomed to the familiar habits that fit within their comfort zone, they avoid using newer automation capabilities because they seem complex, unnecessary, or optional. Unless these capabilities are promoted by procedure, currency/practice requirements, and line culture, their acceptance may develop unevenly across the pilot group. One common example is hand-flown, HGS approaches. While the HGS is required for CAT II and III approaches, it is optional at all other times. Some pilots choose to use the HUD extensively, while others only use it only when absolutely required.[2] This affects their comfort level and proficiency. Another example from the low end of the automation spectrum is hand-flying using flight director guidance. Consider the following report from a pilot who had become "rusty" with hand-flying.

BOX 16.1 FO TRYING TO PRACTICE HAND-FLYING MISSES LEVEL-OFF ALTITUDE

FO/PF's report: Climbed out and got task saturated while in a turn [and] hand-flying … I realized that I had reached 15,000′. I corrected the situation but overshot the altitude. The Captain pointed out the deviation and I smoothly reduced power and established a descent to assigned altitude. ATC queried and gave us a higher altitude of 17,000′. The cause was focusing on the procedure while hand-flying. Didn't notice the altitude until Captain called it out. *I'm a little rusty from lack of flying and have little experience hand-flying the plane. I was trying to acquire some more experience, but perhaps this wasn't the best time to do it. Would have been better to use more automation.*[3]

This FO recognized their need to practice hand-flying and tried to follow company policy guidance. The last sentence highlights how practice is important, but also how the flight conditions matter. We don't know whether the flight director was used. Hand-flying with flight director guidance offers useful practice, but also protects us against path errors like this one.

This raises the issue of when and where to practice unfamiliar automation. While developing proficiency, it is better to practice in an ideal and unstressed environment like at a low-complexity airport under daylight and clear weather conditions. After we gain more proficiency, we can add more complicated practice opportunities, like night and IMC.

16.2.2 When Task Saturated, Shift to a Less Demanding Level of Automation

This policy addresses situations that commonly happen while we are trying to learn a new automation system. Training and certification ensure that we know how to perform basic procedures under ideal conditions. They rarely teach us the many nuances that complicate smoothly applying automation procedures across operational settings.

Specifically, this second policy directs us to recognize when we are becoming task saturated and shift to a less demanding mode. This means that we may need to move our level of automation up or down. If we become task saturated while hand-flying, we can engage the autopilot to reduce workload. If the automation is not working as intended, we can disconnect it and hand-fly the aircraft to restore the desired path. Consider the last FO's report (hand-flying altitude bust). They were already at the lowest level of automation when their error occurred. In their report, they accurately concluded that the best course of action would have been to revert to the *lowest workload* level of automation, which in this case would have been an autopilot-coupled, flight-directed mode. So, a more appropriate policy is to select the automation mode that reduces our task loading and frees time for us to rebuild our SA. While not specifically stated, it also implies that if neither of these options reduces our task saturation, we should abort the procedure, reset, and try again. This interrupts plan continuation bias and failing game plans.

The policy also implies that we should maintain proficiency with the process of shifting between automation levels. We need to learn when to shift, the steps to execute the shift, and how to verify that the transfer is complete and effective. This process is highly dependent on pilot technique. One pilot's techniques often differ from another's. This creates CRM challenges since PMs need to understand what their PFs are about to do, how they plan to do it, when they will make the change, and the roles that each pilot needs to fill.

16.2.3 Reengaging Automation When Appropriate

After our task saturation subsides and our SA is reestablished, the final policy statement directs us to reengage the automation at the highest level appropriate for the current phase of flight. This articulates the overall goal of automation policy – to develop the skills and proficiency to accurately employ the most appropriate automation mode for each situation. Pilots who focus only on the first two statements justify remaining within their comfort zone. This limits automation options that we may need in particular scenarios. This policy directs us to actively expand our proficiency through proactive practice. Ironically, the only way to expand our comfort zone is to venture outside of it by practicing new skills.

16.2.4 Automation Policy Summary

There are two Master Class automation skillsets – selecting the most appropriate level of automation for the conditions and mastering our skill of switching between levels. We practice to improve our proficiency while controlling task loading and maintaining SA. It follows that we should select line-practice opportunities that are less complex than the designed limits of the automation. For example, we practice HGS CAT III ILS approaches during day, clear weather, and calm wind conditions. As our proficiency improves, we can complicate the conditions to add cloud layers and light turbulence. From there, we can add night conditions with decreased visibility. When we ultimately find ourselves flying the approach to CAT III ILS minimums, we will fly it accurately and confidently.

16.3 AIRCRAFT AUTOMATION – BENEFITS AND LIMITATIONS

Before we consider automation techniques, let's understand the strengths and weaknesses of automation and how it interacts with our human strengths and weaknesses. There are areas where we excel over the automation and areas where automation excels over us.

16.3.1 The Junior Pilot on the Flightdeck

We treat automation like an extra pilot on the flightdeck. Just like another pilot, we transfer aircraft control back and forth to it. The flight director/autopilot acts like a pilot that flies extremely well, but doesn't think or innovate. It blindly follows what we and the FMS computers direct it to do. On the pro side, it is capable of accurately executing most flying tasks. On the con side, it doesn't detect or correct any errors.

Our relationship with this third pilot is one of conditional trust. When programmed correctly, we trust that it will accurately follow the desired profile. We trust that it will maintain altitudes, courses, and speeds, especially in steady cruise (low dynamic flight and low/green AOVs). We also trust it in climbs and descends (medium dynamic flight and medium/yellow AOVs), but not as much. When it transitions between modes, we monitor it closely because we can't entirely trust it (high dynamic movement and high/red AOVs).

16.3.2 Areas of Strength and Weakness

Automation is well suited to sustain and monitor long-term tasks that we are poorly equipped to monitor. For example, if we had a slow oil leak, we might notice it during our periodic systems scans. In this respect, we are better than the automation, which would never detect the slowly dropping oil quantity. On the other hand, automation is well suited to continuously monitor oil quantity and immediately warn us when it reaches a programmed low-oil limit. The automation will reliably capture the oil leak event at the programmed limit value, but will never alert us before that point. Ideally, we should periodically scan our panels to detect anomalies like engine oil leaks early, but rely on the automation to alert us when a low-oil condition is actually reached.

The same goes for holding our airspeed, altitude, and course while in cruise flight. Automation tracks these parameters accurately and tirelessly. As long as conditions remain stable, the FMS and autopilot will reliably follow the magenta line. Like with the oil leak, however, automation usually fails to detect adverse trends. In mountain wave conditions, for example, significant wind shifts and turbulence can overtax the response rate of the autopilot and autothrottles. The lag in flight control and thrust response can result in overspeed and underspeed events. Because of this, many airlines direct their crews to disconnect the automation and hand-fly the aircraft until clearing mountain wave conditions because we are better suited to anticipate and respond with necessary thrust and flight control changes.

Another example of automation weakness involves flying into warm air masses that exceed aircraft performance limits. Consider a summertime flight from Los Angeles to Denver flown at the highest performance-limited altitude of the aircraft. While near the ocean, we could sustain our maximum altitude with available engine thrust. As we transition over the Mojave Desert, however, the outside air temperature rises and causes our available engine thrust to drop. If we were actively flying the aircraft, we would notice when we reached the MAX thrust limit. We would detect our airspeed decaying long before any automated slow-speed warning. On the other hand, the autothrottles would quietly advance thrust to the limit, but not generate any warning that maximum thrust had been reached. Exceeding our performance capability, our airspeed would slow toward the stall limit.

To summarize, areas where automation performs better than us include:

- Holding set parameters
- Initiating profile changes like transition to descent and programmed level-offs
- Holding the programmed course
- Alerting us that out-of-tolerance conditions have been reached

Areas where we excel over automation include:

- Detecting slowly changing conditions and trends
- Anticipating external conditions that may develop into problematic events
- Reacting more quickly than automated responses

16.4 AUTOMATION-BASED ERRORS

The NASA ASRS database is rich with reports of automation-caused errors. Pilots are highly motivated to report these errors because they typically result in course deviations and altitude busts. Most automation errors include some pilot involvement. Let's consider some of our contributions to automation errors.

16.4.1 Insufficient Knowledge and Practice

We see many events where pilots mismanage automation due to inadequate understanding and practice. Additionally, many automation applications are not extensively covered in our operating manuals. In training, procedures are typically taught using straightforward, ideal conditions. We are not trained in the many ways that the systems may fail or interact poorly under exceptional conditions.

We often lack opportunities to practice procedures in line operations. One example comes from the early rollout of RNP approaches. Pilots were trained and certified in the simulator, but rarely practiced them while line flying. ATC controllers, pressed by traffic saturation constraints, were reluctant to allow us to fly RNP approaches during busy arrival flows. Soon, many of us just stopped asking for RNP approaches. Later, the situation improved as some larger airports began directing all aircraft arriving on particular STARs to use the associated RNP approaches.

The other knowledge/practice shortfalls involve pilot discretion. Some pilots resist new technology and only practice it when their simulator checkride event is approaching. I experienced one particularly memorable night approach with a Captain who grudgingly admitted that he "never used the HUD", but that he ought to practice flying a CAT III ILS approach since he had a checkride coming up. He lowered the HUD and turned the intensity up so bright that I could see his face fully illuminated with an eerie green glow. His display was so bright that he couldn't see the runway through the glass. He nearly initiated a go around until I informed him that the runway was clearly in sight. He slammed the HUD back up to its stowed position, landed visually, and complained about being forced to use the "worthless" HUD.

16.4.2 Mismanaging Practice Opportunities

Pilots who are unskilled with a particular automation should practice using it under ideal conditions. This gives them the best environment to improve their understanding of the system, its nuances, and failure modes. Unfortunately, many pilots wait until marginal conditions force them to use it. Under the added time pressure and stress, their lack of skill can lead to failure. At some airlines, reluctance to use automation acquires a cultural stigma. "We don't use that here." Reluctant pilots rationalize that relying on new technology represents a crutch or weakness.

Unfortunate situations arise when nonproficient pilots are forced by conditions to use unpracticed automation in complex, time-pressured, and stressful situations. An FO related an event involving their Captain who never practiced HGS approaches. That day, a low fog layer forced them to fly an HGS CAT III ILS approach to minimums. The Captain mismanaged three attempts to land and ended up diverting to an airport with more favorable conditions.

16.4.3 Mismanaged Automation Engagement and Changes

When we conclude that something is going wrong with our automation, we sometimes mismanage the steps for resolving it. We either make mistakes trying to program a change or the automation doesn't respond as expected. We become fixated and frustrated. Since the aircraft continues moving forward, the urgency intensifies and a failing scenario quickens. Examples include trying to select an automation mode that won't engage, engaging a mode that doesn't respond quickly enough, engaging a mode that responds differently than we expect, and engaging a mode that immediately shifts to an undesired alternate mode. In their frustration, pilots often choose to disconnect the automation and hand-fly the aircraft. Unfortunately, a common reason why the automation won't engage is because they attempt to engage it while the parameters are out-of-tolerance. For example, the flight director won't engage because the approach is well outside of salvageable parameters.

16.4.4 Mismanaged Raising or Lowering the Level of Automation

Policy guidance encourages us to raise or lower the level of automation to manage our workload. While this is sound guidance, we make errors managing this process. We become fixated on making the change, lose track of how long it is taking, and become frustrated. Problems include trying to engage an automation mode when our path is outside of the engagement limits, confusion with the parameter indications, and confusion with arming a mode versus engaging a mode.

16.4.5 CRM Breakdown While Making Automation Changes

Many reports cite crew communications and verification breakdowns. These occur when one crewmember is busy or has diverted their attention while the other crewmember is making an automation change. In most operations, the pilot making an automation change (PF, in most cases with the autopilot engaged) announces the change so that the other pilot (PM, in most cases) can verify the action. Some changes are self-evident like disengaging the autopilot ("Autopilot Coming Off") or initiating a descent ("Starting Down"). Others, like FMC route modifications require crew verification before executing them. For example, when the PF executes a clearance to proceed directly to a fix, the bypassed route fixes (often displayed with inverse video) are lost. The PM, whose attention may have been diverted, no longer sees the erased points and misses a chance to recover a required transition. So while the PF may follow part of the procedure by announcing the change, they need to honor the intention of the procedure and allow the PM to verify the change first.

16.4.6 Automation Application Exceeds Practiced Conditions

These events emerge when pilots are trained to use the automation under limited, ideal conditions but encounter real-world conditions that exceed their knowledge and training. On every ILS approach that I practiced during my airline new-hire training,

I intercepted the ILS glideslope from below the beam. Each time, the automation smoothly captured the ILS signal. On my very first line-flying ILS, however, I was vectored to intercept from above the glideslope. As the autopilot captured the glideslope guidance, the aircraft abruptly pitched down. The autopilot then overshot the glideslope and aggressively pulled up. What I wasn't taught in training was that the aircraft automation didn't handle intercept-from-above geometry well. My Captain then informed me that we needed to hand-fly glideslope intercepts from above, then engage the automation only after centering the glideslope guidance. When we train under a limited range of conditions, we may not be prepared for situations that fall outside of that range. That is why we should gradually increase the complexity of conditions during our line practice events so that we can achieve full proficiency.

16.4.7 Automation Glitches

The range of automation glitches runs from small deviations to full system dumps. The following narrative reports two high-speed events that occurred when automation did not respond as expected.

BOX 16.2 AUTOPILOT CAUSES TWO HIGH-SPEED EVENTS DURING DESCENT

PF's report (ERJ-145 aircraft): … experienced two high-speed events during the descent into ZZZ1. While descending from cruise altitude it became necessary to expedite descent and slow the aircraft. As the PF, I retarded the throttles and extended the airbrake, configuring the aircraft for a 250-knot descent. *Upon doing so, the aircraft responded with an unexpected and unusual nose low attitude of approximately 15 degrees which caused rapid acceleration. My immediate reaction at that time was to retract the brake in order to clean the configuration of the aircraft and interrupt the pitch with TCS (Touch Control Steering) and pull carefully back into a less-accelerative attitude. While the aircraft was being slowed, the high-speed warning annunciated and was active for approximately 8 seconds. … During coordination with ATC and at approximately 12,000′, I selected a FLC descent with speed brake to attempt to bring the aircraft through 10,000′ with proper speed. The aircraft responded again with unusually steep downward pitch, and expecting this, faster action to apply TCS and reconfigure the aircraft for a hand-flown descent was enacted by both crewmembers. Even with swift corrective action, the aircraft did accelerate to approximately 280 knot for a period of roughly 5 seconds.*

When faced with an unexpected and unusual response from automation, react immediately with manual control. Afterward, *be extremely clear [when] detailing to the PM what was entered, what was received, and verify clarity of the situation in order to develop a most-conservative response and closer monitoring of possible human or automation factors at play.* Trust the automation, but be prepared to verify any and all factors at play.[4]

In this event, the crew chose to write up the automation. Notice the PF's emphasis on communicating with their PM. In the next report, the crew experienced a total dump of the autopilot control system.

BOX 16.3 COMPLETE AUTOPILOT AND FLIGHT DIRECTOR FAILURE

FO/PF report: Departing ZZZ, LNAV, and VNAV armed on ground. Passing roughly 2,000′, VNAV failed. Cleared to 6,000′, used LEVEL CHANGE, tried to restore VNAV. Cleared to 15,000′, Captain set 15,000′ in MCP window. *Sometime after that, the altitude alert window reset itself to 50,000′, LNAV failed, both Flight Directors failed, Auto Pilot disengaged, and speed window on MCP would not respond to inputs.* Notified ATC of automation failure and requested vectors to the east to stay VMC. Thunderstorms in the vicinity. I believe I disconnected the autothrottles myself as a precaution to avoid uncommanded throttle movement. So at this point we are level at roughly 12,000′, slowed to 280 knots and getting vectors to avoid weather. *Hand-flying, raw data. No annunciated warning or caution messages to pursue in QRH. Captain checked the Un-annunciated section [of the QRH] for MCP failures or anything relative to our situation and found nothing.* Captain sent ACARS message to Dispatch, tried reaching local maintenance at ZZZ for possible solutions, but received no response. With sunset looming and thunderstorms approaching ZZZ we decided it was best to immediately return to ZZZ.[5]

The Captain's report also noted that they had an aural warning tone that sounded from the automation failure until touchdown.

16.4.8 Changes Made during Mode Transitions

A common source of errors involves anomalies that arise while we make changes during automation mode transitions. An example is when we make a change to the target altitude after the autopilot/flight director has entered an altitude capture mode from climb/descent to level-off. Consider the following example involving an early level-off while raising flaps.

BOX 16.4 AUTOMATION RESPONDS UNPREDICTABLY DURING TAKEOFF

Captain/PM's report: Basically empty jet (takeoff weight 223,000 LBS) with a low level-off during the departure (3,000′ MSL, 2,300′ AGL). We had discussed and decided to leave the power setting in Climb 2 due to the low level-off. At 1,000′ AGL the PF called for VNAV. *I pressed the VNAV button, but I believe the aircraft had already gone to altitude capture mode and did not take the VNAV selection – which I didn't immediately catch because I was busy raising the flaps as the jet was rapidly accelerating. The FO commented that "VNAV was doing something strange."* I looked at the MCP and FMAs and saw that 128 KTS was the commanded speed and we were not in VNAV, however the autothrottles were trying to advance to climb power. *The PF was hand-flying and overriding the autothrottles.* The momentary confusion and light aircraft weight caused us to miss our 3,000′ level-off. We ballooned to approximately 3,500′ before getting back down to 3,000′. ZZZ Tower never said anything to us and switched us to Regional Departure. They queried "what happened back there?" I explained that we were light and had some issues with our automation. … *I should have done a better job of monitoring the FMAs when I initially pressed the VNAV button. This would've helped me be a better PM and point out what was happening earlier. I also don't think I was quite mentally prepared for how low the level-off was and how much excess performance the jet would have at our light weight.*[6]

Clearly, this was not a situation that was common, trained, or anticipated. The Captain anticipated some difficulty with the early level-off, but didn't foresee how the low aircraft weight would exacerbate their climbout workload. Here is another altitude capture anomaly.

BOX 16.5 AIRSPEED RESPONDS UNPREDICTABLY DURING CLIMB PROFILE

PF's report: While climbing out of ZZZ, initially we were assigned to climb and level at 4,000′ which changed to 7,000′. I engaged the autopilot at 5,000′ and maintained 250 knots per the SID. *Approximately 6,800′ ATC assigned to climb and maintain 19,000′. The power setting and speed I had was maintaining 250 knots. The aircraft in "Alts Capped" and the speed shot up to approximately 285 knots. The power was reduced, but not enough. I disengaged the autopilot, reduced the power, corrected the airspeed deviation and continued to climb as assigned. … Cause of this event was due to me being behind the aircraft in climb and getting too behind the automation. Another contributing factor was due to me not flying approximately 80 days leading to a lack of proficiency.*[7]

It appears that the aircraft had entered into the altitude acquire mode. From the report, it is unclear whether the pilot attempted to engage a climb mode or ever recognized the cause of the problem.

16.4.9 Mis-programmed Values

Another category of automation error is generated by mis-programmed data entries. The automation computes the aircraft response based on the input values. When values exceed programmed ranges, unexpected outcomes emerge. Following is an extreme example caused by a mismatch between actual aircraft weight and balance loading and the values programmed into the FMS of a Boeing 767.

BOX 16.6 MIS-PROGRAMMED MAC RESULTS IN DIFFICULT LEVEL-OFF AFTER TAKEOFF

Captain's report: At XA+00z the flight crew arrived [in] the hotel lobby. When leaving to board the van we noticed the loadmaster was also leaving the hotel. Upon arrival to the aircraft, the crew noted that it had already been loaded with the main deck and lower belly doors opened. The weather had changed to active precipitation before block out with LLWS advisories in effect, and reported 20-knot loss on final and approach ends of runway XX. The crew reran the numbers for the OPT with max thrust calculations. It gave us flaps 15 takeoff for a weight of 271,985 lbs. The TOW MAC was 22.25% and a trim of 2.5.

On takeoff at about 3,000′ we noticed the speed began to rapidly bleed off, we then disconnected the autopilot and autothrottles, applied MAX thrust available and lowered the nose in an attempt to increase airspeed and stabilize. During the maneuver, the First Officer had to fight the aircraft in order to bring the nose down. We recovered at about 3,000′ and 118 knots. We hand-flew to 5,000′, which was the ATC assigned level-off, before re-engaging the automation. We then continued the flight without any further incidents to ZZZ. *Upon arrival in ZZZ the Load Supervisor, noticed the cans had been loaded incorrectly and the FAK (Freight of All Kinds) had not been accounted for properly. [Load Supervisor] then re-weighed all the cargo aboard and noticed the weight on some of the ULD (Unit Load Device) sheets were incorrect.* The weight and balance the crew had received did not match the final paperwork that ZZZ station had received from ZZZZ.[8]

In hindsight, the Captain accurately traced the chain of errors back to the hotel when they notice that the loadmaster was departing from the hotel with them. This is uncommon since the loadmaster typically departs the hotel much earlier to monitor the freight loading and paperwork verification. We also surmise that adverse weather conditions diverted the flight crew's focus of attention toward their takeoff profile and away from verifying the load documentation. Approaching 3,000′, the autopilot was unable to correct for pitch and accelerate (because the aircraft was too tail-heavy for the computed trim settings). The pilots disconnected all automation to hand-fly the aircraft to 5,000′ before reconnecting automation.

16.4.10 Misdirected Attention between Automation and the Outside Environment

Another aspect is how we divide our attention between inside and outside while using automation. Managing our automation requires that we crosscheck back and forth. For example, on an instrument approach in IMC conditions, we expect to transition from inside-instruments to outside-visual for landing. Following is an instructive account from a pilot who became overly focused inside while relevant information was clearly available outside.

BOX 16.7 FO LOST SA WHILE FOCUSED INSIDE INTERPRETING AUTOMATION DISPLAY

FO/PF's report: While on a heading of 150° to intercept final, at 2,300′ and inside of ZZZZZ, we were cleared for the visual for XXC. We were using the ILS/LOC XXC as back up. My intent was to stay in NAV mode, white needles to intercept, and then hand-fly the remainder of the approach. *As we approached the final course, the white needles were not moving as I expected and while "heads down" to confirm that the automation was set correctly, the Captain [directed me] to turn. I looked up and realized that we flew through the final course. I believe I turned off the autopilot, but hesitated in turning because I was unfamiliar with the airport.* The Captain took control and corrected, ATC advised us of the course deviation, I replied that we were correcting. Once the Captain got us reestablished on final. He gave me control back. We were above 1,000′ AGL and landed uneventfully.

Cause*:* Scan breakdown, my failure to look outside when distracted by the automation. This was also my first time flying into ZZZ so when I looked up, it took a few seconds to orient myself. *During our debrief, both the Captain and I realized neither one of us sequenced the FMS when ATC started to vector us for the approach. This caused the issue with the white needles not moving.* This trip was to maintain landing currency, I have not flown since the early April. I was also ZZZ1 based [flying] the Aircraft Y type. This trip was on the Aircraft X type, which I have not flown in a while.[9]

This report includes many of the elements of automation errors – failure to verify the intended mode of the automation, lack of currency/practice, confusion when automation doesn't respond as expected, fixation on the problem while the aircraft continues moving forward, and confusion between models of aircraft with different automation configurations.

16.4.11 Selected, but Not Activated

This error category involves cases where pilots select or press an automation button expecting the desired mode to activate or arm, but it doesn't. With extremely reliable systems, we sometimes relax how well we verify or monitor. This is especially prevalent when we get busy. Stopping to monitor the automation's response after quickly pushing the button interrupts the pace of our workflow. Instead, we assume that when we press a button, the desired mode is engaged, the automation will do what we expect, and that we can move on to our next task. We may even make a mental note to revisit the action in a few moments to ensure the correct automation response, but this is one of those fragile prospective memory tasks that we easily forget.

A common version of this error happens when selecting a path mode, like initiating a descent through the FMS. Under path automation, the aircraft should initiate our descent when we reach the computed profile point. With the lower altitude set and path descent armed, we shouldn't need to make any further pilot inputs. Now, imagine that ATC alters the plan by directing us to begin our descent early. We typically choose to initiate a shallow descent until intercepting the profile descent path. We select the mode, but it takes a few moments for anything to happen. So, we divert our attention toward a different task. Soon, ATC asks us why we haven't started our descent. We realize that the automation failed to start us down.

16.4.12 Unexpected Mode Transitions

Engineers attempt to design system software to match desired performance. They apply a set of assumptions that match typical conditions to generate operational profiles. When our actual conditions exceed their assumptions, anomalies emerge. Consider flying through an abrupt wind shift during descent. The FMS assumes a steady transition from the actual wind at top of descent to the expected wind at bottom of descent. As we pass through this wind shift, the FMS may exceed its descent profile limits and switch to an unintended mode that won't satisfy a crossing restriction. A similar problem can occur during climbout while we are climbing in a strong headwind and entering an even stronger jet stream headwind. This condition can cause the aircraft to exceed its high-speed Mach limit. These problems are insidious because they can happen during otherwise stable climbs and descents. Unless we are actively monitoring for anomalous events, we might miss them. Typically, we engage the automation and trust that it will "do its job".

As we gain proficiency, we develop techniques to notice subtle changes that may foretell these path shifts. We notice increases or decreases in pitch – either through visual changes or through subtle "seat of the pants" changes. Unfortunately, turbulence and IMC conditions often mask these effects.

16.5 AUTOMATION TECHNIQUES

Following are some ideas and considerations for developing and using automation techniques.

16.5.1 Selecting and Verifying

There are two considerations for selecting and verifying automation changes. The first is for the PF, or the pilot making the change to the automation. The second is for the PM as the pilot monitoring the effect of the change.

- *PF selecting and verifying:* Anytime we select a new automation mode, we need to verify that the mode has engaged as intended. Treating the automation as we would treat another pilot, we require that pilot to verify that they have assumed aircraft control. The autopilot is skilled (flies well), but lacks judgment (does what is told to do whether right or wrong).

  Most training techniques direct us to select the desired automation mode, see that the panel light has turned ON or OFF, and then verify the commanded mode via an annunciator panel display. This is important because some automation modes generate immediate effects while others arm modes that will activate at some future time.

  The PF should select the mode and verify that the aircraft responds as intended. This may interrupt other workflows. For example, if we select a descent mode from level flight, it may take a few moments for the aircraft to actually begin the descent. If we press the button, see the light change, and expect everything to go as planned, we risk missing an automation error. Even if we make a mental note to check it later, it is common to divert our attention toward other tasks and forget to come back and verify the automation response.
- *PM verifying and monitoring:* Airlines train crews to articulate their automation changes. These include verbal callouts, touching switches, and pointing at displays to support the select, verify, and monitor process. All of these are useful as long as they don't become habitual and automatic. We need to apply mindful attention toward our actions. Problems emerge when tasks become too repetitive and familiar and our attention to detail fades.

  As PMs, we need to expand our perspective to confirm not only that the PF has executed the mode change correctly, but that the action *makes sense.* We verify that the PF has done what they just announced doing, confirm that the action fits within the shared mental model, and judge that the game plan remains appropriate. To satisfy these verification steps, we need to extend our SA at least one step ahead of the PF. While the PF is appropriately focused on the immediate task and the *next* task, the PM needs to verify that *next* task plus the future *next-next* task. This ensures that the current change makes sense in that moment and aligns with future SA.

16.5.2 The Verified or Verifiable Standard

One foundational standard is that any action made by one pilot should be verified or verifiable by the other pilot. The details for this process are governed by specific company procedures. We are typically given latitude with how we communicate many of our actions.

- *Specific automation changes:* Specific automation changes are straightforward. When the PF announces, "Autopilot Coming Off", it clearly informs the PM that the autopilot is about to be disengaged. This is a useful courtesy call since autopilot disengagement generates an alert alarm or tone.

- *Transfer of aircraft control to and from the autopilot:* When changing automation, announce the engagement plus the selected mode. For example, "Autopilot Engaged, LNAV/VNAV Path". This allows the PM to verify that the automation has assumed the announced modes. This also helps with delayed path modes – cases where we arm an automation mode that will execute in the future. For example, if a crew is assigned, "Descend Via the [named] Arrival, Cross [fix] at 12,000 Feet". The process proceeds as follows:
  - o The PM responds to the ATC call, "[Callsign], Descend Via the [named] Arrival, Cross [fix] at 12,000 Feet"
  - o The PF states the automation programming as they spin the altitude window on the ACP and states, "Descending Via the [named] Arrival, 12,000 Set, VNAV Path"
  - o The PM visually confirms that:
    - 12,000′ is set in the ACP window.
    - FMC shows the correct routing to the next fix on the assigned arrival.
    - FMC shows the correct routing to the next-next fix on the assigned arrival.
    - VNAV is armed.
    - The computed top of descent point is accurately displayed on both the FMC display and the map display.
    - The profile and game plan make sense (qualitative assessment).
  - o Both pilots structure their workload to return their attention to monitor the automation when it initiates the descent (high/red AOV).
- *Single pilot automation or system tasks:* Some tasks are performed by one pilot at their discretion. For example, the Captain may decide to balance fuel between the wing tanks. On one extreme, announcing every step of the process is cumbersome and unnecessarily detailed. On the other extreme, changing aircraft systems without informing the FO removes an error mitigation safeguard. A balanced compromise is to simply announce, "Balancing Fuel". This alerts the FO that we are about to move a series of switches. The FO silently watches our actions, verifies the correct steps, and corrects any errors.
- *Personal automation adjustments:* Personal automation adjustments that won't affect the aircraft are treated as discretionary and don't normally require verbal notifications. Still, we learn new techniques by watching how other pilots perform their work. We are natural innovators. Anyone can develop a technique that we may find useful.

BOX 16.8 THE ALTITUDE SET KNOB, OR IS IT KNOBS?

An amusing story involves a senior Captain flying with their FO in a Boeing 737-200 series aircraft. To change the altitude alerter value, we had to spin a dial that was engineered with two concentric dials. The wider/recessed dial was geared to allow for high-rate changes – intended for large changes in our altitude clearance, like from 8,000′ to 17,000′. The narrower/extended dial was geared to allow fine adjustments for small changes – intended for minor changes like 11,000′–12,000′. Typically, when assigned a new altitude, we twisted the wide dial to get close to the assigned altitude (like spinning from 8,000′ to 16,400′, for example), then used the narrow dial to accurately set the assigned altitude (fine tuning it to exactly 17,000′).

While flying one day, the FO used this technique to quickly set the assigned altitude. The Captain perked up in his seat and asked, "What did you just do?" The FO replied, "What do you mean?" The Captain pressed further, "How did you set the altitude so quickly?" The FO was now extremely perplexed, "I spun the big dial, then the small dial." The Captain then reached over and moved the wide dial, then the narrow dial. A look of amazement came over his face. He then admitted that he had only ever used the narrow dial and never realized that there was a second dial. For many years and over thousands of flights, he laboriously turned only the small dial to set assigned altitudes. For some reason, on that one fateful day, he happened to closely watch his FO to realize that there was a better way.

16.5.3 Verifying Dynamic Changes in the Path

Setting and verifying changes in automation prevent most problems, but some errors still slip through. Automation is quite reliable, until it isn't. We need to apply our techniques to ensure that it continues to follow the desired flightpath. The techniques we adopt should align with AOV and sample rate priorities.

- *High/red AOVs:* Even when the automation is fully controlling the aircraft, we need to closely monitor our flightpath during high/red AOVs. Examples include, level-offs (transition from medium/yellow AOV climbs and descents to high/red AOV level-offs), entering and exiting turns (low/green or medium/yellow AOVs to high/red AOV transitions in and out of turns), and approaching the FMC-computed top of descent point (from low/green AOV cruise to high/red AOV transition to descent). Essentially, anytime the path changes in pitch or roll, our sample rate should be high.
- *Medium/yellow AOVs:* Sustained climbs and wings-level descents are examples of medium/yellow AOVs. While the automation tracks the computed path, conditions can sometimes cause the aircraft to exit the programmed path mode. The first indication may be a subtle change in aircraft pitch. This unexpected transition is sometimes masked by turbulence and changes in the relative wind component. Our mitigation is to maintain an appropriate sample rate to detect undesired deviations quickly.
- *Low/green AOVs:* Sustained cruise flight is a low/green AOV. Use this time for rest, eating, conversing, and other discretionary activities. Maintain a low sample rate of aircraft automation and systems. Check the FMS route

display to track the route. Select an appropriate map range to monitor upcoming waypoints, flight progress, and potential diversion airports.

16.5.4 Monitoring Automation Displays

As a general rule, select the most usable display presentation that supports building and maintaining SA.

- *Map display range:* Set the map display range to show the most useful presentation of upcoming waypoints and/or the arrival airport environment. As PFs, we extend or retract our range display to clearly show at least one waypoint ahead. PMs should consider selecting a range that displays that next fix plus at least one follow-on, or next-next waypoint. In the low-altitude environment, set the display that optimizes the TCAS presentation.

  In IMC conditions or when weather is a consideration, display radar information on one pilot's map display and EGPWS (terrain and obstruction) information on the other. Typically, PFs choose their preferred mode while PMs display the alternate mode. For tilt-selectable radars, choose the best angle that minimizes ground clutter and still maximizes SA of the most significant convective cells.
- *HGS displays:* HGS displays are a special case because we need to balance our ability to read the projected symbology while maintaining our ability to "see through" it to the outside environment. As display intensity is increased, the lines get thicker and the display becomes harder to see through. As it is decreased, they get thinner and the display becomes easier to see through. One useful rule-of-thumb is to adjust the display intensity until the lines become "hair thin". This balances our transition from monitoring display symbols to seeing the runway environment.

  The HUD auto-brightness setting can also cause problems. Many airports raise the intensity of their approach lighting with low cloud heights. Clearing the clouds, a HUD set to AUTO may suddenly become too bright to discern runway features. Runway sequenced flashers and runway threshold flashers can cause the HUD display to oscillate between bright and dim as the ambient light sensor tries to accommodate for the flashing. The manual (MAN) brightness setting may be preferable.

NOTES

1. Sheridan (2010, p. 31).
2. For perspective, while I used HGS automation routinely to maintain proficiency, I only needed it for actual CAT II or III conditions less than ten times over 20 years of line flying.
3. Edited for clarity and brevity. Italics added. NASA ASRS report #1747234.
4. Edited for clarity and brevity. Italics added. NASA ASRS report #1752883.
5. Edited for brevity. Italics added. NASA ASRS report #1752841.
6. Edited for brevity. Italics added. NASA ASRS report #1755997.
7. Edited for brevity. Italics added. NASA ASRS report #1750923.

8 Edited for brevity. Italics added. NASA ASRS report #1749544. MAC is a measurement of the aerodynamic balance of the aircraft. Specific trim values are computed and set before takeoff to generate a consistent pitch performance.

9 Edited for clarity and brevity. Italics added. NASA ASRS report #1746772.

BIBLIOGRAPHY

Active Pilot Monitoring Working Group. (2014). A Practical Guide for Improving Flightpath Monitoring. Alexandria, VA: Flight Safety Foundation.

ASRS – Aviation Safety and Reporting System – National Aeronautics and Space Administration (NASA) Online Database. (n.d.). Retrieved from Aviation Safety and Reporting System: https://asrs.arc.nasa.gov/search/database.html.

Sheridan, T. B. (2010). The system perspective on human factors in aviation. In E. Salas, & D. Maruino, *Human Factors in Aviation – 2nd Edition* (pp. 23–64). Burlington, NJ: Academic Press.

17 Communications Techniques

17.1 COMMUNICATIONS ENVIRONMENT

Policies and procedures assign roles and responsibilities that encourage open and proactive communications. How we achieve this is heavily influenced by our personal choices, conversational skills, and airline culture.

17.1.1 UNDERSTANDING ROLES AND COMMUNICATIONS PROTOCOLS

Captains exert the strongest influence on the communications environment – not just on the flightdeck, but across the operation. Harsh words or indifferent undertones from one Captain can damage the communications environment for future interactions with other Captains. To promote a resilient operation, every one of us needs to support an open communications environment where we all feel empowered to share information, ideas, and concerns.

Effective communications start with each team's assigned leaders. The lead flight attendant leads the cabin crew. The operations agent leads the station crew. The lead ramp agent leads the ground support crew. Teams can function effectively internally, but perceive barriers that block information flow with other teams. For example, a junior ramp agent may feel comfortable sharing information with their team leader, but feel reluctant to speak directly with the Captain. If the ramp team leader is busy, distracted, or also feels reluctant to share the information, the information may never reach us.

Communications barriers can be disguised or subtle. As Captains, we should assume that latent communications barriers exist and take proactive steps to remove them. We start by demonstrating how open and receptive we are to hearing everyone's inputs. With each conversation, we need to use encouraging words and project positive non-verbal undertones. Opening a dialog, we commit to listening to their concerns, no matter how trivial they may seem. If a team member expresses a minor concern, we use it as an opportunity to demonstrate our commitment to promoting an open communications environment. Consider a case where an inexperienced ramp agent expresses concern about "fluid leaking under the aircraft". When we walk down to the ramp and examine it, we discover that it is just water dripping from the air conditioning pack. First, we thank the agent for bringing it to our attention. We reiterate how important it is that they serve as our eyes and ears to detect any aircraft conditions that may affect safety. Then, we use the opportunity to educate them about what is actually happening and how it doesn't affect aircraft safety. We communicate with a smile and attentive demeanor. Otherwise, we may contribute to an adverse communications environment.

DOI: 10.1201/9781003344575-19

BOX 17.1 THE DOWNWARD SPIRAL TO ABSOLUTES

The average person tends to retain associations with bad events more readily than good events. A series of unfavorable exchanges can build upon each other and generate a deteriorating spiral until a person forms absolute opinions. It starts with one bad experience. A flight attendant comes to the flightdeck to express a concern. An indifferent Captain dismisses it as trivial while using a condescending tone. The flight attendant interprets it as a personal affront and develops a reluctance to share information in the future. "I'll not talk to THAT Captain again." This starts the downward spiral. Barriers begin to form. Later, that flight attendant has a second unfavorable exchange with a different Captain. They begin to generalize their attitude toward all Captains. "Captains don't care about our problems back here in the passenger cabin." The barrier hardens. After the more bad encounters, the barrier becomes absolute. "I'll *never* bring cabin problems to the pilots anymore."

These extreme attitudes develop in pilots, too. Perhaps a Captain experiences a frustrating exchange with their dispatcher. The dispatcher is having a bad day, vents their displeasure, and makes some ill-chosen comments. That Captain becomes reluctant to call that dispatcher in the future. "They challenged my authority. Just see if I call them next time." Next, a second encounter with a different dispatcher goes poorly. The barrier hardens. "Dispatchers don't understand what it is like out here flying the line." The spiral continues until an absolute barrier forms. "I *never* call dispatchers unless I absolutely have to." While this seems extreme, I have encountered Captains who have used these exact words. In one case, the Captain's complaint originated from a single difficult encounter experienced over a decade prior – a long time to maintain an absolute barrier.

Each team develops communications conventions that they use within their group and different conventions that they use with associated groups. Barriers can form between any teams. Consider our example of the ramp agent who discovers a fluid leak under the aircraft. Ramp agents know that the procedure directs them to "inform the pilots", but not realize how important it is that they do it quickly. Flightdeck access protocols sometimes block or delay timely notification. Perhaps their sense of urgency is adversely affected by a past event when a pilot dismissed their report of a similar leak as unimportant.

Additional barriers are created by our physical separation. The ground crew's primary mode of communication from the ramp to the flightdeck is via the intercom. Most of us don't don our headsets or monitor the intercom until preparing for gate departure. The ramp agent may try one early attempt to contact us, but since we aren't listening to the intercom, we don't hear their call. Rather than coming to the flightdeck to speak with us, they move on to their other tasks and make a mental note to tell us in the future (a fragile prospective memory task). Their next opportunity comes right before pushback. This causes a significant delay.

To overcome these barriers, we take proactive steps to open the lines of communication. The pilot performing the exterior walkaround inspection can take a few moments to connect with the ramp agents. By projecting a friendly, open attitude, we invite information sharing. They will feel more welcomed to initiate conversation. Another technique is to monitor our intercom audio with the overhead speaker ON. If they happen to call us, we'll hear them right away.

17.1.2 Reducing Unpredictability within the Communications Environment

Consider a generic flight operation where the Captain makes little or no effort to promote open communications. Each member of the operation enters that environment with their unique and personally biased mindset. Perhaps the lead flight attendant has recently had a difficult conversation with a previous Captain. Perhaps the FO finds it difficult to express themselves in English since it is not their first language. Perhaps the Captain speaks very quietly which some crewmembers misinterpret as appearing unapproachable. Perhaps the operations agent is overloaded and fatigued from working a double shift. All of these people are assembled to form a team to launch a flight. Their predispositions may interact in unpredictable and possibly undesirable ways.

We counter this multiple-barrier environment by proactively engaging with each team member. As we discover barriers, we make special efforts to overcome them. Speak face-to-face with the lead flight attendant and form a connection that wins them over with friendliness. Practice active listening with the FO to quickly detect any gaps in understanding. Offer to help the operations agent in any way that reduces their workload. These measures reduce the occurrence of unpredictable outcomes and reinforce an open communications environment.

17.1.3 Opening the Communications Environment

When we take the time to interact with our team members, we accomplish three goals. Our first goal is to open the lines of communication between the flightdeck and the other operations teams. Interactions conducted via intercom or radio lack the emotional context conveyed by facial expressions. If meanings aren't clear, we fill in the gaps with our own interpretations – sometimes inaccurately. This is why face-to-face contact is much more effective. Consider some examples. If the Captain doesn't meet with the operations agent, the agent might assume that the Captain is aloof or uninterested in the ground operation. If that operations agent subsequently encounters a problem with the ATOG weight limit, what are their options? They could try to solve the problem by themselves, coordinate directly with the dispatcher, or direct the ramp supervisor to remove freight. All of these options can occur without ever informing the Captain. Maybe freight is offloaded, passengers are denied boarding, or jumpseaters are turned away. Compare this with an alternative scenario where the Captain takes the time to converse face-to-face with the operations agent and establish a cooperative rapport. In this improved communications environment, the operations agent informs the Captain as soon as the ATOG problem arises. The Captain discovers that the dispatch release can be amended to raise the ATOG limit.

A quick phone conversation with the dispatcher and the problem is resolved without inconveniencing passengers, losing freight, or turning jumpseaters away.

Similar situations can arise with the cabin crew. Sterile flightdeck rules restrict when a flight attendant is "allowed" to contact the pilots with passenger cabin problems. Many NASA ASRS reports reveal the conflict that flight attendants feel with determining whether to "break sterile" versus waiting until 10,000′. Consider how this changes if the Captain spends some time clarifying this process with them before the flight. "If you encounter a problem in the cabin that we should know about, please call us – even if we are still in sterile flightdeck environment. If possible, check to see that the flaps are up so we don't interrupt our takeoff procedures. When you call, I may ask you to wait or I may address the issue right away." This clarifies the communications process and helps with their decision making. This is especially important for regional jet operations employing a single flight attendant in the passenger cabin. Larger flight attendant teams can discuss problems between each other, but solo flight attendants lack this mutual support.

Our second goal of opening the communications environment is to assess the mindset of each team member. For example, during our first conversation, we learn that all of our flight attendants just got reassigned to our flight. In their rush to make it to the aircraft, they didn't have a chance to acquire food. Sensing their frustration, we can try to turn their negative moods around by making a quick trip into the terminal to get them some meals. This will go a long way toward easing their disappointment, establishing good rapport, and opening communications.

The third goal is to convey our intentions for how we want people to handle unclear situations. Gary Klein calls this communicating our "executive intent". Not every situation is clearly guided by procedure. Team members need to apply their judgment to the many gray areas that crop up during line operations. Our face-to-face meeting conveys what we expect team members to accomplish when we've assigned a task, promotes their independence, supports their initiative to improvise, and expresses our permission to act.[1]

This also reduces communications barriers. Consider a situation where the lead flight attendant unilaterally decides not to inform us about a cabin problem. If we have clearly conveyed our executive intent, the other flight attendants can remind them of our instructions. This gives them the authority they need to override the lead's reluctance and share the necessary information.

17.2 COMMUNICATIONS – SENDING

The communications process involves the sending of a message, the receiving of that message, and the feedback that verifies that the message is accurately understood. Sending incorporates the environmental conditions, the tone of our voice, non-verbal undertones, and the quality of the transmission.

17.2.1 Environmental Conditions

Environmental conditions affect how well the receiver can accurately hear our message. Noisy environments include loud ambient/background noise, speaking too quietly, or any conditions that make it difficult for the receiver to discern our spoken words. Our solutions apply common sense. If the environment is noisy, reduce the noise, move to a quieter location, don ANR headsets, or speak more loudly. The bottom line is that noisy environments are the responsibility of the sender. If people can't hear us, we need to take proactive steps to make ourselves heard.

The same goes for distracting environments. Trying to speak with someone while they are overloaded interrupts their work. If they place more importance on completing their task, they won't direct enough of their attention toward understanding what we are saying. We should either wait until the person is available to listen to our message, assist them with reducing their workload, or get them to stop what they are doing while we share our message. We need to make sure that the message recipient is available to hear and process our message. Again, sending the message in a distracting environment is the responsibility of the sender.

Another problem is environmental complexity. Consider a situation where we learn of a conflict between two of our flight attendants. Imagine how our attempt to resolve it would go if we tried to mediate the disagreement while in front of passengers. Compare this with a conversation conducted in an empty jetway. The empty jetway environment improves the effectiveness of conversation and our ability to resolve the conflict.

Another related example is the aircraft handoff between pilot crews while standing in the jetway. If the off-coming crew wants to share some information about the aircraft, imagine how differently the conversation would go if we conducted it in front of passengers versus a private setting. Perhaps we could move the conversation to the relative privacy of the flightdeck to escape the noise and complexity of the crowded jetway. Again, reducing environmental complexity is the responsibility of the sender.

17.2.2 The Tone of Voice Used

The tone we use affects how well the recipient interprets our message. For example, if we sound angry, our message will be received differently than if we sound concerned. The same message is also affected by the receiver's mindset. Consider a situation where we are deadheading in the cabin when the Captain delivers their preflight PA in a quiet, monotone, bored, robotic tone of voice. From our perspective, we dismiss it as unimportant. They are just filling a square. Make the passenger PA – check. Now, consider the mindset of the nervous, first-time flier sitting next to us. Since the Captain is speaking, they immediately assume that the PA is highly important. The Captain is an important person, so their message must also be important. They strain to understand the mumbled words. Lacking context and clarity, they begin to superimpose their own interpretations to the Captain's lethargic tone. They sound tired. They sound ill. Should we be concerned? They turn to us, eyes wide, and ask if everything is okay. We assure them that everything is fine and do our best to allay their fears. Notice how two different people arrive at two widely different interpretations based entirely on the tone of the delivery. This is why it is so important to speak clearly and confidently whenever we address our customers. They assess meaning based on our tone and the clarity of our words. When we deliver a clear, unambiguous PA in a confident tone, it matches their expectations. They don't have to construct a background story to explain an apparent mismatch.

17.2.3 Non-Verbal Communication

Non-verbal aspects, like physical proximity, body language, and emotional expressions, affect how our message is interpreted. Consider an example of introducing ourselves to the Captain while requesting the jumpseat. Consider three different cases of non-verbal communication.

- *Case 1:* The Captain stays facing forward in their seat while they speak over their shoulder.
- *Case 2:* The Captain turns their head to view us out of the corner of their eye.
- *Case 3:* The Captain turns their body around so that they can converse with us face-to-face, offers a warm handshake, and smiles.

How do each of these conversations feel? Which one would we prefer? The first case feels neutral or negative. The second falls in the middle. The third case clearly conveys that the Captain values our presence. We feel welcomed. Turning to make face-to-face contact, shaking our hand, and conversing with a smile present an unambiguously positive experience that matches our expectations. As with all message-sending considerations, non-verbal communication is the responsibility of the sender.

17.2.4 Clarity of the Message

Assuming that we have solved for noise, distraction, complexity, tone of voice, and non-verbal aspects, our next concern is on how understandable our message is for the person receiving it. The clarity of the message depends on the simplicity of the words used. For example, if the sender lacks language proficiency, they may struggle to select the best words to clearly convey their meaning. Speaking face-to-face helps because we can read their facial cues that reflect their understanding.

If we can't converse face-to-face, we'll need to ensure that our messages are clear and complete by asking for feedback. Consider flightdeck briefings conducted while we each reference our EFBs. Since most EFBs are mounted on our respective side rails, both pilots turn away from each other as we speak and reference our displays. For complex briefing points, consider removing the EFB and holding it over the center console so both pilots can view the same display as we point at specific details and speak face-to-face.

Another aspect of the quality of transmission is the meaning behind our words. Consider a situation where a flight attendant comes to us to make a tough decision for them. They are reluctant to speak openly, so they hint at their meaning or use codewords. For example, FARs clearly state that we cannot carry intoxicated passengers. Anytime a flight attendant calls a passenger intoxicated or drunk, we are directed to have the passenger removed. Consider a case where a flight attendant chooses to inform us of their concerns, but won't speak clearly. They might say, "We have a passenger who has had a lot to drink." We conclude that they were trying to express their concerns, but they don't want to be responsible for removing the passenger. Before guiding a solution, we need to discern their underlying message. Are they concerned about behaviors the passenger is exhibiting? Are they unsure of what to do? Do they want us to make the decision for them? Do they want a customer service supervisor to make it? What do the other flight attendants think? One technique is to encourage them to decide whether they feel the passenger will become a problem or whether they are clearly intoxicated. This encourages them take responsibility for their passenger cabin while supporting whatever course of action they choose to take.

Messages are sometimes disguised with subtext and hinting. We may need to dig down for the accurate meaning. In cases like this, we should focus on defining the points of the procedure or regulation. In FARs, "intoxication" is treated as a clear yes-or-no decision. The reality is that many passengers drink before they fly. Also, many airlines offer alcohol on the aircraft. At what point does a particular passenger cross the line to be classified as intoxicated? Since we are not qualified to make medical determinations, we should guide and support those who are trained to make that determination.

17.3 COMMUNICATIONS – RECEIVING

Assuming that the sender has followed good practices to effectively send their message, the responsibility for effective communication now shifts to the receiver. The main aspects of reception are ensuring that we can accurately hear what is being said and applying our attention to understand the message being sent.

17.3.1 Understanding Rapid or Abbreviated Speech

One barrier arises from the fast pace of flightdeck operations. When we are busy, we tend to accelerate and abbreviate our speech. We have so much to communicate that normal conversation can't keep up. When we are both functioning at peak performance, our rapid-fire, back-and-forth communication is highly effective. Each pilot simultaneously speaks while completing different tasks to stay ahead of the fast-moving pace of the situation. We become rather good at it.

Consider what happens when something interrupts this flow. We may feel reluctant to stop and clear up the confusion because it may disrupt the pace of our workflow. We resist slowing down because we value staying ahead of the situation more than clearly understanding every message. We may even rationalize that we caught the gist of the message, that subsequent exchanges will clarify the garbled message, or that emerging information will clarify meaning. In any case, the receiver is responsible for understanding the message.

17.3.2 Focusing Our Attention

The second barrier involves our attention focus. This, too, is the responsibility of the receiver. High workload encourages us to attempt multitasking. As we have previously discussed, we can't actually multitask. Instead, we rapidly shift our attention back and forth between various tasks and concerns. Every time we shift, we need a moment to reorient our attention to restore context. The more rapidly we switch, the less focused our attention becomes.

Our habit patterns also affect this process. The more experience we amass, the more habit patterns we develop. Proficient pilots can complete aviation tasks while devoting very little attention toward them. Take an example of completing a required checklist while we are particularly busy with other tasks like flying a challenging final approach in moderate turbulence. We call for the Landing Checklist. Because of the turbulence, most of our attention remains focused on maintaining our flightpath. As the PM reads, "Landing Gear", we quickly glance toward the gear lights, see some green lights bouncing around, quickly return our attention back outside, and respond "Down, 3 Green". Did we actually see three green gear-down lights? Did we mentally process that we actually saw or did we just recite the familiar response because we "knew" that the gear was down? Investigations reveal events where the crew initiated the checklist, read each step, and responded to each step, but failed to detect configuration errors that the checklist step was designed to capture. When we are busy, we can regress into our habit patterns and habitually recite checklist responses. We don't accurately verify because our attention is focused elsewhere or constantly flitting back and forth between other important things. The verification step can feel like a distraction, or at least an interruption to our workflow. The following report reveals a case where one pilot sends clear messages in a challenging situation, but the receiver was either too busy or too distracted to process the information.

BOX 17.2 NUMEROUS DISTRACTIONS ADVERSELY AFFECT CREW COORDINATION

Captain/PM's report: I have never [before] set flaps incorrectly, or departed, with an incorrect flap setting. This was the First Officer's takeoff. *Both he and I departed with the incorrect flap setting today.* After flying 4 tough, weather legs between ZZZ and ZZZ1 the day prior, *I believe we were both still tired from dealing with the thunderstorms the day before. I didn't sleep well. Earlier, we discussed some serious stressors regarding challenges of a forced commute from a mandatory displacement and we could no longer rely on securing a passenger seat to [commute to] work. My First Officer is being furloughed and was very concerned about the future and our Company. I had not been into ZZZ1 for a very long time.* They had numerous taxiways closed and under construction. Initially, I thought I understood the taxi clearance, then realized I had it wrong. *While doing the Taxi Checklist on the way out, I was preoccupied with my EFB trying to verify taxiway names. My EFB was (and remains) very slow to respond to screen manipulations. So, I was a bit frustrated looking back and forth between my device and the called checklist items. Given the glare of the sun against the flap gauge, and my difficulty reading it (I had not yet switched to sunglasses), I simply was distracted and misread the gauge. On the before takeoff checklist, I guess I simply "saw what*

I expected to see". I have a hard time believing I missed it. Once airborne on takeoff, I directed further acceleration in the climb to account for the flap discrepancy. The flight proceeded normally without further event.

What I could have done differently: I should have stopped the taxi when I could not get the EFB to respond. I should have put on my prescription sunglasses earlier.... Other thoughts: Why not add physically touching the flap handle and looking at it to confirm its position in addition to current practice. I will do this myself. I think it's easier to see than the MCDU or gauge. Since FOQUA (a flight data analysis program) monitoring captures errors, is it possible to program this system to alert the pilots when the flap setting disagrees with the programmed takeoff data?[2]

There wasn't one profound distraction that caused this error to happen. Lots of small, individual distractors added up to prevent this crew from recovering. Notice all of the stress factors that complicated this crew's communications environment and how their expectations affected their perception. Also, notice all of the personal corrective techniques the Captain vowed to adopt to prevent similar occurrences in the future.

17.4 FEEDBACK

The best way to ensure the effectiveness of our communications is through feedback. To explore the facets of feedback, consider the following examples from a crew beginning taxi from a crowded ramp area. As they prepare to move the aircraft, both pilots focus their attention outside. While clearing to the right, the FO detects an approaching ground support vehicle towing a line of bag carts. They state, "Ground vehicle coming up on our right". Consider the range of Captain responses by the quality of feedback that each one offers.

- *Case 1:* The Captain says nothing. The FO continues watching the vehicle.
    - This is a feedback failure since the FO receives no feedback that the Captain has either heard or understood the callout.

- *Case 2:* The Captain says nothing. The FO glances at the Captain and observes that they are looking toward their left and making no attempt to acquire the vehicle approaching from the right.
    - This provides slightly better feedback. The Captain hasn't provided any verbal feedback, so the FO looks at the Captain. They discover that the Captain isn't trying to acquire the vehicle. The FO concludes that their callout was neither received nor understood.

- *Case 3:* The Captain replies, "Okay". The FO continues monitoring the vehicle.
    - The Captain verbally responds to the FO's callout, so we can conclude that they heard it. However, this minimal level of feedback doesn't tell us whether

they understood the callout, whether they replied habitually, whether they see the vehicle, or what they intend to do about it. The FO might inaccurately assume that the Captain has received the message and that they are assuming responsibility for the ground vehicle threat. This ambiguous feedback actually elevates the level of risk from the safety threat.

- *Case 4:* The Captain responds, "Okay". The FO glances at the Captain and observes that they are still looking toward the left.
    - The FO acquires slightly better feedback. Seeing the Captain focusing their attention to the left, they know that the Captain isn't looking toward the threat and probably doesn't see it. We don't even know if the Captain comprehends the severity of the hazard. In fact, it supports the opposite – that the Captain has is either responding habitually, doesn't see the ground threat, and has no plan for mitigating it. The FO assumes the worst, actively monitors the vehicle, and assumes responsibility for alerting the Captain in case the vehicle does not stop.

- *Case 5:* The Captain responds, "I don't see them."
    - This minimal response provides some useful feedback. We conclude that the Captain understands that there is a ground threat approaching from the right, that they have looked for the vehicle, and that they haven't acquired it. The feedback gives the FO some useful SA. They know that they need to provide directions or information. They might state, "Hold taxi. I'll tell you when the vehicle stops or is clear."

- *Case 6:* Captain replies, "I don't see them. Let me know if they become a problem."
    - This response provides useful feedback. The Captain has acknowledged the hazard, has looked for the vehicle, has not acquired it, and has assigned a specific responsibility to the FO. While operationally useful, it falls short of an ideal response. Since the FO isn't physically controlling the aircraft, they'll need to predict both the vehicle's actions and their Captain's actions to decide whether a hazardous event is evolving. If the Captain elects to start taxiing, the FO has to anticipate how the Captain might maneuver the aircraft, predict whether clearance might become a problem, and speak up if it does. This is a common scenario with ramp mishaps.

- *Case 7:* Captain says, "I don't see them. I'll hold our position. Tell me when they are clear."
    - This is effective feedback. The Captain fully understands the FO's callout and understands the taxi hazard. We don't know whether the Captain has looked for the vehicle or not, but it doesn't matter. Since the FO can monitor the vehicle, the Captain assigns them a task that is completely within their control – judging when the vehicle is clear.

Since the FO cannot control wingtip clearance, the Captain decides to wait until the driver is clear of the aircraft's path before starting to taxi.

From this exercise, we see that effective feedback includes five parts. It confirms with the sender that the receiver has:

- *Heard* the message
- *Understood* the message
- Accurately assessed the *importance* of the message
- *Formed a plan* for how to apply the information
- *Assigned* roles or tasks to mitigate threats.

As a minimum, feedback should satisfy the first two. The remaining three components aren't always necessary, as with informative or routine messages, but are useful to align the shared mental model between both pilots.

17.4.1 Feedback between Pilots on the Flightdeck

Our easiest communication events are between each other on the flightdeck. On the plus side, we can speak face-to-face and hear each other clearly, especially when using ANR headsets. We also have similar aviation backgrounds and perform the same tasks, so it is easier for us to verify each other's actions. This familiarity allows us to convey complex messages using very few words. On the minus side, our common frame of reference biases us. We assume that when we send one of our brief messages that the other pilot will accurately receive it. We don't routinely expect or solicit feedback. Especially when we are stressed, our message meaning and importance may become garbled or misinterpreted. In mishap debriefs, it is common for one pilot to report, "I told you about that problem." The other pilot replies, "I never heard you."

Most of the time, we work quickly and efficiently as a team. When this efficiency breaks down, our feedback begins to degrade. Consider the following three effects – when we reply abruptly, when we clip sentences, and when we stop replying altogether.

- *Replying abruptly:* When we get really busy, we talk faster, speak fewer words, and use less-meaningful words. As the pace of workload pushes against our limits, it feels like every moment should be used by either saying something or doing something. The first communications component we drop is meaningful feedback. Instead of replying with words that reflect our understanding, our replies become more abrupt and reflexive. Replies like, "I see that", "You take care of that" and, "I'll take care of that", become "Yeah", "Copy", and "Check" – none of which provide meaningful feedback. We use these automatic responses to fill the gaps within the conversation. As our situation races along, we don't have time to circle back and confirm shared meaning. Instead, we just keep moving forward because falling behind feels so much worse.
- *Clipping sentences:* Many CVR transcripts document the clipping of sentences – where pilots began saying something, stopped themselves,

restarted, and then stopped again. It suggests that their minds were operating faster than they could form and speak the words. Usually, the PF is the first to become overloaded, often without realizing it. The PM is less susceptible to overload, so they can use their PF's clipped sentences as a warning sign. While there may be little that can be done to remedy the task overload, it should cue the PM to continue making directive callouts and anticipate switching to a contingency backup plan. When both pilots start clipping sentences, it indicates mutual task overload.

- *Stopping replies altogether:* In the advanced stage of task overload, we stop speaking altogether. For example, while the PF is struggling with an unstabilized approach, the PM makes a deviation callout. The PF doesn't acknowledge it. The PM looks at the PF and sees them staring intently at the touchdown zone as they struggle to salvage the approach. PMs should assume that their PFs have completely tunneled their attention, abandoned SA-building, and suspended their deliberative decision-making process. Again, PMs should continue making directive callouts and anticipate switching to a contingency backup plan.
- *The flightdeck authority gradient:* In the early days of airline aviation, the Captain held godlike authority. The flightdeck authority gradient, the virtual power slope between the Captain and the other crewmembers, was practically insurmountable. It was like standing next to a sheer cliff with no way to overcome it. Few First or Second Officers would dare challenge their Captain's judgment or question their decisions. Unsurprisingly, many avoidable accidents resulted. This led to demands for change. Procedures were revised to empower the non-flying pilots to speak up. While the slope of the authority gradient eased slightly, resistance and reluctance remained. Junior flightdeck pilots would typically hint at deviation parameters. Captains weren't required to comply or acknowledge these callouts. A Captain who refused to answer and a Captain who was too overloaded to hear or respond behaved exactly the same. Feedback was unreliable. Procedural changes codified deviation callouts and required verbal acknowledgment. "Correcting" or "I Got It" were considered acceptable responses. While these procedures eased the gradient even further, the decision to comply or not was still left up to the Captain. Eventually, the fully empowered PM role emerged. The PM now has clear responsibilities to make deviation callouts, ensure compliance, and intervene when their PFs fail to comply. The flightdeck gradient has finally eased to a manageable slope.

  Recent improvements with the PM role include specific guidance on how many times to make deviation callouts before directing a recovery action (twice in most cases), a policy that the PM's directive callouts must be followed (removing the Captain's ability to override a "Go Around", for example), and specific guidance for intervening and taking aircraft control.
- *Absence of any feedback – incapacitation training:* A rare scenario is when one pilot stops providing any feedback whatsoever. We practice these incapacitation events in the simulator. Typically, the Captain flies an instrument approach. When the FO makes the "Approaching Minimums" call, the

instructor taps the Captain on the shoulder signaling them stop flying the simulator, remain silent, and cease all verbal feedback. This simulates incapacitation or extreme tunnel vision. When the FO detects the incapacitation, they intervene and announce, "I Have The Aircraft." Some airlines have added FO incapacitation exercises so Captains can practice their incapacitation procedures. The policy is that if the PF can't or doesn't reply when required, incapacitation is assumed and the PM takes over.

- *Learning from feedback failures during task overload events:* As our task workload rises and feedback falls, we compensate by shedding mental workload. We focus more intently on parameters that seem important at that moment. Our tunnel vision can lead to either successful or unsuccessful outcomes. On the beneficial side, it helps us filter out extraneous details so we can concentrate on important details. On the detrimental side, some of the details that we filter out may actually alert us to important counterfactuals. The trap is that while we are overloaded and in-the-moment, both scenarios look exactly the same. It is only afterward, in hindsight, that we can see that our tunneled focus either helped or hurt us.

  After we conclude a tunnel-focused event, hindsight reveals how close we came to the edge of the safety margin. We may have a, "that worked out, but let's not do that again" type of realization. The opposite epiphany rarely happens. We don't look back on events that ended poorly and conclude, "I should have focused harder and I might have pulled it off." Using this revelation, we conclude that we should avoid judging tunneled-focused events while they are happening. Anytime we recognize that we are becoming overloaded, that feedback is dropping, that our attention begins tunneling, or when we begin shedding required tasks, we should abort the game plan.
- *ATC communications and flightdeck verification:* When ATC calls us with an instruction, we read it back. This is important because ATC sending us an instruction doesn't ensure that we accurately understood it. Our feedback allows ATC to correct any misunderstandings. In HF, we call this feedback verification loop, *read-back, hear-back.* We read-back the clearance in a way that communicates our understanding. ATC hears-back our response to verify that we fully understood it.

  We also engage in read-back/hear-back on the flightdeck. Take a complex ATC taxi instruction as an example. "Trans American 209, taxi to runway one-two right via Bravo, Bravo 2, Mike, hold short of runway three one." Assume that the FO understands the instruction and reads it back to ATC word-for-word. How do we confirm that the Captain also understands the instruction? Was the Captain distracted at the time? Did they hear it, but misunderstand it? We don't know unless we get useful feedback. Let's compare a range of Captain's responses.

1. [Silence – the Captain says nothing]
   - The Captain says nothing and begins to taxi. In many airline operations, this is procedurally acceptable. Since the Captain starts moving the aircraft, the FO assumes that the Captain heard the taxi

instructions. Lacking feedback, they can't verify understanding. Without any further clarification, the FO needs to be especially vigilant to monitor the taxi movement. Hopefully, the FO detects any misconceptions by the Captain before they make a wrong turn.

2. "Got it"
   - The Captain acknowledges hearing the ATC instruction. Their "Got it" implies that the instructions made sense and that they have a mental picture of the taxi route and clearance. If they are mistaken, they won't know it until the Captain makes a taxi error.
3. "Runway one-two right via Bravo, Bravo two, Mike, hold short of runway three one"
   - This is a literal repetition of the ATC instruction. The FO concludes that the Captain heard all of the parts of the taxi clearance. They don't know whether they understand what the instructions entail. Maybe the Captain has a very good short-term memory, but subsequently fails to convert it to useful working memory. Their shared mental model isn't verified.
4. "Runway one-two right, taxi straight ahead and turn left onto Bravo, then right on Bravo two, then an immediate left onto Mike and hold short of the crossing runway, three one"
   - This response demonstrates that the Captain has heard the instructions and that they have constructed a mental picture of each taxi turn. Notice that they have added practical detail by stepping through each turn, the direction of each turn, an understanding that taxiway B2 is a short side-step onto taxiway Mike, and that they must stop short of runway 31. This response is somewhat verbose, but it demonstrates an accurate understanding of the taxi clearance.
5. "Standard taxi – follow that regional jet ahead"
   - Option #5 is a much shorter response, but also demonstrates an understanding that their aircraft will join an on-going taxi flow. The Captain understands that the regional jet ahead has been assigned the same taxi routing. While it doesn't verbalize specific details of the taxiway transitions, it demonstrates familiarity with the typical operation used at that airport, an acknowledgement that the regional jet just received the same taxi routing, and an understanding of how their aircraft fits into the taxi flow.

The Captain's feedback needs to demonstrate both accurate detail and situational understanding. Options #1 and #2 fail these standards. Option #3 is technically accurate, but doesn't communicate situational understanding.

Both options #4 and #5 achieve effective communications feedback – although in very different ways.

17.4.2 Feedback to Team Members

Feedback with other team members is affected by proximity, complexity, and workload. Additionally, since other team members have different backgrounds and work processes, we lack the ability of using abbreviated speech and flightdeck jargon.

- *Proximity challenges:* We prefer face-to-face communications so we can watch facial expressions for evidence of understanding. More often, however, our communications are limited to intercom or radios. Noise and transmission quality interfere with quality feedback. Many airlines compensate by requiring face-to-face briefings for rare-normal procedures like pushback without intercom (using only hand signals) and external air cart engine starts (when the APU is unavailable). If we need to conduct detailed conversations via intercom and radio, we should invite feedback to ensure understanding.
- *Complexity challenges:* When we communicate about familiar tasks, feedback is easier. Both parties share common references and understand their work processes. As complexity increases, we move further away from familiar scripts. Feedback becomes more difficult and more important. Especially when particular team members have decision-making authority, we need to clearly understand how the decision will be made and who will make it. A challenging example is an inflight passenger misconduct incident. Here, the decision of whether to continue, divert, or call for law enforcement resides with the Captain who relies on intercom communications from the flight attendants. Information is often relayed to the Captain by one flight attendant who may not be directly involved with an on-going incident or may be observing it from a distant location in the cabin. We need to listen and ask questions to determine the proper course of action. What is currently happening? Is there a threat to passengers, crew, or to flightdeck security? Is the information first-hand or second-hand? Is it calming down or escalating? What can we do to help? Feedback is vital to accurately handle the situation.
- *Workload challenges:* Another problem with intercom or radio communications is that we don't know each other's current workload. We regularly have situations where the flight attendants call us while we are busy with other tasks. Do we ignore their call, ask them to standby, or split one pilot off to converse with them while the other pilot finishes with the pressing task? Each option has advantages and disadvantages.

  The higher the workload, the less time we devote toward acquiring feedback. We tend to rely on shortcuts and assumptions. Ideally, we can call back after the workload subsides, but this becomes one of those prospective memory challenges that are so easy to forget.

NOTES

1 Klein (2003, p. 210).
2 Edited for brevity and clarity. Italics added. NASA ASRS report #1759362.

BIBLIOGRAPHY

ASRS – Aviation Safety and Reporting System – National Aeronautics and Space Administration (NASA) Online Database. (n.d.). Retrieved from Aviation Safety and Reporting System: https://asrs.arc.nasa.gov/search/database.html.

Klein, G. (2003). *The Power of Intuition: How to Use Your Gut Feelings to Make Better Decisions at Work.* New York, NY: Currency Books.

18 CRM Techniques

Crew resource management (CRM) and similarly named programs are designed to improve our effectiveness on the flightdeck. Early airline flightdecks followed an autocratic command structure with Captains exercising absolute control. The effectiveness of the flight crew was highly dependent on the effectiveness of the Captain. When they led well, it worked. When they erred or missed warning signs, team members felt reluctant to verbalize errors or challenge the Captain's decisions. I recall a flight very early in my airline career. The Captain was exceeding stabilized approach limits, so I made a deviation callout as I had been taught in training. He snapped his head toward me and gave me a look that clearly communicated that he was surprised and offended by my callout. On the ground, we discussed it. I stated that I thought I was making a callout as it was procedurally directed. He responded, "Well, we don't do that here."

The early culture of the airline industry evolved from military aviation where co-pilots arrived with minimal experience. Straight from pilot training to the right seat, they were new to the aircraft and mission. The procedures and culture were designed to help these novice pilots learn by watching and following their experienced Captains. The unfortunate side-effect was that it concentrated responsibility at the top. If the Captain made an error, the First and Second Officers had to weigh whether the error was serious enough to warrant speaking up. If they did, some Captains felt that their authority was being challenged.

Early generations of CRM programs were intended to model collaborative/team leadership styles. They distributed responsibility and encouraged open lines of communication. The training and exercises were designed to convince Captains that they were better served by using the full experience and error mitigation capability of their crews.

Since most new-hire pilots at major airlines arrive with past crew experience from their previous carriers, few situations arise that require authoritarian leadership. Instead, Captains are expected guide the operation as team leaders using open communications, pilot monitoring, deviation callouts, and safety interventions.

18.1 CRM LEVELS

CRM has evolved to fill essential, supportive, and enhanced operational needs.

18.1.1 Essential CRM

Essential CRM directs us to take actions to ensure safe operations. No matter what the levels of experience, types of personalities, or extent of team cooperation, essential CRM procedures charge everyone to take the necessary steps to break an accident's error chain. When one pilot falls short, other crewmembers step in. At no

DOI: 10.1201/9781003344575-20

time should the crew's performance expose vulnerabilities that allow an accident to develop. Examples of essential CRM behaviors range from a simple deviation callouts like "Airspeed", to directive callouts like "Go around", to interventions such as "I have the aircraft, go around thrust." CRM programs must effectively and unambiguously train these procedures to ensure predictable, reproducible, and reliable outcomes. CRM training teaches these skills through situational discussions, assertiveness training, simulator scenarios, modeling, and role-playing.

Many accidents involve an execution/decision-making error committed by one pilot (typically the PF) and a CRM/crew failure from the other pilot (typically the PM). Investigations often showed that erring pilots didn't detect their own errors until too late. For this reason, accident prevention cannot rely on the PF's self-detection and correction.

Applying essential CRM, PMs need to detect apparent or evolving errors, announce the deviations, select trigger points to intervene, initiate their interventions, and ensure safe outcomes. These goals require us to develop the requisite skills to recognize when a chain of events is deteriorating, choose which actions to take, decide when to act, and intercede effectively. Individually, we need to understand the vulnerabilities in our own decision-making process, the effects of complexity and stress on our decision making, and how our biases influence our choices. It is only when the concepts become interwoven through all aspects of the flight operation and line culture that an airline can achieve its goals. This makes CRM challenging to teach in a classroom. Line-flying events often occur unexpectedly, under stressful and time-pressured situations, and involve a wide range of personalities. Exactly when should we intercede? Will our procedures support the most challenging situations and encourage the most reluctant pilot to intervene effectively with most resistant pilot during a rapidly deteriorating situation?

18.1.2 Supportive CRM

Supportive CRM centers on the interaction between the PF and PM roles. Each pilot performs duties that are monitored and verified by the other. PFs clearly communicate their game plan and how they intend to operate the aircraft. PMs monitor that game plan, detect errors, and verbalize information that their PFs might be missing. Together, we correct errors early before adverse situations develop. Supportive CRM training promotes teamwork by promoting clear communications, approachability, and receptivity.

18.1.3 Enhanced CRM

Enhanced CRM actively promotes teambuilding and group synergy. While supportive CRM ensures effective crew performance, enhanced CRM actively promotes qualitative aspects that stimulate resilience from system complexity and unpredictability. Experienced crewmembers mentor each other to share wisdom and techniques. Our highest goal is to not only to detect and mitigate errors, but to improve personal, professional, and team skills.

18.2 ACHIEVING AN EFFECTIVE CRM ENVIRONMENT

To demonstrate the irony of extreme CRM environments, consider two cases. In the first case, we are the FO flying with the worst Captain imaginable. They have berated us repeatedly over the past few days. They are the PF trying to salvage an unstabilized approach. It just isn't working. As PM, we know that we should call for a go around. We fear that if we do, they will probably launch into a tirade and make the rest of the trip unbearable. Since the runway is long and dry, we choose to remain silent. We don't say a word as we taxi to the gate. We would rate this CRM environment as "bad" since it allows an undesirable, although safe, outcome to happen.

In the second case, we are flying with a best Captain imaginable. Everyone is getting along extremely well. We place a high priority on teamwork and getting along. This Captain is currently flying an unstabilized approach. They already know that it is a mess and have commented on it. We see that they are working as hard as they can to salvage it, but it just isn't working. We know we should call for a go around, but we rationalize that the runway is long and that we don't want to ruin team rapport, so we remain silent. After pulling clear, they continue to berate themselves. We console them. "It's okay, we all have bad approaches sometimes." Prior to this event, we would rate this CRM environment as "good". Ironically, like the previous case, it allows an undesirable, although safe, outcome to happen.

Our two cases reflect the extremes of CRM environment, yet they both generate the same undesirable outcome – our failure to direct a go around for an unstabilized approach. The common flaw is that they both unintentionally raised barriers against required callouts and interventions.

The caustic Captain CRM environment is unquestionably bad. Nothing useful comes from it. PMs in this environment learn to avoid conflicts that might ignite their Captain's volatile behavior. The PM's threshold for intervening shrinks to a point where only the most hazardous profiles feel scary enough to warrant action. On the other hand, the friendly Captain case starts with an open and cooperative environment. This is desirable. Sometime during the formation of this cohesive team, however, the PM elevated their priority of team cohesion over their PM role. Many other features of this friendly team remain desirable including open communications, mutual support, and the desire to succeed. The feature that they lost was their willingness to make callouts and intervene.

Here is where the Captain's briefing becomes so important. While Captains are setting the tone for an open, supportive, and team-oriented CRM environment, they need to emphasize how highly they value following procedures, making required callouts, identifying counterfactuals, and interdicting failing game plans. These expectations lay the groundwork, but they aren't enough. When FOs actually make callouts, Captains need to support, acknowledge, and complement them. Thanking and encouraging callouts or interventions reinforce the desired behavior. Simply stating the expectation during the Captain's briefing, but ignoring it when it actually happens during the flight undermines the effective CRM environment.

18.3 CONFLICT RESOLUTION

Interpersonal conflicts pose a particular challenge to effective CRM between team members. Unresolved, they can raise barriers that undermine error mitigation safeguards. As we converse while resolving our disagreements, our statements tend to operate across three levels – factual, motivational, and personal.

18.3.1 Difference of Opinion

In factual discussions, both parties present the facts as they see them. Statements follow a format of, "Because [this factual event] is happening, we should follow [this procedure]." For example, if a Captain wants to ignore deice holdover time limits and proceed with the takeoff, the FO can disagree with the statement, "We have exceeded our holdover time and snow is accumulating on the aircraft. The procedure requires that we return to the deice pad for another treatment." This FO's statement presents both the facts and the required procedure that resolves the situation. When both parties conduct their discussion on the factual level, the stronger facts prevail and we usually reach an acceptable outcome.

18.3.2 Personality Conflicts

When discussions intensify, statements often transition from facts to motivational or personal levels. Egos emerge. Facts fade in importance as the parties perceive the need to defend themselves against personal attacks. This makes conflicts much harder to resolve. Even when one person tries to keep the discussion factual, the other person questions their motivations or underlying intentions. For example, a frustrated flight attendant might claim, "Gate agents don't care about letting passengers board with oversized bags as long as they can push the aircraft on time." The inference is that agents intentionally allow passengers to board with oversized bags because pushing the aircraft on time is more important to them. This further implies that the agents choose to shirk their responsibility and let the cabin crew resolve the bag storage problem themselves. Notice how the fact (oversized bags reaching the cabin) is used to support their version of the operations agent's motivations (they don't care).

The next escalation of personality conflicts resorts to direct personal attacks. At this level, discussions move beyond motivations and assign unfavorable personality traits or stereotypes to individuals and groups. "Operations agents don't care about gate-checking oversized bags anymore." "That Captain doesn't care about our safety." These conflicts are much harder to resolve because absolute statements devalue facts, inhibit consensus, and prevent compromise. The discussion either stops or escalates to an argument.

18.3.3 Resolvable Personality Conflicts

To resolve personality conflicts, we need to recognize the escalation and steer the discussion back toward the facts. Once a discussion returns to a fact-based perspective, procedural and social standards can guide compromise and resolution. Try the following steps.

- *Accept the situation:* Accept that a conflict exists. When disagreements become personal, our egos and emotions fuel a deteriorating spiral. By pausing and accepting the conflict, we can arrest the spiral. Then, we can engage the situation without emotional distortion and identify common ground for consensus.
- *Select an appropriate setting:* If the problem cannot be resolved quickly, determine an appropriate location and list of who needs to attend the discussion. Consider a situation where two of our flight attendants are conflicted about how to handle a problematic passenger. One thinks the passenger should be deplaned. The other thinks the passenger should stay. They reach an impasse and come to us for resolution. We decide to huddle in the relative privacy of the flightdeck. This keeps every crewmember on the aircraft, yet clear of passenger view. Another option after all passengers have boarded would be to gather in the empty jetway near the aircraft entry door with the remaining flight attendants fulfilling passenger cabin manning requirements. Both options allow for face-to-face discussion in relative privacy.

  Another consideration is whether the discussion can wait until a later time. Perhaps a personality conflict arises, but a functional crew working relationship remains intact. For example, we have a personality conflict with the other pilot, but we both agree that we can still perform our duties and fulfill our roles. We can postpone the discussion until the overnight or during a sustained low-AOV phase.
- *Look for a resolution opportunity:* When we begin the conversation, acknowledge that a conflict exists and that everyone wishes to resolve it. This keeps us focused on the goal and away from egoic and emotional barriers. Perhaps the other person doesn't realize how they have offended us, or vice versa. Simply shining a light on it raises awareness and opens resolution opportunities.

  Avoid trying to determine who is "right" and who is "wrong". This promotes a winner versus loser strategy. Since no one wants to lose, this shifts the focus of the discussion toward the person and away from the behavior or action. If the conflict stems from a misapplied procedure, reference the manual to clarify the correct procedure. If the conflict is rooted in societal norms such as racism, sexism, religious orientation, offensive language, or political affiliation, we can share our personal reactions to such references and establish clear boundaries regarding off-limits topics.
- *Remain focused on the facts:* We attempt to tactfully express our positions without triggering defensive reactions in another person. We limit our statements to facts and our personal reactions. Instead of, "Don't you dare speak to me like that", focus on the event or the behavior, "I don't respond well when you raise your voice with me." Avoid labels, categorizations, or name calling. Remember, it is about facts, ideas, and behaviors, not about motivations, personalities, or people. We can still share our feelings and reactions, but always try to steer the discussion back to the facts.
- *Argue fairly:* Conduct discussions respectfully. Let each side have an opportunity to fully state their position. Use summary statements to refocus back on the facts. "I understand your frustration with the delay, but the procedure

states that we need return to the gate to process this mechanical discrepancy." Avoid using position or seniority to dominate the discussion. If the conflict is about procedure or some aspect of the operation, clearly state what the procedure is, how it applies in this situation, and what we need to do to comply with it. Again, our goal is to resolve the conflict and retain or rebuild an effective CRM environment. If the CRM environment fails, it is the Captain's responsibility to rebuild it. If the conflict remains unresolved, perhaps the best we can do is to restore a functional working environment where everyone agrees to fulfill their duties while tolerating each other as individuals.
- *Listen:* If our fellow crewmember has demonstrated the courage to express their concerns, we should listen respectfully. Their concerns are important to them. We need to understand their perspective. In the end, we may not agree, but we should always listen.
- *Work toward professional common ground:* We all want our workday to run smoothly. Somewhere within the conflict, we all share that common goal. After all parties have expressed their positions, look for professional common ground – that shared space where we can all agree to fulfill our duties, work as a team, and contribute to the safe continuation of the flight. Steer the discussion toward that common ground and build the resolution around it.

18.3.4 Unresolvable Conflicts

If we reach a point where we can't resolve the conflict, then we'll need to bring in outside help.

- *Bringing in outside help:* If the conflict is between two people from the same workgroup, bring in mediators or supervisors. If the conflict is between passengers or concerning issues of compliance, bring in customer service supervisors. The advantage of using supervisors is that it orients opposing parties toward reaching operational solutions. Other combinations may require that the Captain step in, but try to use outside expertise first.
- *Consulting the resolution committee for assistance:* If the airline or union has a department or committee empowered to help individuals with resolving conflicts, use them. Often, two people get too close to a conflict and cannot detach their emotions enough to focus on the facts. It becomes too personal and upsetting for them to resolve by themselves. Bringing in a resolution specialist can offer a trained, outside perspective. They can accurately assess the conflict and locate common ground. If everyone agrees that the conflict is irreconcilable, then supervisors can facilitate rescheduling individuals to different workgroups.
- *Calling someone we trust:* Another option is to call someone we trust for a fresh perspective. This can be a fellow pilot sharing the same overnight, another pilot friend, or a union representative or committee volunteer. Consider the following report of a crew conflict.

BOX 18.1 CAUSTIC CAPTAIN ADVERSELY AFFECTS THE CRM ENVIRONMENT

FO's report: … I was the Pilot Flying (PF). *At around 3,000′, the Captain unilaterally went into the FMS box, pulled up the first fix of our flight plan, hit ACTIVE, and then CONFIRM, and then armed and activated NAV on the flight guidance panel without me calling for anything.* Additionally, ATC had not cleared us direct the fix. … I asked the Captain what she was doing and she replied, "The Clearance". This indicated to me that the deviation from the SOP was willful and intentional. We also hadn't been given direct the fix by the Center. During the flight the Captain, who was the PM, did not appear to keep a fuel score. If she did, she did not inform me if we were on target for fuel. *This was a further breakdown of CRM.* I kept a fuel score on my iPad when I hadn't heard her brief it in the air. *I was afraid to say anything to prevent her from another outburst. She had cussed at me in the flight deck the previous day and when I had objected to that, she told me, "Report Me". I was trying to avoid further outbursts.*

During descent into ZZZ the First Officer (FO) floor was peeling up around the rudder pedals and emergency extension gear box. I told the Captain it needed to be written up. When we blocked in, the Captain handed me the maintenance can and told me if I wanted it fixed, I had to write it up. *I asked her why she was acting so hostile and deviating from the SOP (Standard Operating Procedures).* I told her that I had not called for anything in the box during our departure, she replied that she was the Captain and it didn't matter. When I returned from the walk around, the Captain was exiting the aircraft with the maintenance can on the center console, nothing written up about the floor peeling up around the rudder pedals. There was a known maintenance issue and the Captain had refused to write it up while exiting the aircraft. I filled out the maintenance can, called maintenance, and signed the logbook with my name signature and employee number.

The Captain in the day prior to these incidents had spoken to me in an inappropriate manner repeatedly, cursed at me, spoke in a condescending manner and made it increasingly difficult to maintain a CRM environment. I had twice tried to speak to the Captain to restore a CRM environment and was subject to her outbursts, raised voice, demands, and inappropriate behavior. This report is being submitted to encompass the complete breakdown of CRM, and willful deviation of the SOP from the Captain. … This Captain exhibited a sense of impunity that her seniority allowed her to act in this manner.

Suggestions for improvement: *In training, the company discusses the importance of CRM, but doesn't actually train on CRM techniques or iterations.* Training on techniques of CRM, how to get crews back on track, and discuss issues on a trip is needed. Training is absolutely needed to teach new FOs on how to deal with abusive and inappropriate behavior by Captains. While rare, it's important to know how to handle it in flight and on a trip pairing. Where is the line between a Captain who is rude, and when that rudeness affects CRM and as a result affects safety of flight …[1]

This is an extreme and rare example of CRM breakdown. What other steps could the FO have used to resolve the conflict? Was professional common ground available?

18.3.5 Irreconcilable Differences and the Need to Separate

What if we can't find professional common ground and can't work together? These are cases where the conflict interferes with required procedures or we become so upset that we can't focus on the work. To ensure flight safety, we need to break up the crew.

- *Immediate rescheduling following a CRM breakdown:* A CRM breakdown is an unacceptable safety hazard. If we are unable to function as a crew, we need to break up the team. Most operations have a procedure for elevating these situations to a decision-making authority. Chief pilots, flight attendant supervisors, and station managers can intervene to shuffle personnel around to reform teams. We need to be clear with them that we have a CRM breakdown and that certain crewmembers need to be rescheduled. Most supervisors will coordinate the crew swaps so we can get the operation moving again. The immediate emphasis is on reforming functional teams and getting the operation moving. Expect some follow-on meetings and sessions to investigate the situation.
- *Personally or socially offensive behavior:* Maybe the other pilot is engaging in socially offensive behavior that we cannot condone. What can we do? If it is offensive, but not unsafe, we can clearly express our position. Hopefully, the offensive behavior will cease. If they don't agree, bring in a supervisor to clarify acceptable boundaries and limits.
- *Procedural violation, but not an unsafe act:* When we reach a debriefing opportunity where we can discuss the event, we need to clearly state that the action didn't follow procedure and that we do not support it. The standard here is safe compliance. Hopefully, the other pilot will cease the contentious behavior. We may wish to file an appropriate report to document the event and protect ourselves from sanction due to the violation.
- *Unsafe act:* If an unsafe act has occurred, we need to clearly voice our concerns. If the offending pilot disagrees or pressures us to accept their unsafe act, we need to make a choice. If we don't feel comfortable with continuing to work together, we should use established procedures to remove ourselves from the crew. File an appropriate report to document the event and protect ourselves from sanction due to the violation.
- *An unsafe act is about to occur:* We may encounter a situation where the other pilot chooses to continue with an unsafe act. Consider an example where one pilot shows up for work unsafe to fly. We'll need to intervene to stop the flight from continuing. First, try to convince the unsafe pilot to remove themselves from the flight. If this fails, we can leave the aircraft and contact our supervisor. Keep statements factual. "I've removed myself from the aircraft because my Captain is unsafe to fly in their current condition." Use available assets within the union or company to process the event.

18.4 CRM CASE STUDY – THE IRRECONCILABLE CONFLICT

Consider the follow scenario. You are Captain on a flight preparing to depart from a small station. During boarding, a passenger enters with what appears to be an oversized bag. The lead flight attendant, Sam, states, "This bag is too large, I'll need to gate-check it." The passenger replies, "I have a tight connection, can I please try to find a place for it?" Sam abruptly counters, "No!". They pull the bag aside and affix a claim tag to it. As the passenger receives the bag check receipt, they mutter an expletive and proceed down the aisle. Sam is upset by this, enters the flightdeck, and demands that the passenger be removed for using offensive language with a crewmember. We did not witness the event. How would we proceed?

18.4.1 Call for Outside Expertise

Since this is a passenger issue, we call for a customer service supervisor. The supervisor arrives and interviews the passenger who admits what they said, apologizes, and offers to apologize to Sam. Sam refuses and insists, "One of us is not staying on this aircraft. Either remove that passenger or I'm getting off." The customer service supervisor reports the impasse to us. They don't support removing the passenger. What can we do?

18.4.2 Team Meeting

At this point, we decide to become directly involved. We gather available flight attendants near the forward entry door to fulfill regulatory manning requirements with passengers onboard. With the remaining flight attendants and customer service supervisor, we interview Sam. We allow Sam the opportunity to fully describe the encounter and the impasse. They won't change their mind. There are no replacement flight attendants available, so we must either resolve this conflict or the flight will be cancelled. How do we proceed?

18.4.3 Guide the Discussion

See if the other flight attendants can establish some professional common ground to sway Sam's position. Coming from coworkers, Sam may begin to see the extremity of their position and relent. Guide the discussion to keep it factual. Sam still won't agree. What next?

18.4.4 Clearly State the Facts and Intended Actions

At this point, we can see that Sam isn't willing to budge. Emotions are running high. We ask Sam to step deeper into the jetway for some relative privacy. In case the discussion goes poorly, we ensure that another witness is positioned to observe the conversation. We calmly acknowledge that we see that Sam is upset and validate that passengers using offensive language with crewmembers is unacceptable. We restate a factual summary of the conflict. We confirm that while the passenger's reaction was wrong, they have offered to apologize. We state that the passenger is not seated in Sam's section, so Sam will not need to interact with that passenger. We remind Sam that the customer service supervisor doesn't support removing the passenger. We offer to call the arrival station for a different customer service supervisor to meet the flight to speak with the passenger. If these steps are still not acceptable to Sam, we state that our only remaining recourse is to contact company flight attendant supervisors and share the facts that we have just listed. We inform Sam that our actions are procedurally required as Captain.

When the facts are clearly stated, the path forward is fully up to Sam. Throughout it all, we strive to present ourselves in a calm, mature, and professional manner. Sam should connect the dots and realize that the supervisors will not support their choice to delay the flight. Hopefully, they agree and return to their duties.

NOTE

1 Edited for brevity and consistency. Italics added. NASA ASRS report #1606768.

BIBLIOGRAPHY

ASRS – Aviation Safety and Reporting System – National Aeronautics and Space Administration (NASA) Online Database. (n.d.). Retrieved from Aviation Safety and Reporting System: https://asrs.arc.nasa.gov/search/database.html.

19 PM Role Breakdowns, Callouts, and Interventions

19.1 THE PM ROLE

The pilot monitoring (PM) role is one of the more recent innovations of airline aviation. In a sense, it has always been around, but until recently we haven't given it the level of emphasis, specific procedures, or clarity that it needed to become fully effective. We do quite well when conditions are clear-cut and our workload remains manageable. However, when stress, complexity, and personality conflicts push us toward our vulnerable edge, the PM role sometimes breaks down or fails to interdict failing game plans. Our challenge is to infuse strengths of the PM role into the remaining nooks and crannies of airline culture.

19.2 CAUSES OF PM ROLE BREAKDOWNS

PM role breakdowns occur most often when the Captain is flying. Mishap FOs report that they were either reluctant to speak up or were overruled by their Captains. There are several factors that contribute to these events.

19.2.1 Quickening Pace

One contributor is the quickening pace of failing game plans. We fall into this trap because both salvageable and unsalvageable situations look very similar while we are immersed within them. We become so focused on working harder and faster that we lose SA. This is when PMs need to assert themselves. Consider this following scenario where the crew became distracted and failed to select landing flaps or complete their Before Landing Checklist.

BOX 19.1 QUICKENING PACE LEADS TO PLAN CONTINUATION BIAS

Captain/PF's report: I was flying my second Captain trip. I was practicing my first HGS CAT III Approach. So, inexperience is definitely a risk factor. We were on approach to land flying a HGS CAT III Approach. The weather was calm and clear. *Somewhere around the FAF, I became distracted and forgot to call final flaps 30 and the Landing Checklist. I allowed myself to become completely*

DOI: 10.1201/9781003344575-21

engrossed by the procedures and callouts. Distraction was the second risk factor. Somewhere below the 500′ callout, I heard, "TOO LOW, FLAPS" [an automated EGPWS callout]. I looked at the flap indicator and saw that the flaps were still at 15. I immediately called "Flaps 30, Before Landing Checklist!" The First Officer complied, and by the time we had completed the checklist, the radio altimeter was making the "100, 50, 30, 10" calls. *Things happened so fast that I did not think to go around.* Being rushed was the third risk factor. I landed, and realized on landing rollout that "TOO LOW, FLAPS" is not a caution, but a warning. *I was in violation of Go Around/Missed Approach requirements.*

FO/PM's report: During final approach as Pilot Monitoring (PM), I failed to recognize that we were not properly configured from flaps 15 to flaps 30 for landing. … *During the final approach segment I became internally distracted trying to simulate as best as possible the CAT III conditions, as it has been some time since I have conducted those procedures.* … This distraction caused a lack of SA and a lack of proper configuration. Somewhere below 400′ AGL we got the, "TOO LOW, FLAPS" auditory warning and immediately recognized our error. The Captain called for flaps 30 and the Landing Checklist, which I complied with. We made an uneventful landing. *In retrospect, I should have called for the go around upon hearing the auditory warning for flap configuration.* Personally, I need to have the internal discipline to study and remain comfortable with CAT III procedures rather than trying to think about hand placement and callouts during the final phase of the approach.[1]

Notice how the crew detected the problem only after the EGPWS announced, "TOO LOW, FLAPS". Procedurally, this required a go around. Instead, the crew reverted into a "detect the problem, fix the problem" mode – exactly what we normally do for most salvageable errors. This is a common feature of quickening scenarios as pilots revert to a fix-it mode instead of recognizing that they are past the procedural limit. From another NASA ASRS report, a crew tried to compensate for a deteriorating approach by attempting a 360° turn at low altitude.

BOX 19.2 RUSHED APPROACH LEADS TO UNWISE MANEUVER AT LOW ALTITUDE

PM's report: …Rather, we made a snap decision to take what seemed like a reasonable and safe course of action. In retrospect, a go around would have been the obvious best course of action. *If I had it to do over again, I would go around. S-turns or 360s on final have worked in the past, but never to configure and never at or below 1000′.* It was not intentional, and I certainly learned a good lesson. *I need to be more spring-loaded to go around [rather] than to try to find a way to "make it work."* Also, as Pilot Monitoring, I need to be more inclined to intervene verbally.[2]

19.2.2 Short-Circuiting the PM Role

In some events, the PF detects that the profile is not acceptable, but says something that causes the PM to hold back. Instead of following established procedures, they move the goal posts to something that is "good enough". They follow a, "I know what we are supposed to do, but we are going to do it this other way" line of reasoning.

BOX 19.3 CAPTAIN OVERRIDES FO/PM TO CONTINUE UNSTABILIZED APPROACH

Captain/PF's report: ... There was a tailwind at altitude. ... *The approach ended up being very compressed and stabilized approach requirements at the FAF and 1,000′ were not met.* Flaps 30 and +15/–5 knots wasn't met until 500'. *At 1,000′ we were fast, not completely configured for landing, with no landing checklist complete. ... I stated that I will go around if not meeting all stable approach criteria by 500′. The FO seemed to agree. At 500′ all stabilized approach criteria were met, and we landed without incident. We were in VMC conditions with the runway in sight from the FAF inbound.*

The approached was debriefed at length. Both of us were extremely uncomfortable, after the fact, that we had indeed violated the stable approach criteria SOP. *While flying, and in-the-moment, I knew a go around was very possible and stated that fact, but felt I had until 500′ to get the remaining criteria stabilized. Post flight, it became more apparent that the FO didn't agree with my assessment and interpretation of the stabilized approach SOP. I didn't adequately recognize how behind the FO was. ... In essence, I missed what Captains are expected and required to do such a situation.*

Bottom line is that we found ourselves in a difficult state, and even though we at least partially recognized it, I didn't properly apply CRM skills and [Error Management] as well as a Captain of my experience should have and could have. We both regret our performance but have learned a great deal from it. The debriefing has been extensive, we will both grow better as a result of the experience.[3]

The Captain recognized the unstabilized approach, but decided that it would be acceptable to continue as long as they were stabilized by 500′. The Captain specifically declared that the go around determination would be made at 500′. We can imagine how the Captain's confident statement neutralized the FO/PM's intention to make the required callouts. Why bother? The Captain clearly knew the parameters. It is also relevant that their runway was 12,000′ long – plenty of pavement to accommodate a long or fast landing. From the FO's perspective, the approach was technically unstabilized, but the Captain/PF had it well under control and was performing all the right steps to fix the problem.

19.2.3 Assertively Suppressing Intervention

Another nuance is when Captains are so sure of their decisions that they choose to shut the FO/PM down entirely. They overrule their FOs callouts and continue. Consider the following reports documenting a significant role breakdown between the pilots. The approach was to El Paso International airport which is surrounded by high terrain.

BOX 19.4 CAPTAIN OVERRIDES MULTIPLE GO AROUND CALLOUTS FROM FO/PM

FO/PM's report: I was operating flight as pilot monitoring (PM). At first, I didn't notice anything out of the ordinary. In fact, the entire flight was normal up until landing time. … After being cleared [for] the visual approach and being at a safe altitude of 9,000′ (mind you, the field elevation is roughly 4,000′), *the Captain puts in an altitude of 6,000′ [in the MCP] and didn't communicate to me that he [had] set a new altitude.* After the short amount of time passing, I told him, "Sir, I show 6,000′ set – confirm?" Of course, as pilot monitoring, I'm watching very carefully what the Captain is doing so that nothing serious happens to risk the safety of our approach and landing. I give the Captain 10 seconds to see if he will tell me of the new altitude set in the MCP. *I hear nothing, just silence*, as the aircraft descends into mountainous terrain. We are now descending through the highest terrain which is 7,176′ and the Captain puts in 4,500′ in the MCP we are still descending closer and closer and closer towards the highest terrain which, mind you, the MSA in that quadrant states 8,400′ for a minimum safe altitude. *Now, I know something seriously is wrong. Something just doesn't feel right. I know we're not supposed to be this low especially over mountainous terrain. Now … the very loud and serious "TERRAIN, TERRAIN, PULL UP, PULL UP" warning flashes red and sounds in the flight deck. As I immediately begin to perform a recovery procedure for this uncomfortable undesirable situation, the Captain says, "I got it." Unfortunately, he did not recover properly which scared me even more and showed me that he was playing chicken with mountains and with our lives.* This was [not a] profile approach or visual. ATC kept us at 9,000′ for a reason, so that we can be clear of obstacles and mountainous terrain. Yet, for whatever reason, he decided to set 4,500′ in the MCP and descend to this dangerous altitude in mountainous terrain. Instead of recovering properly he began a left base turn within close proximity to the mountains.

He was "cutting corners" way too close and there was no APP mode or LOC or LNAV selected. I mentioned to the Captain where the runway was and pointed it out clearly as the visibility was greater than 10 miles and the winds were calm. You couldn't ask for better weather during this approach. I've flown many flights to and from [this area] and am very familiar with this airport as it can be tricky if pilots don't pay attention to detail and don't brief or execute an approach as briefed correctly. *Now, with no CRM from my Captain into the equation, terrain warnings, flying toward the wrong runway, [let] alone wrong airport, me yelling "GO AROUND", and no response from the Captain continuing the very unstable approach, I was put in a situation I would never want anyone to ever experience.* The Captain was hand-flying from the time being cleared for the visual to the runway. *Now, the aircraft [was] still in a left turn towards final 1,000′ above touchdown, still no gear down, and still no flaps configured for landing.* This put me in a very uncomfortable situation. Not to mention we weren't even heading towards the airport, the Captain was flying the aircraft blindly toward Biggs [Army Airfield], which we even discussed during the approach briefing – that it can be easily mistaken for El Paso International airport. … *At this point I am so uncomfortable to the point that my training kicks in and in the name of safety I say, "Sir, we are way off course, go around." Unfortunately, he does not respond and does not execute a go around procedure.* He is flying the aircraft closer and closer to the ground. I keep trying to point out Runway 4 which is 12,020′ long and it is very clearly visible. … *To my disbelief the Captain now is doing a series of unstable erratic maneuvers and is now setting himself to land on Runway 8R – a much shorter runway that we were not cleared to land on. … Even the Tower makes a remark and asks if we're okay. Meanwhile, I am telling the Captain [that] we must go around. I now raise my voice even more and say, "Sir this isn't right, GO AROUND". He claims that he has control and he "can do it".* Yet we were way off course and never were [on] course in the first place for a proper stable landing … Because of the descent he initially performed into the high terrain, as well as aiming towards the wrong airport and wrong runway, I [tried to guide] him towards the correct airport and correct runway once again. *Then, it was my third and final yell for, "GO AROUND". He said, "No, we're landing, everything is under control" when in fact the aircraft was 100% not in control. It was fast, sloppy, behind the curve, and very unstable all the way until we made contact hard with the runway.* The Captain barely put the aircraft on the runway with extensive abnormal high power settings in the flare to keep the aircraft from hitting extremely hard.

I hope and pray that nobody ever has to go through what I have gone through this morning. We are not Cowboys of the sky. We are professional aviators and are set to high standards and have standardization for a reason. *Either pilot can call for a go around. And if an approach does not seem right then it most likely is not right. And unstable approach deserves a go around, period – especially one with pull up terrain warnings, wrong airports, and runways in front of the pilots windshield.* We preach callouts for a reason – to follow them and to be safe so we can live to fly safely another day with many blue skies and soft landings and so we can be role models to the generation in future of aviation safety.

Captain/PF's report: [This narrative starts while being vectored for a visual approach to Runway 4 at El Paso.] We were cleared to 9,000′ heading 270° about 7 miles north of ELP. We were cleared for visual approach to Runway 4.

> We continued descent toward 6,000′ and heading westbound. *Once clear of the mountain ridge to the west of ELP, I began further descent when we had a momentary terrain warning. I could immediately determine that terrain clearance was not a problem as we were in the clear and visually [clear of] all terrain prior to descending, I stopped the descent and the warning ceased.* I … had difficulty in reacquiring the landing runway. There was some confusion looking into the sun at Biggs [Army Airfield] and ELP [International]. I continued towards the airport and realized I was north of the position I planned. *The First Officer suggested a go around, which I considered. I elected to continue since I still had adequate area to maneuver for the landing.* I continued toward the runway and the First Officer assisted in confirming the landing runway. The First Officer again called for a go around, I told him I had the Runway and was confident the approach could safely be completed. I made a right turn to a modified left approximately 2-mile base for Runway 4. During the turn, the First Officer called for a go around. The aircraft approach speed was about 125 knots and during this approach nothing seemed rushed. At no time during this operation did I feel I was pushing the aircraft or my capabilities.
>
> Having thought many hours about this flight, I can see where my CRM skills were far from my beliefs. I do believe in the company's policy on go arounds. *I do believe that the approach, although not pretty, was safe. But the big thing was that the First Officer was not comfortable and his judgement should not have been overlooked.* I will not do this again. As for the terrain warning, although not actionable could have been avoided by planning the flightpath either higher or over a different ground track.[4]

Was the Captain non-compliant, arrogant, overly aggressive, or willfully negligent? We can't say. In many ways, he appeared to be a typical line pilot. His most concerning flaw was that he made his aviation decisions based on his personal standard of safe operation versus airline policy. Policies and procedures are expressly intended to prevent this. While ignoring warnings that the approach was too close to terrain (the "PULL UP" warning), to the wrong airport (Biggs AAF versus ELP), to the wrong runway (8R versus 4), while unconfigured at low altitude, and unstabilized, he chose to continue.

These reports describe a Captain who appeared to be standard and compliant, but suddenly became nonstandard and noncompliant when falling into a Recognition Trap. What caused his latent vulnerability to emerge when his game plan began falling apart? One possibility was the desire to succeed. Successfully landing every single time establishes a habit pattern that relies on *always succeeding.* It is one thing to be sent around by ATC (external, outside cause) and quite another to have to go around because we personally mismanaged a visual approach (personal, internal cause). Consider each pilot's mindset as this scenario unfolded. Starting with clear weather, calm winds, good visibility (except for a low sun angle), no traffic, and cleared for a visual approach, neither pilot expected any difficulty with this arrival. The terrain warning and mistaken runway caused the game plan to fall apart very quickly. The

Captain maneuvered unpredictably and abruptly in an attempt to recover from his errors. The FO directed three go arounds – all of which were ignored or overruled. The Captain even termed the first go around callout as "a suggestion". We don't know what the FO decided at this point. While clearly uncomfortable with an approach described as fast, sloppy, behind the curve, etc., they sat silently through the landing. The Captain's actions had completely shut down this FO.

While this appeared to be a harrowing experience, it apparently didn't reach a point where the FO/PM felt justified to intervene and take control of the aircraft. It would be interesting to interview this FO to understand how their mindset progressed throughout this event. At which point would the approach cross the FO's personal decision trigger point to assume aircraft control? Did they even have a trigger point to intervene? Did they feel shut down by the Captain's repeated denials? It would also be interesting to interview the Captain to understand their mindset throughout this scenario. What was the Captain's go around threshold? The Captain recognized that the scenario was poorly flown, but judged that everything was "safe enough" to warrant overriding all PM callouts and go around directives – decisions which he appeared to regret in hindsight.

The Captain became so committed with completing this landing that following procedure and upholding professional standards were abandoned. It probably didn't feel like that at the time. It probably felt like he was working hard to solve an aviation problem and maneuvering the aircraft to land. This Captain probably recalls this incident as a regrettable one-off within his otherwise compliant aviation career.

19.2.4 Reluctance to Intervene

Another nuance is when PMs feel reluctant to intervene. I participated in a mishap investigation of an aircraft sliding off of a wet taxiway. Interviewing the crew, we learned that the FO felt reluctant to intervene even though they knew that the Captain's taxi speed was excessive. What led up to this decision was quite instructive. The FO reported that the Captain had been commenting on every single error and failure that the new-hire FO had made during the trip. They felt beaten down by the Captain's relentless corrections and abrasive tone. As the FO flew the approach, the Captain micro-managed them all the way down final. They landed, but didn't slow sufficiently to make the typical mid-field taxiway exit. The next available exit wasn't until the end of the runway. The Captain angrily took control of the aircraft and abruptly added thrust to expedite to the end. The FO prepared themselves for another tongue-lashing. For perspective, while the FO was new to the aircraft and airline, they had been a Boeing 727 Captain with their previous carrier. They knew that the Captain's speed was excessive, but felt reluctant to say anything. Also, the airline had no prescribed PM callouts for excessive taxi speed. The FO felt that speaking up about the excessive speed would trigger another angry tirade, so they said nothing. Racing toward the end of the runway, the Captain finally realized that they were too fast, hit the brakes, and yanked the tiller. The aircraft began sliding uncontrollably, spun 180° around, and departed the pavement.

During the debrief, we asked the FO what they thought about the airline's CRM training. They replied that the training had been great. They had completed all of the CRM exercises and vowed that they would never become "that pilot" who failed to intervene when a Captain was operating unsafely. Yet, there they were. They had become "that pilot".

There are many reasons why PMs feel reluctant to intervene. Some are based on fear of consequences. Others feel reluctant because of cultural norms within the line operation. Whatever the case, PMs must overcome their reluctance and act to ensure safe outcomes.

19.2.5 Being Overly Helpful

This case happens more with Captains wanting to help out their struggling FOs. They want to preserve the good, cooperative rapport of the flightdeck. It feels less confrontational to let the minor errors pass or to help the PF to salvage a failing profile.

BOX 19.5 OVERLY HELPFUL CAPTAIN CREATES AIRCRAFT CONTROL CONFUSION

Captain/PM's report: My day started in LGA [LaGuardia], day 5 of 5 on reserve. I flew the first leg which was a non-event. The First Officer (FO) seemed very sharp. I gained some confidence in him. The FO flew the return leg to LGA and the flight was uneventful from takeoff all the way through the descent which again instilled confidence for me in the FO. *We were vectored for the visual approach to Runway 22 and ATC brought us in kind of tight. There was a crosswind from the right, which delayed our intercept with the localizer. The FO did not notice this and I realized we would be high if he didn't start down. This is where the approach completely fell apart.* After I vocalized that the FO needed to set a lower altitude and start descending, it took him 5–10 seconds to process what I said. By that time, we were high. *At that point, the FO completely shut down and I realized he became very flustered, overwhelmed, and tunnel vision set in. I began calling for all his gear, flaps, checklist, and speeds.* We were configured by 1,000′ but when I looked down, we were doing 1,400 feet per minute [sink rate] at around 700′. *I said, "Watch your descent rate". The FO then froze again so I nudged the controls to bring us within limits.* The FO then completely blew through the glide slope on the bottom end, the aural glideslope alert sounded to which the FO promptly responded, "Correcting." With his prompt response, I thought he would level off to get back on glideslope. That is not what happened as the aural glideslope alert continued to sound. *I then nudged the controls to level the aircraft and bring us back onto the correct glideslope.* We were visual the whole time and I had the runway right in front of me so I allowed the approach to continue. The FO then landed the plane pretty hard and side loaded. I took the flight controls at 60 knots and taxied to the gate without incident.

... As a new Captain, this the first time I've flown with an FO ... who completely fell behind the airplane. After the first leg, I believed this FO was sharp and this approach caught me completely off guard. It was a beautiful day outside and a completely normal operation which added to me getting caught off guard. *In-the-moment, I believed the safest action was to coach the FO down to landing and nudge the flight controls, as needed. After reflecting on what happened, I was completely wrong. I should've called a go around and allowed the FO to restart the approach from scratch.* Considering the state the FO was in, I don't believe he would've done the required items on a go around, but regardless I should've called it and coached him through the go around, as needed. This was a big learning experience for me and one I will take with me going forward.[5]

We have all mismanaged an approach and know how embarrassing it can feel. This was an extreme example with the Captain coaching, guiding, and nudging controls most of the way down final. Speculating on the FO's mindset, they might have even assumed that the Captain had assumed aircraft control as evidenced by their failure to make an appropriate flight corrections between the first nudge and the second nudge. At what point did *helping* become *taking over*? An interesting detail is that the FO acknowledged one callout by immediately stating "Correcting", but didn't appear to make an appropriate correction. This indicates how automatic and habitual our replies can become.

19.2.6 Rationalization

Another nuance is the crew rationalization. In this case, both pilots agree with the nature of the problem and fashion a nonstandard compromise to solve it.

BOX 19.6 CREW RATIONALIZES DECISION TO CONTINUE UNSTABILIZED APPROACH

FO/PM's report: During approach to PIA, in night IMC, the Captain was pilot flying (PF), with the autopilot engaged. After being vectored onto the approach, the Captain did not engage approach mode on the FCP. *Apparently, NAV was selected during intercept but not APPR mode, possibly due to distraction with the new checklist procedures.* As PM, I was busy reading the new Landing Checklist and failed to notice the approach mode [was] not engaged. *I finished the checklist and noticed that we were off-scale glideslope (GS) deflection high, due to not capturing the GS. I brought this to the Captain's attention.* He immediately started a descent to capture the GS from above. *As we started descending, I commented that this is likely going to be a go around due to the possibility of being unstable thru 1,000′ AGL. We discussed the go around option briefly, but as we descended thru 1,000′ we were still 2-dots high and*

1,600 feet per minute descent. I said ok, that is a go around due to our descent rate and GS deviation. We discussed the go around requirement briefly and as I prepared to initiate a go around, I could see that we were decreasing descent rate and capturing the GS. *By 750′ AGL, we were stabilized, spooled up, on speed and that a go around seemed to be unnecessary at this time.* A safe landing was made in the touchdown zone, on speed.

[I should have exercised a] better vigilance of pilot monitoring (PM) duties especially when distractions are present. This would have prevented the entire situation. *However, the real error here was an unwillingness to go around at 1,000′ due to sink rate and the failure to go around even when it was requested by myself. The real issue was my, the PM's, failure to initiate a go around at 1,000′ which should have been automatic not a polite request, followed by a discussion of stabilized approach parameters, another request and then conceding that at 750′ AGL there was no longer an excessive sink rate or GS deviation and therefore no current need for a go around.*[6]

The FO/PM felt that a go around was procedurally required and "politely requested" it, but the Captain/PF countered with a compromise that shifted their stabilized approach decision altitude from 1,000′ to localizer approach minimums. This rationalization applied an arbitrary limit that was closer to their personal safety standard. This happens because we know that we can safely land from approaches that become stabilized much lower than 1,000′. As standards drift, safety margins can become discretionary zones that pilots feel that they are allowed to manage. It is more difficult for FOs to oppose their Captains on a "logical" point like this since it does not compromise safety. Of course, the problem here is that limits are not starting points for further negotiations. Limits are limits.

19.2.7 Informing, but Not Acting

Another PM role breakdown occurs when FOs believe that their role is to inform the Captain of deviations, not make flight decisions for them. These FOs recognize the Captain as the final authority for conducting the flight. After they announce the deviations, these FOs feel that they have fulfilled their PM responsibilities. This misconception is a reoccurring theme in mishaps. FOs make the initial callouts, their Captains acknowledge them, and nothing further is said. When asked why they didn't make additional callouts when the instability was not corrected or worsened, the FOs replied that the Captain already knew of the deviation and that any further callouts wouldn't have helped the situation. First, this tacitly justifies the noncompliance and virtually guarantees that they will accept the failing game plan. Second, this overlooks the vulnerability of tunneled attention. When we asked mishap Captains if they realized how extreme their parameters had become, they rarely did. The irony was that the only pilots who knew how extreme their parameters were allowed the situation to continue. For this reason, many airlines now direct PMs to repeat deviation callouts until the discrepancy is resolved.

19.2.8 Mismanaging Discretionary Space

A wide range of individual practices evolve as pilots manage what they perceive as the discretionary space between procedural limits (like the procedural approach stabilization altitude) and their own personal safety limits (where they would direct a go around or take control of the aircraft because the situation is becoming hazardous). PMs start by making callouts and monitoring for trends. They only intervene when their personal safety limit is exceeded. This cultural norm is strong because it promotes safe operation while preserving a congenial flightdeck environment. Practices vary widely when there is a significant gap between procedural limits and personal safety limits. For example, pilots know that they can safely land from an approach that only reaches stable parameters shortly before landing. Long runways allow them to slide their personal safety limit even further. A poorly flown approach can touchdown fast and long if there is enough pavement to dissipate the excess energy. This is what Clem decided during his unstabilized approach. Over time, success encourages more drift. Add confusion, indecision, stress, task overload, and startle and we reach situations where PMs fail to intervene. This is why we need to honor procedural limits. The gap between procedural limits and personal limits gives us time to manage how we will affect the intervention, but the intervention should still happen.

19.2.9 Different Mindsets and Perspectives

Underlying many of these CRM breakdowns is a mismatch between each pilot's mindset or perspective. We use assumptions about the mindset of the other pilot to make sense of their behaviors. These assumptions may be inaccurate. For example, behavior that *appears* to be willful defiance against following procedure by the PF probably isn't. Instead, PFs may have become so task-overloaded and mesmerized that they have lost sight of the larger perspective. This is why it is so important that PMs follow established procedures regardless of the PF's mindset and perspective. Make callouts automatically and repeatedly. Make interventions decisively and effectively.

19.3 PRESET INTERVENTION TRIGGERS AND APPLYING JUDGMENT

Understanding the vulnerabilities that contribute to PM role breakdowns, let's explore some ways to strengthen PM effectiveness. We'll start with examining how we calibrate our callouts and interventions. Procedural limits define many of our intervention points, but we need to solidify our personal response techniques. For this, we need to develop strong trigger points.

19.3.1 Preset Triggers

When we become rushed and stressed, our doubt and indecision rise. Optimism bias then subdues our doubt and subconsciously urges us to continue shaky game plans. We rationalize our reasons to continue and hope for the best. To counteract this,

PMs need a predetermined plan for when and how to speak up or intervene. If we set and rehearse these intervention trigger points ahead of time, they become much easier to execute while we are immersed in-the-moment. Like a red engine light that illuminates when the engine limit is reached, our interventions should trigger automatically and reliably.

19.3.2 Personal Safety Limits

Judgments regarding personal safety limits are not consistent across situations or between pilots. They vary too much. The unsuitability of this standard is demonstrated by mishap pilot's statements during investigations. When asked, "Why didn't you direct a go around?", many answered, "Because the approach didn't feel unsafe." When asked what makes an approach feel unsafe, they couldn't accurately define their parameters. They described it as a gut-felt assessment. Despite its variability, we've come to rely on our intuitive personal safety judgment is an important skillset. So, it is important that we find ways to reliably calibrate and apply it.

A useful practice for calibrating our intuition is to imagine how possible mishaps *might* unfold. Using a range of nuanced scenarios, we can mentally chair-fly any conceivable event, experience how they might feel, and rehearse the actions that we would take. We start by identifying extreme events where interventions are undeniably required. For example, if the PF is flying an unstabilized approach, diving for the touchdown zone, thrust levels pulled back to the idle stops, and sustaining an excessive sink rate, we would judge this as a hazardous profile that risks a hard landing, bounced landing, or departing the runway surface. We can imagine how this scenario would unfold, what it would look like, and how it would feel. We then set and rehearse our trigger point for when we would call for a go around, the words we would use, and the tone that we would use. We would also rehearse a second trigger point for when we would intervene and take control of the aircraft.

Why aren't all mishap events fully trained by our airline or clearly described in written directives? They aren't because there are too many potential scenarios and nuances. Consider landing mishaps. We may have problems with aimpoint, sink rate, lateral offset, airspeed, thrust setting, bank angle, sideslip, windshear warning, loss of visual landing references, touchdown attitude, over-control, under-control, floating, and touchdown point. There are too many cases to specifically address in writing. As proficient pilots, all of us already have an intuitive sense for how each of these events would develop. Our company expects us to apply our judgment to any event that strays from acceptably safe parameters. Since any event outside of that middle range is one that we can imagine, we can practice what we'll need to do if we experience it.

19.3.3 Applying Judgment

Limits are limits, but we are expected to apply our discretion and judgment. Consider some borderline examples. If the directive sets the stabilized approach limit at target airspeed plus 10 knots at 1,000′, a PM seeing target airspeed plus 11 knots would be procedurally directed to call for a go around. In practice, few of us would actually direct a go around for a 1-knot exceedance. In a related case, if the PF is doing their very best to hold target speed in moderate turbulence, would a momentarily exceedance to target airspeed plus 11 knots warrant a go around? All of us would apply our judgment and classify this as a perfectly acceptable momentary deviation. So, as we apply our judgment with these two examples of limit exceedance, target airspeed plus 10 knots doesn't seem to qualify as a realistic absolute limit.

- *Stabilized approach entry:* One way to apply perspective to procedural limits is to examine the intention underlying the directive. The directive of target airspeed plus 10 knots at 1,000′ is intended to ensure that the PF *enters* stabilized approach parameters at or before 1,000′. If the approach starts out fast, but the speed is trending down and reaches target speed plus 10 knots exactly at 1,000′, we would judge that the approach is stabilized. We would also agree that if the PF reaches stabilized approach parameters at 1,200′, but turbulence momentarily bumps the speed to target airspeed plus 11 knots while passing 1,000′, that the approach still qualifies as stabilized. The momentary deviation to target plus 11 is acceptable as a short-lived deviation caused by the turbulent conditions. The underlying intention of the directive is to prevent continuing profiles never reached stabilized parameters by 1,000′.
- *Subsequent approach instability:* We also recognize that once an approach becomes stabilized, it can subsequently become unstabilized and require a go around. Expanding our interpretation of the company policy, the PF must achieve stabilized parameters by 1,000′ and then maintain a stabilized profile through touchdown. This is also subject to our judgment. If the PF passes 1,000′ stabilized, but maneuvers to dodge a bird at 800′, trends 1-dot high, then applies a smooth correction to restore a stable path by 400′, most of us would agree that the approach should continue. If, however, the PF dodges a bird at 100′, and it becomes clear that the aircraft won't land within the required touchdown zone, then a go around is warranted. These examples highlight how momentary deviations in approach stability are acceptable as long as stability can be regained using reasonable control inputs to land within the required touchdown zone.
- *The sliding scale of judgment – hard and soft limits:* In effect, our judgment recognizes a sliding scale for approach stability. While maneuvering down final approach, reasonable deviations are acceptable as long as we have sufficient time to restore stabilized parameters. Nearing touchdown, our tolerance narrows. We wouldn't tolerate any large instabilities approaching touchdown. This acknowledges that real-world conditions sometimes cause momentarily exceedances. If we prohibit any excursions beyond target airspeed plus 10 knots, few aircraft would be permitted to land at Las Vegas International (LAS) on practically any hot summer day since warm air turbulence typically causes significant airspeed and sink rate deviations down final. In these turbulent conditions, we would expect PFs to enter stabilized approach parameters before 1,000′ (hard limit), correct all deviations back to target speed (soft limit), hold glide path (soft limit), hold crosstrack (soft limit), and land in the touchdown zone (hard limit).

- *Holding the middle:* Can soft limits also trigger go arounds? Absolutely. If we encounter a strong rising thermal that drives us over 1-dot high and 20 knots fast at 200′, our judgment would be to direct a go around. Why? Because it violates an underlying assumption that stabilized approaches should remain *near the middle* of normal flight control inputs. We want the thrust levers to vary back and forth around the middle of the normal approach thrust setting. We want flight control inputs to vary around the middle of reasonable flight control corrections. We want pitch controls to hold the glide path using reasonable corrections up and down near the standard glidepath. We want crosstrack deviations to stay near runway centerline with reasonable corrections left and right. All of these assessments are guided by our judgment of what feels *reasonable.* Anytime that we judge that these reasonable limits are exceeded, the approach is classified as unstabilized. We wouldn't accept thrust level corrections alternated between the MAX and idle stops, for example. They aren't reasonable corrections around the middle. We wouldn't accept a sink rate of 2,000 feet per minute through 200′. It is too far from the middle value of about 800 feet per minute.

19.4 SCRIPTED CALLOUTS, DEVIATION CALLOUTS, AND RISKY DECISIONS

Now that we have explored the judgment that bounds when we will make our PM callouts and interventions, when will we say them and what will we actually say?

19.4.1 Scripted Deviation Callouts

The simplest PM callouts are procedural deviation callouts. Clearly written in our manuals, they reflect direct readouts from aircraft gauges and displays – "Airspeed", "Glideslope", and "Crosstrack", to name a few. Depending on the airline, PMs are encouraged to add descriptive details such as "Airspeed, plus 15", "Glideslope, 2 dots high", and "Crosstrack, 2 dots right." Typically, the PF responds with an acknowledgment like "Correcting", and makes an appropriate adjustment to resolve the discrepancy.

Company procedures may direct how often the PM repeats their callouts. They may also direct what to do if the PF fails to acknowledge the callout, if an appropriate correction is not observed, or if the adjustment is insufficient to resolve the deviation. Scripted deviation callouts are effective because they are clearly mandated in written procedures. Generally, PMs aren't reluctant about making them and PFs aren't resistant against hearing them. Although, some line cultures discourage them as misguided efforts to promote congenial flightdeck environments.

19.4.2 Unscripted Deviation Callouts

The next class of deviation callouts includes deviation callouts that aren't specifically scripted in the manuals. These deviation callouts alert the PF of adverse trends or impending errors. For example, if the Captain is distracted while taxiing and starts straying from the taxiway centerline, the PM may call, "Drifting right". This would alert the Captain of the undesirable trend and return their attention to controlling the aircraft's trajectory. For another example, consider a PF who is hand-flying a departure SID, becomes distracted by an inside task, and is unaware that they are quickly approaching the assigned level-off altitude. The PM might say, "Approaching level-off". Like the taxiway drift, this highlights an adverse trend and alerts the PF to redirect their attention toward correcting the flightpath before busting the assigned altitude.

- *Unscripted callout verbiage:* Since these callouts cover a wide range of pilotage issues, the specific verbiage is left up to the PM. We are empowered to use any informative statements that accurately communicate the nature of the deviation. These callouts tend to be more welcomed when they factually describe a condition and less welcomed when they imply fault. Consider the example of a PF hand-flying a climbout and approaching the assigned level-off altitude. The callout, "Approaching level-off" factually states the condition. Compare this with a more accusatory callout like, "You're about to bust our assigned altitude." This option communicates the same information, but with blame. Ratchet it up one step further to get, "You're about to bust our [expletive] altitude." On top of the accusation, this callout implies that the PM is upset about the impending error – a "you're about to get us both in trouble" message.
- *Adding corrective directions to the callout:* Revisiting our previous example of the taxiway drift, the PM might state, "Come left, you're drifting." If the PF is highly distracted, a callout like this eliminates potential confusion. Otherwise, it might take them a few moments to look up, realize what is happening, and apply an appropriate correction. The directive/deviation format streamlines this process. For the impending altitude bust, the callout might be, "Level off now – 300′ to our assigned altitude." This directs the corrective action, identifies the problem, and communicates the urgency. Make callouts unambiguous and free of accusation or negative undertones. If the PF displays frustration with their error, make an effort rebuild flightdeck rapport. For example, if they successfully correct the trend and smoothly level off, consider adding, "Nicely done".

19.4.3 Risky Decisions

Calling out risky decisions is more difficult because they highlight the differences between the PM's opinions and the PF's decisions. It's no longer just about factual parameters. By questioning the quality of their decisions, we risk appearing confrontational. We need to resolve these issues skillfully. Start with facts. Each pilot can perceive similar facts, yet draw different conclusions. Consider a situation where our flight is navigating around some towering thunderstorms. The PF has selected a course that flies, in our opinion, too close to a menacing buildup. We are uncomfortable and want to say something. How can we proceed?

- *Assess the conditions:* Recognize that our decision is based on our experience, our perception of conditions, and our risk assessment. The PF has selected their course based on their experience, their perception of conditions, and their risk assessment. Something has generated a mismatch between our mental models. If we can uncover the source of this mismatch, we can address it directly. Perhaps the PF lacks experience with maneuvering around towering buildups (inexperience). Perhaps they are overlooking important conditions that would otherwise change their assessment (missing information). Perhaps they perceive the same conditions that we do, but weigh their importance differently (risk assessment). Maybe they are just more risk tolerant than we are (risk management). At first, we really don't know. To restore our shared mental model, we need to initiate a discussion.

  A good place to start is to assemble the facts and understand how we each formed our particular mindsets. If the PF is an experienced Captain, the selected course is based on their wealth of experience and past success, so they might be right and we might be wrong. As we investigate deeper, we notice a strong crosswind component blowing toward the closer buildup. Given that additional piece of information, the selected course makes sense. We see that the PF has chosen a course that shaves closer/upwind to one buildup and further/downwind from another buildup. The selected path will probably have the smoothest ride while avoiding any hail that the further/downwind buildup may be generating. Given this information, we change our opinion and accept the PF's decision. If, however, their chosen course is actually downwind from the close buildup, then a different consideration must be driving the PF's decision. We notice that their chosen course tracks closer to the magenta-line flight planned course. We conclude that the PF may be trying to shorten the flight distance to save time, even though it risks a turbulent ride. Their chosen course accepts the increased risk from the buildup to shorten the air miles flown. In our opinion, this decision is riskier, so we decide to speak up and offer a safer alternative.
- *Clearly present our reasoning:* We clarify the facts as we see them and organize our reasons why a different course might be better. We frame the scenario by stating that we are uncomfortable with flying so close to the buildup because of the downwind hazards. "This course seems too close to that buildup. I'm concerned about turbulence and hail. Also, the flight attendants are still up finishing the passenger service." Ideally, the PF agrees with our reasoning and alters course.

  If the PF doesn't share our concerns, we expand the discussion. We may need to add more facts to convince the PF to alter course. Perhaps their risk management has drifted over time to accept higher levels of risk. Discussions like this recalibrate our thinking and return us to a safer standard. Let's ratchet up the scenario further. After voicing our concerns, the PF still elects to continue with the risky course. If we are the Captain, we can override the FO's decision and direct the safer course. If we are the FO, our options are more difficult.

- *Offer a safer alternative:* Assuming the Captain/PF won't accept our concerns, suggest a safer alternative. One tactic is to highlight better options. "Flying to the left of that buildup keeps us upwind of the turbulence and any hail that the buildup may be throwing out. However, if you want to stay on this heading, I'd like to have the flight attendants immediately take their seats." If our airline is using a risk management system like Risk and Resource Management (RRM), we can include a color code. "This course puts me in the Red. About 20 degrees to the left feels safer to me." Notice in these two statements, we stick to either factual statements or our personal opinions. This is what *I* feel and this is what *I* need. Most Captains value a cohesive flightdeck team. Most will bend to our concerns when they understand that the issue is important to us.

  Avoid accusatory statements and undertones. "You are flying us directly toward the severe turbulence and hail from that buildup." "You are about to hurt some flight attendants when we hit the turbulence from that buildup." Statements like these link their decision with bad outcomes. The implied message is, "Bad things are about to happen and they will be your fault." If necessary, this option is still available to make our point. If the Captain is particularly headstrong and resistant to our input and we have tried persuasion without success, this may be a useful tactic to get them to do the right thing.

19.5 INTERVENTIONS

The majority of interventions involve Captains assuming aircraft control from their inexperienced FOs. Generally, these active interventions happen for the right reasons and lead to successful outcomes. The rarer scenarios involve an FO/PM intervening when their Captains are flying. Assuming that we have tried callouts and persuasion to reverse the PF's failing game plan, our final option is actively intervening. In these cases, PMs are potentially the final safety barrier that can stop an accident trajectory. Consider the following report of an FO that tried unsuccessfully to get their Captain to go around.

BOX 19.7 CAPTAIN LANDS DESPITE FO'S REPEATED GO AROUND CALLOUTS

Captain/PF's report: We were on the RNAV approach into ZZZ. For half an hour leading up to this, we were experiencing a bumpy ride and were extended out over the lake (off the approach) for sequencing into ZZZ. Gusty winds were reported (180/23 gust 35) on final and aircraft were also reporting plus-or-minus 20 knots. We were stabilized before 1,000′, but shortly thereafter, we experienced a plus 15 knot gust. Still stabilized, but a possible flap overspeed entered my mind. On short final, I received an aural "PULL UP". We were not low on the glidepath, nor experiencing an excessive sink rate and over airport property, so I continued and landed without incident in the touchdown zone.

FO/PM's report: It was apparent to me that a stable approach was not possible from this position so I called, "Go Around". The Captain said, "I got it" and nosed the aircraft over towards the runway and continued the approach while attempting to regain the flightpath. At 100′, I called, "Go Around" for a second time. Soon after my second call, we received the audible "PULL UP" warning. The Captain continued the unstable approach and landed. *There was so much going on at that point that, instinctively, my primary focus became stopping the aircraft on the runway because it was clear that a landing was imminent. I felt a third, "Go Around" call would not be heard and it would have been more dangerous to assume control of the aircraft at that critical point than to land and aggressively slow the aircraft.* I cannot verify if the landing was in the touchdown zone or not. We rolled to the end of the runway and had adequate braking to make the last turn off at the end of the runway. I never felt that we didn't have enough runway to stop once on the ground.[7]

The first thing we notice from these reports is the stark difference in risk assessment mindsets between each pilot. The Captain found the approach challenging, but well within what they considered as safe. The FO thought it was unstabilized, but not hazardous enough to assume aircraft control. So, they experienced this event from mismatched mindsets.

The FO/PM was concerned about approach stability and directed a go around twice. From the Captain/PF's report, we perceive a pilot working very hard to land the aircraft in windy and turbulent conditions. The Captain might have recognized the instability, but seemed to judge it as remaining within acceptable stabilized approach parameters. The pilots didn't agree on this point. After two go around calls, the FO/PM made a safety judgment assessment. Resigned to the fact that the Captain could not be swayed from their decision to land, the FO decided that the Captain would succeed as long as they got on the brakes to stop in time. While this event ended favorably, consider if they had failed. The accident investigation would have faulted the Captain for not following stabilized approach criteria, overriding the FO's go around callouts, and ignoring the EGPWS "PULL UP" warning. The FO would have been faulted for not actively intervening and assuming aircraft control to execute a go around.

From the FO's comment, "…it would have been more dangerous to assume control of the aircraft at that critical point…," we witness how fragile intervention triggers are when they rely on personal safety standards. In mishap interviews this sometimes comes out as, "I didn't think [the PF's actions] were going to kill us." This is a rather severe touchstone to use for aviation judgment and is clearly exaggerated, but this kind of statement is fairly common with events like this. PMs attempt to follow their training and procedures, make callouts, are rebuffed, and ultimately allow their PFs to continue. They base their decision on their judgment that the PF can still achieve a favorable outcome. In the heat of the moment, when all normal procedures have failed, we often settle on a consequence-based, safety assessment. As long as we conclude that we will achieve a favorable outcome, we allow the situation to continue. It is when we aren't confident in the outcome that we are moved to actively intervene.

19.6 INTERVENTION STRATEGIES

Intervention strategies range from procedural deviation callouts to actively taking control of the aircraft. How to act and when to do it spans a wide range of situations and circumstances.

19.6.1 Due Diligence

Our society is biased to assume that bad consequences arise from poor decisions. This logic is flawed. Undesirable consequences can emerge even when we make all of the right decisions. Because of this, safety philosophy is moving away from punishment-by-consequence and toward a *just culture* which evaluates the quality of decisions within the context of the conditions.[8] The quality of a decision depends on how much effort the pilot devotes toward acquiring information, understanding its meaning, and selecting a reasonable choice within the time available. This means that decisions are judged against a standard of *due diligence*.

- Did they detect indications that they reasonably should have detected?
- Did their interpretation of those indications make sense?
- Did they make a reasonable effort to handle the situation using available directives?
- Did they select reasonable choices given the time available?

Consider the following example. A chief pilot witnesses one of his crews landing from what appeared to be an unstabilized approach and calls them in for an interview. The Captain admits to flying the approach, knowing that it was unstabilized, and choosing to land anyway. The FO also agrees that the approach was unstabilized. The Captain was clearly at fault and receives administrative sanction. The chief pilot applies the just culture process to rate the quality of six possible FO scenarios.

- *Case 1:* The FO reported not making any callouts and claimed, "The Captain was flying. It's the Captain's aircraft."
  - *Analysis:* This is troubling since the FO didn't make any required callouts. Further, the FO didn't seem to fulfill their PM responsibilities or exercise an appropriate level of care. The standard of reasonable due diligence was not met. The chief pilot imposed administrative sanction and scheduled remedial training.

- *Case 2:* The FO reported not making any callouts and stated, "I was uncomfortable, but I deemed that the Captain would land and stop safely."
  - *Analysis:* This was slightly better. The FO appeared to understand their PM duties, but didn't exercise reasonable due diligence with performing them. As a minimum, they should have made the required callouts. Instead, they appeared to revert to their personal safety standard. The chief pilot asked the FO why they didn't make the callouts. The FO

shrugged and stated, "I don't know." The chief pilot issued a letter of counseling. The FO was assigned remedial training with a company CRM instructor to review required callout procedures and CRM intervention strategies.

- *Case 3:* The FO reported that they told the Captain that the approach was too steep and fast, but the Captain said, "It's okay, I've got it." The FO didn't say or do anything further.
  - *Analysis:* The FO communicated that the approach was unstabilized, but didn't make required callouts. On the surface, it appeared that the FO considered that alerting the Captain of the approach instability satisfied their PM duties. The FO was counseled on their procedural shortcomings and assigned remedial training.

- *Case 4:* The FO reported making one required callout. The Captain replied, "Correcting". They didn't say anything further during the remainder of the approach and landing.
  - *Analysis:* This is the first example where the FO followed at least some of the written directives. They should have followed their first callout with another. If the approach remained unstabilized, they should have directed a go around. The FO admitted that they knew the procedure, but felt that the Captain was doing all that they could to handle the approach and land safely. This reflected that they were holding a PF mindset instead of a PM mindset. This was concerning since this mindset alignment can lead to Recognition Trap errors. Doing all we can to salvage the approach is not justification for continuing a failing game plan. The chief pilot had both pilots role-play the approach to review the required callouts, when to make them, and when to determine that a go around is warranted. While the FO was judged to have exercised some due diligence, their failure to follow through was debriefed.

- *Case 5:* The FO reported making one required callout and that the Captain replied, "Correcting". The approach remained steep and fast, so the FO made a second callout. The Captain replied, "It's a long runway. We'll stop just fine." The FO agreed and didn't make any further callouts.
  - *Analysis:* This reflects a common type of scenario where pilots land from unstabilized approaches. PMs often make the required callouts, but when rebuked or reassured, they typically revert to a personal safety assessment to decide whether to allow the landing, direct a go around, or intervene. The chief pilot reiterated the importance of directing and executing a go around anytime that stabilized approach parameters are not met. The FO was encouraged to be more assertive in the future. While not reaching the desired level of due diligence, the FO was acknowledged for initially following directives. The chief pilot reinforced the need for the PM to direct a go around and for the PF to comply.

- *Case 6:* The FO reported making one required callout and that the Captain replied, "Correcting". The FO made a second callout. The Captain again responded, "Correcting". The approach remained steep and fast so the FO directed, "Go around". The Captain replied, "It's a long runway. We'll stop just fine. I've got it." The FO agreed that the landing could be safely completed and decided that actively intervening was not warranted in this situation, so they remained silent. The FO expressed their discomfort with the Captain when they debriefed the event at the gate.
  - *Analysis:* This represented a reasonable level of due diligence by the FO. All required callouts were made, including directing a go around. When it became clear that the Captain was intent on continuing the unstabilized approach, the FO assessed that the landing would be completed safely. The FO followed up by debriefing the event with the Captain after reaching the gate. The FO was released from the interview while the chief pilot discussed the proper handling of unstabilized approaches with the Captain.

These cases show steadily improving levels of due diligence up to a point short of actually taking control of the aircraft. In the end, none of them succeeds in preventing a landing from an unstabilized approach. Additional callout and intervention techniques are covered later in this chapter.

19.6.2 Barriers to Making Interventions

Most of us practice interventions during CRM training or simulator events. Since these training events are preplanned, briefed, and anticipated, they can feel uniquely different from situations we actually encounter while line flying. Added to this artificiality, most written guidance remains vague on exactly when to intervene. This policy is well intended to give PMs some discretionary space to balance whether to allow a profile to safely continue or to actively intervene. Unfortunately, these factors can erect a psychological barrier which leads to hesitation and rationalization during real-world events. It may take a very strong and scary line-flying event to push through this barrier.

In daily operations, many pilots tend to use their safety assessment to decide when to intervene. Since safety standards are fluid, this leads to a wide range of situational responses. We optimistically assume that we will be able to detect all relevant information, accurately assess hazards, and make decisive interventions – just like we practice in the simulator. The problem is that the line-flying interventions are never preplanned, the details are rarely clear, flightdeck dynamics are complex, time is compressed, and the negative consequences of intervening loom large. These factors contribute to hesitation, optimism bias, and indecision.

Intervention is easier for Captains. When Captains intervene, it feels like an instructor intervening with an inexperienced pilot. It is much harder for FOs because it can feel like we are overriding the Captain's judgment. Recalling the bygone era of sailing ships, it can feel a bit like mutiny.

19.7 INTERVENTION TRIGGER POINTS

To solve these problems, we need to apply some of the trained skills from the simulator toward our line-flying mindset. In simulator events, we practice using trigger points to mark when to intervene. They work because our actions are linked to specific profile points.

19.7.1 Escalation Following Standard and Deviation Callouts

To allow time for dialog, we need to make callouts before reaching our procedural limit point (for example, 500′ for unstabilized approaches). The earlier we start, the better. If we remain silent until we reach the limit point, we tend to be more reluctant to act. Let's examine a range of scenarios and intervention decision options for a Captain-flown unstabilized approach (assuming that our airline uses a 500′ point to assess approach stability).

1. *Desired callouts and early go around:* Passing 1,000′ the approach is grossly unstabilized and unsalvageable. The FO/PM makes the required deviation callout. The Captain/PF acknowledges the callout, concludes that the approach is unsalvageable, and initiates a go around.
   - *Analysis:* This is the desired outcome for an unstabilized visual approach. It reflects industry data that shows that most go arounds are initiated well above the approach limit altitude (500′, in this example). While this reflects awareness and compliance, go arounds started at 1,500′ and 25 knots fast feel quite different from what we typically practice in the simulator. Mishandling these early go arounds can result in flap overspeeds and altitude busts.

2. *Desired callouts and late go around:* Passing 1,000′ the approach is grossly unstabilized. The FO/PM makes the required callouts. The Captain/PF acknowledges, but wants to keep trying to salvage the approach. At 500′, there is little improvement. The FO calls, "Go around". The Captain/PF acknowledges and initiates a go around.
   - Analysis: This is a typical "wait and see" option for an unstabilized approach. Assuming that this complies with the company's policy, this choice is acceptable. Of course, if there was no chance of achieving approach stability, continuing to descend to 500′ is wasteful and unnecessary. The positive features of this example are approach recognition, active discussion, and compliance with company directives.

3. *Marginal approach – desired scenario – FO/PM calls for go around:* Passing 1000' the approach is fairly unstabilized. The FO/PM makes the required callouts. The Captain/PF acknowledges and wants to keep trying to salvage the approach. The approach steadily improves, but by 500′ it is still unstabilized. FO/PM makes the required "Go around" callout. The Captain/PF acknowledges and initiates a go around.

- *Analysis:* This differs from the previous scenario since the approach isn't grossly unstabilized and is steadily improving. If allowed to continue, it might even reach stabilized parameters before landing. In this case, the crew follows procedures and goes around. This is the desired sequence for an unstabilized approach flown to the 500′ stabilized approach limit.

4. *Marginal approach with landing – common scenario:* Passing 1,000′ the approach is fairly unstabilized. The FO/PM makes the required callouts. The Captain/PF acknowledges and wants to keep trying to salvage the approach. The parameters steadily improve, but by 500′ the approach remains unstabilized. The FO/PM makes another callout and the Captain/PF replies, "Correcting". The FO assesses that they will reach fully stabilized parameters before landing, continues to make callouts, but does not direct a go around. The Captain/PF lands.
   - *Analysis:* This approach doesn't follow directed procedures, but is typical of what we witness in line-flying situations. Pilots recognize that there is a significant safety margin between the designated approach stabilization point (500′, in this example) and reaching stabilized parameters before landing. The fact that the approach is steadily improving, that it will reach stability before landing, and that the Captain/PF continues to respond to the callouts all fall within this FO/PM's personal safety judgment. The decision to land is counterbalanced against the added complexity of going around, the added cost of flying another approach, and arriving late at the gate. This reflects one of the more persistent problems in airline aviation – low compliance with go arounds from marginally unstabilized approaches.

5. *FO/PM calls for go around – Captain/PF overrides – FO/PM allows the landing:* Passing 1,000′ the approach is fairly unstabilized. The FO/PM makes the required callouts. The Captain/PF acknowledges with, "Correcting". Parameters steadily improve, but by 500′ the approach is still unstabilized. The FO/PM directs, "Go around." The Captain/PF disagrees and says, "No, I've got it." The FO/PM determines that it is safe to land because conditions are favorable (long, dry runway). The Captain/PF lands.
   - *Analysis:* The FO/PM fulfills their required PM duties and directs a go around. The Captain/PF resists. Rather than escalating the intervention, the FO/PM decides that conditions are favorable for a safe landing and chooses to remain silent. The crew should conduct a detailed debriefing after arriving at the gate.

6. *Captain/PF committed to an unsafe landing – FO/PM repeatedly directs the go around:* Passing 1,000′ the approach is unstabilized. The FO/PM makes the required callouts. The Captain/PF acknowledges,

"Correcting". By 500′, the approach remains unstabilized. The FO/PM calls, "Go around." The Captain/PF disagrees and says, "No, we're good." The FO/PM determines that the landing conditions are unfavorable (short, wet runway) and assertively directs, "GO AROUND". The Captain/PF initiates the go around.

- *Analysis:* The FO performs their required PM duties, but the Captain initially rebuffs the directed go around. The FO assesses that landing is not safe and employs the two-callout technique by repeating the directive and assertively making the second "GO AROUND" callout. This snaps the Captain out of their tunneled-attention and achieves the desired result. The crew should conduct a detailed debriefing after arriving at the gate.

7. *Captain/PF possible tunnel-fixation – FO/PM intervenes and goes around:* Passing 1,000′ the approach is unstabilized. The FO/PM makes the required callouts. The Captain/PF acknowledges with, "Correcting". By 500′, the approach remains unstabilized. FO/PM calls, "Go around." The Captain doesn't respond. The FO/PM determines that the landing conditions are unfavorable (short, wet runway). They assertively direct, "GO AROUND" a second time. The Captain doesn't reply and continues the approach. The FO transmits on Tower frequency, "Trans American 209, going around". On the flightdeck intercom, they announce, "I have the aircraft" and takes over control of the aircraft to initiate the go around.
   - *Analysis:* The FO/PM performs the required PM duties and makes deviation callouts. When they call for a go around, the Captain/PF fails to respond. The FO/PM assertively directs the go around a second time. Again, the Captain/PF doesn't reply. The FO/PM assesses that landing is unsafe and transmits their intention to go around to ATC. Tower cancels their landing clearance. The FO/PM takes control of the aircraft and initiates a go around. Whether the Captain is intentionally non-compliant, is unaware of the hazard due to tunneled-attention, or is incapacitated, the FO/PM ensures the safest course of action. The crew should conduct a detailed debriefing after arriving at the gate.

8. *Captain/PF non-compliant – FO/PM warns of intention to intervene:* Passing 1,000′ the approach is unstabilized. The FO/PM makes the required callouts. The Captain/PF acknowledges with, "Correcting". By 500′, the approach remains unstabilized. The FO/PM calls, "Go around." The Captain/PF disagrees and says, "No, I've got it." The FO/PM determines that the landing conditions are unfavorable (short, wet runway). They assertively state, "If you don't go around, I'm going to have to take control of the aircraft". The Captain complies and initiates a go around.
   - *Analysis:* The FO/PM assertively declares their opposition with landing and presents the Captain/PF with a clear choice. If the Captain thinks they can impose their will on an unassertive FO, this confirms

otherwise. An additional option is to transmit, "Trans American 209, going around" to ATC.

9. *CAT III approach – Captain/PF incapacitation – FO/PM intervenes and goes around:* The Captain/PF is flying a CAT III approach to minimums using the HGS. The FO/PM calls, "Approaching minimums". The Captain does not respond. The FO calls, "Minimums". Again, the Captain does not respond. The FO announces, "I have the aircraft, going around" and executes a missed approach.
   - *Analysis:* This mirrors the trained Captain incapacitation event practiced in the simulator. If the runway is in site, the FO can either land or execute a missed approach.

10. *Captain landing – strong microburst crosswind gust – FO/PM takes over for a go around:* Captain/PF is flying a visual approach in gusty winds with convective weather in the area. ATIS warns of LLWS. In the flare, the aircraft is hit by a strong crosswind gust. It abruptly pushes the aircraft well off the runway centerline. The Captain/PF adds power and makes a significant right turn while attempting to correct back to runway centerline. The aircraft is floating past the touchdown zone. Concerned about dragging a wingtip, landing long, and departing pavement, the FO/PM directs, "Go around." The Captain announces, "I've got it." As the aircraft continues drifting downwind, the Captain/PF increases the bank angle. The FO/PM announces, "I have the aircraft", takes aircraft control, and transmits, "Trans American 209, going around."
   - *Analysis:* This event reflects a safety intervention during quickly developing hazardous situation. The risk factors are building quickly. Lacking time to make a second callout or to discuss the problem, the FO/PM takes control, adds thrust, and executes a go around. The FO/PM bases their intervention trigger on event quickening, rising risk factors, and gut feeling. Warned of LLWS, the FO/PM has predetermined that they won't tolerate much deviation from an on-centerline, in touchdown zone landing. The float, crosswind drift, and excessive bank angle all exceed their intervention trigger criteria and raise warnings flags. Any one of them would have triggered a go around callout. The presence of all three makes this go around intervention imperative.

These ten scenarios vary greatly in risk, time available, and compliance. The successful outcomes benefit from the FO establishing strong trigger points for their intervention callouts and actions. Pilots who have predetermined their intervention triggers tend to act decisively. They resist succumbing to indecision, optimism bias, deference to authority, and inaction. Consider the following example from the old days of airline aviation when the authority of the Captain was rarely questioned.

BOX 19.8 FO-DIRECTED GO AROUND AND CAPTAIN CONFRONTATION

The Captain mismanaged a visual approach resulting in a seriously unstabilized approach. Well above approach speed and diving toward the runway, the FO directed a go around. The Captain refused. The FO keyed the mike, and transmitted, "[Callsign], going around." ATC canceled the landing clearance, thus forcing the Captain to go around. The Captain reported the FO to their chief pilot and an interview was conducted. During the meeting, the Captain asserted that the approach was stabilized, that the FO was incompetent, and that they should be fired. The chief pilot then played the ATC Tower audio recording of their approach and go around. The FO's go around callout was difficult to hear because the ATC's aircraft crash warning alarm (the one triggered by an aircraft's excessive sink rate at low altitude) was sounding loudly throughout much of the final approach. The FO was dismissed and the Captain received some uncomfortable one-on-one counseling from the chief pilot.

Most likely, this was not the first unstabilized approach that this Captain had ever flown. It probably was the first time that a probationary FO had forced them to go around. Since this incident, the airline industry flightdeck environment has changed for the better. FO/PMs are more assertive and Captain/PFs are more compliant. While this scenario features an authoritarian Captain, our trigger points must also accommodate the opposite extreme. Imagine, instead, that our Captain is a really nice person, maybe even a good friend. Do we have predetermined intervention triggers that will work in this environment? Consider a third scenario. Imagine that our Captain happens to be our chief pilot. Taking a day out of the office to fly what might be their first pairing in a month, can we appreciate how non-current and distracted they might be? Are we ready to direct our boss to go around? Logically, we assert that personalities and working relationships shouldn't matter, but in the heat of the moment, they might. If we don't set predetermined trigger points, then interpersonal dynamics can adversely influence our actions.

Intervention triggers need to be calibrated by flight parameters, risk, and time available. Grossly unstabilized approaches with extreme parameters are easy because they exceed safe limits. Marginally stabilized approaches are more difficult because they don't. Consider an example of an approach to a contaminated runway with braking action reported as POOR (RCC 1). We review the landing performance data and see that our stopping distance leaves little to spare. We agree that the approach needs to land on-speed and within the touchdown zone followed by immediate thrust reversers and aggressive wheel braking. What if our PF is 15 knots fast and floating toward the far end of the touchdown zone? In good conditions, we might accept this marginal landing and remain silent. On this slippery runway, however, the unfavorable landing conditions exceed our risk tolerance and warrant a go around.

19.7.2 Intervention Trigger Levels

Since our triggers vary with conditions, how can we set and rehearse them? Start with ideal conditions. We ask ourselves, if conditions are favorable, how excessive must parameters be before we would intervene? This anchors one extreme of our parameter-based intervention strategy. Next, rehearse the chain of events that would typically lead to an intervention. We would start by making procedural deviation callouts. These should be automatic and factual. After procedural callouts, we have several options. Consider the following scenarios.

- *PF opens the discussion:* Given sufficient time for discussion, crews typically open a conversation about the problem before initiating a change of game plan (like a go around). Perhaps the PF makes a comment like, "I really screwed up this approach." This opens the discussion about whether to continue the approach. As PMs, this is our first chance to share our thoughts, opinions, and limits. Consider these four PM responses.
  1. "It doesn't look good. Do you want me to inform Tower that we are going around?"
  2. "What's your plan?"
  3. [Remain silent and say nothing].
  4. "I think you can make it."
     - Option 1 reflects agreement with the PF's opinion that the approach is unstabilized. It cues the next procedural step of the go around procedure. It also communicates that we are not in favor of continuing to land. Option 2 opens a discussion of plans and intentions. What does the PF want to do? Do they want to stick with the approach for a bit longer to see if they can salvage it? Do they want to start a go around? Are they unsure how they wish to proceed? Through discussion, the pilots form a shared mental model. Option 3, while unfortunately common, reveals the PM's indecision. It does nothing to help the PF and leaves them with the impression that they will have to make the call of whether to continue or go around. Silence also implies our tacit support with continuing the approach. Option 4 implies that the PM supports the decision to continue. This is fairly common in line flying, especially with Captain/PMs. Many FOs have continued approaches that they would have otherwise gone around from only because their Captains encouraged them to continue.

- *PM opens the discussion:* The alternative scenario is when the PM says something to open the discussion. Here, the PM voices an opinion or shares their concerns about the approach. Consider these statements.
  1. "We are too tight to make this approach work."
  2. "We need more drag right now."
  3. [Remain silent and say nothing].
  4. "What's your plan?"
  5. "I think you can make it."
     - Options 1 and 2 are common with right-turning visual approaches when the Captain/PF's vision of the runway is blocked while looking across the flightdeck. It is common for the FO/PM to make qualitative

statements of approach geometry and energy. Option 1 shares our opinion that the current profile is too tight to achieve stabilized approach parameters. The PF can widen the turn, increase drag, or ask for an S-turn to salvage the approach. Option 2 specifically shares that we feel that the only way to salvage the approach is by immediately increasing drag – like lowering gear and flaps. Silent option 3 leaves the impression that the PF will make the call of whether to continue or go around. Option 4 shares that we are concerned and that we need them to communicate their intentions. This is an important step toward achieving a shared mental model. Option 5 acknowledges that we feel that the approach is marginal, but salvageable.

- *Intervention warning:* The next stage of intervention intensity includes some form of statement that alerts the PF that we expect them to switch to a backup contingency plan. This is more decisive and communicates our opinion that the current course of action is unwise. Consider these statements.
  1. "When do you want to start our go around?"
  2. "I think we need to go around."
  3. "We need to go around."
  4. "Go around."
     - Option 1 shows that we expect to go around, but leaves the timing up to the PF. If they want to continue trying, this opens the discussion to form a shared mental model of the game plan. With option 2, the PM takes some ownership with the go around decision. It still stops short by only expressing an opinion. Option 3 demonstrates crew ownership of the go around procedure, but implies the need to go around fairly soon. Option 4 directs the immediate go around.

- *PM-directed actions and taking aircraft control:* Any PM-directed change-of-aircraft-control procedures should follow the required verbiage from our manuals. We train with precise words and actions to form a connection between directive callouts and expected behaviors. The PF should immediately comply. If not, and time allows, repeat it again. If the PF still doesn't comply, select the intervention option that fits the conditions. Use the exact verbiage from our manuals and don't soften or modify these statements. Otherwise, the PF might assume that we are asking for permission, showing indecision, or making suggestions.

- *Different trigger points for interventions:* Each level of intervention requires its own trigger point. We have a point where we open a discussion, a point where we state our intentions, a point when we direct actions, and a point when we take control of the aircraft. Each of these is vulnerable to hesitation or indecision. That is why it is so important to mentally rehearse them. At the critical moment, we won't consider intervening differently with one Captain that we like or another Captain that we don't, with this new Captain or that experienced Captain, or with our friend or our chief pilot. The trigger points and actions are linked to parameters, risk, and time available. When we reach each rehearsed trigger point, we just do it.

19.7.3 Pilot Incapacitation

Psychological incapacitation events, while extremely rare, are always possible. We need to be ready to assume aircraft control.

BOX 19.9 CAPTAIN INCAPACITATION – JETBLUE AIRLINES 191

On March 27, 2012, the Captain suffered a psychotic mental episode and began making irrational statements about terrorists and the aircraft crashing. While many of the specific details remain confidential, we know that the FO convinced the Captain to exit the flightdeck and go back into the passenger cabin. When the Captain departed, the FO secured the flightdeck door to prevent reentry. The Captain tried to regain flightdeck entry and was subdued by passengers. The FO diverted the aircraft. The Captain was subsequently treated for mental illness.[9]

We commend this FO for skillfully convincing the Captain to leave the flightdeck and securing it to prevent their return. They prevented this incident from escalating. Other incapacitation events have occurred across the industry. Most involve physiological events like heart attacks, seizures, strokes, and incapacitating illness.

Physiological incidents range from subtle to severe. Our first indication may be when the other pilot does something unexpected. We need to determine if their action is a misguided choice, an error, or something else. Consider a case where we are flying with a pilot who has just returned to flying after a prolonged medical grounding for cancer treatment. They seem a bit rusty, which is understandable. At one point during the flight, however, they seem to freeze up, stop talking, and exhibit confusion over a fairly routine decision. What should we do? Our immediate concern is flight safety. We talk them through the immediate problem and get the flight back on track. When the pace slows down, we express our concerns by stating the facts exactly as we recall them without judgments or emotional labels. Maybe they become defensive or don't recall the details as we do. While we are not cognitive psychologists, we are aware of an effect called "chemo-brain" where people exhibit lingering cognitive dysfunction following chemotherapy. We elevate our awareness and closely monitor their performance. We notice two additional incidents during the remainder of the pairing. Afterward, we contact our union's aeromedical committee representatives and share our concerns. They take over and begin the process for getting the Captain the medical care that they need.

19.7.4 Non-Flight Related Interventions

Consider an example of a situation that arises before the flight. While congregating in the hotel lobby, one of our flight attendants walks over and asks, "Are you flying with that pilot?" We reply yes and ask why. They relate that they saw him in the bar the previous night still drinking as they left for required crew rest. They are concerned that they may have exceeded FAR alcohol limits. What should we do? Since we have been directly informed, reasonable due diligence requires that we act. We walk over to the pilot and open a conversation. We notice that they look tired, that their eyes are blood-shot, and that they seem irritable. We think we detect the odor of alcohol. We inform them that they were observed drinking in the bar at the FAR alcohol time limit and that other crewmembers have expressed concern. To remove all questions and controversy, we suggest that they should call scheduling and get replaced. They claim that they are legal to fly, but didn't sleep well due to noise from the hotel guests in the next room. After we step away, they call scheduling and have themselves replaced.

Whether the pilot was legal or not was beyond our knowledge. We disclosed to them that a crewmember expressed a concern about their fitness to fly. Our confrontation gave them the opportunity to remove that doubt by calling for a replacement pilot. If they resisted the safe option, we might reinforce the facts to them. "I'm not accusing you of anything, but this looks like a good time to choose a conservative option. We are about to get into a crowded van with other crewmembers. If someone thinks they smell alcohol, this could escalate badly. If you call scheduling now, this resolves quietly." By presenting the facts as we see them, we can present a solution that leads to a favorable outcome. Notice that in each of these cases, we limit our comments to factual observations.

If we take this scenario a step further, consider what might happen if the pilot refuses. On the hotel van, it becomes clear that they shouldn't fly the trip. Some airlines use entering the aircraft as the trigger point to make the call. Before entering the aircraft, we clearly restate our opinion that they should remove themselves from the flight. If they refuse and enter the aircraft, we have to act. Refuse to follow them into the aircraft and immediately contact the appropriate company supervisors. We need to avoid any option where we allow that pilot to fly.

NOTES

1 Edited for brevity and clarity. Italics added. NASA ASRS report #1586587.
2 Edited for brevity. Italics added. NASA ASRS report #1274670.
3 Edited for brevity. Italics added. NASA ASRS report #1705125.
4 Edited for clarity. Italics added. NASA ASRS report #1454754.
5 Edited for clarity and brevity. Italics added. NASA ASRS report #1698341.
6 Edited for brevity and clarity. Italics added. NASA ASRS report #1245601.
7 Edited for brevity and clarity. Italics added. NASA ASRS report #1449664.
8 A comprehensive resource on just culture is Dekker (2007).
9 Numerous accounts of the event and follow-up stories are available through an internet search of JetBlue Airlines 191.

BIBLIOGRAPHY

Active Pilot Monitoring Working Group. (2014). A Practical Guide for Improving Flightpath Monitoring. Alexandria, VA: Flight Safety Foundation.

ASRS – Aviation Safety and Reporting System – National Aeronautics and Space Administration (NASA) Online Database. (n.d.). Retrieved from Aviation Safety and Reporting System: https://asrs.arc.nasa.gov/search/database.html.

Dekker, S. (2007). *Just Culture: Balancing Safety and Accountability*. Burlington, NJ: Ashgate Publishing Company.

20 First Officer Roles and Responsibilities

20.1 THE FIRST OFFICER PERSPECTIVE

We'll start by framing the First Officer (FO) role through its cultural legacy, its official roles and responsibilities, and how it has evolved.

20.1.1 The Cultural Legacy of the First Officer

The history of the First Officer role dates back to the earliest naval sailing ships. At the top of the ship's hierarchy was the Captain. As "master and commander", their authority was supreme and unquestioned. Next in command was the first lieutenant – the historical origin of the FO. These second-in-command officers served many roles. Viewed from the top-down, they implemented the Captain's orders and directed the actions of the crew. Viewed from the bottom-up, they relayed information from junior and non-commissioned officers back up to the Captain. Viewed side-by-side, they served as the Captain's confidant to form strategy, weigh concerns, and evaluate conditions. On a personal level, they apprenticed under the Captain who mentored them to prepare for the day when they would be promoted as Captain of their own ship. So, while Captains assumed a singular leadership perspective, first lieutenants had to modify their perspective across a range of leadership and followship roles.

Captains could lead in any way that they wished, but first lieutenants had to conform to the Captain's style and personality. They placed high priorities on maintaining the Captain's comfort zone, accommodating their style, and anticipating their orders. Unburdened from the mundane duties of running the ship, the Captain would pace slowly along the upwind rail while studying the weather, reading the winds, and assessing the sails. The first lieutenant, predicting the Captain's imminent "suggestion", would pre-position the crew out-of-view. The Captain might pause his methodical pacing and casually mention to the first lieutenant, "Mr. Johnson, perhaps we can venture the staysails." The first lieutenant would give the order, which was relayed by the junior officers and mates, sending dozens of sailors racing up the rigging to unfurl the staysails.

In practice, the Captain's duties would involve the fewest operational details and the most strategic planning. The first lieutenant's duties would involve the most operational details, but they also needed to understand the strategy to accurately anticipate the Captain's orders. Additionally, if the Captain became incapacitated by injury or illness, the first lieutenant would immediately assume all of the Captain's roles and responsibilities.[1]

Many of these cultural legacies have carried forward from sailing ships to our modern airships. FOs still accommodate the personalities of their Captains, adapt to their styles, manage many of the details of operational tasks, assist with flight

DOI: 10.1201/9781003344575-22

planning, understand the reasoning behind flight decisions, and remain ever-ready to take over in case their Captains become incapacitated.

One important distinction is crew continuity. The first lieutenants on naval sailing ships would often serve with their Captain for years. Modern FOs, however, can rotate through several left-seaters in a single day. Standardized procedures smooth these transitions, but FOs are still expected to quickly adapt to each Captain's style.

20.1.2 FO's Official Roles and Responsibilities

FAR 121.537(d) states, "Each pilot in command of an aircraft is, during flight time, in command of the aircraft and crew and is responsible for the safety of the passengers, crewmembers, cargo, and aircraft. The pilot in command has full control and authority in the operation of the aircraft, without limitation, over other crewmembers and their duties during flight time, whether or not he holds valid certificates authorizing him to perform the duties of those crewmembers." While this clearly establishes the role of the Captain, the FO's official responsibilities remain rather vague. FARs do not list or assign any specific responsibilities to second-in-command pilots. Some airlines try to fill this regulatory void with some general guidance statements like:

- Exercise Second-in-Command duties.
- Assume, secondarily, all responsibilities of the Captain.
- Should the Captain become incapacitated during flight, assume command of the aircraft.
- Advise the Captain of deviations from established policies, procedures, and/or regulations.

These duties reflect the expectation that FOs will know all policies, procedures, and regulations, advise the Captain of deviations, and intervene when the situation requires. To do this, we expect that FOs should possess many of the same skills that we expect from Captains. Some of these skills are trained, some are assumed, and others we expect FOs to acquire through operational experience.

Combining cultural norms with company directives, we find FOs balanced between roles and responsibilities that sometimes conflict with each other.

- Culturally, FOs fill a subordinate/assisting role under the Captain.
- Practically, they are responsible for completing their own specific tasks.
- Administratively, they monitor the quality of the Captain's work.
- Officially, they stand ready to assume all of the Captain's roles and responsibilities.
- Conditionally, they are expected to intervene and take aircraft control if the Captain pursues a hazardous course of action.

Written guidance is often ambiguous about *when* and *how* to exercise these many roles. FOs are expected to select the proper action across a wide range of Captain personalities, variable conditions, and operational challenges. Does this current situation require a callout, an expression of concern, a discussion, a recommendation, or an intervention? FOs need to know when to assume each role and how to apply just enough pressure to ensure procedural compliance. This becomes even more difficult as they balance their actions between cultural norms and official policy. Like the first lieutenants of old, FOs need to be skilled leaders, followers, confidants, and mentees.

We have a sense of the actions that may be required, but most aren't spelled out in our manuals. This is intentional. We want FOs to respond appropriately to the entire range of operational scenarios and challenges. Detailing a specific action might create an unintended impression that actions are limited to certain listed conditions. We don't want that. Instead, we want FOs to act in whatever way they deem necessary to effectively manage whatever situation may arise. These actions can range from subtle hints to extreme interventions like taking control of the aircraft.

20.1.3 How the FO Role Has Evolved in CRM

Early versions of CRM training focused on resolving flightdeck conflicts, achieving compromise, and improving team effectiveness. They centered on Captains because the underlying training objective was to transform the prevailing authoritarian leadership model into a team leadership model. This transition included FO empowerment. Current CRM programs build on this objective by improving FO/PM assertiveness. Case studies and simulations recreate line-flying scenarios that challenge pilots to accurately recognize problems, resolve conflicts, and employ a team approach to maintaining safe operations.

One recent innovation in CRM training adds a dedicated training module to help newer Captains navigate the transition of their mindset from senior FO to junior Captain. Typically conducted 6–12 months after upgrade, this training deepens our understanding and effectiveness as team leaders and company representatives. While the main emphasis centers on managing the flightdeck team, it includes interactions with cabin crew, ground crew, dispatchers, mechanics, and supervisors.

We are encouraged to act as on-scene representatives of company leadership. Applying company policy and philosophy, we manage emerging problems to resolve them in the same way that senior leadership would. If the company leadership would find a way to continue operations under challenging conditions, then we are expected to manage the situation to keep our flights moving. If the company leadership would shut down the operation due to deteriorating conditions, then we are expected to stop our flights. We are expected to share our observations and recommendations with leadership through the company's operational control center. With every Captain serving as the company's eyes and ears in the field, leadership can project their active management across the operation. This provides the necessary resilience to safely manage thousands of individual flights all over the world. Expanding our previous list, FOs assume the following additional roles and responsibilities:

- Understand company policy guidance and operational philosophy.
- Understand how senior leadership expects us to balance priorities.
- Know when and how to continue operations.
- Know when and how to stop operations.
- Know when and how to coordinate between other work groups.

Like Captains, FO roles and responsibilities have become more nuanced, far-reaching, and challenging. While Captains retain ultimate responsibility for the control and authority of the flight, system complexity demands that FOs act to ensure policy compliance and operational quality. If the Captain's unwise decisions lead to unfavorable outcomes, FOs can be held responsible for what they did or what they should have done to stop the undesirable trajectory of the mishap. While mishap investigations strive to view events fairly, our society continues to apply hindsight bias. "The Captain shouldn't have continued with that failing game plan." "The FO should have alerted the Captain." Even in cases when the FO does speak up, the quality and effectiveness of their efforts may be questioned. "The FO should have made the callouts as they are written in the manuals." "The FO should have repeated the deviation callout a second time." "The FO should have been more assertive."

20.2 FINDING OUR VOICE

We began this chapter with an examination of the challenges that FOs face with balancing their many roles. How do we, as pilots who are organizationally subordinate, direct or override the actions of our superiors? We need to find our voice.

20.2.1 Silence Means Consent

Consider a situation where our Captain is making an unwise decision. We can either oppose it or support it. Opposing it can feel confrontational. Supporting it can feel like we are condoning their ill-considered choice. Torn between these options, some pilots choose to remain silent. While in-the-moment, this may feel like an acceptable compromise. We may even rationalize, "Maybe the Captain sees something I don't", or "I'll just monitor this closely to see how the situation develops."

In crew aviation, silence effectively means consent. Since our process relies on callouts to identify adverse deviations, saying nothing implies that everything is acceptable. Taking this logic a step further, if silence means consent, then speaking up feels like confrontation. What can we do with this dilemma? One technique is to make factual and impersonal callouts. This decouples what is happening to the aircraft from value statements regarding the PF's decisions or actions.

- *Make factual callouts:* Factual callouts keep us focused on the adverse indication. "Sink rate 1500" is a statement of fact. The PF's unskillful energy management probably caused it, but pointing this out doesn't help resolve the problem. Simply state the adverse parameter and allow the PF to resolve it. As the PF replies, "Correcting", they become the fixer of the problem instead of the cause of the problem.
- *State what we see:* Some adverse trends are more qualitative and don't lend themselves to factual callouts. Consider the NASA ASRS reports from the previous chapter where the Captain mistook Biggs AAF for El Paso International. A useful callout might be, "We appear to be lining up for Runway 4 at Biggs. I see El Paso Runway 4 to our right 1 o'clock." This callout highlights the pertinent information without attributing blame.

The callout gives the PF an opportunity to correct their error or communicate their intentions.

- *Make it about us:* We each have different risk tolerances. If the PF is doing something that we consider to be too risky, we should own our risk assessment. "I am uncomfortable with continuing this approach." This gives the PF a couple of options. If they are also uncomfortable and already considering a go around, our statement might be the nudge they need to initiate a go around. If they have a plan that they haven't communicated, it gives them the chance to share it. "I realize it is tight. I'm going to square my turn to final to deplete excess energy. If that doesn't work out, we'll go around."

  If they happen to be more risk tolerant than us, they might say something like, "We'll be okay. There is plenty of runway." This informs us that they acknowledge the approach instability, but are content with exceeding company directives to land. At this point, we know their intentions and can clearly state ours. "I understand, but I still have to make required callouts and direct a go around if we aren't stabilized by 500 feet." This clearly states our commitment to follow procedures and fulfill our PM duties. The PF may not like it, but they can't fault our compliance or integrity.

20.2.2 Separate Directive Callouts from Maintaining Team Rapport

We can choose our response depending on the needs of each event. We can be friendly pilots and respond in a congenial manner. We can be helpful pilots who assist with deviation callouts and suggest game plan modifications. We can be a team builders when it fits or directive when it becomes necessary. We can unemotionally direct a go around. After we complete that go around, we can immediately return to being friendly and helpful. In short, we can compartmentalize events and feelings. Treating the go around as an impersonal aircraft maneuver, we don't need to examine how we got into that position in the first place. It also helps if we downplay the significance of the event. "I'm as surprised as you that the approach didn't work out. We'll get it next time around." Even if it isn't quite true, it feels better to promote a supportive environment.

20.2.3 Use Humor

Another option is to use humor. It assures the other pilot that we don't see the situation as disappointing as they probably see it. We all know when we mishandle the situation and appreciate it when the other pilot lightens the mood. My personal favorite was related by an FO who had jammed their visual approach. The Captain casually mused, "You're about to show me something I've never seen before." The FO laughed, took the hint, and went around.

20.2.4 Be Sincere

A pilot once suggested to me, "You can say anything you want as long as you say it like Gomer Pyle." There is wisdom here. In the TV sitcom from the 1960s, actor

Jim Nabors portrayed a Marine Corp recruit named Gomer Pyle. He was a constant irritant to the always-serious Sergeant Carter, played by actor Frank Sutton. The comedy worked because, no matter how badly Gomer screwed up or how enraged Sergeant Carter became, Gomer still maintained his sunny disposition. He never saw fault in anyone. Even when he highlighted something that Sergeant Carter had messed up, he did it in a sincere way that owned part of the problem. He always made a good-faith effort to learn and improve himself. His sincerity and good nature was infectious. All problems became team problems that everyone could resolve together.

20.3 STAYING OUT OF SYNCH

There is a natural tendency, especially when we are under high workload and stress, for both pilots to align their perspectives. While this does concentrate everyone's attention on solving the problem, it weakens our team effectiveness. When analyzing mishaps, we often question, "How could both pilots have missed something so glaringly wrong?"

Perhaps we have observed this effect from the jumpseat or while observing a simulator training event. The PF struggled to manage a difficult problem, so the PM pitched in to help. Since this is a desirable and expected crew response, it initially looked right. Before long, however, the PM became so deeply involved with helping the PF that they lost their separate perspective. Deeply focused on managing the event, neither pilot noticed that an additional problem remained undetected and unresolved. Consider the example of an engine failure in the simulator. Both pilots can become so focused on managing the engine failure, attempting a restart, and preparing for an engine-out landing that they both fail to notice that the fuel load has become unbalanced.

20.3.1 Becoming Enmeshed with the Problem

As we encounter an aviation challenge, we strive to understand and solve it. We focus our attention on it. As PMs, we are pilots, too. Our natural and familiar inclination is to adopt the same problem-fixing perspective as the PF. Instead of remaining detached, we empathize with their struggle and try to help. To be helpful, we need to understand the problem. As we both work to understand the problem, we align our mindsets. Pushed to the extreme, we essentially become non-flying PFs – mentally flying the aircraft as our PFs fly the aircraft. When this happens, our error detection and mitigation roles fade away. We become enmeshed with the problem and lose some of our ability to accurately *observe* it or *recognize* it. As soon as we synchronize our mindsets, we begin looking for the same things, seeing those same things, not looking for other things, missing those other things, both making the same errors, and both missing that we have made those same errors. Our separate error detection and mitigation roles are lost. As our attention is drawn into the problem, we can become blinded to worsening conditions and counterfactuals. In some ways, as we become *surrounded* by the problem, we lose our ability to accurately *observe* it or *recognize* it.

20.3.2 Maintaining a Detached Perspective

To function as effective PMs, we do need to understand the PF's mindset. So, there will always be some alignment. We just need to remain aware enough and detached enough to continue fulfilling our PM role. Qualitatively assessing what the Captain is doing requires a detached perspective. This detachment allows us to judge the accuracy and appropriateness of their actions and choices. We ensure that the path is managed correctly and that the projected path remains appropriate for the conditions. We can only comprehend this bigger picture by mentally stepping back from the actual flying.

We see this in the different ways that each pilot builds their SA. PFs focus on immediate flying tasks which keep them centered on present moment SA. As they become task-saturated, they lack the time to look ahead and build future SA. By assessing the wider perspective, PMs focus on building future SA. This allows us to intentionally scan for counterfactuals and signs that the game plan may be failing.

20.3.3 Assuming the Flight Instructor Perspective

As PMs, we are not physically flying the aircraft, but we are still responsible for ensuring the quality and outcome of the PF's actions. Does this role sound familiar? It is essentially what we did as flight instructors. When flight instructing, we devoted part of our attention toward our student's flying and part of our attention toward monitoring the big picture. We constantly asked ourselves, "How well is this pilot performing?", and "How well is this flight going?" Both of these are qualitative assessments. We accepted that students become tunnel-focused on their immediate flight tasks as they become task-saturated. In many ways, proficient airline pilots with decades of experience can become similarly tunnel-focused when they become task-overloaded. They can exhibit the same vulnerabilities and errors as novice pilots. By applying our flight instructor skills, we detect and intervene to stop failing game plans.

20.3.4 Looking at the PF

When we become very busy, we tend to focus either outside at an important objective or inside toward a required task. This can interfere with assessing how well the PF is doing. If the situation is becoming stressful, take a look at the PF. If they are leaning forward with their eyes locked unblinking on something straight ahead, we can assume that they are not scanning anywhere else. We can also assume that they are possibly missing everything that is happening outside of their laser-beam focus. Post-mishap interviews reveal that PFs often had little idea of how extreme certain parameters had become. They recalled staring at the touchdown zone, but little else. If we see that the PF has become this intently focused, we can assume that they will need a fairly strong stimulus to snap out of it. We'll need to raise our voice, call them by name, touch them on the shoulder, or use directive callouts. All of these are strong inducements that will force them to break their tunnel-focus and expand their awareness.

20.3.5 Beware of a PF's Automatic Responses

Consider an approach where we make a deviation callout and the PF answers, "Correcting". We can't assume that they have received the message and will correct the problem. They may understand the callout and intend to make an appropriate correction. They may also be so focused that they may not realize how severe their parameters have become. Thirdly, they may not have heard us at all and are just uttering an automatic response. Without feedback, any one of these is possible.

Consider a windshear recovery exercise in the simulator. After the PF initiates the windshear recovery maneuver, we don't stop our callouts. We assume that they are task-saturated flying the aircraft and help them by providing trend callouts.

- "Airspeed 120 knots and holding."
- "Altitude is steady."
- "Airspeed rising."
- "Climbing 1,000 feet per minute."
- "Airspeed 130 and rising."

We can use this same technique to provide trend and qualitative information during a failing profile. Consider the following callouts for an unstabilized approach.

- "Airspeed, target plus 20."
- "Sink 1,500."
- "Airspeed target plus 25 and rising."
- "Approaching flap placard limit."
- "Approaching stabilized approach limit."
- "Prepare for a go around."

In addition to qualitative and trend information, these callouts let the PF know that their corrections are not working. They inform them of upcoming limits and prime the required recovery maneuver.

20.4 TECHNIQUES FOR SCRIPTED CALLOUTS, DEVIATION CALLOUTS, RISKY DECISIONS, AND INTERVENTIONS

While many of these topics were addressed in the previous chapter, some additional nuances and techniques deserve additional consideration to address the FO/PM's perspective.

20.4.1 Making Scripted Callouts

These callouts are the easiest to make because they are procedural, factual, and well-trained. Ideally, the Captain has established an open communications environment and supports us making callouts. A good Captain technique is to brief our FOs (and jumpseaters) that we value them making timely callouts. If we later exceed a parameter and our FO fails to make the required callout, we can make it ourselves, "Airspeed, plus 15. Correcting." When we reach a debriefing opportunity, we reiterate that we expect them to make required callouts.

How we choose to make callouts is often subject to our judgment. For example, if we are in moderate turbulence bouncing down final, the airspeed may momentarily spike beyond a procedural callout limit. Calling "Airspeed" every time it briefly bounces fast or slow is unhelpful. It makes better sense to ignore momentary deviations and reserve callouts for sustained out-of-tolerance parameters or adverse trends.

It may help to recite the callouts in a somewhat mechanical tone, like our automated system callouts. This further separates our callouts from judgmental undertones. Callouts become automatic statements of fact.

20.4.2 Making Unscripted Deviation Callouts

Unscripted deviation callouts are more difficult because they rely on our judgment, they aren't cleared defined in operations manuals, and they require us to select our wording. Also, they tend to highlight the PF's errors and unfavorable trends. A judgmental undertone may be unavoidable. For example, "We are getting high on our descent profile" is a factual statement, but it also implies that the Captain has done something or overlooked something to cause that deviation. Either way, it suggests that they have somehow fallen short of demonstrating good airmanship. Still, most Captains welcome deviation callouts because they allow them to resolve problems before they become unmanageable. The words we use also affect the subtext of our callouts. Compare these five examples.

1. "We are getting high on our descent profile."
   - *Analysis:* This example states the problem and presents it as a team challenge. We are in this together and we are going to work together to solve it.
2. "You are getting high on your descent profile."
   - *Analysis:* This option states the same information, but it attributes the problem solely on the PF. It comes across as somewhat judgmental.
3. "What are you doing? You're getting high on profile."
   - *Analysis:* This option effectively accuses the PF of causing the problem.
4. "Do you want me to lower the landing gear?"
   - *Analysis:* This option bypasses a factual deviation callout because it is clearly evident. It skips ahead to a suggested solution.
5. "We need gear and flaps right now or we are going to have to go around."
   - *Analysis:* This option links a remedy with an undesirable consequence. It offers a last ditch solution to prevent switching to a contingency backup option.

Notice how the accusatory tone varies across the five callouts. Again, these callouts are not scripted in our manuals, so it is up to us to select our words and the tone of our statement.

20.4.3 Making Callouts about Risky Decisions

This category carries a significantly judgmental component. These cases typically involve events where a game plan is failing, but we still have sufficient time to discuss our concerns and work toward achieving a shared mental model. Our statements need to communicate three pieces of information.

1. State the conditions (these are the significant parameters that I see).
2. Clearly present our reasoning (this is what concerns me about them).
3. Offer alternatives (this is what I think would be a better course of action).

As time allows, this process opens a back-and-forth conversation. Consider an example where the PF has selected a course around a convective buildup that we judge to be too risky. We can start by asking the PF for their reasoning behind their choice. "I'm curious how you selected this particular course around that buildup." The PF then replies with their assessment of conditions and reasoning. We can either agree, disagree, or present additional considerations that they might have overlooked. Our conversation strives to achieve a mutually acceptable game plan.

Another option is to treat it as a learning opportunity. "Can you explain to me how you selected his particular heading around that buildup?" This opens a teaching opportunity for the Captain to share their experience and wisdom. If we agree, then we proceed in unison. If we disagree, then we can share our concerns. "Given your process, I would choose a course about 10 degrees further to the right. On our current heading, I think we risk encountering turbulence and even some hail from that buildup." Notice that the statement both validates the Captain's decision making, but also communicates our discomfort and a solution that would ease it. Our suggestion offers an option that we can both support. If the Captain discounts our concerns, we can either rephrase our concerns or suggest an acceptable mitigation. "I understand your reasoning, but I'd like to call back and make sure the flight attendants are safely seated in case we encounter turbulence." Finally, if we remain concerned, we should apply our personal judgment of safety and decide whether to consider a stronger intervention.

20.4.4 Making Interventions

Interventions are rarely required on the spur of the moment. There is typically some advanced warning or incremental buildup. Consider an automated warning from an EGPWS or TCAS. If the PF fails to initiate the appropriate recovery maneuver, we might direct the required maneuver using the same words that the automation transmitted, "CLIMB" or "PULL UP". Remember that automated warnings may elicit a startle or surprise reflex that causes a period of indecision. Consider, "CLIMB NOW" or "PULL UP NOW" to snap the PF back into awareness. If they still don't respond, we may need to intervene and take aircraft control.

Sometimes, a caution/advisory callout precedes the warning/directive callout. This gives us some extra time to make a less directive callout. For example, if the EGPWS announces "Terrain, Terrain", we know that we need to alter the flightpath or the next warning will be "PULL UP". If we receive the "Terrain, Terrain" caution, and the PF fails to climb, our callout may be, "We need to climb." This may be sufficient to resolve safe terrain clearance and proceed. If the PF fails to respond, we can escalate the intensity of our callout, "CLIMB NOW." Also, using the pilot's name helps. "Clem, we need to climb now." If they are momentarily frozen, this may snap them out of it. We can also use an "if, then" ultimatum. "If we don't climb right now, I'm going to have to take control of the aircraft."

If we reach the point where we contemplate actively taking control of the aircraft, we should go ahead and do it. Pilots rarely report, "I shouldn't have intervened." Far more often they report, "I should have taken control of the aircraft earlier." If we do intervene, expect to debrief the occurrence as a crew or with a safety oversight team (like an Event Review Team under the safety reporting program). In hindsight, we may conclude that we overreacted, but we will reach that conclusion under calm conditions versus the stressful heat of the moment.

If we do choose to take control of the aircraft, we should be decisive and use standard terminology. Avoid any less decisive actions like bumping the control column, moving the thrust levers, or rephasing our intervention like, "I'm going to take the aircraft," or "Let me have the aircraft." Additionally, if we take aircraft control, we should retain control until we clearly transfer it back. Occasionally, events like go arounds are mismanaged by pilots who begin an intervention by pulling up, then assume that the original PF will resume control for the remainder of the go around profile. These cases often deteriorate because no one is flying the aircraft.

20.4.5 Getting an Overly Focused Pilot to Comply

Most airlines have defined parameters for stabilized approaches, clear procedures for go arounds, and mandates that FO-directed go around callouts must be followed. Despite these strong countermeasures, we still see events where Captains override the FO's go around callout and land.

Compared with overloaded or tunnel-focused Captain/PFs, FO/PMs are in a better position to assess approach stability. Many airlines are adopting a "Stable" callout announced by the PM to mark when an approach achieves stabilized approach parameters. Assume that procedures direct the PM to make a "Stable" callout at 1,000′. This becomes the first point where the PM can highlight that the approach is unstabilized. It gives them time to make a series of deviation callouts before reaching the procedural limit (for example, at 500′). If the approach doesn't look like it will become stabilized by 500′, they can make a callout to emphasize the unstable parameters and prime the go around procedure. "Sink 1500, still unstabilized, prepare to go around."

If the Captain still fails to go around, the next callout is a decisive, "GO AROUND." Procedures can strengthen this directive by prompting the first procedural step while holding onto the flap lever, "GO AROUND, standing by flaps." This prompts the transition from the go around trigger point to the well-trained go around procedure.

It helps the Captain shift from an indecisive, "I think we can make it" mindset to a decisive, "We need to switch to the go around procedure" mindset.

Escalate the problem a step further as the Captain counters, "I think we can make it." Time permitting, the FO can repeat the directive with a decisive declaration, "We MUST go around, standing by flaps." This affirms that they will not allow the approach to continue.

If the Captain continues to resist, another option is to invite an outside authority to direct the go around. Transmit, "[Callsign], going around" on ATC frequency. Even the most headstrong Captain will realize that this will cancel their landing clearance. They probably won't be happy and the FO can expect an uncomfortable debriefing at the gate, but they will go around.

These measures should be sufficient to achieve a go around. If they don't, the FO has the final option of taking aircraft control. This is the scenario we train under Captain incapacitation scenarios. We need to use trained verbiage to make it unambiguously clear that we are taking aircraft control.

20.5 CASE STUDY – MEETING RESISTANCE FROM THE CAPTAIN

While we have limited our discussion to the scenario of an FO trying to get their Captain to go around, consider the following NASA ASRS report of the FO making multiple attempts to get their Captain to permit them to initiate a go around.

BOX 20.1 FO UNABLE TO CONVINCE CAPTAIN TO ALLOW A GO AROUND

FO/PF's report: Flying into SLP (Ponciano Arriaga International Airport, San Luis Potosi, Mexico)…, we were told to expect the ILS DME 1 to Runway 14. I told the Captain, I fly here a lot; we were going to be too high. They [need] to give us the ILS DME 2. But he did not want to listen to me. So he set the aircraft up for ILS DME 1 on my leg flying. … [The FO then provides a detailed description of the early portions of the approach] … *I said we were unstabilized; we need to go around. Captain said no, we are fine, and continued. While coming down over the runway at 50′, [we] got a ground proximity warning over the runway and we were still fast so I called a missed approach. The Captain told me "no, get the plane on the ground and land." Then we argued and I told him no, we need to go around. [He said to] just put it on the ground. I tried to push the throttles forward and I told him I'm going around and he said no as he pulled the throttles back. I had to explain to him we were halfway down the runway [and] have not touched the wet runway yet. We cannot stop before going off the end of the runway. So, I pushed the thrust levers forward again and pitched the nose up and said we are going around, and up we went.*

The Captain did not give any callouts on the approach as the PM (Pilot Monitoring). Unreported winds after going missed, wind 330/15 [out of limits tailwind]. Wind reported prior to approach was calm. I don't know what to put here. At the final approach fix, I should've disregarded the Captain saying, "you'll be fine" and forced the plane into a missed approach.[2]

As an exercise, let's place ourselves in this FO's seat and examine what options we might use to achieve a go around. The FO's first attempt began when ATC assigned the ILS DME 1. They knew this would be difficult to manage compared to the ILS DME 2. They voiced their concerns, but reported that the Captain "did not want to listen". By this point, they had followed all of our recommendations for countering risky decisions. Even though the FO was the PF, the Captain overrode their decision making and directed them to fly the ILS DME 1. The Captain was asserting their aircraft command authority. This is not uncommon, or necessarily unwarranted, since the Captain bears ultimate decision-making responsibility for the flight. As the scenario progresses, we will see how micro-managing the FO's flying from the left seat introduced weaknesses and safety vulnerabilities.

As the FO predicted, they reached the final approach fix and were not stabilized. In hindsight, they stated that the missed approach should have started there. Perhaps they were not completely convinced of their assessment. Instead of a decisive, "Go around", they advised, "We need to go around." It seemed that they were trying to convince their Captain to agree with their decision. Again, the Captain rebuffed their "recommendation". We can imagine the FO's dilemma at this point. They had twice tried and failed to reach a shared mental model with their Captain. Perhaps they questioned their own judgment. Maybe they thought that the Captain was right about continuing. In any case, the FO had not fully reached their personal safety limit point.

At 50′ over the runway, they received an EGPWS "PULL UP" warning. If the FO still had any reservations, this made it abundantly clear that they should go around. They reported calling for a missed approach. They don't report their exact wording, but they imply that they were, again, asking for the Captain's permission to go around. Following our recommended techniques, they should have used a decisive "Go around, flaps [as required]" callout. The Captain countered again, "…no, get the plane on the ground and land." The FO appeared to be shocked by the Captain's decision to continue. Perhaps this played a part in their decision to "argue" while floating over a wet runway well past the touchdown zone.

The report paints a picture of a Captain committed to land and an FO desperately seeking permission to go around. The FO's indecision creates a discretionary opportunity for the Captain to commit a shocking act of pulling the thrust levers back in the flare without assuming control of the aircraft. This action finally pushes the FO past their personal safety limit. They advance the thrust levers, pitch the nose up, and go around.

To be fair to this crew, we need to envision how this event felt in real time. It is highly unlikely that the FO expected their Captain to do what they did. In their mind, they fully expected the Captain to agree to go around. Consider the mindset created by the Captain as they actively managed the FO/PF's decision making. The FO accepted their subordinate role by repeatedly trying to convince the Captain to agree to a go around. As a result, we have the Captain/PM making flight decisions like the PF and the FO/PF trying to direct a go around like the PM. This cross-matched role breakdown weakened everyone's defined roles and responsibilities.

NOTES

1 Excellent fictional accounts are the Aubrey/Maturin Series by Patrick O'Brian (W.W. Norton and Company, New York) and The Hornblower Series by C. S. Forester (Bay Back Books from Little, Brown and Company, New York).
2 Edited for brevity. Italics added. NASA ASRS report #1701686.

BIBLIOGRAPHY

Active Pilot Monitoring Working Group. (2014). A Practical Guide for Improving Flightpath Monitoring. Alexandria, VA: Flight Safety Foundation.

ASRS – Aviation Safety and Reporting System – National Aeronautics and Space Administration (NASA) Online Database. (n.d.). Retrieved from Aviation Safety and Reporting System: https://asrs.arc.nasa.gov/search/database.html.

Section III

Introduction to Challenging and Non-Normal Operations

DOI: 10.1201/9781003344575-23

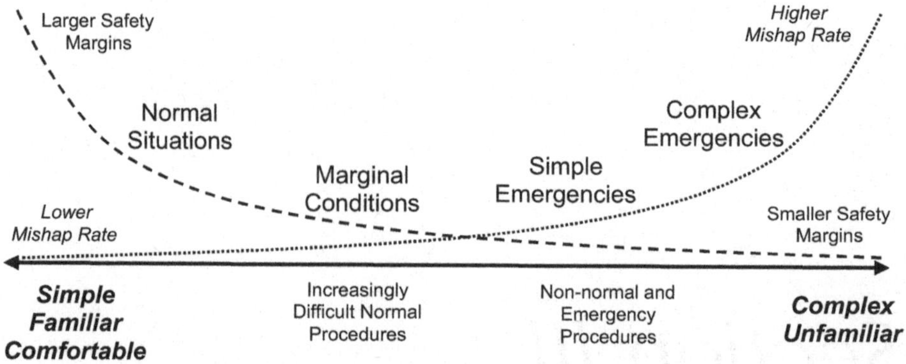

FIGURE III.1 The range of flight events from simple and familiar to complex and unfamiliar.

III.1 CHALLENGING AND NON-NORMAL EVENTS

Visualize a continuum of all possible operational events from simple, familiar, comfortable events to complex, unfamiliar, surprising events (Figure III.1). Situations on the left side of the graph depict the vast majority of line-flying events – normal, familiar flights operated in favorable conditions. As we move to the right, situations become increasingly complex, difficult, and rare. In this section, we'll start in the middle of Figure III.1 with marginal events and work further to the right toward complex, unfamiliar, and surprising emergency situations.

III.1.1 Mishap Rate

Logically, the mishap rate (the rising dotted line) rises with complexity. Conditions begin interacting in unpredictable ways to cause event trajectories to veer off. Mishap events reflect unfavorable outcomes to situations that don't look or feel like failures while we are immersed in them.

III.1.2 Safety Margins

Safety margins are represented by the descending dashed line. As the difficulty increases, our safety margins decrease. Consider landing on an 8,000′ long runway. If the actual landing/stopping distance is 5,000′ in DRY (RCC 6) braking conditions, we would have a 3,000′ safety margin.[1] Compare this with a snow-covered, contaminated runway with POOR (RCC 1) braking conditions which increases our stopping distance to 7,000′. We would have only 1,000′ of safety margin remaining. Let's complicate these two cases. As we react to a wind gust, we add thrust, float, and touch down 1,500′ further than planned. In the DRY runway case, we would stop with 1,500′ to spare. On the POOR braking action runway, however, the aircraft would exceed our stopping margin and slide 500′ into the overrun. Additionally, we have unused braking capability (MAX braking) under DRY conditions that isn't as effective under POOR conditions due to antiskid brake cycling. This means that the standard safety margin on a DRY runway allows us to make extreme errors without encountering undesirable outcomes.

III.1.3 Our Felt-Sense of Safety Margin

Our felt-sense of safety margin is affected by several factors. First, it isn't proportional. We aren't 50% more careful with 50% less safety margin. From the left edge to the middle of Figure III.1, we pay little attention to variations within our safety margin. As we progress to the right, however, our level of concern rises rapidly. Second, each of us measures our level of concern differently. Since we lack consistency, mismatched perceptions between pilots can cause CRM challenges. Third, stress, workload, culture, peer pressure, and ego tend to reduce our felt-sense/concern of safety margin. We see this in post-mishap interviews when pilots make statements like, "I don't know why I continued that approach. I've never landed from an approach like that in the past."

III.1.4 How Our Strategies and Mindsets Change

As we move along the continuum, our game plan strategy and mindset change. On the left side, we apply a small number of familiar, practiced, and reliable game plans. On the right side, emergency and non-normal situations force us to craft unique game plans for each situation. On the left side is confident certainty. On the right side is risky unknown.

Our mindset shifts to accommodate the changes in our situation. Taking the runway, we apply a normal mindset that assumes that we will have a normal takeoff. If while rotating for takeoff, the engine bangs, the aircraft yaws, and the engine gauges start rolling back, we shift to an emergency mindset. How well we manage this shift in our mindset depends on several factors.

- *Training and repetition:* We prepare for emergency events by practicing them in the simulator. If we were test pilots who regularly practiced engine failure events, experiencing one in the aircraft might feel somewhat routine and familiar. To an average line pilot who practices engine failure events only once a year, an engine failure would probably feel unique and unfamiliar.
- *Experience:* Experience helps. A Captain who has been flying the same model of aircraft to the same cities for many years might find a particular event much less stressful than a recently transitioned Captain. The experienced Captain's profile familiarity frees up mental resources to deal with the challenges of the unique situation.
- *Preparation:* Mental preparation makes rarely used procedures easier to execute. If we mentally rehearse a particular non-normal situation, we will find it easier to recognize indications and apply appropriate game plans. Unprepared pilots take much longer to understand the meaning of unfamiliar indications.

III.1.5 Types of Operations

Before we examine marginal and emergency situations, let's define the categories of flight operations.

- *Normal operations:* These are the common situations that typify everyday flying. They are guided by standardized procedures, training, practice, and line culture. Under normal operations, crews act in standardized, predictable ways. Safety margins are ensured through procedural consistency and time-tested success.
- *Rare-normal operations:* Rare-normal operations involve infrequently used procedures required by MEL protocols, environmental conditions, or operational exceptions. For MEL restrictions, crews may need to apply non-standard speed restrictions (required with some flap MELs, for example) and altitude limits (required with an inoperative air conditioning pack or when using the APU to provide pressurization and/or electrical power, for example). For environmental conditions, crews may need to reduce operating limits (required for reduced crosswind limits with braking action POOR (RCC 1), for example). With operational exceptions, crews may need to use non-standard procedures for special cases (for accommodating exceptionally tight gate areas that require ground operations crews to tow aircraft in, for example).

  We need to recognize applicable conditions and apply appropriate rare-normal procedures. We can either perform the required procedure from memory (such as applying modified landing techniques for strong crosswinds) or access our manuals for specific procedural steps (such as configuring aircraft systems to conduct a cross-bleed engine start). Many flight manuals are organized by phase of flight with normal procedures followed by the rare-normal exceptions. For example, the taxi-out section would cover standard taxi and checklist procedures for everyday use. The rare-normal section that follows it would detail procedures for hot weather operations, cold weather operations, taxi-delay, and engine shutdown/restart procedures. Reduced safety margins under rare-normal procedures require us to apply a higher level of attention and care.

  Rare-normal procedures often include prospective memory challenges. For example, while reviewing the station information page during cruise, we note that all aircraft are directed to illuminate their taxi lights when moving and to extinguish them when stopped. Since we won't apply this procedure for another 30 minutes, we'll need to find a way to remember to do it after landing. As we taxi clear of the runway, we won't have distinct reminders to modify our normal habit pattern of leaving the taxi light off during daylight conditions.
- *Non-normal operations:* Non-normal operations are exceptional or emergency situations used to resolve aircraft system malfunctions or adverse conditions. Unlike rare-normal operations, these procedures are scripted in a dedicated manual. For consistency, we will call this the QRH (Quick Reference Handbook), although various airlines and aircraft manufacturers use similar names.

Since we don't memorize the steps or responses, we modify the checklist process to guide all switch movements and system changes. This is because checklist steps involve highly consequential actions, such as shutting down engines or disconnecting generators. We wouldn't want pilots performing an irreversible step, and then referring to a checklist to verify that they did it correctly.

Exceptions are boldface procedures that we perform from memory and then verify using the QRH. The QRH then guides us to deliberately evaluate each step, discuss how it will be accomplished, and verify that it was done correctly. Consider this example sequence of pilots performing one step of the #1 engine shutdown checklist.

- (FO) Reads the checklist step and the required response.
  - "Start Lever (affected engine) – Cut-off"
- (Pilot performing the step – FO) Holds and articulates the action that they are about to perform.
  - FO – holds the #1 (malfunctioning engine) start lever and announces, "I have #1".
  - Captain – guards the #2 (working engine) start lever and announces, "Guarding #2".
- (Captain) Verifies that the action is ready to be completed.
  - Captain – directs, "Cutoff".
- (FO) Performs the step.
  - FO – moves the #1 start lever to cutoff and repeats, "Cutoff".
- Both pilots verify the appropriate system response. In this example, the engine indications show the #1 engine fuel flow dropping to zero.
- The FO moves to the next QRH step.

- *Extremely rare events with no available guidance:* The final category includes extremely rare events from the far-right extreme of Figure III.1. These events are so rare and unanticipated that no guidance or training is readily available. They often generate significant changes across the aviation industry. US Airways 427 (inflight upset accident near Pittsburg International Airport – September 8, 1994) led the FAA to mandate dedicated training blocks to guide recognition and recovery from unusual attitudes. Other examples are United Airlines 232 (engine and flight control failure near Sioux City, Iowa –July 19, 1989) and US Airways 1549 (birdstrike and engine loss leading to ditching on the Hudson River –January 15, 2009).

Organization of the Challenging and Non-normal Operations Section: The following three chapters address deteriorating marginal conditions, QRH emergencies, and time-critical emergencies.

NOTE

1 FAA-mandated landing distance computations include an additional safety margin. For this discussion, we will consider only the actual aircraft landing/stopping performance.

21 Marginal and Deteriorating Conditions

21.1 OPERATIONS UNDER MARGINAL CONDITIONS

Marginal condition operations apply when typical, favorable conditions begin to deteriorate in ways that reduce safety margins and increase mishap rates. Some of the causes of marginal conditions are:

- *Environmental factors:* Deteriorating weather, low ceilings, reduced visibility, storm cells, extreme temperatures, high/gusty winds, night, and low sun angles
- *Aircraft limitations:* Weight, performance limitations, lighting, and system limitations
- *Facility limitations:* Airfield layout, short runways, reduced braking performance, runway contamination, reverted rubber deposits, approach lighting, airfield construction, and confusing lighting/signage
- *Operational constraints:* Traffic saturation, slot/flow restrictions, gate availability, insufficient manning, tight schedules, and misconnections
- *Crew constraints:* Pilot qualifications, crew rest, duty day, fatigue, and CRM challenges

21.1.1 Emergence and Extreme Events

As conditions deteriorate, they interact in unpredictable ways. Most interactions produce consistent and inconsequential outcomes. Sometimes, however, they generate inconsistent and anomalous outcomes that are significantly different from the rest, like statistical outliers. Additionally, they may not follow logical trends. Consider a scenario of a busy airport with continuous landing operations, steady snowfall, and dropping temperatures. We logically predict that braking action reports will trend steadily worse. Following are the landing times and braking action reports from eight aircraft.[1]

- 2200Z – Boeing 737-700 – Reports GOOD
- 2203Z – Airbus 320 – Reports MEDIUM
- 2206Z – MD-11 – Reports MEDIUM
- 2209Z – Embraer ERJ – Reports GOOD
- 2212Z – Boeing 787 – Reports POOR
- 2215Z – Embraer E190 – Reports MEDIUM
- 2218Z – Boeing 737–800 – Reports MEDIUM
- 2221Z – Airbus 330 – Slides into the overrun – reports NIL

DOI: 10.1201/9781003344575-24

Throughout the 21 minutes of this timeline, airport officials monitored the braking action reports to determine when to close the runway for plowing and sanding. The first marginal report was the Boeing 787 reporting POOR at 2212Z. Since the next two crews reported MEDIUM, the managers concluded that they still had some time to prepare. Evaluating the flow of arriving aircraft, they detected a gap in arrivals forming about 10 minutes later. This looked like a good opportunity to close the runway for treatment. They directed ATC TRACON controllers to put all aircraft after the arrival gap into holding. Before they could execute this plan, an Airbus 330 landed at 2221Z and slid into the overrun.

Their plan seemed logical. It was supported by the apparent trend of braking action reports. What did they miss? Their logical flaws were assuming that all aircraft would experience similar braking performance and relying on the accuracy of those braking action reports. While the slipperiness of the runway was steadily worsening, there were additional factors that interacted in unpredictable ways. If they had the same models of aircraft, at the same landing weights, with equally performing braking systems, flown using consistent pilot technique, and reporting braking actions using consistent criteria, the trend would have shown a steady progression from GOOD to NIL. Instead, they had aircraft types ranging from heavyweight wide-bodies to lightweight regional jets. The airfield managers had no knowledge of aircraft weight, quality of tire tread, landing speeds, touchdown points, use of autobrakes, how much reverse thrust was used, pilot technique, or reporting culture. All of these unknown factors interacted in unpredictable ways to produce the wide range of braking action reports.

The failure of the last aircraft to successfully stop seemed to emerge from nowhere. It didn't follow the expected progression. This is one characteristic of emergence as similar conditions give rise to widely variable outcomes. In retrospect, two of the crews reported GOOD and five reported reduced braking action of either MEDIUM or POOR. Expecting a smooth trend, managers thought they would see at least one more MEDIUM TO POOR or POOR report to confirm the deteriorating trend. They even built this into their monitoring plan. "We'll close the runway in 10 minutes unless any crew reports POOR or NIL."

21.1.2 Unknown Actions Disguise Trends

Unknown to the airfield managers, the flight crews were responding to the conditions in ways that disguised the worsening trend. The Embraer crew at 2209Z adapted to the two previous MEDIUM reports and adjusted their aimpoint toward the closest part of the touchdown zone. That, combined with their light aircraft weight and aggressive stopping technique, resulted in an early stop. They reported GOOD. The next aircraft (Boeing 787 at 2212Z) did not compensate in the same ways. Their heavyweight landing and normal touchdown point resulted in a longer rollout and their POOR braking action report. So, we have two aircraft, 3 minutes apart, reporting widely different stopping experiences. Hearing the POOR report, the next two crews (Embraer E190 and Boeing 737–800) compensated by touching down in the first part of the touchdown zone and using MAX braking techniques. Both aircraft experienced antiskid brake cycling, but slowed successfully and cleared the runway. They both reported MEDIUM.

Looking deeper, we discover another hidden factor. Before the mishap, the only two aircraft to roll onto the last third of the runway were the MD-11 at 2206Z (reporting MEDIUM) and the Boeing 787 at 2212Z (reporting POOR). Steady snowfall was accumulating on the last third of the runway. It wasn't piling as deeply on the middle third due to hot gases and jet blast from the thrust reversers of each landing aircraft. This last third also happened to be the portion of the runway with reverted rubber deposits. Dropping temperatures, accumulating snow, and reverted rubber combined to make the last third of the runway much more slippery than the rest of the runway.

Imagine the experience of the mishap Airbus 330 crew. After they switched from Approach Control to Tower, they only heard the two MEDIUM braking action reports. Tower didn't relay the earlier POOR report because the MEDIUM reports were more recent. When the mishap crew of the Airbus 330 landed, they experienced POOR braking action until they reached the last third, where it dropped to NIL. With reversers screaming and antiskid brakes cycling, they slid uncontrollably into the overrun.

21.2 HOW DETERIORATING CONDITIONS AFFECT SAFETY MARGINS

Under marginal conditions, event predictability vanishes and trajectories intensify. We can't anticipate who will experience problems or how each situation will develop.

21.2.1 The Gap between Skills and Capability

We maintain a significant safety margin between our typical, everyday flying and the worst possible situations. Consider landing in a crosswind. The higher the crosswind, the more challenging the landing and fewer opportunities we have to actually practice it. Assume for this discussion that we are certified to land with up to a 45-knot crosswind component. The vast majority of our actual landings fall well within this limit. This means that we have a large safety margin for almost every landing. It is only near the 45-knot limit where our aircraft and personal skill limits are challenged.

Next, consider how this crosswind limit changes under deteriorating weather conditions. While we retain our piloting skills to land up to the 45-knot limit, our tires would lose traction with the pavement. We are still capable of safely executing the landing, but the crosswind would blow us off the side of the runway. A gap forms between our pilot skills and the aircraft performance. This is important because, as conditions deteriorate, we can't rely on our personal flying skills to guide our decision making. Marginal conditions present situations where we *can* perform the landing, but we *shouldn't* attempt it. This vulnerability has emerged in some mishaps as pilots became so focused on completing the landing that they neglected to evaluate whether they should have attempted landing in the first place. To compensate for decreasing aircraft capability, regulators restrict our allowable crosswind component. Under MEDIUM braking action, the crosswind component may be reduced to 25 knots. Under POOR conditions, the crosswind limit may drop further to 10 knots. These limits protect us from confusing our ability to land with conditions that the aircraft can actually handle.

In their book, *The Limits of Expertise: Rethinking Pilot Error and the Causes of Airline Accidents,* authors Dismukes, Berman, and Loukopoulos identify a common theme in accidents that they term the *inadequate response to rare situations.*[2] They conclude that it is unrealistic to expect crews to react quickly and accurately to very novel or rare events.

> No data exist on what percentage of highly skilled airline pilots would be able to execute the most appropriate action in these situations; however, we suggest that these situations severely challenge human capabilities and that it is unrealistic to expect reliable performance, even from the best of pilots.[3]

They describe how the airline industry expects us to execute difficult and complex non-normal procedures accurately, even though these procedures are rarely practiced in training. Additionally, our simulators may not accurately replicate many extreme conditions, thus degrading the quality of the practice that we do receive.

These rare situations are characterized by "surprise, ambiguous cues, high stress, high workload, and the need to respond quickly with a maneuver not frequently practiced". They go on to observe, "Experienced airline crews sometimes recover from these situations and sometimes do not; clearly these situations push the boundaries of human skill."[4] In essence, we practice normal flying every day, but we are still *expected* to produce highly resilient responses and favorable outcomes during rare, ambiguous, and stressful emergencies.

21.2.2 Ego, Experience, and Expectations

We tend to hold unrealistically high expectations of our abilities. We believe that we can perform maneuvers with the aircraft that we actually can't. Because of this, many mishap pilots view their events as personal failings, not as aircraft limitations or training shortfalls. Just because we are extremely good at what we do every day doesn't mean that we will proficiently perform a maneuver that we have never done before while under adverse conditions. Relying on the fact that we have landed in high crosswinds successfully in the past doesn't mean we can successfully land when a 45-knot crosswind gust unexpectedly hits us while in the flare.

21.2.3 Learning the Wrong Lessons from Simulator Training

We cannot practice extreme profiles in the aircraft, so we practice them in the simulator. Unfortunately, the further we move from the middle-of-the-flight-envelope profiles, the less accurately the simulator recreates the experience. For example, when we do upset training, the simulator cannot actually flip us upside down. We ignore the artificial jostling and concentrate on the flight instruments to perform the upset recovery maneuver.

We should carefully consider which lessons we learn from our simulator training scenarios. Consider microburst/windshear training. This training event is typically presented as a dramatic airspeed gain or loss while configured on final approach. When we detect the windshear, we initiate a go around, perform the recovery maneuver, and manage our pitch to fly through the area affected by the windshear. While this is useful training, it does not encompass all microburst phenomena. Since the training is designed to practice the windshear recovery maneuver, it concentrates on slow-speed effects and ignores challenges with crosstrack drift and landing. This can contribute to a biased perspective that assumes that windshear events *always* result in go arounds. Since it is not part of the scheduled training event, we don't practice encountering a crosswind microburst during landing. It turns out that a number of runway departure mishaps have been attributed to exactly this threat. Mishap aircraft touched down while being hit by a strong crosswind. Instead of initiating an immediate go around or a rejected landing, crews attempted to stop on the runway. Mishap analysis indicated that successfully staying on the runway under their conditions was not possible. Their only safe option was to go around. Subsequently, many airlines have added rejected landing training events to address scenarios like crosswind microbursts, bounced landings, and runway incursions.

21.3 LATENT VULNERABILITIES AND MISHAPS

Mishaps occur more often when marginal conditions interact in unpredictable ways to allow latent vulnerabilities to emerge. The complexity of these interactions skews our judgment. Choices that we would normally rate as unsuitable can seem acceptable.

21.3.1 Plan Continuation Bias

After we select a game plan, we become reluctant to change it. With most situations, this makes sense because we may only need to apply a reasonable correction to get it back on track. A wind gust bumps our wingtip, so we move the flight controls to restore level flight. The weather at our destination drops, so we brief and plan for an instrument approach. Making these on-the-fly corrections is an everyday aviation task.

As conditions deteriorate, we reach a point where our corrections fall short. Ideally, we recognize when this tipping point is getting close. When we don't accurately recognize the tipping point, plan continuation bias can encourage us to take steps to preserve our failing game plan.

When we are surprised, overloaded, or unprepared, we rate conditions as being less severe than they really are. This disrupts our future SA. The flight no longer follows our projected path. We should see this as a warning sign, but plan continuation bias adversely affects our judgment. We rationalize that the unexpected indication is probably just a minor disruption that doesn't matter or that will soon disappear. Even if we accurately identify the cause, we reason that we can still make our current game plan work. As task-saturation deepens, we lose the time and mental resources to process rapidly changing information. We tunnel our attention toward a few select parameters. Effort replaces active risk management.

Unpreparedness is the final hook of plan continuation bias. Lacking a backup plan, we force our original, familiar, game plan. We conclude that we don't have time to form a new game plan, so we might as well make the current one work. To avoid this trap, we need to recognize how the warning signs make us feel. Then, when we sense these feelings in some future situation, we'll recognize the need to switch to an exit plan that resets the problem or gives us more time to rebuild our SA. As a technique:

- IF we are surprised by significant changes in conditions,
- AND we are overloaded trying to process what they mean,
- AND we are unprepared with a backup plan,
- THEN we should exit the current game plan, reset, and start over.

Otherwise, we succumb to quickening, tunnel-focus our attention, force our failing game plan, and risk making a Recognition Trap error.

21.3.2 Locating the Failure Zone

Every time we fly, we adjust to changing conditions to successfully achieve desirable outcomes. Some days, it's easy. Other days, it takes all of our skill and experience to succeed. We recognize that shrinking safety margins are moving us closer to a point of failure. How do we reliably locate that point? The truth is that we can't. We typically lack the experience and time to accurately pinpoint it. We do, however, sense when we are approaching it by a rise in our gut-felt feelings of anxiety and anxiousness. When this happens, our first priority is to gauge how much time we have available. If we have enough time to search for the cause of the surprising event, then we should direct our efforts there. If not, we need to switch to a safer game plan.

Our next priority is workload. If we discover that an airport has dropped to IFR while we are still in cruise flight, we'll have plenty of time to program and brief the necessary instrument approach. The workload remains manageable. If, however, the weather drops to CAT III minimums while we are joining final for a planned CAT I ILS, then the workload exceeds the time available. We'll need to go around, brief up the CAT III ILS, and try again.

Finally, given enough time, we should prepare a contingency option. "If that storm cell doesn't move clear of the airport by the time we reach the FAF, we will break off the approach and divert." "If I can't get the airspeed within limits by 500′, I'm going around." Not knowing where the point of failure lies prompts us to shift our perspective back toward more conservative options. Not knowing where the failure point lies shouldn't promote continuing. That is a product of plan continuation bias. Not knowing where the failure point lies should promote either generating more time or switching to a safer game plan.

21.3.3 Optimistic Assessments of Marginal Conditions

As experienced pilots, we confidently believe that we can accurately judge how much our safety margin has shrunk. In hindsight, however, we discover that we can be rather poor judges of our ability to assess the causes of deteriorating and marginal conditions.[5] We underestimate the severity of the problem and overestimate our ability to handle it. Mishap pilots often report that they detected the adverse conditions, thought they fully understood them, and felt that they could handle them. Afterward, they concluded that they didn't truly grasp the magnitude of their problem. They lacked the time, mental resources, and perspective to understand the adverse conditions. They were too busy flying the aircraft and hoping that their efforts would be enough.

21.3.4 The View from Inside the Tunnel

When we study an accident or an incident, we ask many "why" questions. Why did the PF do this? Why did the PM allow the PF to continue? Why did they ignore those warning signs? We might even apply labels to explain the crew's actions – loss of SA, task-saturation, tunnel vision, or non-compliance. If we move beyond these labels and project ourselves into the real-world, real-time flow, we discover that these situations don't feel like loss of SA, task-saturation, tunnel vision, or non-compliance while we immersed in them.

In hindsight, we clearly see all of the options that were available, all of the indications that were present, and all of the better choices that they should have made. We see the whole picture. To the pilots within the situation, it seemed more like moving along the inside of a tunnel.[6] Their perspective was limited to the sides of the tunnel that were immediately visible around them – a restricted picture that quickly moved and changed.

To learn from their mishaps, we try to recreate how they felt as they moved through their tunnel. Imagining ourselves flying their aircraft, we refine our ability to recognize our own gut-felt sensations of overload, quickening, and loss of SA. When we later experience these feelings while flying, we'll recognize that we are in the Red and that we need to take decisive steps to return to the Yellow or Green.

21.4 MANAGING RISK IN MARGINAL CONDITIONS

As conditions deteriorate, manufacturers, regulators, airlines, and pilots all attempt to manage rising risk.

21.4.1 Operational and Regulatory Boundaries

Operating limits are imposed by the aircraft manufacturer and the regulator. Through testing and certification, they set Aircraft Flight Manual (AFM) limits for aircraft operation. These include parameters like maximum altitude, wind limits for takeoff and landing, weight limits, and operating temperature limits. For example, the aircraft may be limited to a 45-knot crosswind component to guard against dragging a wingtip or over-stressing the landing gear. These limits preserve our operational safety margin inside of the engineering or structural limit.

The airline may further narrow regulatory limitations. For example, they may drop the allowable crosswind component to 35 knots. This safety buffer recognizes that real-world conditions may be too dynamic to manage while flying in-the-moment. Pilots following a 35-knot crosswind limit may miss a transient 45-knot wind gust while landing, for example.

The regulator also imposes limits on marginal conditions. A familiar category is instrument approach minimums. Successive approach increments require additional aircraft capability, crew training, currency, and airport facilities. As ceilings and visibility drop, requirements become more stringent. Airlines may impose additional operational restrictions to manage their risk exposure. These include Captain-only landings, currency requirements, experience requirements, special airport/route certifications, and specific crew procedures.

21.4.2 Written Guidance

Marginal operations are so variable that written guidance can't detail how to handle every situation. There are just too many nuances and combinations. Instead, companies form policies to guide how we should manage risk and operate safely. Conflicts arise when scheduling priorities encourage us to continue forward while safety policies encourage us to stop. Written guidance directs us to balance the two by forming game plans that keep the operation moving forward as much as is reasonably possible while preserving desired safety margins.

21.4.3 Personal Boundaries

As pilots, we make further adjustments to written limitations based on our personal experience and risk assessment strategies. A common pilot request is to add additional fuel for marginal conditions. An example is summer weather in desert environments where fast-building, intense thunderstorms form quickly and move unpredictably. Airports can swap from unrestricted VFR operations to indefinite closures with little notice. Normal divert fuel assumptions may prove inadequate as dozens of aircraft simultaneously divert to limited diversion airports at the same time. For example, a mass-diversion from Phoenix to Las Vegas can generate significant landing delays since Las Vegas may already be operating at its arrival capacity limit with scheduled flights.

Between normal procedures and imposed limits lies a range of pilot discretion, also called discretionary space. Unable to manage every flight from their central operations centers, companies rely on each of us to proactively manage risk by assessing the unique conditions affecting our flights. They delegate risk management authority to us because they trust us to find ways to succeed. This works because we can uncover opportunities that might remain hidden from central operations. For example, if an airport is affected by convective weather moving through, we can evaluate real-time radar to locate an opportune window to safely take off and depart.

Sometimes, we encounter conflicts between regulatory limits and our personal limits. Consider a significant ground delay. As long as we remain within legal crew rest and duty day limits, we could stay with the aircraft until the flight is permitted to depart. As long as we "remain legal", we are considered good-to-go. What managers may not consider is whether we previously had a long, challenging duty day following a very short overnight. While technically legal to fly, we may find ourselves suffering from cumulative fatigue. This is why each of us is empowered to override regulatory hourly limits by declaring ourselves fatigued.

21.4.4 Continuing Operations – The "Go" Mode

Airline organizations instill a "go mode" bias into the operation. Whenever disruptions occur, the airline diverts resources to resolve the disruption and keep the operation moving. In a sense, the airline system can apply the same plan continuation bias as a flight crew continuing an unstabilized approach. Consider an airport experiencing continued snowfall and dropping temperatures with only a few remaining flights scheduled to land. Priorities can shift from, "When should we shut down operations?" to "What do we need to do to get these last three flights in?"

21.4.5 Communicating Information Back to Central Operations

If we are delayed for something like a mechanical problem, we need to keep our company informed so they can coordinate with down-line stations for gates and passenger connections. If the mechanic estimates that a repair will take 2 hours, the operation may need to evaluate swapping aircraft, replacing crews, moving passengers to other flights, or cancelling the flight. Another example is a fog bank slowly moving toward an airport. The weather reports may not register the threat as quickly as pilot observations. Our timely notification allows the airline to increase fuel loads for future flights and prepare for the possible diversion of enroute aircraft. In essence, we help the airline fine-tune its game plan.

21.4.6 Reaching the Stopping Point

As marginal conditions deteriorate, we will eventually reach a point where the operation should stop. When conditions reach the regulatory limit, stopping is automatic. The decision is made for us. It becomes more difficult when the decision relies on our personal judgment. Our choice may be resisted by station personnel, central operations, or supervisors. Following is an account from a Captain whose recommendation to cancel the flight was opposed by their dispatcher and chief pilot.

BOX 21.1 CAPTAIN ENCOUNTERS RESISTANCE WHEN RECOMMENDING FLIGHT CANCELLATION

Captain's report: *Upon review of the flight conditions in ZZZ, I decided that the conditions exceeded the limits of the aircraft and elected to not do the flight.* The weather was winds 310° at 19 gusting to 30 knots with snow and marginal VFR, at best. Moreover, taxiways were reporting braking action MEDIUM leading me to believe that the runway conditions were likely suboptimal. As such, we would not be able to land on the long runway due to crosswind limitations on a contaminated runway. As such, we ran numbers for [the shorter runway]. This showed that we exceeded the distance required for flaps 22 and would have to do [flaps] 45. *We ran the numbers and discovered that we would have an unacceptably high likelihood of overspeeding the flaps*

> *and additionally the numbers were marginal at best. This was running them without engine anti-ice which likely would have exceed the distance available.*
>
> *When I called Dispatch to look at the numbers the Dispatcher began yelling at me and talking over me. It felt like I was being pushed to go. I elected to cancel. I then went to the Chief Pilot's office and explained it at which point [they] attempted to explain why I was wrong and that I could have gone and attempted to convince me to go. Feeling pressured, I elected to go with expectation of diverting.*[7]

This event became a dilemma for this Captain. On one hand was the assertion that operational limitations already preserved an acceptable safety margin. On the other hand, operational limits don't compensate for adverse interactions between marginal conditions. The Captain's analysis did. The dispatcher and chief pilot focused on legality. Since the flight was legal to operate, the Captain was encouraged to accept it. The Captain, however, concluded that it was unwise to continue due to unfavorable scenarios that might arise from contaminated runways, crosswinds, and required flap settings. In the end, the Captain succumbed and took the flight "with expectation of diverting".

Another example was presented at the beginning of this chapter with the eight aircraft landing on the snow-covered runway. Every aircraft was legal to land. Airport operations had a proactive game plan for plowing the runway. Despite everyone's efforts to actively managing risk, an aircraft still slid into the overrun. Who could have, or should have, stopped the operation to prevent that mishap? One possibility was the crew of the Boeing 787 that reported POOR at 2212Z. They were the last aircraft to roll to the end of the runway and experience the braking conditions along the entire runway length. Was it really POOR, or did that crew hedge their assessment because they didn't want to be the ones to close the runway and cause subsequent aircraft to divert? Pilots have reported this line of reasoning with similar past incidents. If that crew actually experienced some NIL braking, they might have felt some reluctance to report it. It's common to rationalize, "Most of the runway was MEDIUM and POOR, except for the small patch near the end that might have been NIL. Since it was mostly POOR, we'll report that." Another complication is when pilots modify their assessments to soften the severity and to avoid using consequential words like "NIL". "The last part was REALLY slippery."

Perhaps ATC could have asked the crew for more details about their landing experience. If the crew then reported that braking was MEDIUM until they reached the last third, then dropped significantly to POOR, it would have provided airfield managers with a better understanding of the deteriorating braking performance. Knowing that the Airbus 330 was due within 10 minutes and would need that slippery portion of the runway to stop, airport managers would have chosen to act earlier. Another option would have been to station an observer near the end of the runway to closely observe snow accumulation and aircraft stopping performance. They might have noticed the Boeing 787 slipping as they began their turn off of the runway. In any case, this example demonstrates the challenge with making decisions under deteriorating marginal conditions using limited information.

We promote the go-mode as long as conditions allow, but we need to be equally ready to engage the stop-mode. Stopping an operation is not a bad thing. We should see ourselves as the guardians of the safety margin for those pilots following us. Holding for runway treatment is a common winter event. Diverting because we don't have time to hold is another. We would rather hold or divert a hundred aircraft before allowing even one to slide off of the end of a slippery runway. If our report can prevent that mishap, then we become an asset of proactive risk management.

21.4.7 Practicing Failure in the Simulator

Sometimes we need to fly failing events to experience them and witness how they feel. I recall when we first began aircraft upset training at my airline. My simulator partner was a highly experienced Captain. On his first recovery attempt, he completely mishandled the recovery and crashed. As the instructor reset the scenario, he admitted that he had never actually been upside-down in an aircraft before. His second event went better. By his third, he was successful. The simulator is our best place to practice extreme conditions. We can recreate an out-of-limits condition and experience how it feels when our best efforts fail. A useful scenario might be to attempt a simulator landing on a POOR or NIL runway with out-of-limit crosswinds. As we slide uncontrollably off the side of the runway, we'll experience how that kind of failure feels. Later when flying the aircraft, we'll find it easier to recognize and counteract our plan continuation bias. As we practice extreme conditions, we should take care to learn the right lessons. Practicing an extreme profile is not intended to give us the confidence to push limits. Instead, we use it to practice our threat recognition and rejection trigger points.

21.5 PLANNING AND SELECTING GAME PLANS UNDER MARGINAL CONDITIONS

Following are some ideas and considerations for flight planning under marginal conditions.

21.5.1 Select Options that Reverse Rising Stress

Imagine flying an approach under challenging conditions. As we struggle to hold our flightpath against growing turbulence, we feel our stress level increase. Continuing our failing game plan requires an uncomfortable level of effort and force. As soon as we switch to an appropriate contingency option, we resolve the disruption and restore the feeling of normal and familiar flying. The stress falls away. So, instead of using our perception of rising stress as motivation to work harder, we use it as a trigger to switch to a more-manageable option.

21.5.2 Always Preserve an Escape Option

An antidote to plan continuation bias is actively preserving an escape option. For example, if we see the runway ahead, but a thunderstorm looms immediately off the departure end, we would conclude that we lack a safe go around option for windshear on short final. Lacking a viable escape option, we would go around early, turn away from the storm, and wait until it moves off. Committing to the approach and hoping that we won't need to go around sacrifices our only available safe escape option. A resilient game plan maintains at least one safe escape option until we touch down and extend the thrust reversers. Only then are we fully committed to stopping on the runway. For example, the crew of American Airlines 1420 (Little Rock windshear accident) apparently had a visual path to the runway, but no viable go around path. Perhaps this contributed to their decision to land in what turned out to be level-5 thunderstorm precipitation.[8]

21.5.3 Plan an Approach for the Next Lower Available Minimums

In some incidents, crews planned their approach based on the ATIS despite indications that the weather was deteriorating. Expecting CAT I approach conditions, they were surprised by lower ceilings and decreased visibility on final. A better technique for deteriorating conditions is to plan for the next lower approach category. For example, if ATIS is calling marginal CAT I, plan for CAT II. If they are calling CAT II, plan for CAT III. Lower options add more operating margin and increase our chances of completing the approach.

21.5.4 Trust Experience and Intuition

Every one of us brings a wealth of experience, insight, and intuition to flightdeck decision making. When we assess a situation, we weigh it against numerous similar past experiences. The experiences that went well comprise our repertoire of workable game plans. Those that didn't go well identify unviable game plans. When our gut-feeling tells us that a game plan is workable, it probably is. Conversely, if our game plan feels uncertain or risky, we would be better served selecting a more conservative plan.

21.5.5 Run Mental Simulations

We use our imagination to visualize how our situation should unfold. If the path to our desired outcome looks workable, we form and execute a game plan to follow it. When obstacles emerge along our mental simulation, we discard that plan and try something else. Gary Klein calls this seeing the big picture or the power to see the invisible.[9] Evaluating mental simulations as a crew is even better since each crewmember runs their own mental simulation. Together, we can identify additional strengths and weaknesses.

21.5.6 Look for Leverage Points

Gary Klein notes that experts have the ability to spot leverage points – angles that make a workable solution possible.[10] For example, low sun and haze make a late afternoon approach to runway 27 especially difficult. ATC may be following the momentum of their plan continuation bias by continuing operations to runway 27. From our vantage, landing on runway 9 would be much easier because the low sun would be behind us. By informing them of the deteriorating conditions on runway 27 final, we offer the better option of landing on runway 9. It is difficult to turn the airport around, but it is much easier than switching to IFR approach spacing and accommodating missed approach aircraft.

21.5.7 Rank Reasonable Goals and Discard Unreasonable Goals

Mishap pilots often report statements like, "I knew I shouldn't have tried that approach." Their gut-feeling was telling them to stop, but plan continuation bias urged them forward. Our experience guides us to select reasonable goals and to discard unreasonable ones. One decision-making flaw is overemphasizing long-range consequences. "We should divert, but if we make it in, we can stay on schedule. If we divert, it will disrupt the rest of our duty day." It is often better to use the more conservative plan with its ample safety margin over an aggressive, idealistic one that requires everything to go well.

21.5.8 Select the Best Plan for the Current Conditions, Not for Future Consequences

Sometimes, the most appealing short-term game plan generates undesirable long-term consequences. If it is reasonable to hold or divert, but we know in the back of our mind that doing that will create future problems (longer duty day and passenger connection issues), we may bias our choices toward the riskier, more expedient game plan. "I know that thunderstorm is approaching the field, but if we can land before it hits, we can stay on schedule." Under marginal conditions, our game plan needs to follow what is actually happening, not what we hope will happen. If the conditions favor holding or diverting, then hold or divert. Deal with the operational consequences later.

21.6 MONITORING GAME PLANS UNDER MARGINAL CONDITIONS

After we select a game plan, we need to monitor both for signs that it is working and signs that it may be failing.

21.6.1 Expand Situational Awareness

We build future SA to predict how events are likely to unfold. As time allows, we extend our SA to anticipate changing conditions. For example, approaching LAX near sunset, we notice that the ATIS shows using visual approaches to runways 24R and 25L. On downwind, we see haze caused by the coastal marine layer. Taken together, we predict that the low sun will interfere with our ability to fly a visual approach. We brief and plan for an instrument approach even though the ATIS still reports VFR conditions.

21.6.2 Accept Conditions As They Are, Not How We Wish They Would Be

We have to accept our actual parameters. We can't pretend that we have parameters that we don't. Similarly, we can't ignore parameters that don't match our expectations. Mishap pilots often remember being "fast", but they either didn't check their speed or they couldn't recall their actual airspeed. Wishing and hoping accompany plan continuation bias. They allow us to pretend that our parameters are good enough. This creates a false impression that we are actively monitoring while we really aren't.

21.6.3 Apply Aircraft Knowledge

Every model and category of aircraft exhibits strengths and weaknesses to particular deteriorating conditions. Some aircraft are sensitive to airframe icing. Others are vulnerable to crosswinds on slippery runways. We need to learn the particular nuances of our aircraft. This information may not be available in our manuals. Much of it can only be learned through our own experience or from other pilots. An example is checking "critical surfaces" for icing during a holdover time inspection. The procedure allows us to exceed deice fluid holdover time if we perform a visual inspection of critical surfaces to ensure that they are free of frozen precipitation. This visual inspection is typically made by examining the tops of the wings from the passenger cabin. Our manuals tell us which surfaces to check, but may not specifically describe the details. Which areas tend to accumulate ice first? Where is the best vantage point for inspection? How do we accomplish the inspection at night? How does wind-blown snow affect icing? Know where to look first. It may be the trailing edge of the ailerons. It may be the wing root on the FO's side. This doesn't remove our requirement to inspect all affected surfaces, but if we know where to look first, it saves time and improves compliance.

21.6.4 Understand the Subtle Details Embedded Within Procedures

Every year, a working group examines the lessons learned from the previous winter season and proposes regulatory changes for the following winter. Changes are made to deice procedures, holdover times, and other frozen precipitation nuances. Sometimes, subtle changes are buried in the footnotes of our charts. We need to know how to apply these rules to the actual conditions. When would this particular requirement apply? What statement in the ATIS report would require us to apply that exception? How are we going to accomplish this new procedure as a crew? How will we remember to do this delayed task when there is no procedural anchor point to remind us? If there is a timing requirement, who will run the clock and monitor the elapsed time?

21.6.5 Plan Continuation Bias and Gray-Maybe

We are results-oriented people who are strongly motivated to find ways to land at our scheduled destinations. To counter this go-bias, directives and limitations define limiting parameters that bar us from continuing. When we reach any of these regulatory limits, the decision is made for us. We must stop. Marginal conditions, however, may not quite reach these regulatory boundaries. Between legal-to-go and illegal-must-stop, there is a zone of gray-maybe. Decision making in the zone of gray-maybe depends less on pursuing desirable goals and more on balancing risk probabilities. The push and pull between our desire to continue and our concerns about risk often manifests in the use of conditional language. "Braking action was MEDIUM TO POOR down most of the runway, but the end was really slippery." Is this pilot waffling because they don't want to be the ones to force following aircraft to divert? Is "really slippery" actually NIL? We may also resort to conditional language that overemphasizes hopeful outcomes. "That storm cell *appears* to be moving off. We *should be* able to make it in." When entering the zone of gray-maybe, we need to counteract our go-bias by actively evaluating possible unfavorable outcomes. If this reasoning process causes us to feel unsure about continuing, then we should switch to a conservative contingency backup.

21.6.6 If We Need to Fly Our "Best Plus", Then We Shouldn't Continue

One pilot technique for contaminated runways is to shorten our normal touchdown point. While this sounds logical and even desirable, is it an appropriate technique? I recall a mishap where a line of aircraft continued to land under deteriorating winter conditions. Eventually, one aircraft slid off of the end. During the investigation, we discovered that crews were increasingly shortening their touchdown point and exaggerating their braking technique to compensate for the deteriorating conditions. Some even attributed the mishap to that pilot's unwillingness to "land on brick one". This reasoning begs the question, if we think the only way to land safely is to touch down on the first brick, then should we be attempting to land in the first place? We all agree that we need to fly our best under deteriorating conditions. This means that we should fly a stabilized approach, not float in the flare, de-rotate promptly, get on the brakes and reversers, and slow straight ahead to taxi speed. If we feel that we need to do all of these things *plus* shave knots on approach speed, land on brick one, and immediately apply MAX braking, then we should divert or hold until the runway is cleared and treated.

21.6.7 Guard against Reverting Back to Normal Habit Patterns While Still in Marginal Conditions

Marginal conditions often require us to employ rarely used procedures. When stressful events ease up, we have a natural tendency to revert back to normal habit patterns. For example, procedures direct us to delay flap retraction in snowy conditions until we have an opportunity to inspect the flap tracks for ice buildup. If we fly a challenging approach and successfully stop on a snowy runway, our stress level eases as we taxi clear. Feeling that the exceptional part of the flight is over, we may revert back to our normal habit pattern and raise the flaps. We inadvertently replace the exceptional procedure with our normal practice. Remembering not to do something or to do it differently can be a challenge, especially if we haven't recently practiced it. Marginal conditions require that we act mindfully and deliberately.

21.6.8 Search for Counterfactuals

Counterfactuals are indications that contradict our current game plan. They are the warning signs that may suggest that our game plan is failing.

- *The optimistic perspective:* If we are optimistic about our game plan, we tend to look for and see indications that it is succeeding. These *profactuals* support our game plan and encourage us to continue.
- *The skeptical perspective:* If we view our plan skeptically, we actively scan for indications that it might be failing. We see the *likely* progression of events, not our *desired* progression of events. Under marginal conditions, wise pilots apply a healthy dose of skepticism. If we look for flaws, we'll see flaws. This perspective helps us predict not only how our plan might fail, but all of the counterfactuals and warning signs that precede that failure.

  Whenever we encounter marginal conditions, we need to wear our skeptical glasses. This motivates us to seek out and detect counterfactuals. If they don't materialize, then our game plan should succeed. If we do detect warning signs, it may be a good time to abandon our game plan and switch to a contingency backup. A little bit of skepticism proves to be a good thing.

21.6.9 Monitor Abort Triggers

If we are in marginal operations and are following a workable game plan, actively watching for counterfactuals, and preserving our escape options, then we are following a *fail-safe* strategy. This ensures that our game plan must either succeed safely or fail safely. A failing instrument approach must result in a safe missed approach. A failing landing must result in a safe go around. If we have rehearsed our fail-safe contingencies, then we will quickly and accurately execute them.

21.6.10 Avoid Bumping against the Limits

The closer we operate near a limit (operating manual, aircraft, skill), the more unpracticed, unknown, and unpredictable the situation becomes. The prudent, proactive pilot avoids operating near these limits. If we choose to bump against operational limits, then we should carefully set our abort/reject triggers and become even more sensitive to deviations than usual.

21.6.11 Don't Debrief during the Event

While their adverse event was unfolding, mishap pilots often wasted precious time trying to understand or diagnose their problem. "What is happening?" "What is this thing doing?" "Why isn't this working?" None of these questions improve our in-the-moment handling of the situation. It is much better to save debrief analysis for later.

21.7 MARGINAL CONDITIONS IN WINTER OPERATIONS

Winter conditions generate a significant portion of marginal condition mishaps. Between low ceilings and frozen precipitation, our challenges accumulate. Following are some specific cases, considerations, and techniques for winter operations.

21.7.1 Make Braking Action Reports

In 2016, the FAA instituted the Takeoff and Landing Performance Assessment (TALPA) system. It incorporated a revised list of braking action measurements in the Runway Condition Matrix (RCAM). Runway Condition Codes (RCC) from 6 to 0 are designated for braking action between GOOD and NIL. The Pilot Runway Condition Assessment Matrix defines the values based on both braking action and directional control.

While this system represents a significant improvement over past versions, there are some notable limitations. Braking action reports can be highly variable depending on the aircraft, pilot assessment, and the portions of the runway that aircraft used while stopping. Consider two extreme cases. The first case is a lightweight turboprop RJ. By reversing propeller pitch, stopping performance is especially effective on contaminated runways. This RJ crew stops early and pulls clear at a midfield turnoff. Their short landing roll and narrow wheelbase track along clear pavement near the runway centerline. They report braking action GOOD (RCC 5). The second case is a large wide-body aircraft crew landing near their ATOG landing weight. Using their best braking effort, they still roll down the entire length of the runway. Their widely spaced wheels extend onto snow-covered pavement far from the runway centerline and result in antiskid brake cycling. They report braking action POOR (RCC 1). They base their report on the poor braking and directional control that they encountered over the last portion of the runway – runway the RJ never used.

There are also pilot misconceptions based on runway length. Many pilots believe that if they can slow and make their typical turnoff taxiway exit, that the braking performance is GOOD, regardless of other performance considerations. Following one mishap, we interviewed crews that stopped successfully before the mishap flight. One pilot reported that they had significant antiskid brake cycling during rollout, but didn't report POOR because they "made the high speed exit". They said that they would only report POOR if they couldn't slow down until the last available turnoff.

Another significant variable is directional control. Before the RCAM, the guidance didn't address directional control challenges. Under RCAM, it is now possible to have significantly reduced braking action reports due to wet runways and strong crosswinds. While this represents a significant improvement over past rating systems, pilots aren't consistent with their reports. Years after TALPA's release, many pilots still don't consider directional control in their braking action assessments.

A useful practice is to make a braking action report anytime the airport is reporting braking action less than wet-GOOD (RCC 5). Our normal habit pattern is to switch over to Ground Control frequency as we clear the runway. If we transmit our braking action report to them, there may be some delay as the Ground controller relays it to the Tower controller who then relays it to subsequent landing aircraft.

A more timely technique is for the PF to rate the braking action after clearing the runway. The FO then transmits the report on the ATC Tower frequency. That way, all aircraft on final approach immediately receive our report.

BOX 21.2 POOR BRAKING ACTION REPORT IS NOT ACKNOWLEDGED BY ATC

I had an experience where I was the first pilot to land on a runway after airport support crews had finished plowing it. The weather was moderate to heavy snowfall with temperatures near freezing. As I tried to slow, I experienced full antiskid cycling and difficult directional control. I called "braking action POOR" over the Tower frequency. The Tower controller did not acknowledge my report. I called it a second time. Again, they did not acknowledge my report. The next aircraft on final queried, "Did you just say braking action POOR?" I waited for the Tower controller to reply. They didn't, so I transmitted again, "Yes, braking action is POOR." The controller finally joined the conversation and tried to argue that it couldn't be POOR because they had just plowed the runway. I informed them that I experienced continuous antiskid brake cycling and degraded directional control as I exited the runway. After a long pause, they acknowledged with a disappointed, "Roger".

21.7.2 Monitor Temperatures and Snowfall Rates

When we are scheduled for a destination forecasting dropping temperatures and snowfall, we should start monitoring conditions before departing from our originating gate. Compute landing performance data using worst-case conditions. Verify the suitability of the assigned alternate airport.

In cruise, continue to monitor the conditions. Be especially vigilant when temperatures begin to approach 0°C. If ATIS is unclear about snowfall rates, consider calling station operations to have someone subjectively assess the snowfall, whether it is sticking to the pavement, and the condition of the ramp. Pilots who have recently arrived can relay useful information about runway and taxiway conditions. This is timely information that is often slow to reach ATIS or Field Condition reports. Knowing that temperatures are dropping and that snow is falling, we can assume that braking action may degrade to POOR (RCC 1) or even NIL (RCC 0).

21.7.3 Anticipate Uneven Snow Accumulation

One flaw is assuming that snow accumulates evenly on the runway. Certainly, it falls evenly, so why shouldn't it accumulate evenly? There are several reasons why it doesn't.

- *Single runway is used for takeoffs and landings – continuous operations*
  - *Analysis*: As aircraft depart, crews use full engine thrust. This generates heat and jet blast that either melts the snow or blows it off of the side of the

runway. It is common to see the approach-end and centerline portions of the runway free of snow while it is accumulating beyond those portions. Additionally, landing aircraft touch down and engage reverse thrust. This also heats the portion of the runway from touchdown until they stow their thrust reversers (approximately 2,000′–5,000′ down the runway). The furthest portion of the runway (beyond 5,000′ in this example) is only used by heavyweight and large aircraft. As they roll over a particularly slippery portion with accumulating snow and reverted rubber deposits, they typically experience dramatically reduced braking action.

- *Departure runway is used only for takeoffs.*
  - *Analysis:* Many airports use separate runways for takeoffs and landings. The early portion of the takeoff runway will be clear due to engine heat and blast. Problems are only discovered by aborting aircraft. Consider a crew that rejects their takeoff due to an engine failure. Their heavyweight rollout will extend over pavement that hasn't been used because every departing aircraft achieved liftoff before rolling over it. This snow-covered portion may prove to be quite slippery.

- *Arrival runway is used for landings only.*
  - *Analysis:* Landing-only runways generally have clearer surfaces in the first and middle third due to engine heat and jet blast. Pilots may experience GOOD braking during these early portions. As they ease off on their thrust reversers and braking, they may experience a drop in stopping performance as they transition onto the snow-covered portion.

- *Single runway is used for takeoffs and landings – low arrival volume*
  - *Analysis*: Braking action reports may be old and inaccurate since no similar aircraft have recently taken off or landed. Assume the next worse braking action while evaluating performance.

21.7.4 The Deformable Runway Surface Problem

As snow accumulates on pavement, it no longer behaves like a hard surface. As our tires roll over it, the snow compresses under each tire and squishes through the tread grooves. We classify this as a deformable surface. We typically associate deformable surfaces with snow and ice, but standing water from a heavy downpour also behaves in similar ways. Regardless of the cause, deformable surfaces may cause our braking experience to be significantly different from what previous aircraft have reported. This is because our aircraft and tires may respond differently than theirs.

21.7.5 The Tire Tread Problem

The deeper the tire tread depth, the more effectively it can evacuate the snow and water away to maintain contact with the pavement. Consider an aircraft landing on a snow-covered runway with pilots reporting MEDIUM TO POOR (RCC 2) braking action. We compute our landing performance numbers. The results direct us to use autobrakes MAX and full thrust reverser to achieve an acceptable stopping margin. These results are based on a new aircraft with new tires. Our older aircraft with minimal tire tread cannot match that ideal performance. For this reason, our performance calculations include an FAA-mandated additional safety margin.[11]

21.7.6 The Different Aircraft Problem

Different aircraft deliver different levels of braking performance. We accept that the braking/stopping experience for an RJ on a snow-covered runway is probably different from that of a medium-sized jet or a heavy-jet. That is why braking action reports include the aircraft type. Even within aircraft type, reports can be significantly different depending on tire tread quality, aircraft model, weight, landing technique, braking technique, touchdown point, stopping distance, and pilot assessment.

21.7.7 The Deceleration Problem

Sometimes, pilots report, "The first part of the runway was GOOD and the later part was MEDIUM."[12] Let's consider what might have motivated the crew to make this mixed report. They probably felt GOOD braking deceleration during the early, higher-speed portion of their rollout because their thrust reversers and aerodynamic drag provided the majority of their deceleration rate. Both thrust reverse and drag are effectively immune to slippery surfaces, so the initial deceleration probably felt normal. As they slowed, the effectiveness of thrust reverse and aerodynamic drag diminished. Wheel braking became their main source of deceleration. Runway contamination became relevant. As the brakes engaged, their antiskid brakes started cycling. So, they called "...later part was MEDIUM."

Autobraking affects our perceived braking experience. The autobrake system attempts to deliver a constant deceleration rate. At higher speeds, reversers and aerodynamic drag provide most of our deceleration, so hydraulic brake pressure meters back. Braking effectiveness feels normal. As the aircraft slows, the thrust reversers and aerodynamic drag become less effective, so the autobrakes system increases hydraulic brake pressure to the brake pads. As we lose traction, antiskid cycling begins. This initially feels like a loss of deceleration. It surprises us. Some pilots respond by deselecting autobrakes in an ill-advised attempt to restore the feel of normal braking through manually braking. This can actually result in less deceleration, increased landing roll distance, and directional control problems.

Another nuance occurs while using autobrakes MAX. This is a braking mode we rarely use in everyday flying. After we touchdown on clear pavement, the autobrakes apply maximum hydraulic pressure to the brake pads. This generates significant deceleration. As we roll onto snow-covered pavement, the antiskid system starts cycling. Pilots have misinterpreted this as autobrake failure, disengaged autobrakes, and attempted to brake manually.

21.7.8 What GOOD Braking Action Means to Us

What does a braking action report of wet-GOOD (RCC 5) really mean? For most of us, it implies that we experienced what felt like a typical experience of deceleration and directional control. Rarely, do actually we apply the full braking capability of the aircraft. We use just enough braking to decelerate in a comfortable, familiar way. So, GOOD actually means that the brakes engaged as expected, that the deceleration felt normal and familiar, and that we didn't notice antiskid cycling or directional control issues.

Let's assume that we experienced some antiskid brake cycling as we slowed. Do we still call it GOOD? It's certainly an indication of increased slipperiness, but if we still slowed normally, we might still call it wet-GOOD (RCC 5). According to the RCAM matrix, it only becomes MEDIUM (RCC 3) when, "Braking deceleration is noticeably reduced for the wheel braking effort applied OR directional control is noticeably reduced."[13] We could rate it GOOD TO MEDIUM (RCC 4), but pilots rarely use this option. As a result, we have a wide range of pilot reporting discretion between wet-GOOD (RCC 5) and MEDIUM (RCC 3).

21.7.9 Effective Braking Techniques and Tire Alignment

Aircraft tires brake most effectively when the aircraft is tracking straight ahead. When we attempt to turn and brake at the same time while on slippery surfaces, we can begin sliding uncontrollably. Mishaps involving turning and sliding are relatively common. Following are some causes.

- The pilot employs differential braking to facilitate turning. This causes the heavy-braking wheels to begin antiskid cycling. The remaining normally braking tires pull the aircraft out of alignment. Then, all tires begin skidding.
- The pilot attempts to turn too aggressively. The nosewheel tires lose alignment and begin to skid.
- The pilot attempts to take a high-speed taxi turnoff at excessive speed. Often, these turnoffs and taxiways don't receive the same care and frequency of plowing as the runway.
- The pilot loses traction crossing a snow berm left by the plows between the runway and the taxiway or left by the plows as they make U-turns at the end of the runway.
- The pilot turns aggressively from grooved pavement onto un-grooved pavement. Some runways only have grooves cut in the center portion.
- While turning, the pilot transitions from the heavily trafficked center portion of the runway to the unused, snow-accumulating sides of the runway.

The bottom line is that all tires need to remain oriented straight ahead while braking until reaching a safe taxi speed. This is especially important for FO landings as they transfer aircraft control during rollout. A useful technique on slippery runways is to slow straight ahead on the runway centerline until reaching a manageable taxi speed. Then, transfer aircraft control to the Captain.

21.7.10 Pavement Temperature near Freezing

Runway braking action is highly dependent on pavement temperature. If snow is falling on a warm surface, it quickly melts. Our braking action will feel like wet-GOOD (RCC 5). As the pavement begins to cool or as snowfall increases, the snow begins to accumulate. The deformable surface and dropping temperatures degrade our potential braking performance. When the runway temperature approaches a narrow temperature range between +2°C and −2°C, the braking action quickly deteriorates to POOR or NIL (a condition often coincident with wet snow or standing slush). This doesn't follow a straight-line, predictable progression. Braking performance trends down, but then rapidly plummets. Following is a pilot report of this quickly developing NIL effect.

BOX 21.3 NIL BRAKING FROM A "FLASH FREEZE"

Captain's report: We set up for the ILS in ZZZ3. … no indications of any runway contamination made this ILS seem like a perfectly rational choice. *Per landing data, we could land with autobrakes 3, but opted for MAX just for the extra cushion. … the approach was very stable and the landing was on the numbers and on speed. In the rollout, we both quickly realized and the FO announced that we weren't slowing down. I saw he had thrust reversers fully engaged and speed brake was deployed, so I commanded, "MAX manual brakes, MAX manual brakes!" I looked again to ensure the FO was fully braking. At this point we were decelerating some but approaching the end of the runway. I jumped on the brakes to no avail and we slowly slid off the end of the runway into the grass where we soon came to a stop in a rather smooth, non-violent manner.* We took stock of the situation, having passengers remain seated and FAs check their condition. No unusual lights or indications in cockpit and cabin, all looked very stable, immediate evacuation not required. We fired up the APU, secured the engines and began coordination with ATC, company, etc. No reported injuries, everyone calm. Ground crew reported no immediate signs of aircraft damage. One of them took a measurement from the tail cone back to runway threshold and it was 115′. *Another responder called this a "flash freeze", saying that it happened the day before as well.*[14]

The airport probably had conducted successful landing operations prior to this event, but as the runway surface temperatures continued to fall, the pavement reached this +2°C and −2°C danger zone. Notice that the crew took every reasonable precaution. The aircraft was lightweight, they selected the highest autobrake setting, they flew a fully stabilized approach, and they maximized thrust reverse. None of these precautions prevented them from sliding off the end. We might fault them for not diverting, but this crew had already diverted from their intended destination and from their first alternate. This second alternate was considered to be their best option. Given the reported conditions, they should have stopped normally. This is also an example of a crew doing everything right, but experiencing an unfavorable outcome.

21.7.11 Conga-Lines into High-Volume Airports

If we are part of a line of aircraft landing at a busy airport, we'll have the advantage of receiving continuous, real-time runway reports after each aircraft lands. If the calls are consistent and accurate, we can monitor the trend of braking performance. On the other hand, we can become lulled into a false sense of security. "Everybody else is getting in, I guess we'll get in, too." If crews are calling MEDIUM (RCC 3) while temperatures are dropping and snow is falling, we can expect that the braking action may get worse before we land. Be prepared with performance computations for the *next-worse* braking report. For this example, we should analyze our landing data for MEDIUM TO POOR (RCC 2) and POOR (RCC 1). We check MEDIUM TO POOR (RCC 2) because it is the next worse braking action. We check POOR (RCC 1) because many pilots don't make the transitional reports of GOOD TO MEDIUM (RCC 4) or MEDIUM TO POOR (RCC 2). If we discover that either of these reports place us out-of-limits, we'll be ready when a crew reports POOR while we are on short final. We'll be spring-loaded to go around and divert. Conversely, if we know we are legal to land with that next lower report, we can continue and land.

21.7.12 Infrequent Flights into Low-Volume Airports

At smaller airports, our challenge is that the last report may be untimely. Either no one has landed in a while or the last braking action report was from a dissimilar aircraft. What should we do if we break out of the weather and see the runway as an unbroken snowfield bordered by edge lights? What if we don't have reports measuring the snow depth? What if the runway is half clear and half covered with drifting snow? What if we have a stiff crosswind? If temperatures are dropping and snow is falling, the braking action might be significantly worse than the last report. Predict which conditions might exist by extrapolating trends in temperature and snowfall.

21.8 MARGINAL CONDITIONS IN SUMMER OPERATIONS

While winter conditions are universally accepted as challenging, summertime rain and wind can also present significant challenges.

21.8.1 Reduced Visibility from Rain

We typically make our instrument approach decisions based on the ATIS (current weather) and the TAF (Terminal Aerodrome Forecast – weather forecast for the airport area). If they report mostly clear with isolated rain showers, we typically expect to fly a visual approach. Most of the times, we'll be right. Now, let's assume that we are on downwind with the field in sight. We see a thin rain shaft across the final approach path. We can still see through it to the runway. No problem. When we turn onto final and actually enter the rain shaft, however, the distortion of the rain sheening across our windshield causes us to lose sight of the runway. We turn on the wipers, but they fail to clear enough water away to restore contact. We start our go around and subsequently fly clear of the rain shaft into clear air with unrestricted visibility.

Rain can cause our visibility to drop rapidly and unexpectedly. For this reason, many airlines direct crews to complete full instrument approach briefings anytime rain is present or forecast in the terminal area. This may seem like an excessive precaution until we experience this scenario for ourselves. In our rain shaft example, we could continue flying the instrument approach until reaching approach minimums. By then, we'll probably clear the rain shaft and land normally.

21.8.2 Windshear

Windshear is another transient event associated with thunderstorms and warm weather. Fortunately, many airports have excellent microburst detection and prediction equipment. Even so, the hazards of windshear can occur with little advanced notice or predictability. Consider the following report from an RJ crew landing in Denver.

BOX 21.4 CREW DELAYS GO AROUND DECISION IN WINDSHEAR CONDITIONS

FO/PM report: I was the Pilot Monitoring. *On our descent into DEN, we had gotten the ATIS and noticed that it was starting to get pretty windy down on the ground and that moderate turbulence was being reported by multiple aircraft.* But [otherwise] everything remained normal. Approach told us to expect Runway 34R. *While we were north of the airport, approach had reported an increase of 50 kts in airspeed and wind shear by aircraft and that they were going around. We quickly briefed and reviewed the wind shear escape maneuver if we were to encounter it. We also both noticed what looked like a dust devil had formed right over or near the airport. Then, approach changed our runway.* In the meantime, we kept getting new ATIS reports via ACARS as new ones were coming quick.

Around our base turn, approach announced that there was a microburst alert for the airport. We elected to continue the approach. The aircraft in front had landed without any incident. As we headed down on final, we got a wind shear caution and a jump of about 25 knots. We both verbally acknowledged the caution and continued. I noted a dust front that was moving across the field as we approached. As we got close to the ground, around 100', the plane got tossed side-to-side and the wings rocked up and down. I called "UNSTABILIZED, GO AROUND" as I felt very uncomfortable trying to land with the plane being pushed all over the place. The Captain applied go around power and made the call outs and I performed those tasks. The go around was still a bit hairy as the plane was still being tossed to where the stick shaker came on for a second even though we hadn't pitched up that much nor that hard and the flaps had not been retracted yet since we hadn't attained a positive rate of climb yet.[15]

Following their windshear encounter, this overloaded crew then busted their assigned altitude, which motivated them to file this report. Notice how variable the reports and experiences were for this crew and for other aircraft. The changes were happening too fast for ATIS to keep up. Their personal observations and reports from other aircraft proved more useful.

As a discussion of hypothetical options, should they have gone around following ATC's "microburst alert"? In hindsight, we could argue, yes. Evaluating their in-the-moment, inside-the-tunnel perspective, we understand why they chose to continue. Microburst sensors are arrayed all around the airport. The sensor reporting the microburst may have been many miles away from the landing environment. It was useful information, but not enough to warrant an early go around. Additionally, the previous aircraft had landed without incident. Signs supported that they were in a safe window of opportunity to attempt landing. They increased their vigilance for additional environmental and aircraft indications. Next, they got a windshear caution and a 25-knot jump in airspeed. Let's assume that these parameters fell within their allowable windshear limits/procedures. Given the other reported indications, however, their decision to continue seems shakier. The FO then reports noting, "a dust front that was moving across the field as we approached." At this point, they have the ATC microburst warning, a 25-knot airspeed bump, and visible dust front. All of these indications supported going around. Still, they continued until they encountered a strong gust at 100′. Do we see signs of plan continuation bias? Does it appear that they lacked a pre-rehearsed go around trigger? It appears that they needed an actual windshear warning or significant aircraft control challenges to trigger their go around.

21.8.3 Hydroplaning

Hydroplaning is an effect where our tires ride atop a layer of standing water. Aircraft tires lose traction with the pavement and slide uncontrollably. Hydroplaning is affected by speed, tire tread depth, and water depth. Runways aid water evacuation through crowning and grooves.

Sometimes, conditions prevent water evacuation from the runway surface. On some runways, the cut grooves are limited to the center half, leaving the remaining runway portions smooth. Also, as thousands of aircraft land on runways over many years, the original centerline crowning begins to flatten out. Another complicating factor is the effect of strong crosswinds.

BOX 21.5 CROSSWIND-CAUSED STANDING WATER

I was on final following a brief, heavy rainfall. The preceding aircraft reported significantly degraded braking action. I noticed that the entire right side of the runway appeared to have standing water while the left side only appeared wet. This effect was caused by a strong right crosswind and runway crowning which slowed the evacuation of water toward the right edge. Rainwater pooled on the right side of the runway making it appear like the smooth water surface. That same right crosswind assisted with the draining of the water off of the left side of the runway making the pavement appear only wet. I noticed that the previous aircraft had landed on the runway centerline, so their right-main tires probably hydroplaned on the standing water while their left-main tires rolled normally. This explained their controllability and braking difficulties. I informed my FO that I planned to offset slightly to the left side to keep all of our wheels on clear pavement. We experienced a smooth rollout with GOOD braking.

Because it is a transient effect, predicting the presence of standing water and the threat of hydroplaning is difficult. ATIS is unreliable because the reports don't keep up with a condition that may only last for a few minutes following heavy rainfall. Heavy thunderstorms may be reported in the vicinity, but what really matters is the rainfall during the few minutes before we land. Often the most reliable information comes from the aircraft immediately preceding us.

Consider an event when a heavy rainfall falls on a runway, then stops. We would expect the center portion of the runway to clear quickly as the water flows through the grooves and away from the crowned slope. The sides of the runway would evacuate water more slowly as water from the center backs up against water already pooled there. This would cause standing water to pool along the sides for longer periods. We rarely notice this effect because we strive to land on the centerline.

What if, as we are in the flare, a strong crosswind gust pushes us toward the side. As we drift, our downwind tires enter the standing water and begin hydroplaning. This adversely affects our directional control. The further we drift, the deeper the water becomes and more hydroplaning we experience. While this is a rare situation, it follows the profile of several runway departure mishaps following heavy rain. Pilots reported being blown downwind by a gust and then encountering uncontrollable hydroplaning. We can counter this threat by making centerline alignment a high priority. If we experience a strong gust during landing and are unable to hold centerline, we can reject our landing, go around, and try again after the standing water clears. Most manufacturers use thrust reverser activation as the point that separates this go around option with continuing the landing. As long as we haven't initiated thrust reverse, we can reject the landing and go around.

21.8.4 Heat and Tires

High ambient temperatures, heavy breaking, and fast/long taxiing generate high tire temperatures. This appears to increase the probability of tire failure and tread separation events. These can result in airframe damage, hydraulic leaks, and fuel leaks.

- *Preflight the tires carefully:* Trends seem to indicate that the probably of tire failure increases in tires that have been recapped many times. Look for evidence of splitting through the tread pattern or around the edge between the recap and the original tire.

- *Release the brakes at the gate:* Evidence suggests that releasing the brakes slows the transfer of brake pad heat to the tire. Ensure that the tires are securely chocked before releasing the brakes. A useful technique is to keep the brakes set while the ramp agents connect the tow bar. This protects against unintentional aircraft movement and prevents potential injuries and damage to the aircraft entry door. After the tow bar is connected and the wheel chocks are installed, release the brakes.
- *Respect rotation speed:* Virtually all tire failure events occurred at or just after aircraft rotation. Don't delay rotation unless precautionary profiles require it.

21.9 BALANCING APPROPRIATE OPERATIONS IN MARGINAL CONDITIONS

We are entrusted with wide discretion when managing marginal conditions. The flight schedule and our desire to complete each flight on time push us forward. Countering these go-forces, we have risk management, our safety mindset, and our intuition that encourage us to slow down or stop. We need to find the balance point between these influences. The most useful standard is *appropriateness.* The most appropriate choice acknowledges the go-force of completing the flight, but balances it against the stop-force generated by risk management. Like a mediator in a dispute, we weigh both sides and find the appropriate middle ground that maximizes safe operation and minimizes threats. Following are some practical techniques for balancing and choosing the appropriate game plan.

21.9.1 Manage Stress and Keep the Pace Deliberate

The common advice we hear is to slow down, but that doesn't get the job done, does it? As experienced aviators, we can operate quickly when we need to. The hazard we need to guard against is rushing. When does moving quickly become rushing? When we accurately perform all the necessary steps of each task, we are working quickly. When we start taking shortcuts, combining steps, omitting steps, or compromising our crew communication, we are rushing.

21.9.2 Actively Manage Risk

We need to actively manage emerging problems and apply viable solutions. Hoping that our situation will somehow work out risks allowing it to veer off toward undesirable outcomes. Sometimes, active management means slowing down. Other times, it means speeding up. The key is to modulate our operational pace so that we can detect critical conditions, form effective game plans, communicate them with all team members, and carry them out.

21.9.3 Manage Rest for a Potentially Long Day

Manage rest for the longest legal duty day. Just because we are scheduled for a short day doesn't mean it will stay short. Scheduling can reroute us and extend our workday. Begin the trip rested and continue to manage rest for the longest possible duty day. If, despite our efforts, we end up fatigued, we need to make the fatigue call, get pulled from the flight, and file a fatigue report.

21.9.4 Follow Procedures to Control Complexity

Procedures are designed to control complexity. Each procedure is an integral thread of the operational system. Woven together, they form strong barriers against mishaps. Born from events that we've never heard about, crafted from engineering objectives that we've never seen, and woven into the tapestry of all other procedures, they protect us against latent vulnerabilities. When we bypass a procedure, rush a checklist, or skip a briefing, we loosen that protective fabric. Maybe nothing will happen or maybe the whole thing will unravel. Procedures are repetitive and cumbersome, but following them is the best way to protect against the many underlying threats that we don't know about. Fulfill the standard of *reasonable due diligence*. Take a reasonable amount of time and apply a reasonable amount of effort to effectively assess conditions and select workable options.

21.9.5 Detect and Communicate Warning Signs

Someone usually detects the warning signs before a problem becomes uncontrollable. Where we sometimes fail is effectively communicating that information with other crewmembers. Sharing concerns is what makes teams resilient. If we see something and aren't sure what it means, talk about it. Get it out in the open for everyone to evaluate.

21.9.6 When Things Go Wrong, Start with the MATM Steps

When the unexpected happens, employ the four basic steps of MATM – Maintain aircraft control. Analyze the problem. Take appropriate action. Maintain situational awareness. Make enough time to discuss the situation and agree on the action plan before launching into a remedy. Effective teams working together rarely make big mistakes.

21.9.7 Ask for Outside Help

Successful crews seek outside help to deal with unique or difficult situations. Certain individuals within our organizations routinely deal with marginal operations every day. Over the span of weeks, they encounter more rare events than most of us will experience over our entire careers. Their advice can clarify the situation and answer our questions. They can inform us about previous events, how the crews handled them, what went wrong, and what went right. Ask for their input. Then, assess the appropriateness of their advice. Remember that company organizations tend to have a go-bias. In some cases, crews felt pressured to continue even when they personally felt that it was too risky. Ultimately, we are in the best position to judge rapidly changing, marginal situations.

21.9.8 Don't Be Misled by the Success of Others

Beware of the precedent set by the success of others. Consider a situation where we are within a long line of arrival aircraft. When everyone before us is landing successfully, it promotes a herd mentality that implies that we will also succeed. In steadily deteriorating conditions, however, someone will be the first to fail. While we are encouraged by the success of others, we still need to assess the conditions for ourselves. Perhaps we are heavier than other aircraft. Maybe our approach speed is higher. To appropriately managing risk, we cannot become overly swayed by their success. Someone needs to be the first to say stop. Someone will be the first to divert. Today, it might be us.

21.9.9 Ask Questions to Gauge the Appropriateness of Continuing

Following are some questions to calibrate appropriate decision making in marginal conditions.

- How are conditions trending?
- What are the probable consequences if our decision to continue goes poorly?
- What factors are working in our favor?
- What factors are working against us?
- Which risk factors are trending worse?
- How are these risk factors interacting?
- What can we do to improve our safety margin?
- What is our gut-felt intuition telling us?

NOTES

1 For reference, the FAA Runway Condition Assessment Matrix (RCAM) uses the following Runway Condition Codes (RCC) values: dry-GOOD (RCC 6), wet-GOOD (RCC 5), GOOD TO MEDIUM (RCC 4), MEDIUM (RCC 3), MEDIUM TO POOR (RCC 2), POOR (RCC 1), and NIL (RCC 0).
2 Dismukes, Berman, and Loukopoulos (2007, p. 294).
3 Dismukes, Berman, and Loukopoulos (2007, p. 294).
4 Dismukes, Berman, and Loukopoulos (2007, p. 294)
5 Dismukes, Berman, and Loukopoulos (2007, p. 43)
6 Dekker (2002, p. 77).
7 Edited for brevity and clarity. Italics added. NASA ASRS report #1596532.
8 NTSB (2000) – American Airlines MD-82 accident in Little Rock (LIT) on 1 June, 1999. For a comprehensive human factors analysis, reference (Dismukes, Berman, & Loukopoulos, 2007), Chapter 19.
9 Klein (1999), Chapter 10.
10 Klein (1999), Chapter 8.
11 FAA (2014) – FAA Advisory Circular 91-79A, dated 9/17/14 directs "… an additional safety margin of at least 15 percent should be added to that distance. Except under emergency conditions, …"
12 We discourage the practice of giving mixed braking action reports, so this report should be, "Braking Action MEDIUM".
13 FAA (2014) – From the Pilot's Runway Condition Assessment Matrix (RCAM).
14 Edited for brevity and clarity. Italics added. NASA ASRS #1521830.
15 Edited for brevity and clarity. Italics added. NASA ASRS report #1689503.

BIBLIOGRAPHY

ASRS – Aviation Safety and Reporting System – National Aeronautics and Space Administration (NASA) Online Database. (n.d.). Retrieved from Aviation Safety and Reporting System: https://asrs.arc.nasa.gov/search/database.html.

Dekker, S. (2002). *The Field Guide to Human Error Investigations*. Burlington, NJ: Ashgate Publishing Company.

Dismukes, R. K., Berman, B. A., & Loukopoulos, L. D. (2007). *The Limits of Expertise: Rethinking Pilot Error and the Causes of Airline Accidents*. Burlington, NJ: Ashgate Publishing Company.

FAA. (2014, September 17). *91-79A - Mitigating the Risks of a Runway Overrun Upon Landing*. Retrieved from FAA Advisory Circulars: https://www.faa.gov/documentLibrary/media/Advisory_Circular/AC_91-79A_Chg_2.pdf

Klein, G. (1999). *Sources of Power: How People Make Decisions*. Cambridge, MA: The MIT Press.

NTSB. (2000). *PB2001–910402, NTSB/AAR-01/02, DCA99MA060 – Runway Overrun During Landing – American Airlines Flight 1420 – McDonnell Douglas MD-82, N215AA – Little Rock, Arkansas – June 1, 1999*. Washington D.C.: National Transportation Safety Board.

22 Non-Normal, Abnormal, and Emergency Events

The conditions preceding emergencies interact in ways that challenge the limits of our knowledge, experience, decision making, and time management. In this chapter, we will examine how our emergency mindset differs from our daily line-flying mindset, how mismanaged events lead to mishaps, and techniques for skillfully managing these events.

22.1 HOW SIMULATOR TRAINING AND LINE-FLYING EMERGENCIES DIFFER

Training in a new aircraft, we learn to use its emergency procedure guides. These are titled Non-normal procedures, Abnormal procedures, QRH (Quick Reaction Handbook), QRC (Quick Reaction Card), EAC (Emergency and Abnormal Checklist), and others.[1] The guides are divided into categories like Engines, Hydraulics, Air Systems, and General. Each procedure references a title, a summary of indications that identify the event, and a sequence of remedy steps intended to resolve or stabilize the malfunction. The procedure may direct us to complete additional procedures needed to secure the malfunction. For example, after we complete an inflight engine failure checklist, it directs us toward follow-on procedures for inflight engine restart or landing with an inoperative engine.

As we learn a new aircraft, our instruction starts with simple non-normal events. A warning light illuminates, we identify the required procedure, access the appropriate checklist, perform the remedy, and the light extinguishes. After mastering these straightforward malfunctions, our training scenarios increase in complexity and difficulty.

22.1.1 Non-Normal Events Interrupt Our Operational Flow

Whether flying the aircraft or the simulator, non-normal events interrupt our operational flow. Consider a simulator training event where the instructor gives us an engine overtemp during start. We detect the rapidly rising temperature, recognize the indications of an impending engine overtemp, abort the start (often using boldface memory steps), reference the checklist, complete the remaining remedy steps, and announce, "Checklist complete." The instructor checks the box on our grade sheet, resets the simulator, and states, "OK, that's cleared up." Since the malfunction is completely removed, we immediately return to our normal operational flow. We rarely accomplish all of the real-world tasks like calling station operations, returning to the gate, shutting down, making passenger PAs, calling dispatch, calling maintenance, documenting the discrepancy, cancelling the flight, getting rerouted by scheduling,

DOI: 10.1201/9781003344575-25

or scrambling to another aircraft. All of these operational considerations are significant line-flying events, but we don't give them much attention in the simulator.

When we enter the simulator for training, we expect to experience many non-normal events. This leads us to form an emergency simulator mindset. While line flying, we don't expect to encounter any rarely occurring, non-normal events. This leads us to form a normal line-flying mindset. The difference between our emergency simulator mindset and our normal line-flying mindset affects how we perceive and manage actual non-normal events. Different mindsets form different perspectives and different perspectives guide different decision making.

22.1.2 Differences between Simulator Events and Line Events

Simulator training events lack the variability and nuance of line flying, non-normal events. Each actual line-flying emergency emerges from a unique operational flow. These nuances are difficult to recreate in the simulator. Consider the ubiquitous V_1 cut training event. Typically, it starts with an engine failure, fire, windshear, or a "bang". Real-world events prove far more complex. Following is a summary of reported V_1-related non-normal events from the NASA ASRS database from a one-year period (November 2019 – November 2020).

- The ATC Tower controller queried the crew during takeoff prior to V_1 (crew had initiated takeoff without clearance). The Captain elected to continue the takeoff. The FO was returning from a 7-month absence (Report #1769610).
- A highly experienced Captain rejected their takeoff due to a "sudden veer" to the right. They neglected to make a required "Reject" callout. Both the FO and the jumpseater were confused, but monitored and supported the Captain's actions (Report # 1761771).
- Both pilots noticed a flock of birds on the runway approaching V_1. They elected to continue. They hit about 30 birds which inflicted damage to both engines. They returned for an emergency landing (Report # 1759404).
- The crew experienced numerous engine anomalies during takeoff roll (second incident for the same aircraft on the same day). With too many indications to analyze, they decided to reject at V_1 due to the accumulation of unknowns (Report # 1758495).
- The pilots became confused during taxi-out and made an intersection takeoff when full length was planned. Approaching V_1, they detected minimal runway remaining, but continued their takeoff. They would not have been able to stop if they had rejected at V_1 (Report #1751568).
- The EECs (Electronic Engine Control computers) reverted to ALTN during taxi-out. The crew coordinated with maintenance, reset the EECs, and were cleared to continue. The autothrottles disengaged during takeoff as the EECs again reverted to ALTN. The crew reported startle effect, but continued their takeoff. When airborne, they experienced airspeed indication problems (Report # 1748317).
- The crew received a windshear warning at V_1. They successfully rejected the takeoff (Report #1746586).

- The crew experienced anomalous airspeed discrepancies near V_1 and rejected their takeoff. Maintenance discovered mud dauber wasp blockage in pitot static system (Report #1740194).
- The Captain/PM became tunnel-focused on possible engine underperformance and missed both the V_1 and V_R callouts. The FO/PF made his own callouts and rotated. The Captain, who was heads-down, called for a rejected takeoff. The FO informed, "Negative, we are passed V_1." The Captain pushed engines to full thrust and continued the takeoff (Report #1739089).
- The crew reported that multiple avionics blanked during takeoff and rejected 50 knots below V_1 (Report # 1720363).
- One turboprop engine rolled back, but didn't auto-feather after V_1. The crew continued their takeoff and returned for an emergency landing (Report # 1715079).
- The crew experienced multiple anomalies during rotation. While returning for an emergency landing, they had unsafe gear indications and multiple confusing electrical system anomalies (Report # 1704467).
- A spoiler warning at activated at V_R. The crew performed a high-speed reject. The same spoiler warning from the previous flight had been signed off (Report #1702333).
- The crew struck a large (10′ wingspan) bird approaching V_1 causing a very loud bang. They rejected the takeoff. All tires subsequently deflated (the proper functioning of a tire overheat protection feature designed to prevent tire explosion) (Report #1700045).
- The FO lost attitude indications during takeoff into a late afternoon sun. They estimated normal pitch until airborne, then transferred control to Captain who had normal indications (Report # 1699712).

Analyzing these events, we conclude that when crews experienced events that were similar to the V_1 cuts that they practiced in the simulator, they accurately followed the trained procedure. When scenarios strayed from the profiles presented in typical simulator events, their decisions and actions became less consistent. Despite how often they had practiced V_1 cuts in the simulator, many of these crews made significant procedural errors, experienced startle, or became confused by the indications that they saw.

Most engine failures occur well outside of the few seconds near V_1. The more that these events deviate from trained profiles, the more we need to modify procedures. For example, an engine failure that occurs during climbout passing 500′ requires that we modify the typical V_1 cut takeoff profile. So, while simulator training provides excellent practice, real-world events challenge our expectations of timing, causes, and indications.

Non-normal events require focused attention and discussion. This means that we'll need to hold a train of thought as we diagnose and remedy the problem. While flying, operational tasks constantly interrupt us. ATC asks us questions or the FAs call up wanting to know what is going on. Simulator training typically reduces this real-world complexity. These differences unintentionally promote a mindset where we view simulator training as a fundamentally different experience than line flying.

22.1.3 Using Debriefing to Reframe Our Mindset

Ideally, our simulator training prepares us to handle actual line-flying emergencies. Our debriefs need to bridge the gap between the two mindsets. We need to recreate how the event felt while we handled it and how it would feel in the aircraft while line flying. Some useful questions are:

- What was the first indication we detected?
- Were we surprised or startled? How did it feel?
- Did it interrupt ongoing tasks?
- What were our first thoughts about the cause of our problem?
- What other indications did we investigate to support or contradict our first impression?
- Who saw critical indications first? What did they say?
- How effective was our CRM?
- Did we make any procedural or technical errors?
- Was there anything that I did to trip you up?
- Was there anything that you did to trip me up?
- As a crew, how well did we manage the operational flow and workload?
- Did we feel rushed? If so, what caused it? How did we respond?
- What real-world conditions would have complicated this event?

Answering these questions, we gain a greater understanding of how our mindset affects our perspective and work process. We refine our non-normal event handling skills and improve how we would apply them in actual situations.

22.1.4 Advantages of Advanced Qualification Program (AQP) Training

AQP is a simulator training innovation that emphasizes non-normal event handling within a flight profile. It recreates emergency events within a realistic operational flow. Problems challenge us to complete the flight by modifying aircraft systems/landing configuration, diverting to more suitable airports, managing fuel, and completing flight tasks from pushback through shutdown. Throughout the training event, we need to balance priorities, manage separate timelines, handle distractions, consult with outside agencies, and coordinate our efforts as a crew.

22.1.5 Recreating Unexpectedness and Context

Real-world emergencies are both rare and unexpected. Flying year after year without experiencing any noteworthy events builds an expectation of normalcy. We subconsciously develop the mindset where bad events "never" happen in the aircraft. When an actual failure unexpectedly occurs, we can experience strong startle and surprise effects. It takes more time to recover, interpret indications to understand what is happening, choose a resolution strategy, and coordinate the steps to resolve it.

In the simulator, we accept that we are in an artificial environment without enduring consequences. Many operational tasks are simplified. For example, if we need to call our dispatcher, the instructor quickly answers, "This is dispatch." In the aircraft, establishing this communications link may involve a lengthy process. We may encounter delays with reaching our dispatcher, difficulty communicating information, and longer conversations. Even after we reach them, they may put us on hold while they coordinate with their supervisors or research options. During this delay, other operational tasks and distractions disrupt our flying. The AQP experience strives to present a more realistic experience, but even so, many of the coordination steps are still streamlined or quickly resolved.

Flying the line, we encounter far more operational challenges than aircraft malfunctions. This shapes our problem-solving perspective. Reviewing reports from the NASA ASRS archives, we see many cases where operational demands complicated the crews' efforts to manage their non-normal events. These operational demands repeatedly distract us, present conflicting priorities, and push us forward.

Our desire to quickly resolve disruptions can promote rushing. To counter this, we need to apply methodical, event resolution processes from our simulator training while maintaining our operational flow. Consider the following fuel leak emergency encountered by a Boeing 737-800 crew.

BOX 22.1 CREW SURPRISED BY A QRH PROCEDURE DURING FUEL LEAK EVENT

FO's report: During climb around 10,000′ to 15,000′, the PF first noticed the aircraft required an unusually high amount of left rudder trim. … [The crew then details their process of diagnosing a fuel leak event]… Once I retrieved the checklist, the PF transferred control and radios of the aircraft to me for the duration of the flight. We then weighed our options. Because we were by now 2,000 pounds light in our left tank and knowing there are not many airports between ZZZ and ZZZ1, we decided to return to ZZZ. *Somewhere in the process approach asked if we wanted ARFF (Aircraft Rescue and Fire Fighting) standing by and I replied, "affirmative." Approach then asked for number of souls on board and fuel remaining. Meanwhile, the Captain was busy with completing checklists, notifying flight attendants, and notifying company. They informed me that the flight attendants had smelled fuel on departure. The flight attendants smelling of fuel helped us confirm that our decision to return to ZZZ was correct. Then on about a 5–10 mile left base for Runway XX, the Captain came to the part of the Fuel Leak Engine checklist that required us to shut down the left engine. We weighed our options. There was 5,000 pounds of fuel in the [leaking] left tank, the fuel leak had subsided at the low thrust setting, engine indications were normal, there was a lot of traffic in the terminal area (a missed approach wasn't out of the question). We were close to landing and we didn't want to do a single engine missed at a high altitude/high terrain airport. We decided to leave the engine running and complete the Descent and Before Landing checklists.* We landed without event and I stopped the aircraft and transferred control of the aircraft to the Captain.

... The entire flight from takeoff until touchdown lasted approximately 15 minutes. The Captain did an amazing job of delegating responsibilities among the crew while still keeping everyone in the loop. Although there was a lot to do in a short amount of time, I never once felt rushed or in the Red. I attribute this to the Captain setting a great tone on day-one of our trip that promoted great teamwork and communication.[2]

This crew handled their situation and landing expeditiously. Their surprise with discovering the checklist step directing an engine shutdown implies that they were performing the QRH checklist one step at a time. The portion of the checklist directing the shutdown happened to be on the top of the second page of this QRH procedure. We can envision them juggling all of the operational requirements for an expeditious return and working back and forth through each step of the checklist as time allowed. The Captain then turned the page and discovered that it directed shutting down the engine. Surprise! Clearly, they didn't have enough time to continue the approach while shutting down engine, recomputing landing data, finishing the fuel leak checklist, and completing the inoperative engine landing checklist. Going around might have increased their fuel imbalance and further complicated their situation. The Captain made the decision to land with both engines running. While this decision worked out for them, we don't know the underlying reasons why Boeing included this engine shutdown step in the fuel leak checklist. It might have been to prevent streaming fuel from igniting during landing rollout. Had they practiced this event in a simulator setting, they would probably have reviewed the entire checklist first or gone around when they discovered the need to shut the engine down. Within their operational flow, however, they chose to leave it running and land. We can empathize with their sense of urgency to land and how it influenced their decision to continue versus going around.

22.1.6 How Consequence and Responsibility Affect Our Mindset

When we are in the simulator, we know that it is a training event. If we totally botch a procedure and even crash, we just reset the scenario and try it again. There are no enduring consequences. Also, we are expected to make errors in the simulator. Committing and prolonging errors deeply into a practice scenario is useful training. For example, in the aircraft, we would execute a windshear recovery maneuver as soon as we detect strong exceedances in airspeed and flightpath. In the simulator, however, we intentionally push past those warning signs until we are deeply into the windshear conditions and the automated windshear alert activates. This is intended to give us practice with recovering from a worst-case type of windshear event. In this way, we use the simulator for windshear escape practice instead of calibrating our real-world windshear detection and decision-making skills. Some instructors modify the exercise by asking us to tell them when we would initiate our go around if we were actually line flying. While this helps, it illustrates the inherent differences between simulator training and line flying. In the aircraft, safety overrides practice. In the simulator, practice overrides safety.

In the aircraft, we embrace our responsibility to safely move our passengers, crew, and freight to their destinations. In the simulator, we just have a couple of pilots, an instructor, and maybe some observers. This difference in responsibility affects our mindset. Also, few consequences endure following our simulator event. There aren't any public records of our mistakes. A serious simulator outcome becomes a learning opportunity. If we actually skid off the runway at a real airport, it makes the evening news.

Peer pressure is also a factor. We don't want to embarrass ourselves in front of other pilots. Imagine how many low-fuel situations have occurred because the pilots were unwilling to openly declare minimum or emergency fuel over the radio. Consider how many severe turbulence encounters have been reported as "really bumpy" so we don't have to admit that we just flew through conditions that we probably shouldn't have.

22.1.7 The Effects of Noise and Distraction

It is difficult to recreate the line-flying experience of noise and distraction in the simulator. There is less radio chatter, fewer operational interruptions, no smoke or fumes,[3] less wind noise, and less challenging weather encounters. During one AQP training cycle, check pilots were instructed to simulate more background noise and distraction. Most of us found it to be very disruptive to be constantly interrupted as we tried to coordinate and resolve our emergency event. Restoring our operational flow was especially challenging. Many of us count this training event it as the most frustrating simulator experience of our careers.

22.1.8 Recreating Complexity

Instructors present simulator training emergencies using fairly uncomplicated conditions. While they are able to simulate weather and day/night contrast, it is difficult to simulate all of the nuances of distractions, anomalous indications, and confusion. The following report demonstrates how conditions surrounding a single malfunction and its interaction with MELs can create more complexity than the crew ever experienced in training. As you read it, note the compounding factors of:

- Late arrival at the jet and long duty day
- Air conditioning pack and APU deferred by MEL
- Crossbleed start due to no APU
- Pressurization malfunction during takeoff
- QRH didn't cover their particular MEL-directed pressurization panel configuration
- Level off below 10,000′ to diagnose the problem
- Procedure required reduction of engine to idle (effectively single engine)
- Conflicting guidance between continuing to the destination or diverting back
- Different information between MEL and QRH
- Both packs off, the cabin temperature rising to an uncomfortable level – passenger concern
- Priority handling recovery to departure airport
- Heavyweight, single engine landing with reverser inoperative
- Option offered to continue flying which would have pressed duty day limits

BOX 22.2 HIGHLY COMPLEX INFLIGHT EMERGENCY EVENT WITH UNANTICIPATED CONSEQUENCES

FO's report: [assigned to flight last minute after scheduled FO no-showed] … Arrived … as the last of the passengers were boarding. The Captain helped out the situation by completing a big portion of normal FO duties to help expedite the operation. Once complete with my duties and prepared for the flight, we conducted normal preflight briefings and cockpit set up. We started engine #2 at the gate prior to push back with a crossbleed start on [the taxiway].

[The Captain was flying and I was monitoring] With deferred items of Pack 1 and APU, an ECS-off (Environmental Control System) takeoff was required and performed per the takeoff data we received via ACARS. Acceleration, takeoff roll, and rotation were all normal. Shortly after takeoff on climb out, we received the EICAS message for BLEED 2 OVERPRESS. Since we were still in a critical phase of flight, we elected to continue the climb out and departure till we could get the airplane cleaned up and at a safe altitude heading back toward ZZZ.… We elected to level off at 10,000′ and requested delay vectors as we worked the malfunction. Captain maintained PF while I broke out the QRH (Quick Reference Handbook) and ran (BLEED 1(2) OVERPRESS) procedures. The [error] message failed to go out after securing affected bleed button along with the APU and XBleed (Crossbleed) ones which required the associated throttle to be reduced to idle and completion of the Single Engine and Approach and Landing Procedure. Completed the procedure to set up for return to ZZZ on Runway XXR for landing as we were above MAX gross weight for landing with one reverser available. *Of note with both packs now inoperative, the cabin temperature quickly rose to above 90°F for the duration of the flight. [Requested priority handling] with ATC, landing and taxi-in were uneventful,* … Captain called and briefed the on-call Chief Pilot who gave everyone the option to continue on or be done for the day. The original pairing had us landing back in ZZZ at XI:25 (if on schedule) and only gave me 35 minutes until I timed out per FAR 117 (duty day limitation) rules.… I was replaced as I was no longer going to have the duty day required with the delay.

While the experience was similar to what we receive in training, we did all procedures called for with the QRH, and had effective CRM with the flight crew, Dispatch, and ATC, and safely landed the aircraft. There were some lessons to be learned. First would be some limitations with the QRH. *With deferred items and MELs, individual malfunctions can quickly become complex and with switches in non-standard positions it will limit some messages on EICAS. The QRH in this case assumes that the other bleed/pack system is working normally.* Per the QRH if we were able to get the message to clear after securing the affected bleed, APU, and cross bleed buttons, it would direct us to check fuel requirements and continue with normal operations. In our case, we would still have both packs secured (OFF). This would make the cabin extremely hot, but more importantly, we would lose our cabin pressurization.

By securing the switches, we did not have any pressurization messages since we did not exceed 9,700′ cabin altitude by leveling off at 10,000′ with initial pressurization prior to securing the second pack (of note, climb check done at initial level-off at 10,000′ showed normal). Going back to the MEL for 21-XX-XX gives guidance for both packs inoperative, which was our current situation after completing the QRH procedure. *It talks about limiting flights with passengers to 10,000' among other requirements for operation. Nowhere in this QRH procedure is this limitation documented. The QRH also directs the crew to reference single engine approach and landing procedure. During most, if not all of our training scenarios we would execute this procedure after securing the engine due to fire or failure and not have it available. In our case, the affected engine was at idle and available to us, if needed in extremis.* We launched to the west and turned south of ZZZ where we don't have much terrain or high elevation airports to be concerned about performance. *If this had occurred in a number of other high elevation or terrain threat airports the situation would be much different.* It would have behooved us to brief the times when we would use the affected engine if aircraft performance was in question (GPWS warning, terrain clearance on missed approach, etc.). At the end of the day as a crew we discussed most of these issues/possibilities and completed all procedures as published with a positive outcome it is a good reminder that the QRH is not all encompassing and as pilots we need to take into consideration all conditions our airplane is in with both EICAS messages as well as MELs.[4]

Unexpectedness, context, responsibility, noise, and complexity work against us. They expose a number of human biases that affect how we interpret, process, and choose decisions. How we think and act under high stress and time pressure does not mirror our behavior under unstressed conditions.

22.2 UNDERSTANDING HOW EXCEPTIONAL EVENTS BECOME MISHAPS

As we categorize mishaps, we rate some as handled *well* while others as handled *poorly*. These are both labels that we apply in hindsight. Neither label gives us much insight into the chain of events, decision, and choices that the crew experienced during the incident. Viewing events from the crew's viewpoint helps us learn more about it. This helps prevent us from repeating their mistakes.

22.2.1 The View from Inside the Tunnel

Sidney Dekker describes the progression of an event from the perspective of moving along the interior of a tunnel.[5] When we are *inside* of the situation, we only see what is immediately around us (the sides of the tunnel). Unseen indications and unconsidered choices only exist outside of the tunnel. This makes them effectively invisible to us even though they are obvious to observers reviewing the event from their hindsight perspective.

This hindsight bias also affects many mishap crews during their post-incident interviews. Pilots admonish themselves for making "bad" choices and overlooking "good" choices. They regret losing control of a situation even though they felt that they were handling it appropriately and safely at the time. Our challenge is to recreate their inside-the-tunnel perspective. By focusing on what they perceived at each moment, we begin to visualize the reality that existed for them, the indications that seemed relevant, and the choices that seemed logical. Using their event perspective, we study our reactions and improve our own skills. Consider an unstabilized approach where the PF elected to land. The investigators download all of the parameters from the DFDR. They conclude, "The approach was clearly unstabilized. They should have gone around." They interview the PF.

- *Investigator:* "Why didn't you go around when you were 40 knots fast?"
- *Pilot:* "I had no idea that I was that fast."
- *Investigator:* "How could you have missed that?"
- *Pilot:* "I don't recall ever looking at my speed. I was focused on the touchdown zone."

Just because the information was *available* doesn't mean that it was detected or understood. Learning that they were tunnel-focused on restoring a familiar glidepath and achieving a touchdown zone landing, we begin to understand how they never considered a go around. Their choice to continue their approach made sense to them at the time even though it was clearly "wrong" in retrospect.

22.2.2 Our Subjective Perceptions

One factor that contributes to unwise choices is how we subjectively assess our options and available time. Consider an unexpected wind shift that makes us 20 knots fast on final approach. As soon as we see this, we instinctively reduce thrust. Next, we assess several factors. Was our instinctive thrust reduction appropriate? How long will it take to slow back to target airspeed? Do we have enough time to restore stabilize approach parameters? None of these questions are answered with numbers. We measure them subjectively. If we decide that our initial thrust reduction was appropriate, then we leave the thrust back and choose when to restore thrust to hold target speed. If we decide that the initial reduction was too much or too little, we make a second correction. If the 20-knot exceedance happened on 6-mile final, we would conclude that we have plenty of time to regain target airspeed. The same exceedance event on ½-mile final would concern us much more. In neither case would we assign an actual number value to our correction. In the flow of aviation, our decisions, actions, and time assessments all acquire subjective qualities.

22.2.3 Understanding Accident Event Timelines

Studying a mishap event, we survey the scenario along its timeline. We read the CVR transcript and plot their communications and actions. Using hindsight, we identify the moment when they made their questionable decision. We calculate that it took them 3 minutes and 25 seconds to discover their error. While this is a factual measurement of time, it doesn't help us understand why they selected that ill-fated decision, why it took them so long to detect their error, or how the passage of time felt to them as they moved along inside of their tunnel.

To understand their perspective, we recreate their mindset starting from a point well prior to their unwise decision. We imagine their workload, why they selected their game plan, how they thought it would play out, and which parameters would have seemed important. We discover the flaws in their SA that shaped their mindsets. Viewed from their mindsets, we appreciate how their errant decision might have seemed reasonable at the time. In their post-mishap interview, they reported that time felt like it was moving much faster than the 3 minutes and 25 seconds that we calculated. They remember being very busy, doing lots of different tasks, considering alternatives, and struggling to understand their situation. The gaps that we see along their timeline may look like they were wasting valuable time. In fact, these gaps were filled with wondering how the problem developed, why they didn't realize it sooner, searching for the problem's cause, and weighing whether to continue or change their game plan.

22.2.4 Calibrating Our Personal Awareness and Abort Triggers

It is difficult to recognize flawed mindsets while under stress and high workload. Once mishap pilots become immersed into their biased, inside-the-tunnel, limited perspectives, the operational momentum carries them forward. To avoid falling into this trap ourselves, we refine our skills at identifying the precursors to unwise decisions that sustain flawed game plans. We develop the ability to recognize what an ill-fated situation looks like and feels like *before* it deflects our intended trajectory.

Using the mishap crew's experience, we recreate how their event would have felt to us. Working from this in-the-moment perspective, we improve our ability to detect deteriorating parameters, counterfactuals, and warning signs. We notice how our own habits and mindsets might contribute to similar failing scenarios. Next, we construct safeguards that trigger us to switch our game plans to contingency backup options. To practice, we mentally rehearse scenarios where we reach these trigger points and automatically switch to safe backup plans. This refines our personal awareness and perception.

Consider a range of aviation deviations. As minor deviations arise, we resolve them quickly without giving them much thought. As deviations become more severe, complex, or consequential, our tolerance level begins to factor in how much time we have available. As time runs short, our deviation tolerance shrinks. It would take a severe deviation (like a windshear warning) on long final to trigger our go around. Our abort trigger tightens as our available time shortens. At 1,000′, we might accept momentary deviations up to 20 knots. Approaching the flare, it shrinks to 10 knots. In the flare, we would go around for any significant deviation that threatens a controlled landing in the touchdown zone.

In the same manner, we establish triggers for confusion, indecision, or quickening. For example, if something unexpected happens when we are close to landing, we would immediately switch to our go around contingency option. Rather than risk continuing a possibly failing scenario, we initiate a go around, make time to evaluate what happened, and reattempt the approach. Mishap crews often lacked this discernment. Absent a premeditated rejection trigger, their default plan was to work harder, force their existing plan to work, and continue.

22.2.5 Operational and Personal Priorities

A blend of operational and personal priorities influences our decision making. If our approach is unstabilized but we have a passenger suffering from a heart attack in the cabin, we would continue to land as long as we believed that we could stop safely. Afterward, our company might praise our choice even though it violated stabilized approach criteria. Personally, we would view this as an appropriate use of emergency authority.

Conversely, if we were flying the same unstabilized approach, with plenty of fuel, no emergency, but had a tight connection to make our commuter flight home, our decision to land would clearly be a violation of company directives. Even though we could complete the landing just as safely as with the medical emergency scenario, operational priorities don't justify the violation. Our personal desire to make our commuter flight home is not an acceptable justification for the use of emergency authority.

22.2.6 Fatigue

Fatigue often emerges as a factor with mishap events. It seems to affect our willingness to choose time-consuming contingency options. We lose our motivation to select the appropriate choice. A go around feels complicated and time-consuming. It's easier to just land and get the aircraft to the gate. Fatigue also affects our ability to detect, understand, and prioritize counterfactuals. Mishap crews often miss critical parameters as their fatigued minds process less, understand less, and settle for less.

BOX 22.3 FATIGUE CONTRIBUTES TO PLAN CONTINUATION BIAS

Captain/PM's report: We were descending into ZZZ and were set up for Runway XX. Approach advised that winds had shifted and asked if we would like Runway XY. We accepted and began to set up for XY. When we had the airport in sight, we were cleared for the visual. As we armed the approach for the autopilot, we were told by Approach that the localizer and glideslope may not be working. *My First Officer turned off the autopilot and began hand-flying the visual. He became distracted by configuring the airplane and was flying through the localizer when I informed him [that] he needed to start his turn. As he began the turn, he started descending, which caused a GPWS*

> *warning for a couple seconds due to terrain.* We were confident of terrain clearance but leveled off until we were back on the localizer. Once we were on the localizer, we were high. We began our descent and continued configuring. *We weren't fully configured until about 800′ AGL.* We proceeded with the landing without further event. *I think fatigue played a factor in our poor decision making to continue the approach. We had a [before-dawn] wake-up call that morning and this was our 4th leg of the day. We had been dealing with thunderstorms and delays all day long.*[6]

Notice that the Captain is the PM in this event. We expect Captain/PMs to interdict failing profiles. Instead, this Captain not only allowed the approach to continue, but actively assisted the FO with continuing. In hindsight, the Captain acknowledged that their event exhibited "poor decision making" due to fatigue.

22.2.7 Difficulty Processing Relevant Information While Overloaded

Complexity and quickening degrade our ability to process relevant parameters. When we become overloaded, we narrow our attention. The parameters we choose to monitor may not be the most relevant, but they are the ones we judge to be most important at that moment. Consider a turbulent final approach in hot weather conditions. The aircraft is buffeted as rising thermals pull us up and sinking air masses push us down. Working hard, many of us narrow our focus to maintaining our aimpoint in the touchdown zone. Monitoring airspeed drops to secondary importance. Monitoring sink rate drops to third. It is not that airspeed and sink rate aren't important. It is just that maintaining our path for the touchdown point consumes most of our attention and effort. On some subconscious level, we choose to accept deviations of airspeed and sink rate as being "close enough" as long as we maintain our aimpoint. After landing, mishap pilots often can't recall their parameters. They remember aiming for the touchdown point and that other parameters *felt* good enough.

22.3 EMERGENCY PROCEDURE STRATEGIES

When emergency events go poorly, it is often because we mismanage the situation (misdiagnose the problem or perform the wrong remedy) or we mismanage our time (rushing, skipping, or abbreviating procedures). Consider that most emergency events progress through five phases. We manage time differently in each phase.

1. Initial reaction and maintaining aircraft control
2. Analyze/Assess the problem
3. Develop a game plan to resolve the problem
4. Take appropriate action
5. Maintain situational awareness

22.3.1 Initial Reaction and Maintaining Aircraft Control

In our standard MATM response, this is contained under the first step, *maintain aircraft control*. Our first priority is to maneuver the aircraft from a hazardous position to a stable, safe, manageable position where we'll have enough time to deal with the problem. This includes all of the steps needed to get the aircraft into that safe position. This phase may be very short, like dodging a bird in our flightpath, or protracted, like controlling a damaged aircraft to achieve stable flight. Initial reactions may apply immediate action and boldface emergency steps performed from memory.

Our initial reaction starts at the first moment we detect the problem. Consider an engine failure during takeoff at V_1. Our list of immediate tasks include tracking straight down the runway, rotating smoothly, safely lifting off, raising the landing gear, climbing to single engine level-off altitude, raising flaps, and accelerating to a safe maneuvering speed. During this early stage of the emergency event, we'll have little time for analysis or planning. Once the climb is established, the PM may go heads-down to evaluate the engine instruments, but as a crew, our priorities center on completing these initial action steps.

Let's complicate the scenario. Assume that this runway requires an engine-out turn procedure due to nearby terrain. Add communicating our intentions with ATC and challenging weather. Complicate it further by making it an engine fire that is continuing to burn or severe engine damage that is violently shaking the aircraft. As we are dealing with all of these complications, ATC is calling, the FAs are chiming the intercom, and the engine is still burning or shaking. Like in an emergency room during a mass-casualty event, we need to perform triage and take care of our most pressing problems first. We tell ATC what we need and that we will call them back later. We direct the FAs to standby. We concentrate on flying the aircraft through the special engine-out routing. This all happens before we open the QRH. During this initial reaction phase, we pay little attention to time management. We are just too busy. This initial phase is all about *doing*. The good news is that we regularly train managing our initial reactions to these kinds of events.

22.3.2 Analyze/Assess the Problem

After we reach a safe, sustainable position, the next phase directs us to analyze and assess the problem. From our MATM emergency steps, this is, *analyze the problem*. From the Risk and Resource Management (RRM) model, it is the "A" (Assess) from ABC*D*. We probably performed a quick assessment during our initial reaction phase, but this assessment is much deeper and involves other crewmembers. The more complex the situation, the more time we devote toward making our assessment.

Our assessment determines how we will measure our time available for the remainder of the emergency event. The more complex the situation, the more time we'll need to devote toward analyzing the problem. For example, if we threw an engine fan blade that damaged the aircraft, we would analyze our engine instruments to assess associated subsystems failures like fuel leaks, hydraulic leaks, electrical failures, and loss of aircraft pressurization. We might call back to a deadheading pilot to have them survey the damage from a cabin window. We may ask airport officials to search for aircraft debris on the runway or to share eyewitness reports of the failure event. This analysis/assessment phase is where we determine the extent and significance of our problem.

This is the phase where some scenarios start veering off toward unfavorable outcomes. One pitfall is latching on to our first assessment and launching too quickly into the QRH. Why do we do this? First, this is how we do it in the simulator, especially when we know exactly what is happening. Having performed engine failure training events many times, we execute our simulator version of an engine failure resolution game plan. Second, the emergency pushes us into the Red. We are highly motivated to do whatever it takes to move back toward the Green. Third, the engine failure completely disrupts our desired operational flow. We want to restore our familiar comfort zone as quickly as possible. In some ways, the emergency feels like a distraction that we need to remove so we can get back to our familiar operational flow. Fourth, emergency events tend to be extremely rare and scary. We are immediately pushed from a familiar, normal situation into an unfamiliar, threatening situation. Securing the engine quickly restores some sense of normalcy.

A more effective strategy is to analyze the problem, determine the tasks that we will need to complete before landing, and choose an initial game plan that will get the aircraft safely on the ground. If we estimate that it will take us 15 minutes to accomplish these tasks, we won't want to immediately maneuver back for final. If the event requires us to land as soon as possible, we would like to arrive near final approach soon after finishing our preparation tasks. So, our analysis needs to accommodate our recovery profile before launching into the checklist. Before we start *doing*, we need to *plan* for the time we'll need to prepare the aircraft.

Back to our example scenario of an engine failure at V_1, now that we have our aircraft in a safe position, we have time to analyze the problem. We examine the aircraft indications, discuss them as a crew, and agree on our assessment of the problem. We envision the list of tasks that we will need to complete, the importance of each, the order of accomplishment, and how long they will all take. We compare this with where we are, where we want to land, and how long it will take to reach a favorable position on final approach.

22.3.3 Develop a Game Plan to Resolve the Problem

The next step is to develop a game plan that balances our priorities. This is the "B" (Balance) of ABC*D* from the RRM process. Our initial game plan may be little more than a rough outline. Don't spend much time on it because it is only a starting point that we will refine along the way. With our engine-out example, we decide that we need at least 10 minutes to complete our checklists. When we add agency coordination and passenger PAs, it rises to 15 minutes. Assuming we want a 10-mile final, we start a timer to remind ourselves to start back toward our intended runway in about 5 minutes. This balances priorities and aligns our flight profile with the expected workload.

A challenge to balancing is over-analyzing or under-analyzing our options. We either spend too much time questioning our choices or too little time evaluating options. Identify the tasks that we definitely need to complete before landing. Resolve them first. If there is more time, then we can address the secondary tasks. The gear incident cited earlier in this book (gear handle stuck in the up position and nosewheel retraction during rollout) is an instructive example. The crew had plenty of time to work on their problem because they wanted to reduce fuel weight anyway. This gave them ample time to perform several fly-bys to allow ground observers to examine the gear. They spent almost an hour working to get the wheels to indicate down and green. When they finally achieved this resolution, they landed without further delay.

22.3.4 Take Appropriate Action

From the MATM process, *take appropriate action* means that we have reached the point where we can begin performing the remedy steps for our emergency ("*D*" for Do in the ABC*D* process). Assuming that we are following a viable game plan, taking appropriate action tests our time management skills. We have an initial game plan, but we also have many demands on our time. Besides all of the flightdeck tasks, we need to consider calling the station, dispatch, the FAs, and making passenger PAs. Do we ask for a 12-mile final or an 8-mile final? How do we want the FAs to manage the cabin and passengers? Do we delay our landing to reduce fuel weight or land promptly? The list of considerations can be lengthy.

22.3.5 Maintain Situational Awareness

Our final emergency step of MATM is *maintain situational awareness*. This is not really a process step since we continuously build SA to provide context to the ongoing emergency procedure resolution process.

- *Guarding against let-down:* As our workload eases, we tend to relax. Maybe we have worked ourselves back into the Green, or at least into the Yellow. Experience shows us that this is not the time to let our guard down. As an emergency evolves, we progress from the adrenaline rush of our initial reaction, through the busy proactive work phase of running checklists and forming a game plan, to executing that plan. Once we complete these steps, it can feel like the pace is slowing. Having resolved the problem, we recognize the feeling of returning to our comfort zone. We can subconsciously fall back into our familiar habit patterns. The problem is that conditions have changed. What appears familiar really isn't. For example, if we have a partial hydraulic system failure, the aircraft may fly normally. Maybe we even have a normal landing configuration. However, certain unavailable or degraded systems (like loss of half of our brake actuators, loss of ground spoilers, and sluggish thrust reverser operation) only manifest during landing rollout. We may finish the QRH checklist and declare, "Checklist complete", but still have a number of lingering conditions and consequences that will need to be applied later.
- *Delayed QRH checklist tasks:* After workload eases, it is useful to revisit prospective memory items. For example, if we experienced a loss of engine oil at cruise and elected to shut the engine down, a significant amount of

time may elapse from completing our checklists until we finally land. Some landing considerations may slip our minds. It is useful to revisit the effects of our modified configuration and degraded systems to include a higher pitch attitude due to a lower flap setting, yaw changes as we change thrust, asymmetric reverse thrust, slower response of systems associated with the loss of an engine-driven hydraulic pump, emergency vehicle response, communications with fire/rescue personnel, whether we plan to stop on the runway or taxi clear, passenger PAs, and cabin crew coordination.

- *Reassessment:* When we find ourselves with extra time, we should use it to reassess our game plan. Under the RRM model and ABC*D* process, we recognize that we may need to rerun the process several times to get everything right. For example, as we assess (A) and balance (B), we discover that we need to change our game plan. We'll need to assess and balance again. As we communicate (C), we discover something that we missed, so we return to assess, balance, and communicate again. Maybe we make it as far as doing (D) and discover that our game plan needs to be changed, so we cycle us back and repeat the process.
- *What-if:* With extra time available, evaluate "what-if" contingencies. Start with an assumption that we missed something. What counterfactuals might show up? What are the earliest warning signs that our game plan might need to be modified or changed? Are there shortcuts or compromises that we made earlier in the process that we should revisit? Remember, the event has disrupted our familiar habit patterns. Maybe this is the first time we have ever encountered an event like this. Also, we were time-pressured and stressed when we first analyzed the problem. Only by reassessing the can we accurately see what we might have missed.

  Discuss options as a crew. Maybe one pilot feels uneasy about a particular aspect of the game plan. Asking, "What did we miss?" or, "How can we improve our plan?" encourages input. FOs often defer to confident Captains. They may accurately detect problems, but feel uncomfortable speaking up. When asked why they held back during mishap investigations, they expressed that they weren't sure about their concerns while the Captain seemed very confident with the game plan. It made sense at the time to defer to the Captain's experience and leadership. Voicing uncertain concerns felt like undermining the Captain's authority. As Captains, we need to actively solicit inputs from our team members and invite them to share their concerns.

22.3.6 Adding Time to Manage an Emergency Event

Some variables complicate our landing decision. Should we get the aircraft on the ground right now or take an extra trip around the traffic pattern to prepare? Should we do a fly-by to have ground personnel examine the aircraft? Should we enter holding and call for expert advice? On one extreme are time-critical emergencies like onboard fires where landing as soon as possible is the best option. On the other extreme are drawn-out scenarios like unsafe gear indications that advise us to reduce fuel weight before landing.

Make time decisions using the *value added* test. If adding time might change how we, our crew, or fire/rescue might handle the emergency, then making extra time might be worthwhile. On the other hand, if it won't change how we would prepare our aircraft or fly our approach and landing, then we shouldn't. Consider two examples. In the first case, the crew cannot verify that all of their gear are down and locked. They deem that a fly-by might provide useful information. During the fly-by, ground observers note that one main gear assembly appears only partially extended. Since this affects which QRH option the crew would apply, taking the extra time for a fly-by proves valuable. Compare this with a different observation where ground observers report that all the gear appear to be fully extended. While this doesn't guarantee that the gear will function normally, it does give the crew more confidence that the landing will be normal. For an opposite case, the crew experiences a tire failure event that results in engine damage. They already know that they will be landing with a failed engine, but would also like to know the condition of the damaged tire. In this case, the fly-by might not be warranted because knowing the tire's condition would not significantly change their landing procedures. The most conservative option is to assume that the tire has catastrophically failed and will thrash around during rollout. If, instead, the tire's recap tread just separated, then the rollout will feel normal. Regardless of the case, the crew would still plan their landing for the worst case. The extra time required to make a fly-by wouldn't change that, so there is no value added.

22.3.7 Unknown Causes

Another decision-making criterion is assessing how much we don't know about our situation. If the only information we have to work with is the symptom, we won't know either the cause or the significance. Consider the following account from a Boeing 777 crew that begins with a "bang".

BOX 22.4 LOUD BANG BEGINS AN UNKNOWN CAUSE EMERGENCY EVENT

FO's report: … After departure, climbing through approximately 15,000′, and assigned normal speed, we began to accelerate. *As we accelerated through approximately 280 knots, a loud sound was heard from rear of aircraft, an aircraft shudder occurred momentarily and then an aircraft vibration persisted for the remainder of our flight. The flight attendants immediately started calling from multiple stations. They advised us passengers in the middle of the aircraft were experiencing loud noises and vibrations from outside the airplane and that the passengers were scared. We performed a quick analysis and confirmed we were pressurized, engines were running normally with no VIBS and no EICAS messages. The Captain transferred aircraft controls to me and discussed the situation further with the Flight Attendants. The Captain and I evaluated and agreed the source of trouble was unknown; there was no associated non-normal checklist to address, but that continuing the flight was*

> *also not safe due to the aircraft vibrations we were experiencing through the flight controls.*
>
> I flew the airplane, sent a divert message to company, and talked to ATC while the Captain further coordinated with the Company, Flight Attendants and passengers for landing. *A collective decision was made to land … and to taxi to the gate. We did prep the cabin for evacuation but briefed that a normal landing was expected. Emergency equipment met us after landing and followed us to the gate.* Maintenance said some large sections of air plumbing in belly of airplane, fed by the Ram Air Inlet, had exploded into pieces.[7]

The crew handled this event well. Some instructive points are worth noting.

- *Immediate reaction:* The Captain maintained aircraft control while both pilots performed "a quick analysis and confirmed [they] were pressurized, engines were running normally with no VIBS and no EICAS messages."
- *Supporting indications:* In addition to the shudder and "loud sound", several FAs called to report cabin indications which implied a source located somewhere under the mid-cabin. The vibration was also felt through the aircraft flight controls, adding to their concern.
- *Transfer aircraft control and divide workload:* The Captain transferred control to the FO to free workload to coordinate with the cabin crew and outside support agencies. The FO flew and communicated with ATC.
- *Access checklist guidance:* Since the cause was unknown and no warning messages were indicated, the checklist was unhelpful. Given the persistence of the noise and vibration, it made sense to return to their departure airport.
- *Access external expertise:* The report does not say whether they contacted external expertise. It appears that they didn't. Instead, they decided to land promptly.
- *Form a game plan:* They decided that an expeditious return for an overweight landing was the best game plan.
- *Plan for worst case contingency:* The worst case would have been a catastrophic structural failure during landing. They had the FAs prepare the cabin for evacuation in case this happened. This allowed the FAs to rehearse their actions and prepare the passengers.

22.3.8 Extreme Complexity

Another characteristic of difficult emergency situations is extreme complexity. Most simulator training events feature singular causes that follow a clear progression from symptom detection, cause recognition, and QRH remedy. Extremely complex emergency events, however, create a variety of symptoms, many of which are not covered in QRH checklists. Resolution requires innovation. Consider the following steps for resolving an extremely complex emergency situation.

1. Perform an initial assessment of the symptoms.
2. Determine if any primary systems are lost or degraded.
3. Assess how much time is available before landing.
4. Identify working systems that can be used in a recovery game plan.
5. Allocate workload to handle coordination and operational tasks.
6. Develop and communicate a game plan and backup contingencies.
7. Execute the game plan.
8. Monitor for counterfactuals.

Notice that "determine the cause of the problem" is not included. In extremely complex emergencies, determining the cause may not be possible and will only divert effort and time better spent toward innovating and executing a viable game plan. Consider the following account from two pilots who encountered an extremely difficult and complex recovery following multiple birdstrikes.[8]

BOX 22.5 MULTIPLE BIRDSTRIKES LEADS TO COMPLICATED EMERGENCY LANDING

First Officer: It was a night departure. I was flying and had just accelerated to 250 knots and turned south on the departure. We were passing about 2,500′–3,000′ when I saw about 15–30 large white geese illuminated by the landing lights. Immediately, I called, "Birds!" and ducked down. *We heard several very loud impacts as they hit the aircraft. Both forward windscreens were immediately covered by bird goo. This dried instantly, leaving us both with no forward visibility.* Noise made flight deck communications very difficult. The most disturbing noise came from a flapping bird wing that had wedged itself under the Captain's wiper blade.

Captain: We needed to dislodge the bird wing to reduce the noise, so I operated the wiper blade once. The wing dislodged and cleared the aircraft. *Our immediate concern was the engines.* They both looked fine and we didn't smell anything, so we assumed that they had missed the engines. Later, the Flight Attendants reported that they did smell burnt feathers. *While the FO continued to fly the aircraft, I informed ATC that we had experienced multiple large bird strikes and needed a return pattern. Additionally, I called the Flight Attendants and the station.* I completed the [landing performance computation] and informed the station of the impending overweight landing. *Since the engines were operating normally, we did not rush our return pattern and took the time to prepare for the approach and landing. We were concerned about our loss of forward visibility and decided to perform a HGS CAT IIIA approach.* [This mode provides accurate guidance for zero-visibility landing and rollout.]

First Officer: As we set up for the approach, *it became apparent that neither ILS receiver was working. We later discovered that additional bird strikes had destroyed the ILS antennas.* We reevaluated our forward windscreens and determined that we *each had a very small section of windshield that was not glazed over. The Captain took control of the aircraft to fly a visual approach.*

Captain: I called for the gear. *The nose and right main showed green. Even though the red lights went out, the left main green lights remained blank. I performed a lights test and the lights were still out. We broke off the approach and asked for vectors back around.* I asked the FO to replace the bulbs. After he did so, the gear down light showed green. We continued for a landing. *We were able to turn off of the runway, but the combination of the windscreen obscuration and the ramp lighting made taxi-in impossible. We called for a tug to tow us in.*

First Officer: After arriving at the gate, we surveyed the damage. We had experienced at least two bird strikes to the forward radome. The radome had shattered and the birds had done extensive damage to the radar and forward portion of the cabin pressure vessel. We had also incurred significant damage to the tail section.

In their process of handling this situation, notice how this crew:

- Retained PF and PM roles during the initial reaction
- Assessed primary systems
- Determined that they did not need to land immediately
- Established a quieter environment by removing the distracting noise
- Assessed their primary limitations – in this case, forward visibility
- Built a game plan based on workable systems and approach options
- Discovered system deficiencies with the loss of their ILS antennas
- Encountered gear status discrepancy and flew an additional pattern
- Diagnosed and corrected the burnt-out, gear-down light bulbs
- Continued for a Captain-flown, HGS-assisted visual landing
- Determined that they lacked adequate visibility to safely taxi
- Arranged for tow-in

22.3.9 Sharing Our Stress Level with Other Crewmembers

In normal operations, we use well-practiced habit patterns and familiar procedures to match proven game plans with past successful outcomes. With emergency operations, we apply rarely practiced and unfamiliar procedures, usually under stressful, marginal conditions, to devise game plans that we may have never used before. Since these events force us out of our comfort zones, we need to acknowledge our stress levels, mindsets, and personal biases. This is a good opportunity to use the colors in the RRM model. Declaring that we are "in the Yellow" or "in the Red" informs the other pilot of our personal stress level and signals that we are in a fundamentally different operational environment that requires fundamentally different techniques.

22.3.10 Using QRH Checklists

Under normal operations, each pilot performs their own independent flow to set switches and systems. Then, we complete a familiar checklist using a read-verify-respond format to verify that they have been properly set. QRH/Abnormal checklists follow a different format. We don't use flows to preset switches and systems. Instead, we deliberately evaluate and perform each step, one at a time.

1. Read the checklist action step.
2. Decide if it applies and agree on the action before proceeding.
3. Direct the action step.
4. Complete the action step.
5. Verify that it was performed correctly.
6. Respond that the checklist action step is completed.

This deliberate process is particularly susceptible to distraction and interruption. Most of the potential sources of interruption (ATC, FAs, station operations) are unaware that we are busy. To ensure checklist accuracy, we should carefully track each step. If we are unsure following a distraction, we should back up one step. If we become completely distracted, then we should start over.

Our first crew challenge is identifying the correct QRH checklist. Once we agree, it helps to scan through it to familiarize ourselves with the sequence of steps and the expected aircraft state that we should have after finishing it. Recall the earlier example of the crew with the fuel leak that discovered (after they turned the page) that the next step required engine shutdown. At that point, they were already on final, so they elected to land. Had they scanned the checklist earlier, they would have accounted for the need to complete the Engine Failure/Shutdown Checklist, Single Engine Landing Checklist, and single engine landing performance numbers.

22.3.11 Temporarily Dividing Workload and Duties

A fundamental objective of flightdeck monitoring is that every action should be verified or verifiable by the other pilot. This helps us to detect and correct errors. Unfortunately, this process is inherently slow. Under non-normal scenarios, we may be forced to complete more tasks than our meticulous verification process allows. In these situations, we find it useful to split roles and tasks for a short period. For example, one pilot (typically the FO) flies the aircraft and handles the operational flow including ATC coordination. The other pilot (typically the Captain) speaks with cabin crew, coordinates with outside agencies, and completes any "heads-down" tasks like ACARS messaging, FMS programming, and reviewing QRH checklists. When all of the preparation work is completed, the crew reforms to verify each other's work, build the game plan, communicate a shared mental model, complete QRH checklists, and prepare for landing.

22.3.12 Studying Our Personal Biases

We improve our personal skillset by studying the vulnerabilities and biases that surface while handling emergency scenarios. While some of us are quite good at handling emergency events, none of us are better at them than with everyday flying. The moment the alarm sounds or the engine bangs, startle and surprise release adrenaline. The limbic system (the emotion-processing portion of our brain) can over-stimulate and inhibit the reasoning ability of our cortex (the logical thinking portion of our brain). We may experience a period of indecision, confusion, and difficulty recalling important information. Training and experience help us to recover quickly. Immediately applying some simple recovery steps like MATM also guide the process. In any case, it may take a few moments to settle our nerves before we can effectively analyze the problem. Work together as a crew. If we need more help, we can bring in outside resources like jumpseaters, other pilots traveling in the passenger cabin, operations center pilots, and maintenance technicians.

Accept that stress will cause our attention level to heighten and our focus to narrow. While this may be exactly what we need to process our problem, it might also lead us to miss important indications by limiting the parameters that our minds can process. This is not a failing. It is a natural human reaction and bias that emerges under stressful situations. If we are aware of our particular personal biases, we can take measures to compensate. Inform the other pilot, "I'm going heads-down to review this QRH procedure. Please take care any distractions until I get this figured out." By acknowledging our personal biases, we can mobilize our strengths and guard against emerging vulnerabilities. To understand ourselves better, consider the following questions.

- Am I sometimes indecisive during emergency events?
- Do I tend to jump too quickly into my problem-solving mode?
- Do I favor familiar game plans that require more effort over modified/innovative game plans that align with the conditions?
- Do I tend to oversimplify my interpretation or rush my analysis of problems to make them more manageable?
- Do I feel anxious when we aren't handling the problem more quickly?
- Do I tend to get frustrated when I think we should be working faster?
- Do I tend to work faster than other pilots?
- Do emergency situations feel like disruptions that need to be quickly resolved so we can get back to a familiar operational flow?
- Do I prefer to manage the QRH checklist while the other pilot flies?
- Do I prefer to fly while the other pilot manages the QRH checklist?
- How do I prefer to handle external distractions while we are completing the checklists and briefings?

As I personally answered these questions, I concluded that I tended to jump too quickly into completing the QRH checklist. I preferred to let the FO fly while I diagnosed the problem and managed the QRH. I felt that this kept the FO in their comfort zone of flying the aircraft and left the unfamiliar, non-normal tasks for me to handle. I clearly communicated what I was thinking and doing so they could back me up, correct my errors, and refine the evolving game plan. Even so, I tended to move more quickly through the steps than my FOs preferred. To compensate, I committed to slowing my pace, moving more methodically through checklists steps, and gaining their concurrence before performing each step. After completing required checklists, I used the extra time to coordinate with the FAs, station, dispatch, and anyone else who had a supporting role. I would make a passenger PA to inform them of the situation, what we intended to do, and to address probable concerns. Finally, I would ask my FO, "What did we miss?" to open a discussion of what-ifs and to guide our monitoring for counterfactuals. As we evaluate our personal strengths and vulnerabilities, we improve our techniques for handling non-normal situations.

NOTES

1 For consistency, we will refer to this type of manual as the QRH.
2 Edited for brevity. Italics added. NASA ASRS report # 1721385.
3 Some simulators do feature smoke generation capability.
4 Edited for brevity. Italics added. NASA ASRS Report #1665327.
5 Dekker (2006, p. 102).
6 Edited for brevity. Italics added. NASA ASRS report #1816970.
7 Edited for brevity. Italics added. NASA ASRS report #1544227.
8 This summary was derived from interviews with the mishap pilots.

BIBLIOGRAPHY

ASRS – Aviation Safety and Reporting System – National Aeronautics and Space Administration (NASA) Online Database. (n.d.). Retrieved from Aviation Safety and Reporting System: https://asrs.arc.nasa.gov/search/database.html.

Dekker, S. (2006). *The Field Guide to Understanding Human Error*. Burlington, NJ: Ashgate Publishing Company.

Dekker, S. (2015). *Safety Differently: Human Factors for a New Era*. Boca Raton, FL: CRC Press.

Dismukes, R. K., Berman, B. A., & Loukopoulos, L. D. (2007). *The Limits of Expertise: Rethinking Pilot Error and the Causes of Airline Accidents*. Burlington, NJ: Ashgate Publishing Company.

23 Time-Critical Emergencies

The vast majority of emergencies and non-normal events, such as the examples from the previous chapter, offer ample time to form deliberate strategies. We can balance priorities, review procedures, and coordinate crew actions. Time-critical emergencies, however, challenge us to respond quickly and decisively, often without the advantages of crew coordination or checklists. Through experience, we know where these time-windows reside. During takeoff, the critical time window is within a few seconds of V_1. Engine failures outside of this narrow window allow ample time to interpret the indications and communicate before we act. During landing, the critical window is between flare and the first few seconds of rollout. Before and after this time-critical window, our urgency to act is much lower.

As we approach these time-critical windows, we should suspend all discretionary and distracting activities and focus our attention. Unfortunately, our minds tend to relax as we settle into our comfort zones. Repetition, comfort level, fatigue, and laxity allow our minds to wander and attention to drift. Since "nothing ever goes wrong", these time-critical windows fade in importance.

We acquire most of our experience with non-normal events through simulator training. Our instructors start with simple malfunctions to teach us how to handle non-normal events. They gradually increase the complexity and time pressure. After graduating to line operations, we rarely see non-normal events. It is a rare day when we need to pull out the QRH. Even then, the vast majority of line-flying events are not particularly time-sensitive. We usually have plenty of time to calm any startle reaction, diagnose the problem, coordinate as a crew, select the appropriate procedure, and methodically accomplish the recovery steps. Only a very small number of time-critical emergency events ever occur. Still, since their outcomes can become extremely consequential, we need to develop the skills to accurately handle every single one of them. This chapter will examine the specific set of skills and techniques for handling these events.

23.1 EMERGENCIES WITH LITTLE TIME TO REACT

We'll define time-critical emergencies as those events where we need to apply at least one initial recovery action step within a few seconds. The aural warning sounds, the light illuminates, the aircraft lurches unexpectedly, or a collision threat appears in our path. With these situations, we have very little time to identify the threat, understand it, and take action. Some examples are:

- Engine failure, fire, or severe damage during takeoff within a few seconds of V_1

DOI: 10.1201/9781003344575-26

- Immediate collision threat (such as birds, other aircraft, or taxi obstructions)
- Uncommanded aircraft attitude upset
- Stall warning at low altitude
- Rapid decompression at high altitude
- Uncontrollable aircraft movement while in the flare (wake turbulence encounter)
- Loss of directional control on the ground during takeoff, landing, or taxi

The following report describes an ERJ-170 Captain's severe wake turbulence encounter while in the flare on short final at DFW.

BOX 23.1 GO AROUND FOLLOWING SEVERE WAKE TURBULENCE IN THE FLARE

Captain/PF's report: I was acting as PF while on approach to Runway 36L at DFW. The approach was typical and stabilized. We were following an Airbus A320 in visual conditions and had the aircraft in sight. The wind at the time of the incident was a quartering headwind. I believe we added 10 knots to our V_{REF} speed. *While in the landing flare at approximately 10′ from the runway surface, we encountered what I would describe as "severe wake turbulence".* The aircraft gained between 20′ to 30′ of altitude and experienced a desire to roll. Realizing that the landing must be rejected, I called aloud "Go Around". Since my eyes were already fixed down the runway during the flare, I kept my sight picture outside, pushed the thrust levers to MAX power, rotated to approximately 10° (takeoff sight picture), countered the roll moment, called for "flaps 2", and then called aloud for "Positive Rate, Gear Up". *This event took both of us by complete surprise. With the speed so low and the attitude of the plane upset I forced the throttles to MAX power without regard for the go around button. … The "startle factor" of a wake turbulence upset so near to the runway was intense. I believe this "startle factor" was what led me to abandon my duty to push the go around button, thus upsetting the normal flow for a go around.*[1]

Notice the immediate onset of this event. They went from flying a typical and stable landing to a flying a barely controllable aircraft in moments. Also, notice how the Captain narrowed their focus on one critical parameter – holding go around pitch using a "takeoff sight picture". This is a common reaction to startle effect. Also, notice that the crew did not spend any time or attention toward diagnosing why the upset was happening or what caused it. All efforts were focused on responding immediately to the threat.

While the previous list of time-critical events was limited to cases requiring *immediate* reaction, several additional events require *expeditious* application of recovery procedures.

- TCAS resolution advisory (RA – as directed by TCAS system)
- Terrain warning (TERRAIN, PULL UP)
- Windshear warning (GO AROUND, WINDSHEAR AHEAD)

These automated warnings give us a few additional seconds of time to react, but still present some of the same challenges as immediate reaction events. Our response needs to be prompt, but not immediate.

Time-critical situations require the full range of non-normal event handling skills covered in the previous chapter plus:

- Recovery from an initial startle/surprise reaction
- Immediate recognition of the threat
- Prompt action to minimize any hazardous consequences
- Controlled transition to deliberate resolution procedures

23.2 THE EFFECTS OF STARTLE AND SURPRISE

Startle and surprise adversely affect how well we handle emergency events. They degrade our performance, confound our decision making, and increase our vulnerability to further distraction.

23.2.1 STARTLE

The startle effect is a natural human reflex reaction following a particularly intense stimulus like a bang, audible warning, or observed threat. It includes both physical and mental effects.[2] The physical effects include a reflexive flinching of the eyes and tensing of most of our major muscles. Any physical motion that we were doing often stops or intensifies.[3] For example, if we were making a left turn, we might immediately stop that turn or jerk the controls further to the left. If we were engaged in a thought process, like performing a checklist or making a decision, it would also stop. The startle effect often erases our knowledge of previous tasks or actions from our working memory. We might not recall which checklist step we had just performed or what we were intending to do next.

The mental effects from startle are caused by our instinctive flight-or-flight response. Our reaction occurs because the primal centers of our brains are hardwired to quickly react to stimuli like loud sounds or scary images. This is largely beyond our conscious control. Along with our initial bodily reaction is a sudden release of adrenaline (all within the first 1.5 seconds). After this initial shock, the logical, thinking part of our brain begins to reassert itself to search for the cause and meaning of the startling stimulus.[4] Following are several general statements regarding startle.

- The more time-pressured, inexperienced, tired, or relaxed that we are when the event happens, the more intense the physical effects and mental shock tend to be.
- The more trained, experienced, ready, and alert we are, the more quickly we recover.
- The more intense, unexpected, or rare the stimulus is, the more powerful the startle effect.
- The more common and expected the startling stimuli, the smaller the effects.[5]

23.2.2 Surprise

Surprise results when something unexpected happens, but also when something we expect to happen doesn't.[6] It is a momentary, "That's not right" or "What's going on here" type of experience. Like startle, the more time-pressured, inexperienced, tired, or relaxed we are, the more intense the surprise effects. The more trained, experienced, alert, and ready we are, the more quickly we sort it out and recover.[7]

23.2.3 Startle and Surprise Comparison

Startle and surprise have many similarities, but there are some important distinctions.

- Startle requires an intense and typically unexpected stimulus like a bang or alarm.
- Surprise arises from an interruption in our expected flow of events. It may emerge following a startling stimulus or from the absence of an anticipated event. This mismatch between expectations and indications interrupts our orderly flow of future SA.
- The startle effect tends to pass fairly quickly.
- Surprise effects may persist, depending on experience, alertness, time pressure, novelty, and complexity.
- Startle affects our body and mind.
- Surprise affects our SA.
- Startle is rare, intense, and powerful.
- Surprise is common and ranges from subtle and short-lived to extreme and long-lasting.
- We can experience only startle, only surprise, or both together.
- The effects of both startle and surprise are intensified by time compression, inexperience, and fatigue.
- The more mentally detached we are when the event happens, the more intense the physical effects and emotional intensity tend to be.
- When something startling or surprising happens, additional unexpected events intensify follow-on problems. This chain of disruptions interferes with our ability to rebuild our SA and recover.

This last point is particularly important. Like a boxer who receives a stunning blow from their opponent, we don't immediately recover. Before we can gather our wits, another blow can knock us back on our heels. This FO reports the effects of a chain of surprising events following a poorly executed go around.

BOX 23.2 SUCCESSIVE STARTLING EVENTS INHIBIT RECOVERY

FO/PF's report: The aircraft was fully configured on glide slope and flaps 50 for RWY 36L. The weather was low IFR with 500′ ceilings and good visibility below the deck. The aircraft weighed 415,000 pounds and was loaded close to the forward CG limit. This resulted in the pitch trim being at 13 units or close to full nose up trim when established on final. *At approximately 1,500′MSL or 1,200′AGL we were instructed to go around, climb to 3,000′, and turn left to 270°. The plane had gone into dual land mode. As such, the go around button had to be pushed.* I took a couple of seconds to think the procedure through because I knew I wanted the airplane out of go around mode as quickly as possible and into level change. This was because we only had to climb 1,500′. *When I pushed the button and the engines went to go around thrust, I was completely taken off guard as to how much effort was required to counter the nose up trim. I was trimming and pushing with considerable effort. As a result, call outs were not correct and the plane stayed in go around mode with the gear down and flaps at 50.* My first attempt to engage the autopilot failed and I continued to push and trim. We leveled at 3,000′ and then began the turn. I again attempted to engage the auto pilot and was successful. *The auto continued to trim and caused a couple +/−100 feet undulations as we retracted gear and flaps.* We flew a normal pattern and landed uneventfully. *Adding to the startle and distraction was the "stabilizer motion" aural sounding continuously.* The cause was due to the aircraft being at MAX thrust, out of trim, and requiring a near immediate level-off. *Go around procedures were not correct as a result of the startle effect.*

… The fact is that if this happens again tomorrow, I'll probably execute the maneuver flawlessly. However, a go around usually occurs unexpectedly, at the end of a long flight, when we're not on our A-game. Although we train to go around using level change, the majority of our training is done at MAX power. In the heat of the moment, an additional threat of a rapid level-off in a turn adds unnecessary complexity to an already elevated risk maneuver.[8]

What should have been a routine go around became quite difficult due to startle effect. The FO anticipated the difficulty, so they "took a couple of seconds to think the procedure through" before selecting the go around mode. Even with this mental preparation, they committed several altitude busts as they tried to get the thrust under control, the trim reset, and the autopilot engaged. Their initial surprise at the MAX thrust engine response was immediately complicated by the abrupt pitch up, delay in recentering trim, and inability to engage autopilot. This rapid succession of unexpected and extreme events affected their ability to recover from the first startling event. Also, notice their comment regarding how training cues the MAX thrust go around even though line practice appears to favor using level change mode to avoid the problems that they encountered.

23.2.4 How Startle and Surprise Affect Our Attention Focus

Both startle and surprise narrow our attention focus.[9] Ideally, this helps us to quickly process and resolve an event. If we respond accurately, then our laser-beam focus is exactly what we need to resolve our problem. Unfortunately, we might focus our attention on the wrong parameter – one that may not provide us with the most useful information. We may narrow our visual attention to a single point, such as the intended touchdown zone of the runway, and stop looking at other sources of information, like airspeed gauges. In extreme cases, tunneled-attention masks our other senses. The following example recounts an accident where neither pilot recalled hearing the gear warning horn or EGPWS "PULL UP" warnings.

BOX 23.3 CONTINENTAL AIRLINES 1943 – GEAR-UP LANDING IN HOUSTON –FEBRUARY 19, 1996

Due to a procedural omission, the crew failed to move the DC-9's hydraulic switches to the HI (high) position resulting in insufficient hydraulic pressure to achieve full extension of landing gear and flaps. Surprised that they weren't slowing down following flap selection, they missed that their gear had not fully extended either. From the NTSB final report, During post-accident interviews, neither pilot recalled seeing "any" landing gear indicator lights; both pilots recalled the gear handle being moved to the down position. The First Officer stated that he did not hear the landing gear warning horn. The Captain stated that he heard the horn sound momentarily and thought that it sounded because he put the flaps to 25° before the gear was down and locked.[10]

From the flightdeck recorder transcript, it was clear that there were repeated gear warning horns and three ground proximity warning system (GPWS) "WHOOP, WHOOP, PULL UP" aural warnings.

23.2.5 How Startle and Surprise Adversely Affect Our Decision Making

Both startle and surprise inhibit our logical thought processes. We can still think, but the speed and quality of our reasoning are degraded. This is because the emotion-processing portions of our brain that trigger the flight-or-flight response temporarily overwhelm the logical/thinking portions of our brain. As our minds attempt to restore order, they might follow this progression.

- What just happened?
- What does it mean?
- What is likely to happen because of that unexpected event?
- What do we need to do?
- How do we rebuild our SA?
- How must our game plan change for us to successfully move forward?

If it was as simple as working through these steps, we would quickly recover. Unfortunately, the road to recovery is strewn with obstacles:

- The startling event,
- is often followed by surprise,
- which leads to confusion,
- but before we can sort it out,
- something else startles or surprises us,
- which makes us start the process over again.

Following is an account where surprise degraded this pilot's ability to recall an aircraft switch location and procedure.

BOX 23.4 NON-CURRENT CAPTAIN STRUGGLES WITH SURPRISE AND OVERLOAD

Captain/PF's report: ATC called and said we had a 100-knot overtake on the Caravan in front of us. So, I selected props 100% and still didn't slow us enough. ATC told us to level off at 6,000′ and canceled our approach clearance. *I was searching for the Altitude Hold button on the autopilot control panel. I have been out of the cockpit for so long, I had forgotten where it was. While I was looking for it, I got the stick shaker alert.* I immediately added power and lowered the nose to about 5,800′ and recovered back to 6,000′. I asked [the FO] what the airspeed was and he said 125 knots. We took vectors back [around] to start the approach again and requested to climb back up to 9,000′.

As we climbed the icing just got worse. … *We are spending too much time in heavy icing conditions. … I leveled off at 9,000′ and turned left to 060° to intercept the localizer and I got the stick shaker again.* This time I did see 125 knots. I recovered again and resumed the approach. From that point on it was a normal approach to landing. When I did my walk around, the airplane was coated in thick ice. The worst that I have seen in 3 years. Situational awareness. Keep the scan moving. Ask for help from the other pilot such as "please select ALT HOLD for me".[11]

This report demonstrates the effect of tunneled attention on one problem (trying to locate the Altitude Hold button) which leads to missing another more serious condition (rapidly deteriorating aircraft airspeed). From the FO's report, we learn that this error was preceded by a steep descent while trying to intercept glideslope from above. As their speed increased, it caused the spacing problem with the preceding Caravan. We conclude that the Captain leveled the aircraft with the thrust back while searching for the Altitude Hold button. This caused their airspeed to deteriorate rapidly until the stick shaker sounded. We also surmise that the FO/PM became equally distracted and focused by the Captain's automation management difficulties and missed the rapidly deteriorating airspeed. This appears to be a case where the PM's focus and mindset became synchronized with the Captain's. They both focused on the same parameters, missed other important parameters, and made the same errors.

23.2.6 How Startle and Surprise Increase Our Vulnerability to Distraction

Startle and surprise can also accompany, and then deepen, the effects of distraction. The following report started with an unexpected EFB reset that dumped the airport diagram display. This led to taxi route confusion, which led to forgetting to start the second engine, which led an embarrassing outcome. Notice how a simple problem snowballed to adversely affect their decision making and problem solving.

BOX 23.5 SURPRISE AND CONFUSION LEAD TO UNWISE DECISION MAKING

Captain/PF's report: After pushing back, ATC instructed us to taxi to RWY 16R via Papa and cleared to cross RWY 10/28. As I was taxiing out, my Jeppesen app crashed. … *My EFB resetting led me to be distracted with the taxi instructions and almost resulted in me taking taxiway Bravo instead of taxiway Papa. Both of these distractions disrupted a normal rhythm and led me to not calling for the delayed engine start checklist.* … Tower told us that we were cleared for takeoff and as we were taxiing onto RWY 16R, I realized that the Engine #1 had never been started. I told the FO and he contacted ATC to tell them we had to taxi off the active RWY due to a situation that arose. We proceeded to taxi-off at Mike and hold short of taxiway Papa as instructed by ATC. *We were trying to figure out a way as to not alarm the passengers, so we decided to shut down engine #2 and then restart them both. Because we were frazzled and embarrassed, … we did not start the APU and when we realized that and tried to start the APU with just battery, the APU would not start. … We decided to shut it down and call Dispatch to let Operations know that we need to get towed in and also let Tower know we had no way of communicating with them.* … [In the future, I should] make sure that I'm more vigilant when it comes to letting distractions affect my performance and safety. I should've stopped the airplane and waited for the EFB to reset instead of continuing taxiing. I will make sure that from now on, there is no reason for me to hurry when things happen and that time constraints are of little concern when it comes to safety.[12]

It appears that surprise from the EFB reboot led to missing the engine start, it may have also contributed to some of the flawed decision making that followed. Surprised by their error of taking the runway before the aircraft was ready, they moved too quickly into their remedy plan and made the second error of shutting down their remaining engine before starting the APU.

23.3 RECOVERING FROM THE EFFECTS OF STARTLE AND SURPRISE

Startling events happen without warning. In the initial moments of confusion, our thinking mind drops offline for a few moments. As we recover, we don't immediately return to full capability. As it reboots, our mind needs something simple, singular, and definitive to guide our decision making. We can't avoid the startle effect since it is hard-wired into our brains. However, we can practice techniques to minimize its adverse effects and speed our recovery. Useful strategies include:

- Rehearsal
- Body positioning
- First look
- First step

23.3.1 Rehearsal

Consider a crew practicing emergency events in the simulator. As their instructor taps busily at the console setting up the next scenario, they anticipate that it will be an engine failure before V_1 scenario that requires a rejected takeoff procedure. They mentally rehearse the procedure.

1. It will begin with an engine fire warning alarm or the sound of an engine winding down.
2. Expect the aircraft to yaw toward the failing engine.
3. Feed in opposite rudder to counter the yaw and maintain runway centerline.
4. If it occurs before V_1, verify the failure and announce rejecting the takeoff.
5. Simultaneously, disengage the autothrottles and retard the trust levers.
6. Ensure wheel braking either from autobrakes or manually.
7. Initiate reverse thrust.
8. Slow to a stop on the runway.
9. Inform Tower of our rejected takeoff.
10. Direct the passengers to remain seated.
11. Call for the appropriate QRH checklist.

Mentally prepared and ready to go, the instructor clears them for takeoff. The takeoff appears normal until approaching V_1. They hear a loud fire alarm and see a blazing red light in the #1 engine shutoff handle. They accurately follow the rehearsed steps and reject the takeoff. They still experience a moment of startle when the fire alarm sounds, but they are ready for it and quickly recover.

In the simulator, rehearsal aligns our mindset for events that we expect will happen. In line flying, rehearsal prepares us for what possibly *can* happen. Reviewing a scenario improves the speed and accuracy of our actions. It takes knowledge of procedural steps that reside deep in our memory, dusts them off, and moves them to the forefront of our mind. When the emergency happens, the necessary knowledge is immediately available. Our rehearsal can be as simple as mentally reviewing a few key words.

BOX 23.6 REHEARSAL OF "THROTTLES, CHUTE, AND HOOK"

In military aviation, we were taught to verbalize the abort boldface steps before every takeoff. Rehearsing those words became automatic, "Throttles, Chute, Hook". Flight after flight, we rehearsed them without needing to use them. One day, that all changed. As I started my takeoff roll, a mis-installed afterburner fuel pump poured jet fuel into the engine bay. Accelerating down the runway, I sensed a lack of thrust from my #2 engine even though the instruments all showed normal. Airflow drew the fuel forward where it was ingested into the jet engine's compressor section. As I reached 100 knots, that fuel ignited. There was a loud explosion, fire lights, and everyone with a radio started screaming that we were on fire. I was startled, but through the confusion, the abort steps automatically kicked in – "Throttles, Chute, Hook". We successfully stopped, egressed the jet, and looked back to see the entire back half and bottom of the aircraft engulfed in flames.

Transferring the rehearsal practice that we routinely use in the simulator to the aircraft requires commitment and discipline. We know that there is very little chance that we will actually experience an engine failure in the aircraft. Still, rehearsal prepares us for the unexpected.

23.3.2 Body Positioning

Another useful strategy is positioning our body to react correctly and quickly to time-critical emergencies. This protects against startle-induced movement errors. For example, it is common for our arms to freeze or jerk following a startling noise like a loud bang. During takeoff, this could cause us to unintentionally push or pull the thrust levers. An appropriate hand position might prevent us from inadvertently moving them in the wrong direction. Consider a situation where we are the Captain/PF performing a takeoff. During initial acceleration for takeoff, we might consider hooking our fingers over the thrust lever knobs. If there is a startling bang, our hand is properly positioned to pull the thrust levers back and reject the takeoff. Later in the takeoff roll, our rejection criteria narrow. For example, we wouldn't want to reject for minor caution lights or distractions. Reaching these higher speeds, we might unhook our fingers from the throttle knobs. If we encounter a problem, our hand position reminds us that we need to deliberately choose whether to continue or reject. Reaching V_1, our procedures emphasize continuing the takeoff except for a short list of extreme cases. Here, we can either remove our hands from the thrust levers or move them behind the levers. This hand position reminds us to favor continuing the takeoff, even following a loud or distracting bang.

The same logic applies for foot positioning on the rudder pedals. While it is common to fly with our toes on the pedals and our heels resting on the floor, it may be more appropriate to position our feet higher on the pedals during taxi movement so that they are positioned to immediately apply brakes during an emergency stop.

This also applies during the early portion of the takeoff roll where our priorities emphasize rejecting the takeoff. As we transition to higher speeds that favor continuing the takeoff (or when autobrakes become armed), dropping our heels to the floor may be more appropriate.

23.3.3 First Look

The first look strategy directs our attention toward the most useful indicator of the most important parameter. For example, if we are rolling down the runway approaching V_1 and hear a loud bang, our first look is toward the primary engine thrust or power indicator (N1, EPR, or torque, for example). By comparison, scanning the entire engine instrument panel while trying to construe meaning would take too long. Some gauges may not be useful (fuel flow and oil pressure). Others may be useful for supporting confirmation, but may contradict or lag behind the primary gauges (engine temperature and N2).

While our minds are still confused from startle and surprise, our first look plan gives us the most useful and immediate information to guide the critical decision that we need to make – for example, whether to continue or abort. We don't even need to process the actual value or number of the readout. For multiengine aircraft, the fact that both gauges are parallel and steady is assurance that we have two operable engines producing expected thrust. If one is significantly lower, especially if it is spinning down, we'll have an unambiguous indication of engine failure. Following is a selected list of first look plans.

- Engine failure, fire, or a loud noise/bang during takeoff approaching decision speed
  - *First look:* the primary thrust/power gauge
- Immediate collision threat – such as birds or other aircraft
  - *First look:* the trending movement or drift of the threat
- TCAS RA (resolution advisory warning)
  - *First look:* the directive display on ADI or HUD (or as TCAS audio directs)
- Uncommanded aircraft upset
  - *First look:* the level horizon reference – either outside or from instruments
- Stall warning at low altitude
  - *First look:* the flightpath reference display
- Rapid cabin pressure decompression or warning horn
  - *First look:* the cabin pressure gauge
- Directional control veer-off on the ground during takeoff, landing, or taxi
  - *First look:* the direction toward the most available pavement

23.3.4 First Step

The next strategy step for time-critical emergencies is to perform the first step of the recovery maneuver. Imagine that we have mentally rehearsed our procedures, that we have our attention appropriately focused, and that our body is positioned to react. When the startling event happens, we look at the best indicator to diagnose the problem. We process what has happened and that we need to react quickly. The next step is to accurately perform the first step of the recovery sequence. Since our minds may not be fully recuperated from the startle effects, knowing and performing the first step help unfreeze our recovery process.

This first step may be the only time-critical task that we need to perform. For example, a quick pull or push on the flight controls to avoid the bird may be the only action we need to take. For more complex procedures, completing this first step activates the trained recovery sequence. If we can successfully initiate the first step of a procedure, we are much more likely to move successfully to the second step, and so on.

Depending on the event, many follow-on steps may be delayed. Consider a windshear warning on short final. The first step may be something like, "Thrust Levers – MAX". As pilots, we naturally make the follow-on step of rotating to a desired pitch attitude. The later steps of raising gear and flaps can wait. Additionally, some remedy steps are linked together. While they may be listed separately in the procedure, our training and practice join them together into a single maneuver. For our windshear example, advancing thrust, disengaging autothrottles, and establishing a climbing pitch attitude mirror the same maneuver that we practice during any go around, missed approach, or stall recovery. Like muscle memory for a physical task, performing the first step engages our body and mind to complete the associated actions.

Many pilots find that it helps to verbalize the first step. First, this reinforces the procedure. Simultaneously saying and doing overcomes the hesitancy often associated with startle effect. Second, it lets the other pilot know what we are doing and thinking. It informs them of our mindset and helps unfreeze their minds if they are still befuddled from startle effect. Third, it alerts them in case we are making an error. They will immediately know if our actions are misguided.

23.4 TIME-CRITICAL EMERGENCIES – EXAMPLE SCENARIOS

To tie this discussion together, consider some specific time-critical emergencies and the applications of rehearsal, body positioning, first look, and first step.

23.4.1 Rejected Takeoff (Immediately Before the V_1 Callout)

Most companies designate a limited number of events for rejecting takeoffs near V_1 (engine failure, fire warning, predictive windshear warning, or the aircraft is unsafe/unable to fly). Unfortunately, the indications that precede these events aren't always clear. Sometimes, we have borderline or deteriorating indications that don't match the simulator training scenarios that we use for practice. Indecision can cause us to hesitate and delay our decision to reject until well past V_1. In most of these cases, we would be better off rejecting the takeoff and investigating the anomalous indications afterward. For this reason, we might consider adding "confusing, borderline, or deteriorating indications" to our reasons to reject the takeoff near V_1.

- *Rehearsal:* Before taking the runway, mentally review the rejected takeoff criteria. Consider the prevailing conditions. For example, if convective weather is a factor, consider how a windshear warning would sound. For favorable weather conditions, review for an engine failure. Mentally review decision speeds and associated actions. Step through the rejected takeoff boldface items while touching each respective control.
- *Body positioning:* Commencing the takeoff, we would position our feet up on the rudder pedals so that we can comfortably apply full brakes, even while fully deflecting one rudder pedal to its forward limit. We position our hands over or on the thrust lever knobs as discussed earlier.
- *First look:* Consider a case where we are rolling down the runway. We are anticipating a "V_1" callout from the PM. Suddenly, an engine fails. We sense the loss of thrust and feel the aircraft yaw. Our first concern is directional control. We reflexively apply rudder to hold runway centerline. Our first look strategy is toward the runway centerline to control aircraft path and alignment. The PM's first look strategy is to glance at the primary thrust indication gauges. These are typically positioned at the top of the engine instrument stack (N1 in most turbojets). If one is dropping or showing automated failure warnings, it confirms the need to reject the takeoff. The PM should provide a useful callout to simplify the rejection decision, "Engine failure, #1 engine".
- *First step:* Most airlines direct the Captain to perform all rejected takeoffs. If the Captain is the PM when the failure occurs, they will need to diagnose the failure (first look at the primary thrust indication), announce the rejected takeoff, assume control of the aircraft, and begin the rejection by reducing thrust.

23.4.2 Engine Loss/Fire Immediately Following Liftoff, But Prior to Gear Retraction

The main challenge here is the startle effect. It may be complicated by environmental conditions like night or IMC, aircraft conditions like heavyweight or high-density altitude, and personal conditions like fatigue and freezing-up from startle effect.

- *Rehearsal:* As with the previous example, review the V_1 reject and V_1 continue procedures before taking the runway. After liftoff, we normally execute the procedural steps successfully because each task is followed by the next in a predictable sequence. An engine loss or fire that occurs within this short time window disrupts our practiced sequence. We might skip tasks that we reliably perform during normal takeoffs. We could, for example, miss raising the landing gear. Additionally, obstructions and terrain may require a specific engine-out departure routing. Transitioning to these exceptional routings may prove especially difficult while we are startled. We should review these special procedures and have them immediately available to reference. Since the aircraft is airborne when the failure occurs,

PFs should concentrate on maintaining the flightpath. PMs should silence any alarms and describe the engine conditions. Which engine is it? Is it still producing thrust? Are there other indications that confirm an actual engine failure or fire?

- *First look:* At night or IMC, the best indicators are the attitude and slip (yaw) indicators. Secondary indicators are airspeed and VVI.
- *First step:* For PFs, the first step is to maintain the aircraft path. Set the trained pitch attitude for an engine failure. With thrust loss, an appropriate rudder input to correct for roll and yaw is also necessary. For PMs, the main threats are missing tasks from the operational flow. Ensure that the gear is raised, evaluate the engine conditions, ensure that the PF has the single engine departure procedure available, handle any ATC calls, and verbalize adverse trends. After these conditions are stabilized, prepare the QRH checklist to assist with handling the engine failure or fire.

23.4.3 Directional Control Problems on the Runway at High Speed

The main threat is the unpredictability of these events. This makes preparation especially challenging.

- *Rehearsal:* If external conditions like high crosswinds or convective weather are present, we can review windshear warning alerts and indications.
- *Body positioning:* PFs should be able to apply full rudder and toe brake deflection. This positions the brake pedals somewhat closer than many pilots prefer. If an event occurs with the pedals adjusted further away, they wouldn't achieve full deflection travel and toe brake extension. To compensate, many pilots adjust their pedals closer for takeoffs and landings and further for cruise flight. Seat belt tension and armrest positioning are also important as many pilots prefer a relaxed lap belt and retracted armrests for personal comfort. A particularly violent loss-of-control event may compromise their ability to activate the controls. During FO takeoffs, Captains need to be positioned to immediately assume aircraft control and apply tiller inputs.
- *First look and first step:* The PF is already looking outside, so the direction of drift will be clear. For PMs, however, this may be especially startling, especially if they are focused inside monitoring the engine gauges and airspeed. As PF, our first priority is to reorient the aircraft back toward the runway centerline. Our first corrections will be instinctive, but their magnitude may vary. Our instinctive correction is to deflect the rudder pedal to deflect the nosewheel. Stronger corrections are available with the tiller or by asymmetric braking. Either of these choices may create further problems. For example, if we overcontrol the tiller, the nosewheels may lose traction and start skidding. Another issue is whether to continue the takeoff or reject. The Captain needs to be clear and decisive.

23.4.4 Loud Bang near V_1

This scenario is particularly instructive for several reasons. First, it encompasses a large range of events from engine failures, compressor stalls, tire failures, mechanical failures, and birdstrikes. Many pilots have rejected takeoffs based solely on hearing "loud bangs". Second, these events are frequently misinterpreted, startling, and distracting. Third, following the bang, most pilots feel the need to do something very quickly. They may reject the takeoff without taking the first step of assessing thrust.

- *Rehearsal:* Mentally review the RTO criteria and procedures before takeoff. Notice that a loud bang approaching V_1 is not a listed rejection criterion.
- *First look:* The most critical indicator is engine thrust. If the engines appear normal, the PM should make a simple, unambiguous callout. Many airlines don't want FOs making continue or reject decisions, so consider a callout like, "Good engines". As the Captain/PF hears, "Good engines", it gives them all the information they need to make their continue/reject decision. Reversing the roles, for Captain/PMs, stating, "Continue" communicates that they have evaluated the indications and want the FO/PF to continue the takeoff. Verbalize any additional essential information to remove confusion created by a startling event. Consider the following reports from a crew experiencing a loud bang during takeoff.

BOX 23.7 LOUD BANG DURING TAKEOFF

FO/PM's report: ... *Shortly after V_R, we heard a loud bang. The Captain says, "I think we blew a tire." I said, "I agree." The Captain kept flying. We left the gear down. Speed was bleeding off and then I said, "Engine Failure." [The FO then requested priority handling with ATC with intentions to return to ZZZ]. We leveled off at engine acceleration height and started to accelerate, thinking the gear was up.* We decided to leave the flaps at 1 to return to the field and not knowing if there was any damage. We completed the ECAM procedure and follow-ups. The Captain talked to the FAs (Flight Attendant) to prepare for a precautionary landing and made a PA to the passengers, per [company guidance]. *ATC asked us to climb to 4,000′. Then, we realized the gear was still down.* [We] raised the gear [and] completed the After Takeoff Checklist. We did the Landing Approach and Overweight Landing Checklists. We briefed the approach to [runway] XXC and completed the Descent Checklist. We configured on schedule, completed the Landing Checklist, and landed uneventfully. After stopping, the Captain made a PA, talked to ARFF (Aircraft Rescue and Fire Fighting), then taxied to the gate. *[The cause of missing the landing gear was] startle effect and the initial confusion of a tire failure or an engine failure.*

Captain/PF's report: ... *Acceleration was normal and rotated at V_R. Shortly thereafter, heard a thump. I stated, "I think we blew a tire." Thought about not raising the gear and continued flying the aircraft and noticed*

> *rapid degradation of speed. FO (First Officer) stated, "Engine Failure." Immediately lowered the nose and started trimming based on rudder trim to level the wings and to center the rudder. Aircraft performance was very degraded but assumed it was due to a heavyweight 321 aircraft.* [Requested priority handling] and stated intentions to return to ZZZ, thinking the landing gear was up. Elected to leave flaps at 1 and return to the field as expeditiously as possible as we were down to 1 engine and 1 generator. Completed Engine Failure ECAM procedure, ECAM follow-ups, notified the Flight Attendants to prepare for a precautionary landing, and made a PA as per the [company guidance] pages. *ATC asked if we could climb to 4,000′ and said we were unable due to weight. At this point I realized the gear was still down. Raised the gear and completed the After Takeoff Checklist.* Accomplished the landing data for overweight landing and checklist, briefed the approach to [runway] XXC, and completed the Descent Checklist. We were then cleared for the ILS XXC approach, configured on schedule, performed the Landing Checklist, and landed uneventfully. After stopping, made a PA, ARFF (Aircraft Rescue and Fire Fighting) checked out the aircraft, and we taxied to the gate. *[The cause of missing the landing gear was] startle effect. Initial confusion of tire failure versus engine failure played a significant role in distraction from raising the landing gear. As pilot flying, my attention was completely absorbed with flying the aircraft with assumed degraded performance and didn't allow for transfer of aircraft control. Additionally, lack of experience on the fleet and as a new Captain also played a significant role. Departing at MAX landing weight with no APU was probably not a good decision.*[13]

This is a particularly instructive event since we are trained not to raise the gear following a tire failure, but we are trained to raise gear for an engine failure climbout. Following the "bang" or "thump", both pilots aligned with tire-failure mindsets. Consider if the FO/PM had employed a first look strategy, they would have detected the engine failure before it became evident by the sluggish climb performance. Notice how their tire-failure mindset contributed to missing the landing gear. Ideally, the FO should have noticed this and alerted the Captain. Viewed from an inside-the-tunnel perspective, the point where they normally would have raised the landing gear was behind them. This explains how it was overlooked. Startle and confusion also contributed to missing the After Takeoff Checklist. It was not until much later when ATC asked them to climb that they discovered that the wheels were was still down. So, because of their startle effect and tire-failure mindset, they found themselves struggling to climb in a heavyweight, single-engine Airbus with the landing gear still extended. We also glean that their simulator-trained procedure usually involved transferring aircraft control to the FO so that the Captain could concentrate on the non-normal event checklist and coordination. The Captain never did this because their "attention was completely absorbed with flying the aircraft with assumed degraded performance". In the end, they recovered from their early misconceptions and did a commendable job returning for a safe landing.

NOTES

1 Edited for clarity and brevity. Italics added. NASA ASRS report #1168197.
2 Two excellent resources are EASA Startle Effect Management report (NLR-CR-2018-242) (Field, Boland, van Rooij, Mohrmann, & Smeltink, 2018) and (Martin, Murray, & Bates, 2012).
3 Field, Boland, van Rooij, Mohrmann, and Smeltink (2018) citing Rivera et al. (2014).
4 Field, Boland, van Rooij, Mohrmann, and Smeltink (2018, pp. 13–14) and Martin, Murray, and Bates (2012).
5 Field, Boland, van Rooij, Mohrmann, and Smeltink (2018, p. 13) citing Koch (1999).
6 Field, Boland, van Rooij, Mohrmann, and Smeltink (2018, p. 15) citing Rivera (2014) and Burki-Cohen (2010).
7 Field, Boland, van Rooij, Mohrmann, and Smeltink (2018, p. 19).
8 Edited for brevity. Italics added. NASA ASRS report #1601223.
9 Martin, Murray, and Bates (2012, p. 389).
10 NTSB (1997, pp. 4–5).
11 Edited for brevity and clarity. Italics added. NASA ASRS report #1698941.
12 Narrative reordered and edited for clarity. Italics added. NASA ASRS report #1394157.
13 Edited for clarity and brevity. Italics added. NASA ASRS report #1818853.

BIBLIOGRAPHY

ASRS – Aviation Safety and Reporting System – National Aeronautics and Space Administration (NASA) Online Database. (n.d.). Retrieved from Aviation Safety and Reporting System: https://asrs.arc.nasa.gov/search/database.html.

Field, J., Boland, E., van Rooij, J., Mohrmann, J., & Smeltink, J. (2018). *Startle Effect Management*. Amsterdam: Netherlands Aerospace Centre: EASA – European Aviation Safety Agency.

Martin, W., Murray, P., & Bates, P. (2012). The Effects of Startle on Pilots During Critical Events: A Case. *30th EAAP Conference: Aviation Psychology & Applied Human Factors – Working Towards Zero* (pp. 388–394). Queensland: Griffith University.

NTSB. (1997). *NTSB/AAR-97/01 and PB97–910401: Wheels-up Landing – Continental Airlines Flight 1943 – Douglas DC-9 N10556 – Houston, TX, February 19, 1997.* Washington D.C.: National Transportation Safety Board.

Section IV

Introduction to Professionalism

DOI: 10.1201/9781003344575-27

IV.1 CAREER PROGRESSION AND PROFESSIONAL WISDOM

Consider our airline career from when we started to where we are today. Recall how hard we worked to develop our skills, gather knowledge, and reach understanding. Notice that it isn't just about accumulating more facts. We have integrated thousands of pieces of knowledge and experience into deeper, wider professional aviation wisdom. Let's examine this wisdom-building process.

IV.1.1 Average Proficient Pilots

Consider a graph that tracks the growth of professional aviation wisdom for average, proficient pilots. While many career paths bounce between seats and airlines, Figure IV.1 depicts a simplified, smooth career path from new-hire, through Captain upgrade, and concluding at retirement.

The horizontal axis reflects the length of a pilot's career at their major airline. The first milestone marks when they are first hired as an FO. With time and experience, they reach the second milestone and upgrade to Captain. After many years in the left seat, their career concludes at retirement.

The vertical axis represents their growth of professional aviation wisdom – how we combine and integrate facts and experiences to deepen our understanding of airline aviation. The first portion of the curve reflects our steep learning curve as we acquire aviation wisdom more quickly early in our career. As the years progress, we encounter fewer unique learning opportunities, so the curve flattens. It continues to slowly rise until reaching retirement.

IV.1.2 Comfortable Pilots

Next, let's add a second curve onto our graph that depicts a comfortable pilots' accumulated aviation wisdom (depicted by the dashed line in Figure IV.2).

The curve flattens because comfortable pilots aren't as motivated to build their professional aviation wisdom as proficient pilots. They view Captain upgrade as the final rung on their career ladder. Having reached the top, there is nowhere left to go. They settle quietly into their comfort zones – content to just fly their line and enjoy

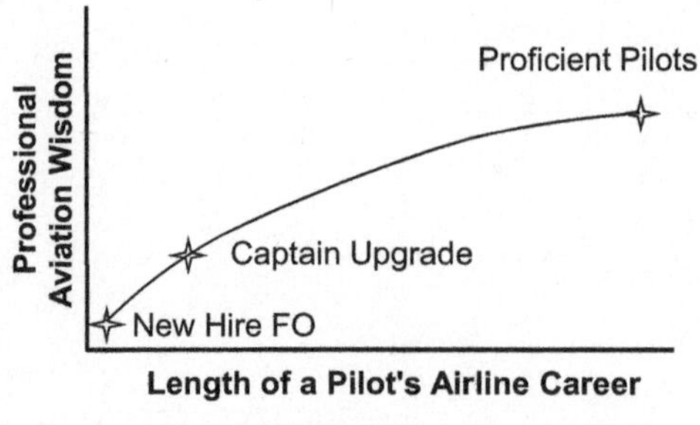

FIGURE IV.1 Proficient pilots' career path.

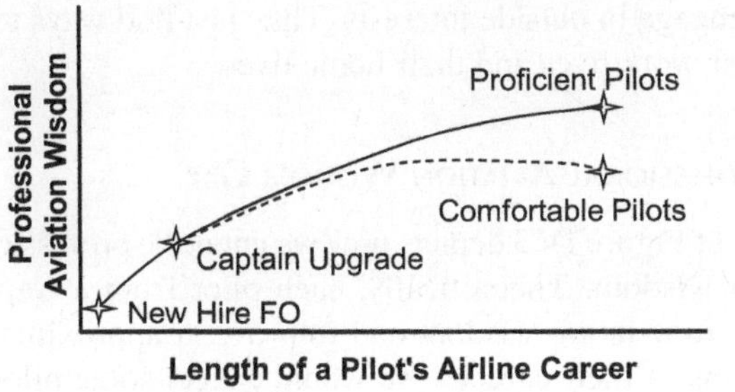

FIGURE IV.2 Comfortable pilots' career path.

the ride. As their habit patterns solidify, their tool box shrinks to hold only a few well-worn game plans. They no longer see aviation as a profession. It is just their job. Their pursuit of professional wisdom stagnates and then begins to fade. They lose more wisdom than they gain. By the time they reach retirement, they are pretty much cruising on autopilot, content to coast across the finish line. These pilots have not lost their passion for learning. As they mark the accomplishment of their aviation career goals, they shift their energies toward outside endeavors. They are still motivated to advance new areas of interest. They just don't see their airline job as one of them.

IV.1.3 Master Class Pilots

Figure IV.3 adds a third line to depict the career path of Master Class pilots (dotted line).

Master Class pilots also find their comfort zones, but they resist settling in and stagnating in them. They retain their passion for learning. From their perspective, there is always something new to learn, more nuances to discover, deeper connections to make, and fresh threads to add to the thick tapestry of their aviation wisdom. Rarely satisfied with cursory understanding, they dig down to study both the underlying processes of airline aviation and the inner workings of their own minds. As they amass wisdom, they become the wise elders of the pilot family. Like comfortable

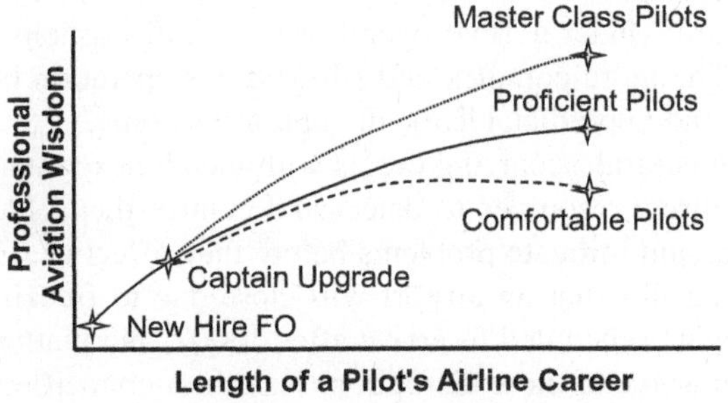

FIGURE IV.3 Master Class pilots' career path.

pilots, they also engage in outside interests. They just find ways to balance their pursuits between their work lives and their home lives.

IV.1.4 The Professional Aviation Wisdom Gap

The early portion of Figure IV.3 depicts how we enter the profession with comparable levels of aviation wisdom. Theoretically, each pilot from a Captain upgrade class should steadily gain aviation wisdom and improve at approximately the same rate. Over the long spans of their careers, we would expect some pilots to excel and rise a bit above the norm and others to progress more slowly and track a bit below the norm. Across the pilot group, however, we would expect a steady upward growth of aviation wisdom in all pilots. Instead, we find some pilots excelling well above the norm (Master Class) and others regressing well below the norm (Comfortable). This is because our steady growth assumption doesn't reflect the faster learning progress with pilots who pursue the Master Class path or the stagnating effects of Comfortable pilots who settle into their comfort zones. By the time our three groups reach retirement, a significant gap forms.

As we scan across any random group of Captains in the pilot's lounge, this gap doesn't seem apparent. All of us perform well enough to handle the complexities of daily line flying. We pass our checkrides, avoid mishaps, and stay out of trouble. We successfully and safely move our passengers and freight to their destinations almost 100% of the time.

The significance of this gap doesn't become apparent until we turn up the heat. It is only under stress, complexity, increased risk, and novelty that the latent vulnerabilities of error, bias, laxity, and flawed decision making begin to surface. These effects rarely emerge during everyday line flying. Well-practiced habit patterns and ample safety margins mask latent vulnerabilities. Errors still occur, but they are caught and mitigated before unfavorable outcomes happen. Like a pot of water on the stove, the water on the bottom starts to boil, but the bubbles don't reach the surface. It is only when the heat rises to the boiling point that the bubbles reach the top of the pot.

IV.1.5 The Pursuit of Resilience

One goal of safety programs is to improve systemic and individual pilot resilience. Resilience reflects our ability to recover from disruptions and to keep the operation moving forward. Under normal operations, most disruptions remain small and inconsequential. The more complex and stressed our operation becomes, the more frequent, intense, and unpredictable the disruptions become.

Disruptions are natural-occurring events within complex systems. On a systemic level, companies direct resources to detect and control them. Ideally, these safety measures intercept and mitigate problems before they affect us. For example, when the company anticipates that an airport will close due to freezing rain by 2100Z, they cancel all flights scheduled to arrive after 2000Z. No matter how reactive the system is, however, some of these disruptions leak through to affect front-line operators like us. Referring to the Swiss Cheese model (Figure 2.1), no matter how many protective safety layers the company creates, some unfavorable event trajectories can potentially slip through. As the last lines of defense, it is up to us to detect and mitigate these elusive problems. By improving our personal resilience skills, we shrink the holes to stop unfavorable event trajectories.

Our personal resilience monitors for unfavorable event trajectories, interrupts them, and redirects them back toward favorable outcomes. Some unfavorable situations arise too quickly for systemic processes to catch. For example, consider departure operations at a busy airport with winds shifting toward a tailwind component. ATC continues to call out the wind speed and direction, but they don't compute the tailwind component or monitor when it exceeds our aircraft's operating limitations. Since it is extremely difficult to turn a busy airport operation around, they try to schedule airport reversals around convenient breaks within the flows of arrivals and departures. This biases their decision making toward traffic flow and operational volume rather than our particular aircraft limitations. We, however, remain focused on our tailwind limits. As soon as the wind exceeds our aircraft limitations, we inform ATC that we need a runway change. This may force ATC to initiate the runway swap earlier than they had planned. Other pilot-initiated changes include transitions from visual to instrument approaches as cloud layers and visibility drops, changes in type and intensity of frozen precipitation falling, unacceptable levels of turbulence, and unacceptable braking performance on slippery runways.

The second area where we affect system resilience is through our personal decision making, actions, and skillsets. Part of developing our aviation wisdom includes learning about ourselves. For example, if we recognize that we exhibit certain biases when we become tired, we can compensate by choosing flight schedules that match our natural wakefulness periods, making more conservative decisions when we are tired, increasing our caffeine intake, or calling in fatigued.

Our Master Class awareness grows from actively debriefing our experiences. Early in our careers, it evolves from "I'll never do that again" types of experiences. Each time we find ourselves operating in the gray zone of increased risk, we resolve to learn from the experience. "While that event ended favorably, I see several ways where we could have failed. Let's discuss it and see what we can learn." As we refine our Master Class perspective, we dig deeper to discover learning opportunities from even the most benign experiences.

Organization of the Professionalism section: The following three chapters will explore the Master Class path, skillsets, and attributes.

24 The Master Class Path

Choosing the Master Class path, we commit to promoting discovery, awareness, discipline, and self-improvement. Pushing against this commitment is a steady headwind of resistance. Human nature prefers the tailwinds of comfort and ease. The more often we encounter easy situations, the more we value feeling relaxed. The critical difference lies between *preferring* our comfort zone and *settling* into our comfort zone. When something disrupts our flight, we work to resolve the problem and return to our familiar, smoothly flowing flight. When we manage disruptions attentively and deliberately, we actively balance threats and conditions. If we conceal problems by over-simplifying them and taking shortcuts, we are just suppressing irritants so we can restore the familiar feeling of normalcy.

Aviation learning is not mathematically cumulative. Experiencing more events more doesn't necessarily mean that we learn more. As we learn some things, we often forget other things. As we acquire one good habit, another one fades into disuse. Our forward progress is interrupted by periods of backsliding. What is important is that we set our intention and resolve to keep moving forward despite the steady headwind. As we commit to the Master Class path, we commit to activities like purposeful practice, life-long learning, standards of excellence, pursuit of perfection, and building conscious awareness. The first step is establishing our Master Class intention.

24.1 ESTABLISHING OUR MASTER CLASS INTENTION

When we form the intention to follow the Master Class path, we adopt a mindset that shapes our future actions and choices. Intention fuels the personal commitment to align our activities toward a goal. Many of life's achievements have clear goal markers like earning an advanced academic degree, qualifying for a professional tournament in a sport, or achieving a black belt rank in martial arts. We divide each goal into increments that mark milestones along the way. Examples would be gaining acceptance into an advance degree program at our desired university, winning a local tournament to qualify for the professional-level tournament, and achieving brown belt rank in martial arts.

24.1.1 Career Advancement by Time versus Merit

Reaching Captain upgrade is dependent on seniority, the retirements of pilots senior to us, and our airline's growth. As we reach the top of the FO list, we can bid for Captain upgrade. Completing upgrade, we become Captains. Even an unmotivated, mediocre pilot will reach this milestone. This is one of the headwinds with promoting excellence in the airline profession. There is little systemic incentive for us to excel. Just stick around, keep passing checkrides, hope that our airline adds aircraft, wait for senior pilots to retire, and we'll all reach upgrade. Even after upgrading,

DOI: 10.1201/9781003344575-28

excelling as a Captain is no guarantee that we'll retain the position. Economic recessions often lead to furloughs and downgrades. These movements, too, are dependent on seniority, not merit. For these reasons, we can't use upgrade as our professional goal or as our motivation to form our Master Class intention.

24.1.2 Our Personal Commitment to Sustain Our Master Class Intention

The Master Class path has no clear markers, no ceremonies, and no visible signs of advancement. There is no badge, epaulette stripe, or pay bump awarded for reaching Master Class. Our motivation depends on our personal commitment. We commit to act in a particular way, to think in a particular way, and hold ourselves to a particular standard – all of which are personally selected, personally defined, and personally measured. Our goals are whatever we choose to set. Our progress toward those goals is as fast or as slow as we wish to make it.

24.2 ENGAGING IN DELIBERATE OR PURPOSEFUL PRACTICE

Having established our intention to pursue the Master Class path, we commit to deliberate or purposeful practice. Simply completing each flight without trying to learn from it leads to stagnation. Instead, we view every flight as an opportunity to discover something new about aviation or ourselves. In his book, *Outliers,* Malcom Gladwell suggests that people who achieved mastery of their area of interest typically required at least 10,000 hours of practice. His examples cited fields like chess (the hours of practice to achieve grandmaster rating) and performers (the hours that the Beatles practiced and performed together before reaching stardom).[1] Many critics inaccurately adopted this number as both a prerequisite and as a guarantee of mastery. Clearly, this is not the case. There are chess players with over 10,000 hours of practice that fail to reach mastery and long-running rock bands who continue to languor in obscurity. Closer to home, many of us log well over 10,000 flight hours or 10,000 flights, but fail to achieve mastery.

In subsequent interviews clarifying his theory, Gladwell detailed the importance of talent, physical ability, coaching, and most notably, that the practice be deliberate.[2] It is only when we combine these attributes with our personal commitment to excel that we can achieve mastery. In aviation, talent and physical ability exert little influence. While they may prove helpful early in our flying career, their importance fades after thousands of flights. Coaching can prove effective, but most line pilots visit the training center only once a year. Much of that training time is reserved for required training events, not individual coaching. While line flying, coaching opportunities prove to be sporadic and inconsistent. Ultimately, if we want a good coach, we will have to become one ourselves.

Gladwell's 10,000-hour reference comes from the work of K. Anders Ericsson and Robert Poole presented in their book, *Peak.* They propose that purposeful practice includes four components: well-defined/specific goals, focused attention, feedback, and getting out of one's comfort zone.[3]

- *Well-defined, specific goals:* We start by actively pursuing self-improvement and self-awareness. This guides us to investigate the underlying conditions at play, how they interact with aviation, and how they personally affect us. By engaging every aviation challenge with a curiosity to learn and improve, we steadily advance our growth.
- *Focused attention:* We keep our attention appropriately focused. We raise and lower the intensity and object of our attention depending on the phase of the flight and the needs of the situation. With familiarity comes habit. With habit comes relaxed attention and stagnated growth. The antidote is to find something new to learn from every situation.
- *Feedback:* We evaluate every learning situation to discover how we can do it better in the future. This requires an honest assessment of the aspects that we can see and feedback from others for those aspects that we cannot see. Since very few of us enjoy consistent coaching from a mentor, we need to rely on introspection and collective feedback from our teams.
- *Getting out of one's comfort zone:* This last point acknowledges the inherently erosive effects from settling into our comfort zone. When we reach our final career step of landing an airline job or upgrading to Captain, there are few external motivators prodding us to excel. We can easily settle back and stagnate. Rising from our comfort zone is an inside job. No one will do it for us.

24.2.1 Purposeful Preparation

Purposeful preparation digs deeper into underlying conditions to uncover potential threats. While comfortable pilots would prepare for the conditions they *expect* to encounter, we use purposeful preparation to anticipate challenges that *might* arise. We actively search for conditions that might affect our flights – many of which are not immediately apparent. For example, a quick review of destination weather reveals that it is currently VFR and forecast to remain so. A deeper dive, however, reveals a marine layer of low clouds lingering off the coast. Applying our experience, we would anticipate that these conditions might favor quickly forming low ceilings and reduced visibility. Preparing for this contingency, we would ensure that we upload adequate fuel to fly an approach followed by a missed approach and a diversion to a favorable alternate. To expand our options for landing at the intended destination, we would verify that NOTAMS allow for a CAT III ILS approach.

24.2.2 Purposeful Briefing

We follow purposeful preparation with a briefing that centers around consequential threats. Using the "threats forward" format of the Briefing Better model, the PM briefs the threat of possible low cloud layers and reduced visibility. The PF integrates this threat into the game plan. As the flight progresses, we monitor for early warning signs. Even if the ATIS is still reporting VFR, we prepare to fly a CAT III ILS approach. Entering the terminal area, we search for low cloud decks or signs of fog forming in low-lying areas around the airport. In most cases, this extra effort will prove unnecessary, but as Master Class pilots, we would rather over-prepare for that one time when unexpected weather moves in.

24.2.3 Purposeful Execution

Purposeful execution is our commitment to complete each flight in the best way we can – accurately, smoothly, efficiently, and safely. This includes superior stick-and-rudder skills, accurate management of automation, effective CRM, superior customer service, and resilient handling of unexpected conditions.

24.2.4 Purposeful Feedback and Debriefing

Purposeful feedback comes from an effective and focused debriefing. After arriving at the gate, we discuss when we first detected the earliest signs of fog, how quickly it thickened, how the visibility changed as our approach progressed, and how well we followed the CAT III ILS approach procedures. We highlight areas where we prepared well and areas where we can improve our future preparation techniques. We might even review how we would have performed a missed approach and diversion to our planned alternate.

24.3 COMMITTING TO LIFE-LONG LEARNING

Life-long learning describes a self-motivated pursuit of knowledge and understanding. Being self-motivated means that we are not dependent on external forces pushing us to learn. Once we start line flying, there are no instructors administering tests or assigning grades. No one will encourage us to pursue knowledge above the required minimum. Once we reach this rung on the career ladder, we are on our own.

The regulatory environment unintentionally promotes a kind of minimalism. As long as we meet the minimum standards and requirements, we are deemed trained, qualified, and certified. The box is checked and we are released to fly for another year. The problem is that this environment sets a low bar for further growth. As the old adage goes, "If you aim for the minimum, you will hit it every time."[4] For those who are content with maintaining minimal standards, the system will support a career of mediocrity.

24.3.1 Growth Mindset

Committing to life-long learning means that we embrace a growth mindset. Acquiring knowledge and understanding is a personal priority. We thirst for it. Take the example of some required reading sent down from the flight operations leadership. The comfortable pilot would stop after reading it once. The growth-motivated pilot would not only read the assigned material, but might seek out associated material to help understand it within its larger context. We learn because we place intrinsic value in discovering new things. It widens our perspective. Consider the points from this list of life-long learner habits.[5]

1. Read on a daily basis.
2. Take courses.
3. Actively seek opportunities to grow.
4. Take care of your body.

5. Pursue diverse passions.
6. Value making progress.
7. Challenge yourself with specific goals.
8. Embrace change.
9. Believe that it's never too late to start something.
10. Share your attitude of getting better with others.
11. Leave your comfort zone.
12. Never settle down.

24.3.2 Pursuing Depth of Knowledge

We seek to understand topics on deeper and wider levels. For example, imagine that we had an interaction with ATC that led to a misunderstanding. Instead of just shrugging it off, we try to understand how the event evolved. We might start by referencing the FAA's organization policy document, Air Traffic Control document (JO 7110.65Y). Next, we might explore an Air Traffic Control forum group like stuckmic.com. The information we gain would help us understand the ATC controller's perspective, how they perceive their role in the aviation environment, and how they perceive ours. Granted, reading policy manuals may seem rather dry and academic. Life-long learners overcome this resistance and press forward.

24.3.3 Integration of Knowledge and Experience

We process new knowledge and experience by integrating it within the existing fabric of our aviation wisdom. This integration guides how we create meaning and judge significance. It helps us to skillfully build SA, manage game plans, and detect counterfactuals. Integrating information reduces the need to remember specific facts. In aviation, we are surrounded by hundreds of numbers, values, and procedures. Remembering individual facts is challenging. Sometimes, we can't recall particular numbers. Other times, we recall them incorrectly. Weaving individual facts within larger concepts connects them to relevant conditions. As experienced pilots, we are skilled at detecting and compensating for conditions. Recalling the facts integrated within those conditions is easier than trying to recall the facts without context. We discover that we don't need to remember the actual numbers. We just need to remember that the numbers exist. For example, consider a case where our aircraft manufacturer imposes altitude limitations for the inflight use of APU for electric load, pneumatic load, or both. Instead of memorizing those three specific altitudes, we simply remember that using the APU inflight has altitude limitations. When the rare event occurs where we'll need to use the APU inflight, we'll remember to look up the altitude limitations. The isolated fact becomes interwoven with the real-world context where it is applied.

24.4 EMBRACING A STANDARD OF EXCELLENCE

What does excellence mean? How do we measure our standard of excellence? Consider the following components.

- *Importance:* Not settling for "good enough", we endow every flight with importance. Every flight becomes significant. Every flight is worthy of our attention. This counteracts the tendency for repetitious flights to become boring and habitual.
- *Accuracy:* We strive to follow established procedures. When conditions require modification, we continue to satisfy the intentions underlying that procedure. We continue fulfilling the higher principles of our company's mission statement.
- *Efficiency:* We select choices that efficiently use resources. This may involve balancing between conflicting goals. We may, for example, exceed the most efficient cruise speed to get our late passengers to their transfer flights knowing that it will assist many other connecting flights to depart on time.
- *Smoothness:* We value smooth aircraft control. If we are a few knots fast, we smoothly reduce thrust and allow the speed to gently return to the desired airspeed. We make our control inputs as gently as conditions allow.
- *Intentionality:* We actively select our choices. Instead of following well-worn, habitual game plans, we make appropriate adjustments for the unique blend of current conditions.
- *Mindfulness:* We focus our attention on the flow of the flight. Instead of dwelling on past mistakes or events, we pay attention to what is happening right now. We also modulate our attention to the appropriate level for the phase of flight. We rest our minds during low task-loaded phases to improve our mental focus during high task-loaded phases.
- *Forward-looking:* We project our SA to anticipate threats and detect counterfactuals. We adjust our game plan to achieve the smoothest transition from present SA to future SA. We prepare backup contingency game plans to respond to deteriorating conditions.
- *Modeling:* When we demonstrate our commitment to excellence, people notice. They also notice when we fail to live up to our high standards. In the public eye and around other pilots, we act and present ourselves professionally.
- *Ownership:* We own our appearance, actions, speech, and choices. They become expressions of our character and personality.

Consider that the most important aspect may be the last one, ownership. Our personal standard of excellence shapes the lens we use to view ourselves and our profession. On the first page of his book, *Blue Threat*, Tony Kern introduces the term *empowered accountability.*[6] We don't just accept accountability after an event, but apply it from start to finish. From arriving to work rested, healthy, professionally attired, and motivated, to executing every facet of each flight according to our standard of excellence, we embody pride and ownership.

24.4.1 Everyday Excellence

We tend to view excellence on a grand scale. Consider, instead, the subtle expressions of excellence – those many simple acts, choices, and words we express every day. Everyday excellence is not just doing a great job when extreme events demand it.

It is also about demonstrating excellence with completely ordinary situations on monotonous, routine flights. It's harder to embody excellence when nothing is happening and no one is looking. Perhaps, this is the measure of excellence that sets true Master Class professionals apart.

24.5 PURSUING PERFECTION

Another component of Master Class intention is the pursuit of perfection. Unlike other professions, we don't attempt to increase the difficulty of our vocation. An athlete, for example, pursues higher levels of expression by performing more difficult maneuvers, increasing their speed, or reducing their reaction time. The closest aviation pursuits might be air racing and aerobatic competitions.

In airline aviation, we strive for a subtle expression of perfection. We strive to flawlessly complete each flight from start to finish by performing ordinary tasks consistently, seamlessly, and smoothly. Our passengers don't want to be awed by our aviation skills. They want to pass the time sitting quietly, reading their books, watching their screens, and arriving safely and on time. We strive for perfect, unexceptional consistency. So, while a performer hopes to achieve a score of "10" on the difficulty of their performance, we strive to achieve 100% successful, desirable outcomes. Following are some of the components that describe our pursuit of perfection.

- *Balanced priorities:* We continuously monitor conditions to anticipate how they might affect our flight. When priorities begin to conflict with each other (for example, trying to satisfy on-time arrival while running late), we balance speed and efficiency with preserving our safety margin.
- *Compliance:* Our standard of excellence compels us to follow procedures. We interpret the guidance both for what it states and for what it intends. When we lack clear procedures, we find a way to complete the tasks while satisfying the guiding policy.
- *Personal skills:* We steadily improve our flying skills. Every flight offers a fresh opportunity to plan better, communicate more clearly, fly more smoothly, extend our SA, anticipate changes, and mitigate errors.
- *Resilience:* We pursue perfection across the entire range of operations. We react intentionally, accurately, and dynamically to each situation. Recalling the Swiss Cheese model, we strive to create an impenetrable final barrier that stops every preventable mishap trajectory. We contribute to our airline's resilience by actively communicating information on deteriorating trends back to our central operations centers.
- *Passion:* On one end of the spectrum are pilots who show up for work and do their job. On our end of the spectrum, we remain passionate about our contribution to our airline and to our profession.
- *Upward trajectory:* While comfortable pilots remain content with satisfying the minimum job requirements, Master Class pilots continue to amass aviation wisdom. Holding steady is not enough. Even after we have mastered the challenges of everyday line flying, we look for new ways to improve our skills, perceptions, self-awareness, and growth.

- *Effortlessness:* Professional athletes describe moments when all of their talent, practice, and motivation come together in effortless, perfect execution. They describe how it feels when they make the perfect dive, the perfect throw, or the perfect swing. We seek this same experience with our flying. We strive for the perfect takeoff, the perfect descent profile, the perfect turn to final, and the perfect landing. Effortlessness means that we are not working harder to achieve something greater. Instead, we seek the perfect groove where everything seems to fall into place.

The pursuit of perfection is the lift that elevates us from comfortable proficiency to achieving mastery. We arrive each day rested and nourished. We keep our outside life in order so it doesn't divert our attention while flying. We build and brief our game plan, but acknowledge that every game plan is dynamically unstable and will require continuous course adjustments. We weave a net of monitoring to catch every relevant condition that may affect our flight. We make well-timed and accurate adjustments to aircraft systems and flight controls. When we get it right – when we get it perfect – everything seems to fall seamlessly into place.

24.6 BUILDING CONSCIOUS AWARENESS

Promoting safety in a high-consequence, low-accident-rate industry like airline aviation is challenging. Across millions of flights, the few errors we make seem small, easily detected, quickly corrected, and inconsequential. To improve aviation, however, we still need to study those few errors, minor incidents, mishaps, and close calls that slip through. As we study these events, we need to carefully choose which lenses to use. The investigation processes that we use for full-blown accidents tend to rely on hindsight. If we eliminate this bias and just examine what the pilots did, we discover that most of their actions and decisions appeared normal, common, and non-hazardous to them while they were in-the-moment. They don't look much different from actions taken by successful crews who achieve desirable outcomes.

To build conscious awareness, we need a fresh pair of glasses. We need to study the seemingly normal work performed by mishap crews from their in-the-moment perspective. Understanding and applying the interplay between normal work, existing conditions, and their evolving event trajectories guides our learning. In the end, what really matters is learning about ourselves, how these scenarios would feel if we were in their seats, and how we can avoid similar errors.

24.6.1 Meta-Awareness

Meta-awareness is our understanding of our own level of awareness – how aware we are of our awareness. Consider an event from our personal experience. Select a particularly familiar and often repeated task like locking our car doors before going into a store. After entering the building, we select a shopping cart and proceed down the aisle. Suddenly, we are hit by the thought that maybe we didn't lock our car doors. Certainly, it is a normal habit for us to press the lock button on our key fob. We recall instances where we did it in the past. We just can't recall whether we actually did it this time.

We return to the car and discover that the doors are, in fact, locked. We conclude that we must have done it subconsciously and that the act didn't create a lasting memory. Let's change the outcome and reexamine this scenario. In this second example, we return to our car and discover that the doors are actually unlocked. We wonder how we could have missed this habitual task. We think back and recall that someone honked their horn across the parking lot as we were exiting our car. We remember looking to see what the horn was about. We conclude that distraction must have interrupted our normal door-locking habit pattern. We further surmise that the break in our habit pattern must have subconsciously planted the seed of doubt that surfaced when we started down the aisle in the store. Compare these two cases and notice how we conduct many of our actions subconsciously or habitually. The more familiar the task, the more of it we relegate to subconscious habit. Now, let's alter the situation a third time and add the presence of some sketchy individuals milling around the parking lot. Recognizing them as potential threats, we pay particular attention to locking our doors. We look at the key fob to ensure that we press the LOCK button instead of mistakenly pressing the UNLOCK button. As we press it, we listen for clunk of the door locks moving into place. We then enter the store confidently aware that we successfully locked the doors. The difference with this third case is that we added conscious intentionality to an otherwise mundane task. We not only paid attention to locking the doors, but consciously created a memory that we had successfully completed it.

Habitual aviation tasks are affected by everything else that is going on in our minds. Monitoring flight conditions absorbs much of our attention, so we rely on habits to complete familiar tasks subconsciously. The more of our attention that we divert elsewhere, the more of our actions we complete subconsciously (or miss subconsciously). This effect becomes evident when we interview mishap crews. We ask them to recall specific actions, many of which they probably completed subconsciously. There recollection is typically quite distorted. The mishap events become intermingled in their minds with hundreds of past subconsciously completed events. Unless the pilots made a dedicated effort to apply conscious awareness to their actions, they struggle to recall the events accurately.

Let's apply this self-awareness exercise to an aviation example. If I promised you a $1,000 cash bonus if you could fly your next leg without mishap, you would certainly accept my offer. After the flight, I could ask you, "How many details can you recall from this flight?" You would be able to accurately recount how you prepared for the flight, the items you covered in your preflight briefing, and how you motivated the crew to be especially vigilant for procedural omissions or errors. You would have read through the weather package more closely, called your dispatcher, made detailed briefings, stayed focused, and spent extra effort expanding your SA. As the flight progressed, you would have remained especially vigilant for changing conditions. You would have actively monitored for counterfactuals and continuously refined your game plan. You would have devoted an extraordinary level of care. Simply stated, you would have done everything you normally do, but with more conscious awareness and focused attention. The reality is that this flight was no different than any other flight, except for the level of conscious attention you chose to apply.

For contrast, examine the opposite extreme. Consider a pilot who is tired, bored, late, in a foul mood, hungry, or ill. How much attention-to-detail would we expect from them? In reality, no one is going to offer us a cash bonus to accurately complete our next flight. Perhaps, however, we may be a bit tired, late from a gate hold, a little bored, grumpy, sniffling from allergies, suffering from acid reflux from the lunch that we ate too quickly between flights, or distracted by an issue from home. How much conscious attention-to-detail would we devote to that flight? This exercise was intended to help us realize that any one of us can reduce our level of attention by succumbing to external and internal conditions.

24.6.2 Attention Drift

The way we focus our attention changes over time. The repetitive nature of airline flying promotes a slow drift in our priorities and practices. We fly the same aircraft over the same routes year after year. Over time, we might adopt shortcuts and techniques that adversely affect the protections built into our procedures. The quality of our work suffers. Our monitoring weakens. The range of our SA shrinks. Our vigilance wanes. It becomes just another day at the office. Add daily fluctuations of fatigue, boredom, schedule disruptions, mood, hunger, and home life distractions and we have the makings of a "pilot error" accident or incident. This is how we become "that mishap pilot".

The repetitive nature of our flying doesn't naturally motivate our upward growth on the aviation wisdom scale. While our operational experience increases through daily exposure, our aviation wisdom can still erode. This slow erosion is a natural byproduct of repetitive work. Mishap pilots are not exceptional individuals. We can become just like them unless we pursue a Master Class level of caring.

24.6.3 Level of Caring

Our level of caring affects how we respond to conditions. By studying our own reactions to familiarity and repetition and by monitoring our moods, decisions, and actions, we can sense patterns in our level of caring. We can study how our mindset changes with different conditions. How does our level of caring change when we are tired? What is our typical level of caring on the last flight of our pairing compared to the first flight? How does the experience level of the other pilot affect us? Do we alter our techniques to compensate for emerging vulnerabilities? Do we employ CRM safeguards to monitor our blind spots? When we allow our level of caring to drop, our growth rate stagnates. Small irritants undermine our motivation. Our attitudes turn negative. As long as we continue caring, our mindset will motivate continued growth. Small irritants vanish. Our attitude remains positive.

24.7 UNDERSTANDING AND OVERCOMING OUR BIASES

Biases can adversely change our mindset and increase our vulnerability to error. Studying events where bias affects our actions, we monitor our personal reactions to particular conditions and situations. We then adjust our techniques and safeguards to control their adverse effects. Let's start with a review of some biases that tend to emerge in aviation environments.

24.7.1 Plan Continuation Bias

This is a common contributor to mishap events. It starts with a familiar game plan that doesn't fit with existing conditions. This causes the situation to become more complicated. Our problems multiply. We make corrections, but they don't work. The pace quickens. We work harder to hold our game plan together. We ultimately accept the error, increase our effort, and force our game plan against a rising tide of resistance. A related bias is called *sunk-cost bias* where we feel like we have invested too much effort into the game plan to abandon it for another.[7] Afterward, we regret our decision to continue. We share comments like, "I didn't see that coming," and "Let's not do that again." These events need to be extensively debriefed. We should schedule a block of time to fully examine the sequence of events, when significant conditions emerged, the indications they generated, when we detected them, and when the game plan should have been abandoned for a safer contingency. As we debrief these events, we need to pay particular attention to our personal reactions. Did we notice our attention tunneling in? Did frustration affect our ability to consider another course of action? How did CRM fail to interdict the failing trajectory? What changes do we need to make in our techniques and mindsets to reduce the effects of plan continuation bias?

24.7.2 Representativeness Bias

This is a precursor to plan continuation bias where we assess the risk of a particular event by subconsciously comparing it to previous similar encounters.[8] The zone dividing profiles that can be salvaged and profiles that cannot be salvaged is fuzzy. Encouraged by past success, pilots may allow their risk tolerance to drift toward riskier choices.[9] Another nuance is *availability bias* where we tend to downplay threats since our previous encounters all ended favorably.[10] It always worked in the past, so it will probably work out fine again in this situation. Another related bias is familiar technique bias (*familiarity bias*) where we tend to revert to our most commonly used game plans when complexity makes improvising seem risky.[11] Familiar game plans provide a sense of comfort by reducing the number of choices to consider. As risk and complexity rise, we need to acknowledge our vulnerability to these biases and increase our vigilance for counterfactuals. When we detect counterfactuals, it should prompt us to switch to more conservative options. "Doing our best" is not enough.

24.7.3 Expectation Bias

This bias leads us to see what we expect to see and hear what we expect to hear. When a stressful event consumes our attention, surrounding events only receive our peripheral attention. We don't fully process verbal statements because we are only "half-listening". Our minds reconstruct these half-heard messages in ways that match our expectations. The same effect applies for visual indications. Focusing intently on a point of interest, we take a quick glance toward a particular gauge. As we return to our main point of interest, we reconstruct the image in a way that supports our current game plan. This bias is also affected by our experience level. As we become more experienced, we become more confident in what we know and less concerned about what we don't know. When something falls outside of our expectations, we tend to discount it as inconsequential. Debriefing these events, we ask ourselves these questions. Where was my attention directed? Did I interpret the warning or counterfactual as a distraction? Did my attention focus contribute to the error? Was I particularly tired, bored, distracted, or unfocused before the event? Countering expectation bias requires mindfulness and detachment. We need to avoid becoming consumed by a stressful situation. As soon as we feel ourselves being drawn in, it should trigger an abort response to exit the quickening situation, reset under better controlled conditions, and reattempt.

24.7.4 Confirmation Bias

Like expectation bias, confirmation bias alters our interpretation of information. When we become committed to a course of action, we tend to increase the credibility of indications that support our plan and minimize indications that conflict against it. For example, if we are threading our way through build-ups, we might increase the significance of previous aircraft making it through a hole. Simultaneously, we might minimize indications that a growing storm cell is quickly filling that hole or possibly spilling a curtain of hail across our path. When we encounter events where we placed too much weight on some information and under-weighed counterfactuals, we should examine how we fell into that trap. Develop quality-rating techniques to gauge the importance of indications and increase the monitoring of counterfactuals. This decouples what we want to happen from what is actually happening. We may need to slow our monitoring sample rate since quick glances tend to favor confirmation bias.[12]

24.7.5 Specialty Bias

As we become more confident in ourselves and our abilities, we tend to minimize the observations and inputs from others who may lack our level of experience.[13] Related biases are professional bias and over-confidence bias. These be particularly strong in some Captains. A particularly amusing example came from an elderly passenger deplaning after a flight. She stopped, looked up at the Captain and stated, "Sonny, there is a wrench on top of the wing." The Captain politely thanked her. He initially discounted her report because of her age and assumed ignorance. Then, he began to wonder. He engaged hydraulics and raised the spoilers and there it was, clear as day, a wrench sitting in cavity under the spoiler. Truth can come from any source. Ramp agents may detect exterior damage or leaks. Flight attendants may detect malfunctioning components by their sound or vibration. Passengers may detect engine or tire anomalies. A corollary of specialty bias is viewed from the opposite perspective. A new FO may give unwarranted confidence to their Captain's opinion, even when they know them to be wrong. Countering these biases requires both accepting truth from unexpected sources and questioning assertions from superiors.

24.7.6 Framing Error Bias

This bias is indicated when we lock in on the first explanation of a problem and fail to investigate other possible causes.[14] For example, if our flaps fail to extend and we don't notice it, the first indication we might detect is that the aircraft is not slowing. We may attribute the high airspeed to a wind shift. If we lock in on this false cause, we might not investigate further to discover the flap malfunction. While this may seem farfetched, this exact scenario has occurred many times. We are biased to attribute common effects to commonly experienced causes. If framing bias leads us to make an error, we should examine our mindset at the time. What led us to lock in on the wrong explanation? How can we improve our monitoring to detect the actual causes behind unexpected effects? How can we improve our CRM so that at least one pilot captures these errors?

24.7.7 Salience Bias

Salience bias is when we focus on the loudest, brightest, and largest indication to the exclusion of lesser indications.[15] Consider a simulator event of an engine fire immediately after takeoff. The warning bell and fire lights overwhelm our monitoring scan. After the pace settles down, we expand our scan and discover that we also have a hydraulic system failure. Another example is when one pilot calls out the traffic, both pilots immediately focus on it. Fixated on that threat, they miss additional traffic converging from a different direction. PMs counter this by ensuring that they remain out of synch with their PFs. Whenever we become focused on a single indication, try to resume the appropriate monitoring sample rate for phase of flight as soon as conditions allow.

24.7.8 Fundamental Attribution Error

This is the tendency to blame others when things go wrong.[16] If we arrive at our destination to discover that the weather has closed the field and we lack a scheduled alternate and diversion fuel, we may blame our dispatcher, even though we had access to the same information that they did. We might even go so far as to attribute their error to personal motivations. "They don't care about what happens to us out here." "They don't know what they are doing." We counter this bias by understanding the various processes within our airline's system. Where are the vulnerabilities within each workgroup? Where do they lack clear information? What can we do to catch their errors before they affect us?

24.7.9 Distancing through Differencing

This bias involves interpreting actions based on our opinions about the source.[17] For example, consider an event where a competing airline's crew elects to go around for turbulence on final. If we hold a low opinion of their pilots, we might choose to continue our approach because, "We are better than them. We know how to get the job done." We counter this bias by focusing on the facts. Rather than attributing their decision based on a stereotype, we assume that they are right and investigate available indications. We could call for other ride reports on final to gain context.

24.7.10 Automation Bias

Automation bias reflects the assumptions we hold regarding automation accuracy.[18] Given a mismatch between an automated result and our judgment, we might defer to the automation because, "the computer is always right". Since a computer can't make a computational error, computer-derived results are assumed to be valid. On the other hand, we've experienced events where the automation fails us. We counter this bias through verification and sensemaking. First, we confirm the accuracy of data inputs. An example is cross-checking the FMS route waypoints with charted waypoint restrictions. Next, we compare the automated results with our experience and rule-of-thumb estimations to verify that they make sense.

24.7.11 My Biases

This class of biases is unique for each of us. They are the particular biases that emerge when we lower our vigilance, often while we are tired, bored, distracted, or inattentive. These are our personal blind spots. To mitigate these, Tony Kern suggests an ICAN method to identify, categorize, analyze, and neutralize our bias blind spots.[19] His suggested steps are:

- Become aware of your blind spots.
- Respect all new situations.
- Look outside of your specialty.
- Recognize changing situations.
- Invite others into your decision-making process.

24.8 COUNTERING THE FORCES THAT DRAG US BACK

Along with practices that keep us moving forward, we need to counter forces that threaten to drag us back into the stagnating effects of our comfort zones.

24.8.1 Countering Cynicism with Proactive Discussion

Many work environments exhibit an undercurrent of cynicism. Some of it is intended as humor. Some of it reflects tribalism. We seem to enjoy complaining about endless procedural changes, policies that make our jobs more difficult, scheduling changes, slow-moving seniority lists, and labor contracts. Cynicism can contribute to a slow erosion of our level of caring. It drives a wedge between accepting responsibility for the operation (ownership) and embracing an us-versus-them mentality (separation).

If we embraced a cooperative perspective, we would *discuss* problems that a new policy is causing. We would focus on its undesirable side effects and look for ways to revise the policy. Conversely, if we assume a separate perspective, we would *complain* about the new policy. Complaining encourages active non-compliance. The more pilots that express their willingness to work around the new policy, the more that a line culture of non-compliance feels acceptable.

The company is responsible for guiding policy discussion. A comprehensive procedure development process starts with thorough research and testing. Before releasing the new procedures, pilots should be given reasons and justifications behind the procedure development process. After implementation, SMS managers need to monitor for unintended consequences, listen to reviews from front-line employees, and correct procedural flaws.

Master Class pilots promote proactive discussion. When we sense discord, we start by deciding whether it is something that can be improved. Not all problems are solvable. Junior pilots complaining about being on reserve or senior pilots complaining about scheduling reroutes are unsolvable problems. It is best to just let them be. On the other hand, pilots complaining about pairing construction that contributes to fatigued flying or policies that create unnecessary barriers are solvable problems. We can forward concerns to leadership through available SMS programs. Consider the following NASA ASRS report and recommendation concerning safety concerns into Hilo Airport (Hawaii – ITO/PHTO).

BOX 24.1 MASTER CLASS CAPTAIN'S SAFETY RECOMMENDATION

Captain's report: ITO Runway 8 Visual Approaches, especially at night or marginal VFR conditions, constitutes increased risk and requires additional resources to mitigate this risk. The ITO airport information pages directs us to "at night, plan on arrivals to Runway 26", but tailwinds and wet runways often prevent this option. The airport information pages also directs us to "follow the shoreline south to Hilo Bay to intercept Runway 8 final". The issue is that the shoreline is only 1.5 NM west of the approach end of the extended centerline of Runway 8, which correlates to a maximum wings level altitude of 500′ AGL upon rollout for a three degree glideslope. To even achieve this minimum, the approach requires "cheating" by offsetting further inland over downtown Hilo prior to the turn to final, to account for the turn radius which would be even greater with increased speeds associated with a flaps 15, single engine approach (such as an ETOPS single engine divert event), or accepting a steeper glideslope and rate of descent. Additive conditions such as the common inclement weather and winds and relatively sparse ambient lighting associated with Hilo, high terrain, (Company) Crews' relative unfamiliarity with the new ITO service, and the short inter-island flight times (which limit time to brief and prepare for a challenging approach) all add to the increased risk. Inter-island flying is often benign, but the ITO Runway 8 Visual is a challenging approach under the best of circumstances and could surprise unfamiliar ETOPS Aircrews without adequate awareness and preparation. I recommend the following steps be taken to mitigate the ITO Runway 8 Visual.

1. Expedite the creation of an RNAV visual to ITO Runway 8 to assist in situational awareness during the approach.
2. Make all ITO Runway 8 approaches Captain Only landings until the RNAV 8 Approach is created. The descending left-hand turn landing is even more challenging for the FO who has to look through the Captain's windows during the approach. The additional experience level of the Captain should help mitigate the risk.
3. Increase awareness through RBF's (Read Before Fly), notes on weather packets and Dispatch Releases describing the risks associated with the ITO Runway 8 Approach.
4. Encourage briefing of ITO Runway 8 Visual Approach procedures on the ground prior to departing for ITO when Runway 8 is in use, due to the short inter-island flight time.
5. Add a module to ETOPS recurrent training specifically addressing the ITO Runway 8 operations.[20]

Notice how clearly the Captain enumerates the specific problems with the existing policy and how it currently encourages "cheating". Imagine how many other pilots felt the same way, but did nothing to change the procedure or just complained about it. This Master Class Captain concludes the report by presenting five action steps to solve the problem in ways that support regulatory requirements and line-pilot compliance.

24.8.2 Reframing the Irritants

Some events will still irritate us. How we choose to frame them will either move us forward or pull us back. For example, at some point during my career, I realized that I was becoming rather jaded with waiting for and riding on airport shuttles. Waiting for a promised hotel van became a point of personal irritation that adversely affected my enjoyment of my airline career. To resolve this, I reframed the experience into a more favorable context. I conversed more with my crewmembers. I looked around to discover details that I had previously missed – families joyfully greeting loved ones emerging from the terminal and happy dogs finally released from their kennels. Instead of viewing the shuttle wait as an irritation, I reframed it as an opportunity to experience and enjoy new things. Sure, I still disliked waiting for the promised hotel van on cold and rainy nights, but more often, it became something new to be experienced.

24.8.3 Embracing Change

"There is nothing more constant than change."[21] There are so many sources of change in our airline career. There are large changes like a learning a new aircraft, crew base reassignment, upgrade, and economic cycles. There are small changes like schedule reroutes, delays, and cancellations. Even after settling in with a particular aircraft and base, we experience changes of line pairings, aircraft systems, and procedures.

The airline experience sometimes feels like shooting the rapids in a raft – constantly bumped and pushed by unseen forces. We ride the waves, sometimes controlling our course and other times just doing our best to hang on. Throughout this steady push and pull of forces jostling us around, we still need to deliver 100% success with reliability moving our passengers and freight from their origins to their destinations safely.

A useful first step is to avoid labeling changes as "good" or "bad". When we assign labels, we affect how we perceive and handle them. Bad changes encourage avoidance. Since many changes are unavoidable, this isn't a useful option. If the weather moves in and closes the field, we can perceive it as bad, but it doesn't help us to cope with its effects. The same goes for changes we label as good. We welcome and encourage good changes. In time, however, even good changes become normalized and melt into normalcy. We forget how good they used to make us feel.

Consider viewing changes simply as altered conditions – neither good nor bad. When flying, we acknowledge new conditions and apply the necessary corrections to restore our intended course. Returning to our river raft analogy, some parts of the river are smooth and easy while other parts are turbulent and challenging. They are all parts of the same river ride. So, handle whatever is happening at that moment. As Master Class pilots, we accept that even the smooth and easy parts are transitory. Enjoy the calm, but prepare for the next set of rapids.

Find ways to embrace change. When a new system or procedure is introduced, actively research the changes to understand and master them. If it is an optional or situationally dependent system such as HGS or Autoland, search for opportunities to practice the procedures under ideal conditions. This way, we'll be confident and skilled when using them under marginal conditions. Declaring, "I don't like using Autoland" doesn't eliminate the need to successfully apply procedures when conditions require. A better strategy is to practice using it under favorable conditions so when the weather drops, we can perform the procedures with ease and confidence.

24.8.4 Affirming Ownership

Another mindset that holds us back is the perceived division between groups – termed tribalism. A common example is the division that evolves between management and labor unions, especially during contract negotiations. Others evolve between us and other work groups (station operations, ground operations, inflight, and maintenance) and between pilots from different backgrounds (military, regional airlines, corporate, and age groups). These divisions promote an "us versus them" mentality. When we experience difficulties, perceived divisions make it easier to blame "them" as the source of the problem. The ill-will spreads. Barriers rise.

A more constructive perspective is to promote both individual and group ownership of the operation. If we all grab an oar and pull together, we all move forward. If one group is struggling, we pitch in and help them. When we are struggling, we allow others to pitch in and help us. When we have information that might help another department, we share it. When we need information that another department might have, we ask. This not only helps the company succeed, but it promotes goodwill by sharing both the satisfaction following success and support following failure.

24.9 CONTINUING OUR AVIATION WISDOM TRAJECTORY

The Master Class path requires work – not hard work, just steady, sustained effort. It is like a long backpacking trip. Every day, we resolve to make forward progress. Every evening, we rest and reflect on what we learned along the way. With the new dawn, we shoulder our pack and move forward again. Sometimes, our motivation wanes. We grow bored with the monotonous landscape. We begin grumbling about small irritants. The lure of the easy comfort zone begins to hold us back. To counter this, we need to remain aware of changes in our mindset and make frequent adjustments to keep our progress on track.

24.9.1 Keeping Our Eyes on the Prize

Our goal is to advance our aviation skills and wisdom to safely and confidently handle anything that happens during a flight. This means that we adequately prepare and brief for expected conditions, that we execute the game plan as envisioned, and that we extend our SA forward to anticipate changes. We strive for smooth planning and execution – trying to fly the perfect flight. When challenges confront us, we demonstrate our resilience by detecting, identifying, understanding, and responding to any challenges. We make appropriate corrections to the game plan, anticipate counterfactuals, set trigger points for discontinuing the original game plan, and confidently execute safer options when they are required.

24.9.2 Balancing Over and Under-Response

One characteristic of highly skilled, Master Class pilots is a finely tuned sense of balance. We are able to react to situations with just the right amount of effort. We know when to ramp up our response to threats and when to leave smaller disruptions alone. Whenever we handle a disruption, we monitor for any aftereffects that might linger. For example, if we overcontrol our correction for a turbulence bump, it may cause an overshoot in the opposite direction. We'll need to apply a follow-on correction. Undercontrolling a problem allows it to intensify. Overcontrolling a problem requires additional corrections to dampen the overshoot. We learn to balance our responses to apply just enough input to arrest the drift and return the trajectory back to our desired course.

24.9.3 Keeping It Steady

Remember that we are engaged in a long career. There are no deadlines, no time-constraints, and no need to stress about how much progress we have made or how far we have left to go. Instead, make a commitment to sustain steady forward progress. Learn something new every day. Improve a skill every day. Instead of worrying about achieving milestones, find satisfaction in achieving inch-pebbles.[22]

24.9.4 Keeping It Smooth

Smooth progress is better than strong bursts of effort. Consider comfortable pilots coming up on their yearly checkride. Knowing that they have let their skills slip and their procedures atrophy, they cram to prepare. They spend hours toiling through the manuals and reviewing procedures. They practice approaches that they had neglected throughout the past year. It becomes an unenjoyable barrier that they must overcome. After passing their checkride, they check off the box for another year and return to their past habits. Instead, strive for easy, smooth, sustained progress. Practice the procedures, review the manuals, and improve proficiency every day.

24.9.5 Keeping Our Professional Aviation Wisdom Moving Upward

With age, experience, and seniority come more responsibility and higher expectations. Our professional experience and wisdom curve doesn't reach its peak until retirement day. Unfortunately, some of us reduce our effort as we round the final turn for the finish line. Content to coast those final yards, we settle back and relax. Perhaps we begin to feel our age. Perhaps we become distracted by post-retirement plans. Whatever the reason, some of us ease off before reaching the finish line. As soon as we redirect our intention away from professional airline flying, our wisdom curve starts trending down. To keep moving upward, consider adopting a strategy of leaning into the tape. Having run a good race, the last few yards become even more important. We sustain our effort and even gain a few more inches by leaning forward as we cross the finish line. This keeps us sharp and focused until the very end. After retirement, we'll have plenty of time to redirect our attention toward other endeavors.

24.9.6 Keeping It Interesting

Any long-running endeavor can become stale and tedious over time. We need to keep our career motivated, fresh, interesting, and fun. Most of our contemporaries won't see our efforts, so we need to find personal satisfaction in our progress. Few people will encourage us along the way. We'll need to generate our own motivation. Take stock at the end of each day of lessons learned or relearned. View each forward step as a positive event. When we discover shortfalls, we study what led to them, identify leverage points where we can handle them more skillfully, and resolve to do better next time. Like a treasure hunt where we experience a sense of accomplishment whenever we discover the next step of the challenge, we reward and encourage ourselves as we advance on the Master Class path.

NOTES

1 Gladwell (2008, pp. 41 and 48).
2 Malcom Gladwell demystifies the 10,000 Rule: https://www.youtube.com/watch?v=1uB5PUpGzeY.
3 Ericsson and Pool (2016, pp. 14–22).
4 Adapted from the Zig Ziglar quote, "If you aim at nothing, you'll hit it every time."
5 This list is an edited version from Oscar Nowik's lifehack.org webpage: https://www.lifehack.org/articles/communication/12-signs-you-are-lifelong-learner.html.
6 Kern (2009, p. 1).
7 Vadulich, Wickens, Tsang, and Flach (2010, p. 188).

8 Dismukes, Berman, and Loukopoulos (2007, p. 81). – citing Tversky and Kahneman (1974).
9 Dismukes, Berman, and Loukopoulos (2007, p. 81).
10 Dismukes, Berman, and Loukopoulos (2007, p. 249). – citing Tversky and Kahneman (1974).
11 Dismukes, Berman, and Loukopoulos (2007, p. 296).
12 Dismukes, Berman, and Loukopoulos (2007, p. 145). – citing Austen and Enns (2003).
13 Kern (2009, p. 154).
14 Vadulich, Wickens, Tsang, and Flach (2010, p. 188).
15 Vadulich, Wickens, Tsang, and Flach (2010, p. 186).
16 Kern (2009, p. 157).
17 Cook and Woods (2006, p. 331).
18 Mosier (2010, p. 165).
19 Kern (2009, p. 151).
20 Edited for clarity. NASA ASRS report #1734822.
21 Greek philosopher Heraclitus, born 535 BCE.
22 Heath and Heath (2010, p. 136).

BIBLIOGRAPHY

Cook, R. I., & Woods, D. D. (2006). Distancing through Differencing: An Obstacle to Organizational Learning Following Accident. In E. Hollnagel, D. D. Woods, & N. Leveson, *Resilience Engineering: Concepts and Precepts* (pp. 329–338). Burlington, NJ: Ashgate Publishing Company.

Dismukes, R. K., Berman, B. A., & Loukopoulos, L. D. (2007). *The Limits of Expertise: Rethinking Pilot Error and the Causes of Airline Accidents*. Burlington, NJ: Ashgate Publishing Company.

Ericsson, A., & Pool, R. (2016). *Peak: Secrets From the New Science of Expertise*. Boston, MA: Houghton Mifflin Harcourt.

Gladwell, M. (2008). *Outliers: The Story of Success*. New York, NY: Little, Brown and Company.

Heath, C., & Heath, D. (2010). *Switch: How to Change Things When Change Is Hard*. New York, NY: Broadway Books.

Kern, T. (2009). *Blue Threat: Why to Err Is Inhuman*. Lewiston: Pygmy Books, LLC.

Kern, T. (2011). *Going Pro: The Deliberate Practice of Professionalism*. Lewiston: Pygmy Books, LLC.

Mosier, K. (2010). The Human in Flight: From Kinesthetic Sense to Cognitive Sensibility. In E. Salas, & D. Maurino, *Human Factors in Aviation – 2nd Edition* (pp. 147–174). Burlington, NJ: Academic Press.

Vadulich, M. A., Wickens, C. D., Tsang, P. S., & Flach, J. M. (2010). Information Processing in Aviation. In E. Salas, & D. Maurino, *Human Factors in Aviation – 2nd Edition* (pp. 175–216). Burlington, NJ: Academic Press.

25 The Master Class Skillset

The previous chapter examined how we form the intention and mindset to pursue the Master Class path. Our intention and mindset guide *how we think*. This chapter examines ways to counter stagnation and drift, followed by proactive measures for improving our skills. Our Master Class skillset guides *what we do*.

25.1 COUNTERING STAGNATION AND DRIFT

Three stagnating and drifting effects are decline of interest, subconscious actions, and procedure shortcutting.

25.1.1 Countering Declining Interest

The opposite of the comfort zone is the discovery zone. It is a mindset where we feel motivated to explore and learn new things. While the comfort zone is relaxed, the discovery zone is energized. While the comfort zone narrows our range of interest, the discovery zone expands it. The challenge of the discovery zone is keeping the learning experience fresh. To do this, we need to notice areas where our imagination and creativity begin to stagnate.

- Identify times during each flight when our minds tend to wander or daydream.
- Notice times when we lapse into habitual actions and automatic task completion.
- Detect areas where our preparation and review practices have become lax.

Even for pilots who love everything about flying, there are aspects of airline aviation that are undeniably repetitive. This repetition can lead to boredom, or at least disinterest. Our minds are naturally attracted toward interesting things. This is the same mental process that leads us to become distracted, only in this case, in a quiet, subtle way. Instead of something shocking like a fire alarm, it is something that we find a bit more interesting than the mundane, repetitious task that we are currently doing. If we don't actively focus on our current task, our minds would happily drift off into a daydream. Our attention to detail fades as we switch on our mental autopilot.

We counter this by performing periodic evaluations of our current practices. We might notice areas where we have allowed our accomplishment of ordinary tasks to become lax. These are early warning signs of drift. We pay attention to them and move them back on course. In a way, we trick our minds to rekindle interest with the drifting tasks. By directing our Master Class intention toward it, we restore its importance.

DOI: 10.1201/9781003344575-29

25.1.2 Becoming Aware of Our Subconscious Actions

When flying becomes common and repetitive, our attention level can wane and our focus can stray. Like driving home from a familiar place for the hundredth time, the part of our mind that drives the car can subconsciously switch to autopilot. After arriving home, we might reflect back and realize that we can't recall a single unique event from the drive. Surely, we made it home successfully. We stayed in our lane, maintained our spacing from other cars, made all the correct turns, stopped at red lights, and continued at green lights. We just didn't pay attention or retain any of it. What happens is that we learn to divide our attention between the repetitive task and some other topic that our mind finds more interesting. We relegate the subconscious aspects of our mind to driving the car while our conscious attention attends to other points of interest.

This same effect can creep into our flying. For example, some of us develop a habit of reciting checklist steps and responses without verifying the switch positions. We were initially trained to complete checklists deliberately. Early in our careers, we dutifully verified every checklist item before responding or moving to the next item. Over time, however, something caused our practice to drift. Maybe the first time, we were distracted. We just recited familiar checklist responses while our minds attended to that distraction. Maybe another time we were just tired and wanted to get the checklist over with. Maybe the repetition of reading the checklist hundreds of times without ever noting an incorrectly positioned switch lulled us into a false sense of security. Eventually, ordinary and repetitive procedures became ordinary and repetitive tasks in ordinary and repetitive flights. They melt together as we complete the steps while cruising along on subconscious autopilot. Thinking back on the flight, we wonder, did we complete the Before Landing Checklist? We must have. We always do, don't we? Yet, we cannot specifically recall completing it. Consider the following examples of warning signs of drifting into subconscious habit.

- Notice how we complete checklist steps. Do we deliberately verify the switch positions and values, or are we just reciting the steps and responses?
- Monitor whether the other pilot is verifying each checklist items or just reciting them.
- Evaluate whether we are actually seeing the switch or gauge or just looking toward it.
- Detect whether one or both of us are accelerating the pace of the checklist – either by reading the next step quickly or by responding quickly.
- Note portions of flightdeck tasks that we perform by habit rather than with conscious awareness.

Once we detect the early indications of drift, we can begin the course correction back toward our intended practice.

25.1.3 Guarding Against Shortcutting Procedures

Another source of drift is shortcutting procedures. Consider a hypothetical preparation flow that consists of 20 items. As our preparation becomes highly practiced and very familiar, we begin to subconsciously decide that only 15 of these items are actually important. The five "unimportant" items just don't seem to matter very much. The switches are never in the wrong position. It begins to feel like we are checking something that really doesn't need to be checked. We start skimming over them. Eventually, we skip them completely. Since they are part of our individual flow, the other pilot won't notice our omissions. Over time, maybe we omit another three items. Eventually, our flow settles on 12 meaningful items. We get through our setup more quickly and nothing bad ever happens. Our new, abbreviated, streamlined, efficient flow has drifted from our original practice. It works well until that one time when something unique happens to reposition one of those rarely used switches. Perhaps a maintenance technician neglected to reset it following a diagnostic test. The latent vulnerability created by our drift combines with a set of unique conditions to generate an error. To counter this drift, we try to detect places in our routine where we have allowed shortcutting to develop.

- Compare our familiar flow pattern with how it is presented in our training manual. Are there steps that we have allowed to drift?
- Evaluate each step in the procedure, especially the rarely moves switches. What would cause one of them to be moved to a rarely used position? Would our habit of glancing at the switch detect whenever it is in the wrong position?
- Imagine what conditions or personal biases might contribute to our missing this misplaced switch? For example, consider cases such as relying on seeing both switches/gauges as parallel (both OFF, versus both ON). Have we become vulnerable to "parallel, but wrong" errors?

25.1.4 Using Drift to Improve Resilience

Drift occurs as a natural result of adapting to changing conditions. Drift isn't inherently bad. It can be a useful early warning sign. For example, a particular latent vulnerability may be undetectable until drift exposes it. After we discover that vulnerability, we can modify our procedures or techniques to remove it. This process improves both personal and systemic resilience.

Problems emerge when people or systems become too comfortably settled or habitually rigid. Our willingness to adapt deteriorates. We rely too heavily on our familiar game plans and stick with them even when conditions warrant changing to backup plans (plan continuation bias). Consider a football team as they line up to execute a play. If the quarterback fails to notice that the defensive lineup is detrimental to the plan, the play will fail. Equally ineffective is when the quarterback does notice the detrimental defensive lineup, but stubbornly refuses to switch to a backup option. A resilient quarterback would read the defense, change the play, or call for a time-out. Resilience demands that we remain attentive, perceptive, agile, and innovative to successfully manage any situation that may arise.

Many organizations and people gradually become less flexible. We complete so many ordinary flights so successfully that our ability to actively adapt begins to atrophy. One adage warns that if you're not moving forward, you're sliding backward. Is this what we are doing when we settle for "good enough"?

To counter the erosive effects generated by our comfort zones, we commit to re-orient and re-center our actions toward mindful and resilient practices. Like our football quarterback, we interpret the threats before we commit to each play. Even though our coaches picked the play (company procedures), it is up to us to react to unexpected conditions (counterfactuals and warning signs), adapt our game plan (actively manage risk), and direct the team to execute the most appropriate play (good CRM). Paying close attention to conditions and indications, we detect emerging problems and constantly adjust our game plans to reach desired outcomes. Few of us instinctively know how to do this. It is a learned skill that we refine through purposeful practice.

25.2 IMPROVING SKILLS BY STUDYING CHALLENGING EVENTS

Not all threats are detectable. Latent threats lurk below the surface of every operation. Our organizations entrust us to discern a safe way to accomplish each flight. If we can execute a safe game plan, they trust us continue. If we cannot find a safe option, they trust us to stop. As Master Class pilots, we refine our ability to accurately discern the difference.

There isn't a clear line between successful and failing scenarios. Instead, we have a deepening gray zone where risk steadily rises, safety margins shrink, and latent vulnerabilities emerge. Our success depends on how well we handle these gray scenarios. While these scenarios contain the most risk they also offer us the most useful learning opportunities. Using our insights from these learning opportunities, we refine our ability to discern how gray our level of risk has become. As we evaluate each gray zone situation, we have three choices.

- *Continue:* We conclude that we can effectively monitor, detect, and mitigate the threats.
- *Adapt:* The situation requires active management. Familiar game plans either don't fit or require significant modification.
- *Reject:* We judge that threats are becoming too unpredictable or unmanageable. We need to abandon the game plan for a contingency backup.

25.2.1 Managing Operations in the Gray Zone of Increased Risk

Between situations we know that we can manage (choose to continue) and situations we know are unmanageable (choose to stop) lies a challenging, fuzzy crossover zone.

Figure 25.1 depicts the operational tightrope where we sometimes find ourselves. Let's examine some specific examples. Consider three proficient, comfortable crews that find themselves on a high and fast final approach profile. They have a range of techniques available for depleting their excess energy.

- *Case 1:* They are confident that they can manage the approach. They reduce thrust, extend gear and flaps, perform an S-turn, and reach fully stabilized approach parameters before their company limit of 1,000′.

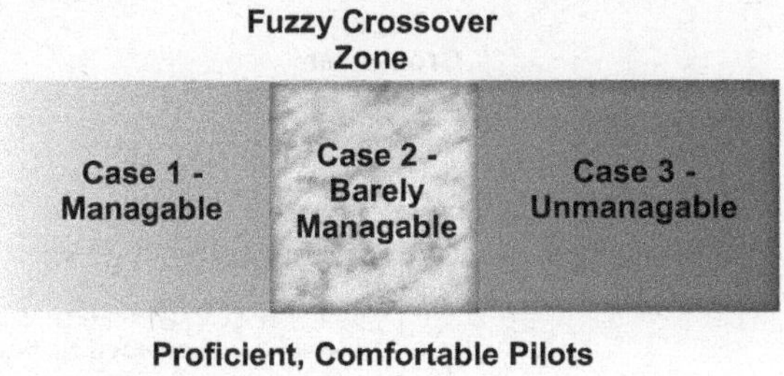

FIGURE 25.1 The fuzzy crossover zone between manageable and unmanageable cases.

- *Case 2:* They detect this as a borderline case, but they think they can still manage it. They use every technique except the S-turn. They fail to reach fully stabilized approach parameters by 1,000′, but do reach them before 500′.
- *Case 3:* They accept that the approach will be unstabilized, but choose to continue it anyway. They only pull the thrust levers back, configure normally, never reach stabilized parameters, and touchdown at idle thrust while still 20 knots fast. They didn't realize when they started the approach that it would turn out so badly.

While each case starts with the same conditions and stops safely, the three game plans generate different flight profiles. In hindsight, they conclude that they probably should have gone around for Case 2 and definitely should have gone around for Case 3. Thinking back to how it felt for them in-the-moment, however, the outcomes weren't very clear. Especially for Cases 1 and 2, they began their corrections with the full intention of reaching stabilized approach parameters by 1,000′. With Case 2, their actions proved insufficient because they didn't reach them until 500′. With Case 3, they thought they would make fully stabilized parameters somewhere after 1,000′, but certainly before landing. It turned out that they were way off with their estimation.

Consider also that Figure 25.1 depicts the three cases from proficient, comfortable crews. Since they haven't devoted the work to refine their recognition and monitoring skills, their central, fuzzy crossover zone appears quite wide. They are often surprised when approaches like these don't work out. This contributes to frequent decisions to continue their approaches past the procedural limit (plan continuation bias).

As Master Class pilots, we refine our ability to accurately discern our situations (Figure 25.2). Three benefits emerge. First, we draw a clear line between profiles we choose to continue and those we don't. This is depicted by the vertical Too Much Risk (dashed) line. We adopt solid decision criteria to sort potential profiles. We leave fewer decisions up to gut-feeling or judgment. This forces a larger portion of borderline-manageable scenarios into our "Unmanageable" category (moves the dashed line to the left). We don't even consider attempting them, even when some of them are possibly salvageable. We recognize which profiles cannot be flown within procedural

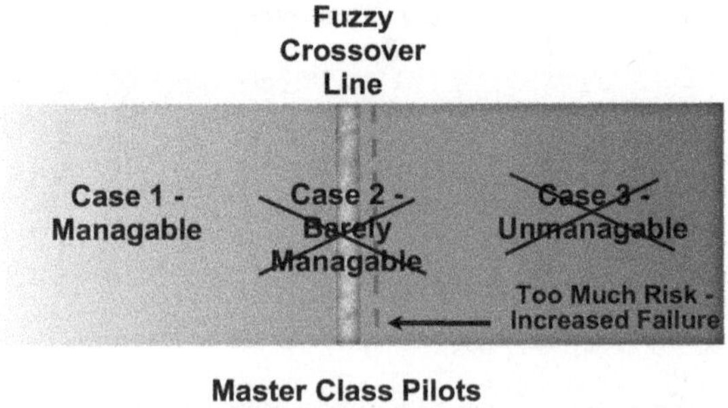

FIGURE 25.2 The Master Class pilot's profile management.

limits, overcome plan continuation bias, and execute a go around. Second, our fuzzy, crossover zone narrows to become a fuzzy, crossover line. This reflects our improved discernment to recognize conditions and accurately predict approach outcomes. Together, the fuzzy crossover line and the Too Much Risk line form a clearly defined barrier. We don't cross it. Third, we abort our Case 2 or Case 3 profiles.[1] We execute go arounds without ever attempting to salvage these approaches. As a result of these combined factors, we aren't surprised by how events turn out. If we can't confidently achieve Case 1 fully successful outcomes, then we go around and try again.

25.3 PROACTIVE DEBRIEFING

While in training, the syllabus is organized to optimize learning opportunities. Every flight began with a dedicated planning, briefing, and teaching period. After flying, we returned to the building for a thorough debrief. This ended when we began line flying. The flight schedule now governs time allocation. When each scheduled flight ends, we either start preparing for the next flight or end our workday. Dedicated debriefs disappear. Even when we resolve to discuss some debriefing points at a later time, we often fail to follow up. The "doing" objectives of line flying replace the "learning" objectives of training. We still value the benefits of debriefing, but the demands of the flight schedule hinder our opportunities. Some airlines have attempted changes to promote debriefing, but the practical demands of the operation override convenient windows of opportunity.

25.3.1 Characteristics of Effective Debriefs

Debriefing is where we identify and examine the learning points from a flight. While we are flying, especially in high-workload flight phases, we need to focus on what is happening and what is about to happen. Our attention flows in real time. For effective debriefing, we need to look back with hindsight. Surveying the timeline, we know when to fast-forward through routine parts and when to slow down through interesting parts. We know where to freeze the timeline and focus on critical parameters. We know what we expected to happen, what we missed, and what actually happened.

Debriefing is where we rate how successfully we achieved our objectives and followed procedures. Without these priorities, we often default to *completion standards* like safely at the gate or on-time arrival. Measuring our performance by completion standards effectively erases errors and unwise decisions. For example, when we successfully land and taxi to the gate for an on-time arrival, our motivation to evaluate our poorly flown approach fades. Debriefing replaces completion standards with learning standards. Analyzing significant learning points compels us to reconstruct our thinking and acknowledge our miscalculations. "I was surprised when ATC turned us in so tight. I thought I would have more time to configure and slow for landing."

Debriefs also create mentoring opportunities in both directions. FOs can share their knowledge and experience about a particular airport's operations. Captains can share their knowledge and experience about unique aircraft characteristics. Together, the crew can examine the effectiveness of their CRM, communications, and monitoring.

25.3.2 The Meandering Course of Debriefs

It is difficult to predict where a debrief discussion might wander. An event that seemed straightforward and simple to one pilot might have been particularly problematic to the other. Consider an event where the PF performs a go around procedure incorrectly. As it is happening, the PM immediately detects the error, but feels that it is inappropriate to speak up at the time. During the debrief, the PM highlights the error and reinforces the correct procedure. The PF learns and improves. Both pilots benefit.

Allow debriefs to follow their own course. A small point may lead to a long discussion. That discussion may, in turn, branch off and meander in new directions. This is how we explore the interconnections between the many facets of aviation. This is the process that weaves new threads of understanding into our professional aviation wisdom.

25.3.3 Scheduling a Debrief Opportunity

Effective debriefing requires a dedicated block of low-workload time. First, it has to be relatively free of distractions. For complex events, everyone needs to pay attention to the evolving line of reasoning. Short and routine distractions are quick enough that they don't break our train of thought. Extended or involved distractions begin to monopolize our attention and interfere with the continuity of the debrief. Second, the time block needs to be open-ended enough to allow the discussion to branch off and explore. Setting a time limit or rushing the debrief so we can move on to other priorities makes debriefs cursory and unhelpful.

The natural opportunities for debriefs are during scheduled ground time segments, during ground delays, while socializing after the workday, or during sustained cruise flight. Each offers advantages and disadvantages. Scheduled ground time segments are free of operational distractions, but tend to conflict with non-flying priorities like meals and personal time. Ground delays can also prove useful, but since they are unpredictable, they might not occur at useful times. Socializing following the workday works as a debrief opportunity, but this time tends to be better suited for positive discussions and storytelling versus dissecting errors and analyzing flawed decision making. Also, the presence of uninvolved individuals like other pilots and flight attendants makes shop-talk inappropriate.

In multi-leg airline flying, the optimal debriefing opportunity is during sustained cruise flight during follow-on flights. Workload tends to be low and distractions are typically routine and quick (like ATC radio frequency changes). The events from the previous flight are still fresh enough on our minds, plus we have ready-access to manuals and reference materials. The disadvantages are short stage lengths with limited cruise segments, when crew changes break up the crew between flights, and after the last segment of the workday. Still, sustained cruise works because most flights are fairly routine and require only a short amount of debrief time. We can accomplish a review of one or two learning points fairly quickly. If a longer discussion is warranted, sustained cruise segments usually offer enough time to explore a range of debrief topics.

25.3.4 Creating a Debriefing Habit Pattern

As long as we treat debriefs as exceptional and unimportant, they won't feel particularly useful. The exchange can feel like square-filling. "You got anything for me?" "No, how about for me?" "No, all good." The debrief is opened and quickly closed without either pilot gaining any benefit. Creating a debriefing habit pattern needs to open doors and invite entry. This starts with honest sharing and open-ended questions.

> Captain/PF: "I thought the last flight went pretty well. One area where I think we could have improved was with our anticipation of arrival weather. The weather product warned of temporary thunderstorm conditions, but I didn't expect it to cause a last-minute change to runway 23 with that complex RNAV RNP approach."

This opens the door. It identifies a learning point and frames a discussion topic. The other pilot then picks the ball and runs with it.

> FO/PM: "Yes, this happened to me once before. We use runway 05 so much here that we expect it every time. Like you, I didn't anticipate the airport swapping directions. I think we did a good job handling the late change, but it would have been useful to review the RNAV RNP 23 before top of descent. I think our scramble to prepare for the approach contributed to our late configuration and excessive speed on final."

Notice that the second pilot takes the opportunity to acknowledge the learning point about not briefing the backup approach and raises another concern about approach stability. This gives the first pilot the opportunity to share.

> Captain/PF: "Good point. The tight arc-to-final segment of the RNP caught me off guard. I also didn't give the high density altitude of this airport enough consideration. I was about here (pointing at the chart) when I realized the problem and called for landing gear. In the future, I'll slow earlier."

Through this exchange, each pilot highlights the problems that they saw, when they detected them, and ideas for personal and crew improvement. Ideally, they would refine their preparation and monitoring to anticipate future weather-related profile challenges.

25.3.5 Questions That Encourage Proactive Debriefing

Avoid simplistic yes-or-no questions. "You got anything for me?" "No, how about for me?" "No, great flying with you." "You too." Exchanges like these help us feel like we are fulfilling the debriefing requirement, but they amount to little more than square-filling. We don't gain any useful learning. Instead, try open-ended questions that invite proactive sharing.

- *Captain:* "I don't like the way I handled that ground delay out of Atlanta. Do you have any ideas or have you flown with anyone who handles cases like that exceptionally well?"
  - *Analysis:* This acknowledges the line-flying reality that Captains don't fly with other Captains. FOs do. This invites the FO to share a useful technique that they witnessed from a different Captain. Since most major airline FOs were Captains or Aircraft Commanders at their previous organizations, it invites them to share their own wisdom.
- *Captain:* "We were tight on our holdover time, but I think we handled that deice operation and departure well. In retrospect, did you see anything that we could have done better?"
  - *Analysis:* This invites the FO to share their impressions and concerns. Perhaps they thought that a cabin check of wing icing was warranted. A useful discussion of when and how to perform a cabin check can follow.
- *Captain:* "That last leg went well, but I have a couple of areas where we could improve as a crew. What did you see?"
  - *Analysis:* This opens a discussion about Master Class improvements following what was already a good flight. The Captain challenges the FO to share their thoughts. This debrief approach relies on an open and supportive CRM environment. Imagine if the Captain was an authoritarian, judgmental sort. Few FOs would feel comfortable sharing their ideas. If, however, the Captain has proved themselves to be truly interested in improving crew performance and welcoming of constructive comments, then this can be a very productive experience.

25.3.6 Debriefing a Significant Event

Every once in a while, some portion of the flight goes poorly. For these events, debriefs need to dig deeper. First, acknowledge that an undesirable outcome has occurred. Next, follow a positive debriefing process. Avoid blame because it just triggers our personal defenses and inhibits learning. Instead, accept the conditions and events as they occurred and focus on how we can avoid making similar mistakes in the future. Following are some useful debriefing topics and questions from Gary Klein[2]:

- Identify the trickiest judgments and assessments.
- Construct a timeline and identify the judgments (versus the events).
- Contrast options using hindsight.
- Minimize debates about details and facts.
- When was the problem spotted?
- What cues indicated the problem?
- What were the differing perspectives on the problem?
- What cleared it up?
- What would have reduced uncertainty?
- How would you have changed your decisions?

Notice that these questions focus on judgments, decisions, perspectives, process, and mindsets. Facts and details take a back seat. The success of this debriefing format depends on us retaining a supportive, inclusive CRM environment. We need to focus on the learning points.

25.3.7 Debriefing Surprising Events

We are especially interested in events that surprise us. Identify indications that preceded the surprising event. Perhaps we detected them, but misinterpreted their cause or meaning. Perhaps we didn't detect them as early as we should have. Perhaps we chose a recovery path that didn't work out well. While comfortable pilots might be satisfied with just getting through the event safely, Master Class pilots choose to dig deeper. After encountering an event where we didn't detect the warning signs as early as we wished, ask:

- What were the critical indications that preceded the emerging threat?
- When did we notice them?
- How quickly did we diagnose their meaning? Was our diagnosis accurate?
- How did we modify our game plan?
- How can we apply lessons learned for future encounters?
- What similar scenarios might evolve where this knowledge may prove useful?

After encountering events where missed the warning signs entirely, we ask ourselves:

- What were the critical indications that we missed?
- How early could we have detected them?
- What secondary indications would have accompanied the primary ones we missed?
- How would they change over time?

Learning to improve our monitoring techniques, we ask ourselves:

- How do we need to modify our personal monitoring techniques to detect these indications more quickly?

- What crew monitoring techniques do we need to improve to detect them more consistently?

After attempting a remedy that failed to solve our problem, we ask ourselves:

- Why did our mitigation fall short?
- What other counterfactuals emerged after we applied our remedy?
- What better game plan should we have chosen?
- When should we have switched to that better game plan?

25.3.8 The Master Class Debriefing Perspective

When we combine our Master Class path to build our aviation wisdom with learning opportunities, we open a range of discussion topics. The debrief doesn't need to be limited to what happened or didn't happen during the flight. It can open topics like:

- Balancing conflicting priorities
- Adapting techniques to limit our exposure to latent vulnerabilities
- Performing concurrent tasks more smoothly
- Understanding the underlying objectives behind certain procedures
- Improving CRM to enhance team resilience.

When we apply our learning perspective, a range of possibilities opens. No matter how routine and simple, every flight contains learning opportunities.

BOX 25.1 "TELL ME WHAT I COULD HAVE DONE BETTER"

I attended a presentation at a soaring club featuring a guest speaker, Don McMonagle, the former NASA space shuttle pilot/commander. He was introduced with a long and distinguished aviation résumé that you would expect from someone who had flown everything from gliders to space shuttles. Before his presentation, a glider club member and close friend of Don's offered a story. They had recently flown together in Don's Cirrus. Upon landing and parking the aircraft, Don asked his friend to debrief him on anything that he could have done better. The friend laughed. Here was a test pilot, astronaut, and Shuttle commander asking him, a glider pilot, what he could have done better. What advice or observation could he possibly offer to such an accomplished aviator? Don turned to him and stated with complete candor, "You're not getting out of this aircraft until you tell me what I could have done better."

Don embodies the Master Class discovery mindset. Even a routine flight in his personal aircraft became an opportunity to refine his professional aviation wisdom. He challenged himself and his fellow pilot to look deeply for learning points, even following the most mundane and ordinary of flights.

25.4 IMPROVING OUR JUDGMENT

All pilots strive to exercise good judgment. For most of us and in most situations, we do quite well. The challenge is that our aviation judgment skills need to work across a wide range of scenarios. The judgment process that we apply for everyday flying is somewhat different than what we would use for a complex emergency situation. Refining our judgment skills across the full range of aviation challenges requires a proactive process. Like the muscles of our body, we need to exercise all of them in every direction to remain strong.

25.4.1 Training versus Line Flying

Early in our flying career, the training syllabus exposed us to an increasingly complex range of situations. In the simulator and during training flights, our instructors created events to stretch our decision-making skills. The wider the range of situations, the more opportunities we had to sharpen our judgment. With each new situation, we experimented with our decision-making process while our instructors ensured that our outcomes remained safe.

After we graduate to line flying, the learning environment fundamentally changes. Instead of seeing a wide range of exceptional events that stretch our judgment skills, we settle into a fairly normal routine. We have plenty of practice applying normal judgment within normal situations. Given the reliability of our aircraft and the resilience of our flight operation, we may not experience a single event that truly challenges our judgment skills until we return to the training center for our annual recurrency checkride.

25.4.2 Practicing Judgment during Everyday Line Flying

Limited to the routine line flying, our judgment skills can become the proverbial inch wide and a mile deep. This is especially true if we don't actively work to expand our skillset. To get the most practice out of everyday line flying, we need to look for opportunities to analyze and discuss what went well, what went poorly, and how we could have handled it better. Instead of shrugging off a problematic flight, we study it to extract useful learning points. We evaluate how well we judged the importance of indications and how well we balanced priorities. Through repetition and steady improvement, we perfect our judgment process for routine line flying.

25.4.3 Imagining Scenarios

The judgment processes that we would need to apply in conflicted, complex, and time-pressured events may not get the frequent practice they need to remain sharp. We just don't experience very many of these events. We need to find ways to practice them while flying our normal schedule.

Referring back to our discussions about the gray zone of increased risk, we observed that pilots who frequently practiced risky, complex, and uncertain scenarios improved their proficiency at managing them. The same goes for developing judgment. The more often we practice near the turbulent edge of excessive risk, the more we improve. Fortunately, there are safer ways to hone our judgment skills without recklessly manufacturing risky events. Imagine that we are flying along on a routine flight and suddenly, we experience a rapid loss of pressurization. It completely surprises us. We would need to react quickly and accurately to handle it. After the trained boldface steps of donning oxygen masks and establishing communications, the remaining remedy steps of restoring pressurization and initiating an emergency descent demand prompt action and coordination. We might do well or we might make some mistakes. Either way, we would certainly learn from the experience. Now, imagine experiencing that same event a second time. We would immediately recognize what was happening. We would recall what we previously did wrong and what we did right. Our decision making would be quick and accurate. Encountering that second depressurization event, we would experience less surprise and startle, know what to look for, know what it feels like, understand what the indications mean, and recall what to do. It would go much smoother. Now, consider that we can recreate these kinds of experiences anytime we choose. By practicing "what-if" scenarios, we can improve our judgment process for handling any complex emergency event. Consider some examples.

- We are landing at a busy airport with a heavy jet landing on the parallel runway upwind of us. We predict that the crosswind component might blow their wingtip vortices toward our landing zone. Nothing bad happens, but after arriving at the gate, we imagine what a wingtip vortex encounter might have entailed.
  - What indications would the wake vortices create as they drift toward us? Dust plumes? Evidence of swirling wind in the grass?
  - How would the vortex encounter feel in our aircraft? Would we experience a strong lift of our upwind wingtip? Would we encounter moderate or severe turbulence?
  - What would our best response be at 100′? At 50′? In the flare?
  - If we had gone around, what would our fuel situation have been?

- Flying over the Rockies with a thick undercast, we consider what we would do if our #1 engine suddenly fails? Since workload is low, we can explore the scenario while cruising along.
  - What are the immediate priorities? Flightpath? Thrust? Pressurization?
  - Assuming that we can't maintain our altitude, what are the steps for letdown? (Use this opportunity to reference the QRH and review the procedures.)
  - What would we say to ATC?
  - How would we configure our TCAS display to display other aircraft around us?
  - When would a restart attempt be appropriate? What are the steps?
  - Where is the nearest suitable field for landing? What is the best use of our moving map display to build SA?

- Following deice, we are in a short line of aircraft for takeoff. Fortunately, the line moves quickly and we depart. Reaching cruise, we imagine some scenarios that might have complicated our departure.
    - What if a snowsquall would have arrived as we were #3 for takeoff? How would it have affected our holdover time?
    - If we had to do a wing ice inspection from the passenger cabin, how would we coordinate with ATC and as a crew? How would failing anti-icing fluid appear on critical surfaces?
    - If we had to return for another deice treatment, how would it have affected our fuel situation?

The more we engage in practice scenarios, the more prepared we will be when something unexpected actually happens. While imagining a scenario isn't as profound as experiencing the real event, it does help us exercise the same mental processes. Like a ball team learning a complex play, rehearsing it on the practice field doesn't guarantee that it will run smoothly in the big game, but it significantly improves their probability of success.

25.5 IMPROVING OUR INTUITION

Intuition reflects our subconscious ability to sense subtle, barely detectable, or unseen factors. It manifests as gut-feelings or flashes of insight. When we view a complex situation, intuition helps us to know what is happening and what needs to be done. Intuition is not an innate skill. It is a complex mental process formed from years of experience and practice. If we passively allow it develop at its natural pace, it grows slowly and eventually stagnates. If we actively and intentionally nurture it, it grows faster and reaches its full potential. Like the intentional process of developing our judgment skill, practicing with hypothetical scenarios improves our intuition.

Consider a professional baseball player up to bat. They don't know which pitch is about to be thrown. It could be a fastball, curveball, changeup, or something else. The speed and variability of Major League pitching prevent batters from consciously detecting the indications, predicting the path, and choosing where and when to swing. The pitch moves much too quickly. If they simply guess and swing for a fast ball, they will certainly miss a curveball. Despite this challenge, professional players manage to successfully connect the swing of their round bats with round baseballs and hit home runs. Their skill does not come from deliberative decision making, innate ability, or luck. It comes from the intuition that they have built from years of focused practice. They study scouting reports, watch videos, note the pitcher's release point, and assess the spin on the ball. They weigh probabilities based on the current pitch count of balls and strikes, the number of outs in the inning, their teammates currently on base, and how the defenders are arranged across the field. Their minds subconsciously combine probabilities with seen and unseen nuances to intuitively predict the path and speed of the pitch. After experiencing untold thousands of pitches, they develop the intuitive skill to quickly process subtle details to guide whether to swing or not. Players who fail to develop these intuitive skills muddle along in the minor leagues and never make the move up to the Major Leagues.

Aviation skills share many of these same intuitive aspects. Consider all of the subtleties affecting how we handle flare, touchdown, and rollout. In the flare, changes happen too fast to accommodate deliberative decision making. Our hands and feet seem to move instinctively to compensate for a host of variables to achieve a smooth touchdown.

While intuition guides our split-second reactions, it also helps us to make sense of uncertain situations. Master Class pilots who refine their intuitive skills understand complex situations far better than those who don't. We detect many variables to assess risk, select appropriate game plans, and develop abort triggers. Pilots who lack these intuitive skills succumb to plan continuation bias, experience indecision, force their failing game plans, and experience surprise when they make errors and end up in mishaps.

25.5.1 Our Intuitive Recognition of Risk

When we encounter a complex aviation situation, our intuition recognizes subtle hazards and generates a gut-felt sense of unease. Many of these types of situations match one of our past "I'll never do that again" experiences. We recognize the threat and know when to switch to a safer contingency option.

We also apply intuition for situations that are completely new to us. We may not understand what the indications mean, but our intuition still generates a sense of unease. Remember that our sense of rightness depends on a smooth transition from past SA, to present SA, to future SA. As experienced pilots, we become quite sensitive to mismatches and speed bumps within this flow. Even when we aren't actively monitoring for counterfactuals, we expect that indications will agree with our future SA prediction. For example, when fully configured on final with the power back, we expect our airspeed to slow. If, instead, we see it holding steady, or even rising, we recognize this as a warning flag. The indication doesn't match our expectation. Something is wrong.

25.5.2 Our Intuitive Sense of Time Available

Our first reaction is to look for the cause behind the disruption. How much time we spend investigating this depends on our intuitive mental balancing of risk, uncertainty, and time. When we feel like we have plenty of time, we investigate. As our intuition discerns the cause, it also spontaneously generates workable options. In this way, analysis and solution are generated simultaneously. When time is too short to allow investigation, we abort the scenario without even trying to determine whether our game plan is failing or not.

25.5.3 Limits of Intuition

The greater the complexity and uncertainty, the more unreliable our intuition becomes. Our intuition looks for patterns. When indications become particularly chaotic, we may detect a pattern that doesn't actually exist. A situation may appear a bit like something we recognize, so we pursue a game plan based on that familiarity.

In hindsight, we discover that we were mistaken. The opposite is also problematic. If we are unsure of our intuition, we may lack confidence in our decision making and hesitate to act. Somewhere between the extremes lies the optimal balance.

25.5.4 Learning from Stories

Our aviation profession enjoys a strong tradition of sharing wisdom through storytelling. This is a vital resource for refining our intuition since our direct exposure to exceptional events is limited. Aside from simulator training, most of us will never experience a severe, time-critical emergency in the aircraft.

Lacking direct experience, we'll need to rely on our generalized process for handling emergency-type events. We'll have to build our understanding about the event at the same time as we decide how to respond to it. This is an important distinction within the aviation profession. Most other professions can pause when they encounter an obstacle. In most cases, we can't. Sometimes, we can create more time by breaking off the approach and taking another trip around the pattern, but we incur a fuel penalty. Some of the considerations we need to make "on the fly" are:

- What are the indications at the onset of the event?
- Do we need to respond immediately or do we have time?
- Do the indications match conditions that we recognize from a past event that we experienced?
- Do the indications match conditions from an event that we have heard about?
- Do we have a workable game plan that can handle the situation?
- What is the safest path forward while we continue to gather information?

Shared stories help us answer these questions. Imagine experiencing an event that we haven't personally seen before. Our SA vanishes in an instant. We need to start rebuilding it quickly. While we lack direct experience with the event, maybe we recall, "This looks like the situation that I've heard about." We recall the main points from the story:

- The cause
- The remedy the crew used
- Some mistakes they made that we should avoid
- What steps they took to get back on track

Their story gives us a starting point for building a workable game plan. In a way, it transfers their hindsight wisdom into our foresight wisdom. Klein observes that it gives us "a package of causal relationships" and their effects.[3] This fits nicely into his Recognition Primed Decision Making (RPDM) process. Our storehouse of personal experiences plus everything we learned from others' experiences drives our intuitive ability to make sense of complex, unique, and confusing events. This sensemaking generates a set of promising game plans that, in turn, simplify our decision making. Even if it turns out that the story we heard is different and proves unusable, it eliminates an ineffective game plan that we know to avoid.

25.5.5 Sources of Stories

Our profession does a commendable job assembling and sharing stories. Most safety departments investigate instructive events and share the details through internal publications, weblinks, and video interviews. Safety departments at various airlines also share event summaries and data between their companies. In the past, these products were published and distributed on paper, making them somewhat perishable. With website and EFB improvements, past events are easily shared, searchable, and preserved.

One of the best sources is the NASA ASRS database. The millions of reports from pilots, controllers, dispatchers, mechanics, and flight attendants form an enduring record of direct reports covering a wide range of aviation events. The database's size can be a bit daunting, but with a little practice, anyone can extract first-hand accounts of events sorted by aircraft type, error classification, and crew position. Additional story sources are available through NTSB accident reports, FAA reports, industry websites, private aviation sites, and blogs.

The easiest supply of stories comes from other line pilots. As an FO, our best source is sitting three feet to our left. Even if they aren't Master Class-motivated Captains, they have experienced and dealt with many worthwhile learning situations. Use some low-workload opportunities like extended cruise flight, a ground delay, or after-hours socializing to invite their stories. As they tell them, notice how they unfold. The way they tell their story traces the event path as it felt to them. Notice the main points during their first telling. These identify significant milestones that anchor their understanding of how the event progressed. Notice inconsistencies or holes and probe for more information. This may uncover details that didn't seem important to them at the time, but are recalled when we ask the right questions. Understand that these subtle details were probably overcome by more prominent indications at the time.

25.5.6 Making the Story Our Own

A story is just a story until we immerse ourselves into it. Using Dekker's metaphor, analyze the event as if we are moving along inside of a tunnel. Understand how their tunnel appeared to them at the time. Ask questions to understand the context and problem-solving process they used. Remember, they are recalling their story from a hindsight perspective. They already know what they did right and what they did wrong. The wrong parts didn't look wrong at the time. They only appeared wrong when the undesirable outcome materialized. Consider the following questions.

- What were their conditions at the time?
- How did those conditions affect their event?
- When did they detect and recognize these influential conditions?
- What conditions did they identify afterward through hindsight?
- How did the conditions interact to allow the event to emerge?
- Was the event predictable?
- What options or game plans did they consider at the time?

- How did the flow of time feel? Did they feel rushed? Did the event quicken?
- Would they make the same decisions if they encountered the same event again?

To fully integrate the lessons from their story, we need to imagine ourselves projected into the event as it happened and see it from the inside of the tunnel. As we suppress our hindsight bias and envision encountering their event in real time, we imagine our perceptions.

- What indications of the problem would emerge first?
- Would our current monitoring techniques and habits have detected those indications?
- How might we misinterpret them?
- How might we select an unworkable game plan?
- How can we improve our monitoring habits to handle a situation similar to this one?
- How can we improve our CRM to work more effectively?
- How can we achieve more resilience for scenarios like this?

Remember, our goal is to improve the accuracy and speed of our own intuition. Every time we engage one of these story scenarios, we exercise our intuitive processes. Like all skills, the more often we practice using them, the better they respond to challenging situations.

25.5.7 Watching, Learning, and Sharing

When we fly, we observe the other pilot completing many tasks. When we see one completed in a particularly unique, effective, or intriguing way, we can learn from them. For example, if an aircraft discrepancy requires an MEL-deferral and our Captain completes the process with exceptional efficiency, ask them about it. "Could you walk me through your MEL process and priorities?" This opens the discussion about how they came to perfect their skills with completing that task. They might share past experiences with that particular MEL, their awareness of the most time-intensive parts of the process, and their own personal checklist for ensuring that all parts of the task are completed. The same goes in reverse. Perhaps we are the Captain flying into an unfamiliar airport where our FO has extensive experience from their previous airline. We solicit their wisdom. "I'm new to this airport, but you operated here extensively. Please share your insights with me."

Master Class pilots look for opportunities to pass their wisdom on to others. If we encounter a particularly difficult situation and handle it well, we can share that wisdom. Consider an example where we have a difficult interaction with our operations agent as they tried to board a passenger that the flight attendants deemed unfit to fly. As Captain, we chose to intervene to help resolve the situation. Through it all our FO watched, but didn't say anything. After we settle into cruise flight, we open a dialog. "What are your impressions of that passenger incident we had back at the gate?" They share their version of what they saw. Ask them some questions.

- If you were the Captain for that event, how would you have handled it differently?
- My intervention worked, but how could it have gone poorly?
- What other options were available?
- Do you think that there are any follow-up steps that I should take with the operations agent or the flight attendants?
- Would you have intervened with that event? When would you have intervened?

25.6 MANAGING UNCERTAINTY

An obvious strategy is to avoid events that fall within the gray zone of rising risk. If we could, we certainly would. There are two problems with this strategy. First, we can't really tell how far into the gray we are at any particular moment. Accurate assessments rely on hindsight. While we are immersed in the situation and flying in-the-moment, we can't accurately predict how conditions will interact or anticipate the consequence of our decisions.

Uncertainty undermines our ability to predict how an event will develop. When conditions are benign, emerging variations remain small and are easily handled. As complexity and uncertainty rise, variations can increase in severity to allow unique events to emerge. These unique events can exceed our imagination or planning. Also, not all conditions are known or even knowable in-the-moment. When investigating mishaps, we often uncover conditions that were never detected by the crew. Had they detected them, they would have chosen differently. Hindsight provides a clarity that isn't available while events are unfolding.

Most of the challenges we face involve scenarios that remain manageable and achievable. However, there isn't a clear line. There is only the range from easy, normal operations toward increasingly risky operations until we reach a defined (usually regulatory) stop-point. Since uncertainty is unavoidable, it behooves us to develop skills to manage it. How each of us handles this gray zone of uncertainty often depends on our personality. Aggressive pilots tend to push into the gray zone more deeply and more often. They confidently believe that their skills and judgment ensure their success. Their frequent encounters with the gray zone tend to make them skilled managers of risky situations. It also biases them toward accepting risk more often than choosing to avoid it. Combined with the effects of comfort zone and drift, they may delve deeper into the gray zone over time. When they experience a failure, they typically correct back toward more conservative choices. Over time, however, their aggressive habits return. They are not reckless. They value safe operations. They just assume that their skill, regulatory limits, and safety margins provide adequate protection against failure.

Conservative pilots tend to hold back from the gray zone. Adverse to risk, they try to avoid uncertain situations. They adopt techniques and habits that keep them well clear of the gray zone. Because of this, they tend to have less practice with handling risky events. They may compensate by increasing their personal safety margins. They tend to add fuel more often, divert earlier, configure earlier for approaches, and go around sooner when parameters exceed their personal limits.

Both aggressive and conservative pilots are aware of rising uncertainty. They just view it differently. Aggressive pilots tend to respond by actively engaging the risk, while conservative pilots respond by actively avoiding the risk. Between these personalities lies a balanced approach that we Master Class pilots pursue. We accept that uncertainty and complexity will push us into the gray, so we refine our skills and techniques to safely manage the tipping point between continuing and rejecting. One important difference is with how we focus our awareness.

25.6.1 Managing Uncertainty with Awareness

We balance our attention between actively managing the risk to continue and monitoring for signs of failure. We accept that we can operate in riskier regimes of tighter safety margins as long as we also remain aware of emerging counterfactuals, the indications they generate, and how much safety margin remains. For mildly risky situations, we typically continue with our game plan and raise our vigilance. Pressing deeper into the gray, we sense the increasing demands on our attention. Instead of becoming drawn into the task saturation of increased task loading, we shift more attention toward our abort trigger points and counterfactuals. This maintains a separation between our focus of attention and the evolving situation. Instead of becoming drawn in by the increased effort needed to force our game plan to work, we use uncertainty as a sign to pull back and reassess. This means that rising uncertainty and complexity encourage us to repeatedly reassess the situation. If we don't have time to reassess, we reject the game plan, move to a safer position, and then reassess. This means that we might abort a small percentage of profiles that may be otherwise workable.

25.6.2 Gauging Trends

The trend toward or away from risk and uncertainty is also important. We can operate safely in marginal conditions as long as they remain stable or are trending better. Consider a line of arrivals into a busy airport in marginal weather. As long as conditions remain steady, we can reasonably predict our success with continuing. If conditions begin to worsen, like with snowfall and dropping temperatures, warning flags emerge. What is causing the worsening trends? Is the trend accelerating toward an unacceptable limit? The trap is assuming that the success of aircraft in front of us means that we will also succeed. The reality in a worsening environment is that someone will either be first to break the conga line of arrivals or follow the conga line, land, and slide off end of the runway. As Master Class pilots, we are ready to be the first to say no, even when everyone else is saying yes. With the example of snowfall and temperatures dropping into the freezing range, we know that conditions can quickly become hazardous. Take the example where ATC informs us that we are the last aircraft scheduled to land before they plan to close the runway for plowing. The aggressive pilot might view this a stroke of luck. We would view it as a possible warning sign. Are they planning to plow it after us because we are at the end of the conga line? Are conditions worsening and we just happen to be the last planned arrival before conditions will be deemed too risky? If we have all of the relevant information and conditions remain workable, we would continue. If too many factors remain uncertain or conditions are trending poorly, we abort our game plan for something safer.

25.6.3 Imposing Tighter Criteria on Acceptability

As risk continues rising, we impose tighter criteria on what we will accept. For example, when flying an approach in convective weather, we typically tolerate a higher level of turbulence on 3-mile final than we would at ¼-mile final. When we experience that strong jolt on 3-mile final, we raise our level of monitoring and search for additional indications of windshear. Until we learn more, we continue. Are there indications of winds and turbulence that are visible on the ground (dust, smoke, water surfaces)? What are the reported landing conditions? Are wind reports changing? Using this information, we set our abort trigger criteria. For example, if we see that the winds are out of limits at 1,000′, but ATC reports marginally acceptable winds on the surface, we might set a trigger point at 200′ to decide whether to continue or go around. The more uncertain our information, the higher we set our trigger points.

BOX 25.2 WHEN WE WERE THE ONLY AIRCRAFT TO GO AROUND AND DIVERT

We were on an approach into a desert airport during summer "monsoon" season. Approaching the field from the South was a wall of dust (called a haboob) from a strong thunderstorm. From visual indications, it appeared that the gust front would reach the field as we were turning base. These haboob events are typically short-lived, but they may close an airport for an extended period or cause ATC to swap runways back and forth. We rolled out on final and experienced moderate turbulence. As we reached our trigger point of 200′, we experienced the strongest gust of our approach. We went around. We didn't have fuel to wait out a sustained haboob event, so we diverted. As it turned out, the winds eased off a short time later. We were the only aircraft to divert that afternoon.

Managing our abort triggers is one of most discerning Master Class skills. Whenever we reach a trigger point, we don't hesitate. Otherwise, we can find ourselves rationalizing and vacillating between continuing and aborting.

NOTES

1 For perspective, airline industry data indicates that most Case 2 crews land while most Case 3 crews execute go arounds.
2 Klein (1999, pp. 60–61).
3 Klein (1999, p. 180).

BIBLIOGRAPHY

Dekker, S. (2011). *Drift into Failure: From Hunting Broken Components to Understanding Complex Systems*. Burlington, NJ: Ashgate Publishing Company.

Klein, G. (1999). *Sources of Power: How People Make Decisions*. Cambridge, MA: The MIT Press.

Klein, G. (2003). *The Power of Intuition: How to Use Your Gut Feelings to Make Better Decisions at Work*. New York, NY: Currency Books.

26 Professional Attributes

Most of us view *aviation professionalism* as a familiar, universally understood concept. Given any particular event, we generally agree whether the crew acted professionally or not. When challenged to list or define the specific attributes, behaviors, and skills that embody aviation professionalism, we struggle. To make progress on our Master Class journey, we need to elevate our personal understanding from having a general sense of what comprises professionalism to clearly identifying its attributes, behaviors, and skills. Then, we can direct our attention toward refining them in ourselves. While this chapter focuses primarily on Captains, we expect the same qualities in FOs.

A good way to understand professionalism is to survey those particular pilots who exemplify the attributes we want to build. Who are the best people to identify these Master Class Captains? It is not other Captains because Captains rarely fly together. It is not training center instructors or flight operations leadership because they rarely observe Captains flying in their everyday line-flying environment. For many reasons, the best judges of Captain professionalism are FOs. There are several reasons for this. Major airline FOs have served as Captains, aircraft commanders, instructor pilots, and supervisors in their previous aviation positions. So, they have direct experience doing the job. They recognize model Captain behavior when they see it. FOs also see Captains within their normal, unfiltered, everyday environment. Unaffected by the "angel behavior" that some pilots display during formal checkride evaluations, FOs see Captains as they handle the full range of line-flying situations.

This chapter contains the results from survey responses to specific questions on professionalism.[1] This "Best Captains" survey was conducted in two rounds. First, FOs were asked to identify three Captains from their domicile that exhibited the finest qualities of Captaincy. The survey was designed to encompass the full range of Captain attributes, behaviors, and skills that surround professionalism. In the second survey, the identified Best Captains were asked to share their views on these topics.

26.1 HOW FOs IDENTIFIED THE BEST CAPTAINS

The survey generated a consistent profile of the professional qualities that FOs considered important and admirable in their Captains. The attributes listed most often were professional competence, personality, team building, and instructing. Following are comments from the FOs about the Captains they identified as exemplary role models of airline professionalism.

DOI: 10.1201/9781003344575-30

26.1.1 Professional Competence

The Best Captains were recognized as being outstanding aviators who knew the procedures and their aircraft.

- "Knows how to fly by the book with excellence"
- "Phenomenal stick skills – passengers think they are riding a magic carpet"
- "Outstanding flightdeck management"
- "Tremendous company asset"
- "Excellent flight management"
- "Deliberate in actions and pace"
- "Knows the book cold"
- "Super SA"
- "Keeps the passengers comfortable with professional PAs"
- "Professional approach to skilled piloting"
- "PhD in Boxology" (management of the FMS)
- "The most standardized Captain I've ever flown with"
- "The consummate professional pilot"
- "Takes the time to project cause-and-effect in everything"
- "Stays ahead of a developing situation and builds plan and contingencies"
- "Smoothest hands on the tiller and yoke of anyone I have flown with"

26.1.2 Personality

The Best Captains were cited for their easy-going and relaxed personality.

- "Calm experience"
- "Puts customer, crew, and others ahead of their own agenda"
- "Natural leader"
- "Laid back atmosphere"
- "Courteous and considerate"
- "Very proactive and an outstanding leader in the jet"
- "Super professional"
- "Egos don't get in the way of their flying"
- "Grace under pressure"
- "Extremely personable and friendly – they enjoy getting to know you"
- "Fun, fun, fun to fly with"
- "Omits personal bias"
- "Their relaxed demeanor is soothing in a stressful situation"
- "Patient and allows everyone to get the job done right"
- "I'd fly every trip for the rest of my career with them"
- "They showed great personal integrity"
- "Attitude and professionalism was contagious"
- "They know their stuff, but show humility"
- "Excellent balance of professional knowledge, even-keeled personality, and airmanship"
- "Not afraid to make tough decisions"

26.1.3 Team Building

The Best Captains made team building a high priority.

- "Exemplifies the golden rule"
- "Briefing at the beginning of the pairing is the best I've heard"
- "Always willing to give a helping hand"
- "Excellent and open communicator"
- "Makes the pairing fun as well as a learning experience"
- "Goes the extra mile with not only FOs, but FAs and whole Team"
- "Takes input with respect and grace, even if it's wrong"
- "Promotes the environment of teamwork the moment you meet them"
- "Takes time to communicate – LISTEN!"
- "Effectively integrates all inputs to make good decisions"
- "Looks after the whole crew – offers to buy food between flights"
- "Flying the aircraft is a team effort and inputs are truly desired"
- "You really want to do the best for them because you don't want to let them down"
- "Exudes humble professionalism"
- "They want input… I rarely observed any errors"
- "Makes sure the entire crew is in-the-loop"
- "Keeps me informed of their intentions"
- "Creates a flightdeck atmosphere that allows any FO to do their best"
- "Cares about the success of each individual and the company"
- "Treats other employees with dignity and respect – buys food for the ramp agents!"
- "Stayed with me in the hospital all night while I was waiting for an appendectomy"
- "No one does a better job creating a relaxed environment"
- "Gets to know all the company people they can (thus, more aware of whole operation)"
- "They are rapport builders"
- "Always gives a thorough briefing to the incoming crew"

26.1.4 Instructing/Teaching

The Best Captains were outstanding mentors and skilled teachers.

- "Able to instruct and have fun at the same time"
- "Awesome mentor with lots of positive constructive inputs"
- "Identified my errors in a non-threatening, non-critical way, but still got the point across"
- "That's the kind of Captain I want to strive to be like"
- "Gave me great debriefings on my flying"
- "Instills confidence"
- "A rewarding experience I will never forget"

- "They were always trying to learn something new from a situation (teaching is a two-way street)"
- "Gives tips to keep me out of trouble early"
- "Benevolent instructor style"
- "Sets a good example"
- "You spend the whole trip in the sponge mode, laughing the entire trip"

26.1.5 Testimonial

One FO offered this full testimonial of an individual Best Captain.

> He is an exceptional Captain because he leads by example and motivates me to be better. After 25 years of Air Force and civilian flying, I've flown with a ton of pilots/Captains. He is absolutely one of the best Captains in my career. He sets an outstanding example in two areas: operational and team building. He really knows the book, and executes the procedures, administrative and operational, but does it with common sense and ease. Throughout the process he consults and advises the FO, checking that his plan makes sense and is agreeable. He listens to the crew, and discusses options; he doesn't dictate. He is humble and will gladly do a walk-around, or a preflight, or just help out during a tight turn. He's always aware of the entire crew's status and is ready to help. But what really sets him apart is his follow-through. If he doesn't know the answer to a question, he'll get it. Through the course of the trip, we came upon 3 or 4 unanswered questions from FARs to training. He did the research, e-mailed training, made phone calls, and got the answers to all the questions in 2 days. Then he e-mailed the details and citations to me!
>
> His real strength that catapults him beyond other good Captains is his team building approach to the whole organization. He always brings chocolate bars for the flight attendants, he helps them put their bags up on originators, he helps clean when he has free time at the end of a pairing. But wait, there's more: He always buys lunch, or dinner, or beverages for the entire crew. He's so generous with his good fortune that it's contagious. Additionally, he is a great people person, always striking up conversations with even the grouchiest of folks. His good cheer and generosity extend to operations agents and other company employees. His good attitude is contagious.
>
> He sets such a great example of being a good pilot, a good leader, and a good team builder that everyone enjoys flying with him. He never preaches; he leads by example. When you see the flight attendants faces and attitudes change when offers candy and puts their bag in the overhead you are motivated to do the same. He makes me want to be better. Because of him I now carry chocolate, make my PAs from the front of the cabin, and spend more time studying the books. Imagine if all pilots were like him. The attitude would always be great, the teamwork would be amazing, the whole team would be tighter, happier, better prepared and more productive. He sets the standard I now aspire to.

26.1.6 Attributes as Ranked by FO Seniority

While fairly consistent across the board, there were some variations of valued attributes based on FO seniority. Newer FOs tended to value attributes like *instructional ability* and *easy-going nature*, while more experienced FOs tended to prioritize *technical expertise* and *flying skills*. Rankings of *CRM environment, personality,* and *professionalism* tended to spread evenly across all FO seniority groups.

26.1.7 Relative Importance of Attributes

The following attributes (numbered 1 through 16) tended to cluster within three main groups.

- *Top group of attributes:* These top three were cited most often.
  1. *CRM environment:* This was the most commonly cited attribute. FOs felt empowered to express their ideas and observations. Their Captains made them feel like important contributors to the flight's success.
  2. *Personality:* This category related how the Best Captains were friendly, relaxed, and psychologically mature. FOs enjoyed working with them.
  3. *Technical expertise:* The Best Captains demonstrated expertise in book and aircraft knowledge. They followed procedures consistently and predictably. They also gave excellent briefings. Standardization was highly valued.

- *Middle group of attributes:* The next grouping identified the following:
  4. *Instructional abilities:* The Best Captains were exceptional teachers. They had a keen ability to identify learning opportunities and employ successful methods for sharing their wisdom in a way that was welcoming and positive. FOs enjoyed learning because it was non-threatening, respectful, and fun.
  5. *Flying skills:* The Best Captains were skilled pilots. They flew precisely and smoothly.
  6. *Easy-going nature:* This was originally grouped with personality, but since there were so many specific comments directed toward easy-going nature, it deserved a separate category. FOs commented how their Best Captains exhibited "grace under pressure" while defusing potentially stressful events.
  7. *Professional:* This category was counted whenever FOs used the word "professional" to describe Best Captain attributes. Comments like "consummate professional" were commonly stated.

- *Group of remaining attributes:* While receiving fewer specific references than attributes 1–7, these were mentioned fairly often:
  8. *Team building:* Best Captains made concerted efforts to build and support the entire crew as a team.
  9. *Open communications:* Best Captains opened lines of communication with their FOs and FAs. This was commonly associated with *Team Building.*
  10. *Company asset:* Best Captains actively promoted the higher ideals of the company in both their words and deeds.
  11. *Flight management:* Best Captains were commended for how well they managed the smooth flow of the flight.
  12. *Sense of humor:* Best Captains didn't take themselves or the world too seriously. Some FOs rated this as their most-valued attribute.
  13. *Mentoring:* This category related how Captains made the extra effort to form a one-on-one teaching connection with their FOs. Like tribal elders, they shared their wisdom with the next generation of Captains.

14-16. *Situational awareness, decision-making, and experience:* These last three areas tallied the fewest mentions, but could easily be combined with *professionalism, flying skills*, and *technical expertise.*

26.2 SURVEYING THE BEST CAPTAINS

Using the FO surveys, several Best Captains were identified within each domicile. Part two of this project asked these Best Captains to share their views on various topics.

26.2.1 Creating a Favorable First Impression

The Best Captains were asked, "FOs stated that the Best Captains created a great first impression that established a rapport of ease and openness. What do you do during your first encounter with your crewmembers to set your tone?"

- "First I try to find them in the lounge to introduce myself instead of them trying to find me. Then, I spend some time talking with my FOs as we walk to the jet and find out what they flew before and where and [about their] family. Usually we come up with some areas of common interests or experiences and we get to know each other a bit. If they are new, I make sure they know we will go at their speed and if they are confused or they don't understand anything that is going on to please ask me. No stigma attached, safe operations come first, schedule second. When I meet the flight attendants, I smile and shake their hands, introduce myself, and make sure I pronounce their names correctly and treat them with respect. After my usual brief…, I make sure they know they can call us as much as they want for temperature control or anything else they may need."

- "I introduce myself, help them stow their bags, exchange hugs (if we've flown together in the past) and ask how long we are together for the trip. I always bring chocolate bars for the first crew of the day. I carry 9 for a 3-day trip and 12 for a 4-day trip. An invitation to visit the flightdeck in-flight is always extended to all flight attendants, especially on the longer legs. I always want them to feel like they can come to me with any problem during our day together."

- "No matter how 'busy' or 'tight' the turn might be, I take the time to walk back and introduce myself to the crew… If we're getting on the aircraft together, I try to talk to everyone individually. I tell them they are 'welcome' up front, ask if they need/want anything and when I run to get food, I ask if they want some. I … just try to have good manners and treat people like people."

- "I feel Captains truly set the tone for the entire crew. The tone the Captain sets will play a big part in how any future situations are handled and resolved. I try to convey mutual respect and friendliness. I make a point to seek out the FAs and introduce myself with a friendly handshake. When I do my briefing, I talk with them, not at them! For FOs, I feel it is very important to put them at ease – especially if they are new. A relaxed tone in the flightdeck ensures that everyone will perform at their best."

- "Put them at ease by getting to know them a little better. Let them know that you know you're not perfect and perfection is not expected from them. Empower them to make deviation callouts and let them know that this is not only 'okay' but expected."

Best Captains valued the first encounter. They made the first minute together special. They strived to make a personal connection and show direct and caring attention to each crewmember. They established a strong foundation for teamwork, open communications, and cooperation.

26.2.2 Open Communications

The Best Captains were asked, "How do you promote open sharing of information? What specific actions do you take?"

- "The tone dictates how much open communication there is on the flightdeck. It starts with mutual respect for each other and our abilities. I tell my FOs if I'm doing something 'dumb, dangerous, or stupid' in the aircraft, tell me before someone outside the aircraft (ATC, Customers, etc.) knows I'm doing it – and I will do the same for them. Listening is as important as speaking!"

- "It's very important to let other crewmembers know that you are approachable and available to help solve problems. Briefings are an important step toward achieving this goal. Solicit and use input from all crewmembers. Make a decision. Explain your decision to others, if applicable."

- "I let them know that what we do is a team effort. I also let them know that I'm not perfect, so don't assume I know what I'm doing. Also, let's not play, 'I've got a secret.' Let me know your ideas and concerns. As far as game plan, I look at the situation, ask for inputs and develop a plan. Then, I run it by the FO for their agreement. More than once, I've had an FO keep me from doing something stupid."

- "I tell my FOs first off, 'Please don't let me do anything to annoy you! If I'm doing something that bothers you, please tell me so I can stop doing it'. Lastly, if they ever need to take the aircraft, do it – but please keep flying because I'm going to give it to you. We'll sort out if it was necessary later."

- "I tell FOs that just because I'm a check airman does not mean that I'm infallible. Everyone can make mistakes, and that we should work as a crew to back each other up."

- "As for the FO, use of humor (sometimes in self-deprecating ways) tends to lighten the atmosphere. Finding commonality (previous flight experience, families/kids, sports or hobbies) always helps to open up the other person. I always fess up to any mistakes I make, and ask for feedback in my flying skills and techniques. It helps to have a thick skin."

Best Captains achieved open communications by actively engaging with their crewmembers and inviting their inputs. All of these Captains either stated or implied that active listening was a vital component. This promoted a willingness by all crewmembers to participate in the decision making process – that their participation in decision making was invited, welcomed, and expected.

26.2.3 Team Building

The Best Captains were asked, "What specific steps do you take to bring your crew together?"

- "Brief them together in the morning or when you begin your trip if you can. See yourself as part of the team, the leader in fact, but allow everyone to do their jobs. Do not tell them how to do their jobs but let them know that you are available to them if they need you for any reason. If they need you to step in and exert your authority, do so without hesitation. Nobody likes an empty left seat. When you go for coffee or food, ask if anybody wants anything and consider buying when it is appropriate. Always communicate with everyone. Let everyone know what's going on with unusual ops, problems, delays, anything out of the ordinary."

- "Mutual respect, tone and open communications. Captains are managers of their crew – not dictators! Have the ability to listen, gather information, consult others and come to a mutually agreed-upon solution."

- "Along with open communications, I ask a lot of questions. When a write-up is required, I'll ask my FO, 'Okay, Captain Smith, how would you write this up, or would you? Who would you call first?' I try to never give my input first, but see what they think – a lot of times, it's way better than what I think. We truly have a wealth of experience here! Soon, they'll be calling the shots – ultimately we want them empowered with this ability."

- "Appreciate each person for their contribution to the team. Give praise when praise is due!"

Best Captains understood that one plus one can equal far more than two and that effective teams generate positive outcomes. Effective teams not only perform better in ordinary situations, but they make fewer errors in exceptional situations. There was a clear understanding that the team didn't build itself. It required specific actions by the Captain to set the tone, build trust, and develop team relationships.

26.2.4 Instructing/Mentoring

The Best Captains were asked, "How do you approach the task of flightdeck instruction? How do you select areas to instruct? How do you get your message across so it is received so well?"

- "When I'm flying with a new pilot, I stress the fact that we have all been new at one point. I convey to them that any mistakes they may make are not new mistakes – they are mistakes that we have all made (myself probably several dozen times). For the most part, I approach the instruction as a discussion, not a directing action. If it is time-critical and I have to direct action, I will always follow it up with a discussion when time permits."

- "This one can be touchy and is often overlooked. Be sure to differentiate between technique and procedure. Focus on nonstandard areas first. Give justifications, if available. Set a good example and use positive reinforcement."

- "I usually don't say anything until a few legs or so to see if there is a definite trend in some area. I'll then mention the item as it comes up by asking them about it. Did they know that the item was contrary to the book? I find that most pilots don't realize that they've let that area slip. I also will instruct in areas where I think an FO can improve. I usually will give my techniques in an area and explain why I think it works (usually drawn on experience and giving simulators). I usually give some of the pitfalls to doing it the way they are now. I always end these sessions with the comment 'technique only'. Then, they can take it or leave it."

- "While doing [check airman instruction], the instruction flows at a higher rate (obviously) than during a regular trip. I try to lead by example, and fly the profiles per the operations manual. You don't want to saturate the new pilot, so take breaks often and remind them that they are doing well. Staying even-tempered, as well as using more humor. This makes it lighter in the flightdeck and allows the instruction to take hold. The instruction tapers off sharply when flying with a seasoned FO. You have to realize that just because they don't fly with your particular techniques, it's OK. As long as they stay within operational procedures and standards, there are 100 ways to fly the aircraft. As an FO gets closer to upgrade, I try to share and discuss any questions they may have regarding procedures or decisions. I always involve them in any MEL's, logbook write-ups, fuel drip-sticking, etc."

Best Captains approached teaching as a process of *sharing wisdom*. They remembered what it was like to be new. They made a special effort to put their FOs at ease. They looked for opportunities to introduce new ideas or correct errors. They wanted their FOs to learn at their own pace. They also viewed teaching as a progressive exercise – how experience and wisdom are built incrementally. There was no attempt to comment on every single error. Finally, the Best Captains saw teaching and instruction as an important part of the Captain's job – something that must be done well and effectively.

26.2.5 Personality

The Best Captains were asked, "Many of the common adjectives used when describing the Best Captains were: selfless, humble, people-oriented, affable, and friendly. Do you have a particular perspective through which you view this profession? What traits are important for you to project or share with others?"

- "All of those are great traits – some come naturally, and some people have to learn them. We all have stuff going on in our personal lives. I find any troubles I have disappear when I'm at work. Common courtesy to ALL employees lends itself to a happier workgroup. This is the core reason most employees enjoy working here. No secret, just the Golden Rule at work! I'm very proud to be a [company] pilot. We've all worked hard, through different avenues, to reach this goal. Many in the company look up to us. We need to be selfless leaders. Leadership involves knowledge, skill, professionalism, courtesy, compassion and humor. Sharing those traits will keep us on top of the industry."

- "Perspective? Yeah! This is the best job in the world but don't jam it down people's throats. Live it. Be friendly, treat others with respect. This is very important. Sometimes we forget who pays our bills/paychecks (our customers). Treat them with respect and civility with which you would like to be treated. Take time to talk to them, ask them about their day. Look for the positive. Most people are favorably impressed if the Captain takes time to be welcoming and people-oriented. They like to see your human side and that you care about them. They already hold you in some awe and mystery, so it means a lot when you become accessible to them."

- "A good Captain should have an open and honest personality. I try to use 'please' and 'thank you' (two very simple concepts!). For example, if an FO makes a deviation call out for me while I am flying, I try to respond with 'correcting, thank you' so that they know that I appreciate their contribution."

- "I try to use humor as much as I can. I think there is a time for 'business' (and that should be upheld), but you can be friendly and people-oriented and still get the job done effectively. I think, sometimes, people get caught up in power and trying to press people too hard instead of pulling them along the with real leadership!"

- "For the most part, I am an optimist, the cup is half full kind of person. This is a great way to make a living and I have a good time when I come to work. I enjoy flying and the people that work here. I try to convey that to the others by attitude and actions."

- "The best leaders I've known have had two outstanding traits – ability and humility. The ego should be kept in the back pocket and rarely make an appearance. There is rarely a perfect day or even a perfect flight. I'm going to make mistakes and so is the FO. The idea is to keep mistakes inconsequential. Never shirk responsibility for making decisions, but it's a rare occasion that a Captain has to order something be done. Asking will suffice."

Best Captains valued selflessness, caring, humility, responsibility. No one described a "perfect personality". Everyone recognized that the ideal personality resulted from the *actions* and *intentions* of people striving for ideals of excellence and following the Golden Rule. Simply hold these ideals as we move through the day and the result is an ideal personality.

26.2.6 Professionalism and Standardization

The Best Captains were asked, "The FOs highly valued standardization, adherence to FOM procedures, and projecting a professional image. How do you view the importance of these traits?"

- "Professionalism and standardization are extremely important. In my humble opinion, professionalism does not have to mean stoic and harsh. You can be professional and still have a good time. I also think it's important to remember that there are many techniques that work for a given procedure. Just because it may not be your technique, doesn't mean that it is procedurally incorrect."
- "These are VERY important traits for a successful career. Flying within the standards and procedures devised by the airline makes it easier on ALL crewmembers. Each employee knows what to expect. I try to project a professional image, while maintaining my sense of humor – it tends to ease jumpy passengers when they see happy, smiling, confident pilots! Don't equate 'professional image' with being cocky, stuck-up, and humorless. Alienating those around you is a sure setup for failure!"
- "I try to fly using company procedures as much as possible – truly important! I think the highest form of professionalism is 'walking the talk!' I'm also open to correction from my co-workers when I fail (sadly, too often). When I am interacting with the customers and coworkers, I think a genuine care and concern for trying to do it right, safely, and yet still enjoying ourselves goes a long way. I think shiny shoes, a clean sharp uniform, and a bright smile offers much in affecting our image as pilots."
- "I tell my FOs that I will attempt to fly by the book. So when they see me not doing that, I tell them to let me know. I let them know this will not hurt my feelings. I want to do it right, and I think as a Captain we need to. We set the tone and FOs are watching. If we fly sloppy or non-standard, this will rub off onto our FOs. I wear my uniform neatly and correctly to set an example to the FO and project a professional attitude to the customer."

Notice how their comments on professionalism were expressed in terms of CRM, team-building, mentoring, personal appearance, and safety. Notice how interrelated they are. The Best Captains weave these qualities, skills, and characteristics into their thick brocade of excellence.

26.2.7 Deliberate and Predictable

While somewhat related to standardization, *deliberate and predictable* are important distinctions. The Best Captains were asked, "The FOs liked Captains who were 'deliberate in actions and pace.' How do you view the importance of these ideas? How do you meter your pace with each FO?"

- "This is an important concept. FOs need to know what to expect from the left seat. We've demonstrated time and again that there is no future in rushing. I let the FO set the pace and adjust accordingly (Assuming the pace is reasonable and safe. Occasionally, I have to slow the FO down.)"
- "I may be a little different here from most – I think the Captain sets the pace and shouldn't try to fly off the FO. That being said, I try not to rush them, but I still pull them along. If they are struggling with a certain area, I'll try and watch them and see if I can offer any techniques. If it's a potential 'safety area' (i.e. taxiing) and they're still hanging on, I won't press. I stop or slow down till they are with me. Later, I'll stress some techniques to keep them up with me so they'll be there when I need them."
- "I have the easy job. They have the hard job. If I set the pace and timing, they have to react to me. I don't get excited when we are behind schedule nor do I get angry when we encounter delays. I do the best I can to expedite and return to schedule within the boundaries set for us, but if we can't catch up, it doesn't ruin my day."
- "I try to make it clear that I'm not in a hurry. You can be efficient and stay on schedule without rushing and making the common mistakes associated with being in a hurry. Being 'deliberate in actions and pace' comes from experience and feeling comfortable and confident on the flightdeck. If you see the FO struggling to keep up, help out if you can. Do the walk-around once in a while, get the ATIS or do performance computations while they get set up. Show them you really want it to be a team effort."

We see a range of ideas on this topic, but the common thread is awareness. All of the Best Captains monitored their FO's workload and pacing. Where they could, they helped the FO's work more efficiently. At all times, they maintained a safe pace.

26.2.8 CRM/Customer Service

The Best Captains were given the following scenario. "The senior Flight Attendant (FA) approaches you with a situation during passenger boarding. They want a passenger removed from the aircraft because they've 'been drinking'. Knowing nothing more than this, how do you begin to resolve this kind of problem?"

- "Support your crew but proceed with caution. Gather information – do the other FAs concur? Does the passenger appear to be intoxicated? Engage the help of a customer service supervisor. Come to a decision with the help of your crewmembers and ground operations staff. Make sure everyone is comfortable with the decision (include the FO). Let the customer service supervisor handle the direct customer contact."

- "I immediately call for a customer service supervisor (CSS) to address the situation. They are trained for these confrontations and know how to handle them. The operations agent will usually assist. Find out if they are travelling with family and inform the CSS – this makes a big difference. I always tend to side with the FA in these situations, but I've had occasions when the FA was dead wrong and totally at fault."
- "A customer service supervisor might be needed. On the ground, I try to step aside, momentarily, to allow people that are trained to deal with this do their jobs."
- "Hopefully, I'll have some sort of rapport with the FA (I'll know seniority, personality, work ethic, etc.). If not, I think we need to support them (and they need to know that we do) in tough areas. I'll call for a customer service supervisor and get the other FAs together and ask them to have a little get-together. Whatever we come with, try to implement it – we are a crew."
- "I say, 'Tell me about it.' Listen with an open mind to the story out of ear-shot of the passengers. Only work with the facts, not emotion. Most of the time, your FA will be right. Have them tell you where the passenger is sitting and observe them yourself, speaking to them if you feel the need. Then, if your FA is right, back them 100%. If they are wrong, and this will be a rare occurrence, pull them aside and talk to them and try to resolve the situation using your best interpersonal skills. This is one of the toughest situations you face as a Captain. Always listen. Never be an empty left seat."

The common themes were to become involved (never be an empty left seat), involve the experts (customer service supervisors, in this case), guide the process to achieve a team consensus (all FAs, FO, and station specialists) and select a good solution (make sure everyone is comfortable with it).

26.2.9 Best Captains Survey Summary

Professionalism is a difficult quality to define. It incorporates mindset, intention, actions, and maturity. Hopefully, we see many of these Best Captain qualities and actions in ourselves. These Best Captains stood out because their actions exceled in quality and consistency.

When we distill the survey results down to a single sentence, FOs valued Captains who are technically competent, psychologically confident, and promote an effective CRM environment. Being technically competent, these Captains knew their aircraft and procedures while demonstrating superior flying skills. Being psychologically confident, these Captains were psychologically mature, comfortable with themselves, and relaxed. By promoting an effective CRM environment, they actively opened lines of communication, encouraged input, and led their crew to form a resiliently functioning team. These results were used to spread many good ideas across the pilot group at the surveyed airline. It encouraged discussions about professionalism across the pilot group. Whenever excellence becomes the cultural norm, everyone strives for it.

26.3 MASTER CLASS PROFESSIONALISM

Promoting professionalism struggles against headwinds. We have systemic obstacles that slow the spread of professional ideals. Captains don't fly with other Captains and FOs don't fly with other FOs. Unless we observe them from the jumpseat, we don't witness outstanding pilots handling operational challenges. Additionally, most companies don't actively promote or model Master Class behavior. They might acknowledge crews who successfully handle difficult emergencies, but not what they actually did to merit recognition. The Best Captains survey gives us a closer look into the specific behaviors and mindsets that elevate superior pilots above the rest. Before we conclude our discussion of professionalism, let's address three additional qualities that weren't included in the survey – moral character, system monitoring, and Captain's authority.

26.3.1 Moral Character

While the Best Captains survey reveals the contemporary standard of Master Class Captaincy, we should acknowledge some foundational ideas from our airline pioneers. In his book, *Aircraft Command Techniques*, Captain Sal J. Fallucco examines the history behind the airline Captain. The transcontinental air carrier business began on July 8, 1929. Transcontinental Air Transportation (TAT – later changed to TWA – Trans World Airlines) chose Charles Lindbergh as their Technical Committee Chairman. The committee's first task was to select the pilot group. Lindbergh reported that they, "exercised the greatest possible care in the selection of its pilots and the final appointments… of those who would occupy the left seat as first pilots and those who would occupy the right seat as second pilots." They evaluated each pilot's experience and assessed their *moral character.* They recognized that these pilots would be, "operating for the most part as an unsupervised work force… and [would be] placed in a position of great trust." During this same period, Juan Trippe of Pan American Airways began using the label "Captain" to designate the commanding pilots of amphibious air ships. The job title stuck. Both Lindbergh and Trippe recognized that the commanding pilot, be they first pilot, Captain, or aircraft commander, must be held to a high standard. Fallucco concludes, "As fragile as the industry was economically, Lindbergh understood that the public would establish and build its trust and confidence in the air transportation system as a direct result of the industry's operating safety record." Nearly 100 years later, this statement remains true.[2]

This standard of moral character remains with us today. In fact, it endures as a stated eligibility requirement for an air transport pilot (ATP) certificate under FAR Sec. 61.153(c) – *Be of good moral character.* The FAA sets a high standard for attributes like professional behavior, emotional maturity, accountability, responsibility, judgment, and decision making. Additionally, since they list this as a minimum requirement for an ATP, it implies that we all begin with an acceptable level of good moral character and improve upon it during our career. What does this mean for us?

Fallucco cites the dictionary definition stating that to be *moral* means one concerned with the goodness of human action. *Character* is the combination of qualities or features that distinguish one person from another.[3]

The requirement of good moral character does not appear anywhere else in the FARs except for air transport pilots and air traffic controllers. We are expected to demonstrate emotional and social maturity. As Captains, we rise above the standard of common pilots. We join a select group of elite pilots within the aviation profession. We are the elders of the tribe – the wise ones.

Our Best Captains demonstrate good moral character through their words and actions. On the opposite end of the spectrum, however, there are those among us who do not. These pilots are often described as self-centered, abrupt, unfriendly, abusive, reclusive, and disrespectful. Some of these individuals even defend their behavior with statements like, "I'm paid to fly aircraft, not to get along with people." They feel that flying aircraft is enough and that there is no requirement to do more or to be more. They reject the notion that good moral character is either required or lacking in their actions. They are mistaken. Our profession only begins with the requirement to fly aircraft well. It also includes exercising sound leadership, building/promoting effective teams, following procedures, training/mentoring future Captains, and proactively managing risk.

26.3.2 Pilots as System Monitors

In the early days of airline aviation, the Captain assumed the sole responsibility for managing the flight. Improvements in computers, radars, and control systems have combined to shift much of our flight management to outside agencies – for example, ATC from the airspace management side and dispatch from the company flight management side. These agencies are not focused on our single flight. They are structured to get all of our flights to all of our destinations. As a "component" within this system, it can appear that our role is to accurately and efficiently execute the system's overall game plan. This perspective adversely influences some pilots to abandon their leadership role.

In a larger sense, we serve as the PM for the system. We monitor the viability and progress of the system's overall plan. As with the flightdeck PM role, this means that we need to make deviation callouts and intervene, when necessary. Consider a scenario where ATC is trying to assign a crossing restriction that we are incapable of achieving. We tell ATC that we are unable – a callout. Consider when ATC is continuing to assign visual approaches while conditions are deteriorating to IMC. We inform them of the need to switch to IFR – an adverse trend callout. Consider when a dispatcher tries to release us without a required alternate or adequate fuel load. We have to stop the process until the errors are corrected – an intervention. Just like serving as PM for a flight, our actions need to be accurate, timely, tactful, and assertive.

26.3.3 Captain's Authority

Some pilots feel that the current airline system has undermined Captain's authority. They feel that they have lost their ability to influence the system. Instead, Captain's authority has actually expanded to influence the larger aviation environment. In addition to our immediate flight decisions, we must evaluate and approve many decisions and systemic plans made by others. From this perspective, our authority has expanded over both our aircraft and the system. Nothing happens without our agreement.

Pilots who believe that Captain's authority has become eroded see themselves as automated cogs in the airline machine – destined to do what they are told. Not only is this attitude unhelpful, but it creates systemic vulnerabilities. The resilience of the system relies on us to monitor, speak up, and intervene – to exercise our Captain's authority. The system may effectively manage large numbers of moving parts, but it depends on the accurate flow of information about conditions and trends. If their assumptions are wrong or if conditions are changing too rapidly for them to respond, the system may either miss important counterfactuals or react too slowly.

Another vulnerability is that the system assumes that all crews share common capabilities. If the weather at an airfield is dropping to CAT III minimums, the system may assume that all crews and aircraft are qualified to fly CAT III approaches. Instead, individual aircraft may lack required equipment. Individual crews may lack qualification or currency. The system relies on each pilot to evaluate the overall plan and certify its viability. Take the example of a fog bank moving toward an airport. The ATIS may continue reporting VMC because of the way that the sensor measures visibility. A proactive pilot would see the fog bank moving in, call dispatch, and inform them that the field is about to drop to IMC. This allows system managers to add takeoff alternates for departing flights and evaluate arriving flights for possible holding or diversion.

Captain's authority hasn't died, but it has changed. In many ways, it has grown. We are the thousands of active monitors reporting back to our central planning organizations. We each evaluate how well the system is managing its overall plan while tailoring our game plan to fit.

26.4 PROFESSIONAL ATTRIBUTES ACROSS THE PHASES OF THE FO CAREER

FO professionalism shares the same attributes as Captains. There are a number of specific differences and nuances that bear consideration. As FOs, we walk a fine line between support and intervention. For the vast majority of the situations, we support our Captains as they lead the flight crew team. We accommodate and adapt to the Captain's style and personality. We endeavor to be "good company" and adapt to the flightdeck environment set by our Captains. On rare occasions, however, we need to act to keep the flight operating within regulations, company procedures, and safety margins. This means that we monitor for errors, make callouts, voice concerns, or intervene to stop an undesirable event trajectory from a failing game plan. Master Class FOs perform these roles with skill and competence. It takes some time and experience to refine these skills, so our early years in the right seat require a heavy emphasis on learning.

Our time in the FO seat is a formative learning period. Flying with many different Captains, we gather useful techniques for the day when we will move to the seat three feet to our left, gain a shoulder stripe, and lead our own flight crew team. Along the way, we will experience a range of leadership styles. Some styles will align with our strengths and personalities. Some will not.

We will notice areas where we will need to devote more study and effort. We will also refine our sense of pacing. How long does it take to process an MEL deferral? What is the best way to assist with a maintenance delay to shorten the process? We will learn to monitor different timelines to understand the pacing of the various components of the line operation. How we learn our role as an FO and how we prepare for our future role as a Captain change over our career.

While FOs are expected to embody all of the attributes of Best Captains, they demonstrate them differently depending on the phase of their career. Consider how FO priorities and attributes change across three phases: new-hire, experienced line-pilot, and nearing Captain upgrade.

26.4.1 Phase 1 – New-Hire/Probationary FO

Congratulations and high-fives all around. We have landed our airline job. Following new-hire and initial operational experience (IOE) training, we take our place at the bottom the FO seniority list. Typically we start out as a reserve pilot. We may fly sporadically or work every available hour within the duty day and hourly limits allowed by FARs or union contract. We may know our assignments ahead of time, have them assigned when we are called out, or be reassigned at the last minute. There is little consistency or predictability, so we must be prepared for anything.

Another challenge is that we may not remain with our crew. We may need to travel to our overnight hotel alone. Finding our way to the right curb and the right hotel van isn't always intuitive. At larger airports and hotels, the pick-up and drop-off spots may be different from the ones used by passengers. This also means that we may need to make our way back to the airport alone. Determining how much time to allow for shuttle transit can be challenging. Add complexity from time zone and international date line confusion and it can become quite difficult. Use airline resources or flight crew phone apps to help.

- *Readiness:* Being on reserve means we need to be ready for whatever scheduling assigns. For major airlines with extensive route structures, this means that we pack our bag for both Alaska and Florida. Nothing says "inexperienced newbie" like standing in the hotel van waiting area, in a snowstorm, wearing summer-weight slacks, and a short-sleeved shirt.
- *Fast learning:* This is an intensive learning phase. Even if we are familiar with airline operations, we will need to learn specific company policies and procedures. The information will probably bombard us faster than we can process it – like trying to drink from the proverbial fire hose. Since our flights' schedules are repetitious, we will quickly accumulate knowledge. Engage this phase with a positive attitude and thirst for learning. Most Captains empathize with the challenge of applying new procedures while learning a new aircraft. Many will adjust the operational pace to reduce our task loading to help us keep up. Sometimes, they will perform some of our tasks for us. Their assistance can be a double-edged sword. On one side, it frees time for us to get our work done. On the other side, Captains who lack recent practice with completing FO tasks may miss important steps. So, accept their help, but verify their work. Maintain a positive learning attitude and ask lots of questions.
- *Wide range of Captain styles:* On reserve, we will find ourselves flying with many different Captains. Within an established seniority list, senior Captains tend to fly with senior FOs on desirable lines, with steady

workdays, fewer aircraft swaps, and longer overnights at more desirable destinations. Consider a senior Captain on a regular international route who always flies with experienced FOs. They expect their FOs to skillfully complete all of their required tasks. When that originally scheduled FO calls in sick, a reserve newbie like us gets assigned to cover the trip. Now, these senior Captains have a new-hire FO who is still learning the aircraft, the airline, and the unique requirements of the route and destination. Expect some frustration as our inexperience pulls them from their comfort zone. Keep a positive attitude, ask questions, and voice concerns.

On the other end of the seniority spectrum, junior Captains tend to fly with junior FOs on less desirable lines, with longer workdays, and to unappealing overnight destinations. Expect earlier starts, later finishes, longer gaps between flights, more aircraft swaps, longer duty days, and shorter overnights. Reserve assignments typically increase during major holidays and during bad weather. So, if a snowstorm is bearing down on the upper Midwest on December 24th, dress warmly and expect to spend Christmas in Cleveland.

On the one hand, junior Captains are more familiar with FO duties since they recently occupied our seat. On the other hand, they are still learning their Captain duties and may be struggling themselves. Expect challenging flying conditions, operational difficulties, and tighter duty day restrictions. Newer Captains may feel more pressure to stay on time and accept borderline flights in marginal conditions. While senior Captains have developed the finely tuned judgment to distinguish between which conditions they will accept and which conditions they will reject, newer Captains haven't. Some may be overly cautious, while others may be overly risky. Whichever the case, we need to assert ourselves to remain within our personal risk tolerance. Mishap FOs found themselves in uncomfortable situations because they were reluctant to speak up when pushed to accept game plans that they thought were unwise.

This last topic is somewhat sensitive. Reserve pilots tend to get assigned to Captains that line-holding FOs try to avoid. Many airlines offer a bid-avoidance option where FOs can designate a few Captains to skip over during the line-awarding process. Additionally, line-holding FOs who are paired with a Captain that they don't wish to fly with can trade-away their pairings or even call in sick. As a result, these bid-avoided Captains tend to fly with junior, reserve FOs. There are lots of reasons why these Captains are avoided, but most tend toward abrasive or unfriendly personalities. The bottom line is that regardless of which Captain we find ourselves paired to fly with, we must ensure that our flights are conducted according to regulations and company procedures.

In the end, we all share something in common. We are all committed to completing our flights successfully. Even pilots with abrasive personalities can conduct their flights effectively. Personally, I have experienced enjoyable flights with highly avoided Captains and unenjoyable flights with popular Captains. We can't expect all personality combinations to mesh. So, keep a positive attitude, follow the procedures, maintain standardization, and work toward favorable outcomes.

26.4.2 Phase 2 – Experienced, Line-Holding FO

As we gain seniority, we begin holding hard lines with more predictability and stability. Depending on how our airline constructs our lines, we may be able to fly a full pairing or a full bid cycle with the same Captain. Flying with the same crew allows us to settle into a routine. We can anticipate each other's actions and adjust to each other's style.

- *Learning:* By this phase in our career, we have mastered a typical flight's workflow. We settle into our comfort zone – for good and for bad. On the good side, the flying gets easier. We feel more confident, detect more subtle indications, understand what is happening, and integrate our tasks into smooth workflows. On the bad side, the quality of our work can begin to drift. We can start shortcutting procedures, relaxing our attention, and succumbing to our biases.

  Periodically, we need to renew our commitment to pursuing the Master Class path. Otherwise, we can settle into monotony and boredom. Again, choosing to reside in our comfort zone is not necessarily a hazardous choice. Many pilots fly successfully within their comfort zone, continue to amass experience, and steadily improve their skills and judgment. Safety margins, talent, practice, and the predictability of the airline environment shield most of them from mishaps. The higher level of expertise we gain by pursuing the Master Class path may never be fully tested. That's okay. We pursue it because it is in our nature to strive for excellence.
- *Lifestyle:* The stability of this phase allows us to redirect more attention toward outside interests. Tracking the growth of the airline and the pace of our advancement up the seniority list, we estimate when and where we should become eligible for Captain upgrade. Many of us predict our likely Captain domicile and decide whether to move to that city or commute. We also focus more on our family life. While this is not incompatible with the Master Class path, we should guard against allowing outside concerns to distract us while we are flying.
- *Line quality:* As we advance, our line quality improves. With larger airlines, we gain the flexibility to choose our aircraft type, route structure, overnights, and pairing quality. We can choose to eliminate variability from our schedule. In the extreme, we can fly the same pairings to the same overnights every week. Limiting our schedule supports our comfort zone, but it limits our practice with unfamiliar situations. For example, if we predominantly fly between Los Angeles and Hawaii and are suddenly rerouted to Chicago in a winter snowstorm, we might lack recent experience with winter operations. In a related problem, narrowing their range of practice sometimes creates problems for international wide-body FOs who upgrade to narrow-body domestic aircraft. As we approach upgrade, it behooves us to practice with the conditions that we will encounter after upgrade.

26.4.3 Phase 3 – Nearing Upgrade

We begin shifting our attention toward the Captain's seat. While upgrade seemed far off, we became quite comfortable within our FO mindset. Upgrade upsets our comfort zone, but in an exciting way. Finally, our prized goal feels within reach. While Captain upgrade represents a significant advancement in our career path, we need to get through upgrade training first. In addition to the physical challenges of switching our flying/thrust lever hands, we need to acquire our Captain's mindset. In the past, we watched our Captains wrestle with tough decisions. Maybe they even asked for our input, but it was still their decision to make and their responsibility for unfavorable consequences. Soon, we will occupy the hot seat and have to make those tough decisions for ourselves.

- *Learning the Captain's perspective:* By this point on our career path, we have become experts at performing our FO duties. We know the books from the FO perspective. Approaching upgrade, however, we need to study the manuals from a Captain's perspective. The same words begin to take on different meanings. The tough decisions that we previously glossed over as "Captain stuff" now become important distinctions to that we need to fully understand.

  FO decision making seems easier. Most FO choices are clear and unambiguous. We are still responsible to speak up when a decision is clearly wrong or hazardous, but those events are rare. Captain decisions require us to balance between potential rewards and negative consequences. Subtle and conflicting goals offer fewer clear decisions and more uncertain ones. We don't often know all of the conditions at play.

  A Master Class strategy is to acquire our Captain's mindset before entering upgrade training. This is not a difficult objective because, by now, we have become skilled FOs. We have plenty of excess time to pay closer attention to the Captain's decision-making process. As line-flying events unfold, we exploit viable debrief opportunities to interview our Captains about their past decisions. If they are not on the Master Class path, they may not be able to offer us as much useful insight since most of their decisions were probably made by habit or subconsciously. When we have occasion to fly with Master Class Captains, we should delve deeply into their accumulated wisdom.
- *The kinds of questions to ask:* Unfortunately, most pilots approaching upgrade ask unhelpful questions. One typical question is, "I'm upgrading soon, what should I study?" This question focuses on upgrade training, not on learning the Captain's role. In truth, few pilots fail with Captain upgrade because they don't know the books or can't fly from the left seat. They fail because they haven't acquired the mindset to think like a Captain. They struggle with Captain decision making and risk management. They either act too quickly before they have all the necessary information or wait too long because they are reluctant to make decisions or lead the flight crew. What to study for upgrade is already listed in the upgrade prep guide. The more important skills are learning how to evaluate conditions and make tough decisions. Consider some of the following questions that we can use when interviewing our Captains.

  - What situations have challenged you most as a Captain?
  - Have you experienced events where you didn't know where to find the information you needed?
  - Have you been pushed by the operation to do something you didn't think was wise?
  - How do you handle conflicts with members of other employee groups?
  - Have you ever had a team member defy your decisions or resist your leadership?
  - When do you refer tough decisions up to higher authorities, like chief pilots?
  - Have you ever had to defend your decisions or actions to a manager who didn't agree with how you handled an event?
  - Have you heard of events from other Captains that really challenged them?
  - How do you anticipate down-line problems during a duty day? How do you prepare for them in advance?
  - How do you decide when to speak up about problems and when to leave them alone?
  - What are some of the systemic holes or traps that challenge you as a Captain?
- These examples outline some negative events that challenge Captains. They explore the process of turning bad into good. Another set of questions focuses on the positive side of building and leading effective flight crews.
  - How do you assess the difference between an operation that is running well enough and an operation that would benefit from your intervention?
  - How do you determine when you should assist others in completing their tasks?
  - How do you encourage various groups to communicate concerns, problems, or ideas to you?
  - When do you contact dispatch to discuss aspects of a flight?
- Questions like these improve an operation that is already running successfully. They explore the process of turning good into better. The following questions examine operational gray areas and when to make tough calls.
  - That was a challenging approach with the moderate turbulence on final. How do you draw the line between continuing or going around?
  - There aren't favorable diversion options for this airport. If you had to divert without planned alternate fuel, like with a runway closure, how would you manage it?
  - If an airport is closed for windshear and we are the only aircraft waiting to depart, what conditions would you consider before departing?
  - Have you ever had an event where you had to exercise your emergency authority?
  - Have you ever found yourself concerned about your remaining fuel?
  - Have you ever had to assume aircraft control from your FO?

Notice how these questions drill down to the root of the Captain's role. The easy stuff won't challenge us. The fuzzy gray area does. Balancing conflicting priorities does. The more stories we hear, the more tools we will have when we are faced with these kinds of conflicts.

NOTES

1 Some quoted survey responses have been edited for grammar, gender reference, and deidentification.
2 Fallucco (2002, pp. 8–9).
3 Fallucco (2002, p. 6).

BIBLIOGRAPHY

Fallucco, S. J. (2002). *Aircraft Command Techniques: Gaining Leadership Skills to Fly the Left Seat*. Burlington, NJ: Ashgate Publishing Company.
Kern, T. (2011). *Going Pro: The Deliberate Practice of Professionalism*. Lewiston: Pygmy Books, LLC.

Glossary of Commonly Used Airline Terms

ACARS: Aircraft Communications, Addressing, and Reporting Systems – a data link system used to send and receive messages and data packages between the aircraft and ground agencies.

ACP: Autopilot Control Panel – The dedicated panel of switches and dials that control the interface between the PF and the aircraft autopilot. Also called the MCP (Mode Control Panel and Autopilot Controller (AC)).

ADI: Attitude Indicator – a primary aircraft flight instrument that displays the aircraft's orientation to the horizon. It is the primary instrument for monitoring bank and pitch within IMC conditions.

AFM: Aircraft Flight Manual – an operating manual that lists the operating restrictions and limits as mandated by the aircraft manufacturer. "AFM limitations" apply to all operators of the aircraft. Regulators and companies may impose more-restrictive limitations.

ANR: Active Noise Reduction – a feature of pilot headsets that eliminates ambient background noise to improve the quality of intercom and radio audio.

AOM: Aircraft Operating Manual – this is the manual that governs the operating procedures for that model of aircraft within that airline. See also – FOM.

APP: Autopilot approach mode – a mode control button on the ACP/MCP (also APPR). Selecting this mode "arms" the autopilot or flight director to "capture" and hold the transmitted localizer and glideslope signals. Pilots typically announce "Approach Armed" when selecting this mode and "Approach Capture" when the automation begins following the navigation guidance.

APU: Auxiliary Power Unit – An internally mounted jet engine that provides an electrical source (to power aircraft systems) and/or compressed air (for engine start or air conditioning pack operation). APUs are routinely used during ground operations, but can provide an electrical power or an air source while inflight.

AQP: Advanced Qualification Program – a training and evaluation simulator event where pilots manage a flight from departure gate to engine shutdown in real time. Events and malfunctions are programmed to challenge the crew with procedures, complexity, distraction, coordination, and CRM.

ASRS: See NASA ASRS.

ATC: Air Traffic Control – a system of ground-based agencies that manage aircraft movement. Interconnected agencies include local ground (Ground) and tower (Tower) controllers, local area approach and departure controllers (TRACON – traffic control, Approach, or Departure), and high-altitude airspace controllers (Center).

ATIS: Automatic Terminal Information Service – An automated airport weather information source available through voice-recorded loop or datalink. Typically updated each hour, it provides pilots with the current weather conditions at an airport. Under rapidly changing or marginal conditions, "special" ATIS reports are issued more frequently.

ATOG: Allowable Takeoff Gross Weight – maximum aircraft operating weights. For departing aircraft, ATOG weights refer to maximum aircraft weight limits for pushback, taxi, and takeoff. For engine-loss scenarios while in certain cruise flight segments over high terrain, maximum cruise weight is designated "enroute ATOG". Maximum landing weight is designated as "landing ATOG".

ATR: The designation of regional jet aircraft manufactured by Aérospatiale of France and Aeritalia of Italy.

Autoland: An automated landing mode where the autopilot flies the aircraft through final approach, flare, and touchdown.

CAT I, II, III: Categories of charted precision (flight-directed course and glideslope guidance) instrument approaches with specific weather minima, pilot qualifications, aircraft equipment, and ground equipment requirements. CAT I approaches use basic ILS localizer and glideslope guidance. CAT II and III require HGS or autoland precision and allow lower approach minima. Subcategories are designated by letters (A, B, or C) and "SPECIAL".

CB: Circuit Breaker – aircraft flightdecks feature extensive panels of CBs to allow pilots to identify and isolate electrical malfunctions. Additional CBs are located in aircraft equipment bays. They are only accessible while on the ground.

CFIT: Controlled Flight Into Terrain – a category of aircraft accidents where an otherwise controllable aircraft impacts the ground, water, or obstructions often due to loss of crew SA or distraction. Events typically happen in IMC conditions with unaware flight crews.

Check pilot: A management pilot who conducts evaluation events for company pilots. These include line checks (to evaluate pilots during normally scheduled flights), simulator checkrides (simulator evaluation events), initial operational experience (IOE – certification of new-hire pilots and recertification of pilots following prolonged flightdeck absences, such as medical grounding), and upgrade operational experience (UOE – initial line certification of newly upgraded Captains.)

Chief pilots: Management pilots assigned to each domicile/crew base. Chief pilots are the direct supervisors of every pilot assigned to that base, including check pilots. Typically, a number of assistant chief pilots also serve at each base.

Commute/Commuting/Commuter: Terms that describe pilot movement between where they live and their assigned crew domicile. Related terms include commute flight (the flight used to fly to or from the pilot's domicile), commuter base (a domicile that is staffed primarily by commuters), and commuter hotel (a hotel close to the airport frequented by commuters before or after a flight pairing).

Counterfactuals: Indications that contradict what we expect to see or happen. They are usually generated by conditions that are adversely affecting the predicted progression of the flight (future SA). Counterfactuals can be warning signs that indicate that the game plan is failing. The game plans either need to be modified or abandoned for a contingency backup game plan.

CRM: Crew Resource Management – the field of study and flightdeck operating philosophy that guides the optimal use of the flightdeck team to operate effectively and resiliently.

CVR: Cockpit Voice Recorder – a continuous loop audio recording system that simultaneously records each pilot, radio transmissions, and flightdeck ambient sounds. Recordings are only downloaded for mishap investigation or as allowed by formal agreements.

DFDR: Digital Flight Data Recorder – see FDR.

Dispatcher: Individuals within an airline company responsible for producing flight release documents and tasked with following the inflight progress of assigned flights. Dispatchers research prevailing and forecast conditions to construct each dispatch release. Captains either accept or amend the release before departure. FAR regulations assign joint responsibility for each flight to both the dispatcher and the Captain.

Domicile: The base or location assigned (or awarded by seniority bidding) to each pilot. Each pilot's pairings begin from and end at their domicile. Pilots either live locally or commute from their residences to their base. Commuters typically arrange overnight lodging (often called "crash pads") at their domicile city.

DOT: Department of Transportation – the cabinet-level, executive branch of the U.S. government responsible for the transportation system, including air travel.

ECAM: Electronic Centralized Aircraft Monitor – An Airbus system that displays engine and system status on flightdeck displays. Following malfunctions, crews check the ECAM display for a list of error codes to guide their diagnosis and handling of abnormal events.

EFB: Electronic Flight Bag – a portable screen device that mounts in a convenient flightdeck location for each pilot. It contains all flight manuals and charts needed for flight operations. It replaces the "brain bag" which previously contained all of the paper volumes of these references.

EFC: Expect Further Clearance time – an ATC-assigned time, typically issued to aircraft awaiting departure or in a holding pattern for arrival. For example, if a weather event affects departure or arrival flow, each flight is assigned an EFC time that designates when they can expect an update that will either extend their delay or resume operations.

EGPWS: Enhanced Ground Proximity Warning System – an automated system designed to alert crews of terrain hazards, obstructions, windshear, unfavorable flightpath trends, and non-standard aircraft configurations. The term, GPWS (earlier version), is still commonly used.

EGT: Exhaust Gas Temperature – a measurement of the temperature of the gases exiting the turbine section of a jet engine. It is typically associated with maximum limits imposed by the engine manufacturer.

EPR: Engine Pressure Ratio – in low-bypass turbojet engines (older jet engine technology – largely replaced by high-bypass engines referencing N1), EPR provides a standard for measuring and setting engine thrust.

FA: Flight Attendants – the cabin crew. Led by the senior or "A" flight attendant, the crew varies from one (on small, regional aircraft) to many (on large, widebody aircraft). The minimum number of FAs is directed by regulation.

FAA: Federal Aviation Administration – administrative and regulatory department of the DOT responsible for the U.S. aviation system.

FAF: Final Approach Fix (also FAP – Final Approach Point) – a designated point on a charted instrument approach that separates the earlier, maneuvering portions of the procedure with the final approach segment.

FARs: Federal Aviation Regulations – regulations governing aviation in the U.S.

FDR: Flight Data Recorder – also DFDR adding "Digital" – an onboard monitoring system that records a wide range of subsystem status and flight parameters. It is continuously recorded for later download and analysis. Most airline SMS departments process this data and analyze it for out-of-tolerance or anomalous events. Events worthy of further investigation are referred to Gatekeepers who are trusted pilots who identify and interview the involved pilots. Crew contacts are protected by contractual agreements. Interview summaries are deidentified.

FMC: Flight Management Computer – the main interface between pilots and the automated guidance systems of the aircraft. System inputs are uploaded automatically via datalink (ACARS) or typed in manually through a keyboard. The FMS (Flight Management System) incorporates all of the aircraft sensing, navigation, and control sub-systems.

FMS: See FMC.

FO: First Officer – the Second in Command pilot on a flightdeck. They occupy the right seat while the Captain occupies the left seat.

FOM: Flight Operations Manual – by this and similar names, it is the operations guide issued by the company that defines policies and procedures. Some companies distinguish between an FOM that covers company procedures for all fleets and an AOM (Aircraft Operating Manual) that covers specific procedures for each fleet type.

Game plan: The plan that each crew forms to accommodate current/expected conditions to guide the flight along a desired path. After detecting conditions and recognizing situations, the selected game plan guides decision making, monitoring, and CRM. Marginal conditions require pilots to prepare and brief backup contingency plans, in case the primary game plan proves untenable.

GND: Ground – It refers to either a switch position or an aircraft sensing mode (on the ground versus in the air).

Hand flying: This refers to times when the PF flies the aircraft without autopilot assistance (also called "uncoupled"). This control mode includes flying with no automation, using only "raw data" (basic navigation instruments), and using flight director guidance.

Hard line: A period of flying (a calendar month, for example) where all pairings of line flying are assigned ahead of time. Pilots typically bid for their lines from a bid package. Lines are awarded by seniority or through choice optimizing software (called Preferential Bidding System – PBS). See also Reserve Line.

HF: Human Factors – the science of the human interaction between each other and machines. It includes the fields of psychology, sociology, engineering, biomechanics, industrial design, physiology, anthropometry, interaction design, visual design, user experience, and user interface design.[1]

HGS: See HUD.

Hold out or holding out: A commonly occurring delay for arriving aircraft when their gate is not available for them after landing – typically due to a delayed flight currently on the gate. Arriving aircraft have to hold out in a designated holding area or "penalty box".

HUD: Heads Up Display – A component of the HGS (Heads-up Guidance System) that projects flight parameters and course guidance onto a combiner glass in front of the pilot – typically projected in green. The pilot looks through the glass to simultaneously view the outside environment and the HUD display guidance. It is a required system to fly HGS approaches.

ICAO: International Civil Aviation Organization – a United Nations agency that guides coordination, safety standards, and practices among participating national governments.

IFR: Instrument Flight Rules – a set of FAR procedures that govern operations outside of VFR (visual flight rules). For example, IFR procedures govern aircraft spacing and the use of charted instrument approaches in reduced visibility and weather-restricted conditions.

ILS: Instrument Landing System – a category of charted instrument approaches using ground-based transmitters that provide both course and glideslope guidance from final approach to the landing runway.

IMC: Instrument Meteorological Conditions – weather conditions that require ATC to manage the spacing between aircraft and the use of published instrument approaches. See also IFR.

IOE: Initial Operational Experience – this is line-flying experience following initial classroom and simulator training. New-hire pilots fly with a check pilot or a training pilot. After completing IOE, FOs are certified for regular line-flying. UOE (Upgrade Operational Experience) is the equivalent for newly upgraded Captains.

Jeppesen: A company (now a division of Boeing Corporation) that publishes navigation, departure, arrival, airport, and approach charts. Historically, they issued paper charts (known as Jepp charts/pages). Now, these products are primarily issued electronically to pilot EFBs.

Jumpseat: One or more additional seats located on the flightdeck that can be occupied by inspectors, check pilots, and authorized individuals. Most often, they are used by pilots commuting to and from their domiciles.

Leg: A common reference to a single flight. Pilots refer to "my leg" for flights when they serve as PF and "your leg" for flights when they serve as PM.

Line flying: Within the airline culture, line flying refers to the everyday schedule of flights moving passengers and freight. The term comes from the process of bidding and awarding a "line" of scheduled trips – typically for a month. Pilots rank-order their choices from a bid package within their domicile and aircraft type. They are then awarded a "line" which consists of a sequence of pairings – each which departs from and returns to their domicile after one or more days. Other common terms are "line holding" (a pilot who can successfully bid for a hard line) and "flying the line".

LLWS: Low-Level Windshear – a potentially hazardous condition cause by wind and ground features. Pilots may experience changes of wind speed, wind direction, and turbulence that exceed aircraft limitations or capabilities. Most modern aircraft are equipped with predictive windshear (PWS) systems that detect LLWS conditions. Certain airports also maintain an array of ground-based sensors to detect microburst events and convective weather gust fronts.

Lobby: A common term for the location and time for a crew to gather in the hotel lobby awaiting their shuttle ride to the airport. Common uses include "lobby time" and "making lobby".

LOC: Localizer – a mode button on the ACP/MCP. Selecting this mode arms the autopilot or flight director to capture and hold a selected localizer course. Pilots announce "Localizer Armed" when selecting this mode and "Localizer Capture" when the automation begins following the navigation guidance.

Local rationality: Behaviors, actions, operating norms, and perspectives used by front-line operators to handle situations in a way that makes sense to them. This is one of the major drivers of drift. Distinct variations can emerge across an airline's system. A benefit is that innovative locally developed techniques can be elevated to become company-wide procedures. A harm is that local practices can unintentionally undermine safety protections designed to mitigate latent vulnerabilities.

LOFT: Line-Oriented Flight Training – a simulator training event that follows a full flight scenario from pushback through engine shutdown. See AQP.

Magenta line: In modern aircraft situation displays, the programmed course is depicted as a magenta line. It appears as connected straight-line flight segments that change direction at each named point, called a "fix".

MCAS: Maneuvering Characteristics Augmentation System – an automated system introduced on the Boeing 737MAX aircraft to compensate for significant pitch-up effects caused by high thrust settings.

MCP: Mode Control Panel – see ACP.

MEL: Minimum Equipment List - a manual that specifies the protocols and procedures for operating with inoperative or degraded aircraft equipment. Provisos ensure that adequate functionality exists with backup systems to allow safe operation.

Microburst: A weather effect associated with convective activity (more common in dry climate conditions featuring "high base" thunderstorms) where intense downflows of air emanating from the thunderstorm core reach the ground and propagate laterally. Microburst events can be extremely intense and exceed aircraft operating capabilities. Microburst events tend to be short-lived and extremely hazardous.

MSA: Minimum Safe Altitude – a designation of the lowest altitude for flight operations due to terrain or obstacles. MSA also stands for Minimum Sector Altitude which is the minimum altitude that ATC can assign while vectoring aircraft through a particular area.

N1: A measurement of the outer fan speed on turbojet (typically high-bypass turbofan) engines. N1 displays the percentage of rated rotation of the outer fan blades. N1 is the most usable measure of engine thrust.

N2: A measurement of turbojet inner fan speed. N2 displays the percentage of the rated rotation of the inner or engine core fan blades. N2 reflects engine speed and performance. While this does produce some thrust, the primary purpose of the N2 fan is to spin the outer fan blades to generate thrust from air bypassing the engine core (N1).

NASA ASRS: National Aeronautics and Space Administration Aviation Safety Reporting System – a NASA agency that manages the voluntary reporting system. Deidentified reports are maintained in a publicly available, searchable database.

NAVAID: Navigational Aids – ground-based sites that transmit signals that can be displayed on aircraft instruments. While still commonly used, ground-based navigational aids are being replaced by aircraft Global Positioning System (GPS) navigation.

NDB: Non-Directional Beacon - a ground-based navigation station used extensively in the early eras of aviation. They are rarely used in airline aviation anymore.

NOTAMS: Notices To Air Missions – Notifications published by the FAA regarding exceptional, hazardous, or abnormal status of facilities, equipment, procedures, or services. Pilots review the NOTAMS affecting their flights before and during each flight.

NTSB: National Transportation Safety Board – the independent federal agency charged by the U.S. Congress with investigating civil aviation accidents in the United States. It also investigates significant accidents in other modes of transportation including highway, marine, pipeline, and railroad.

OPS: A commonly used abbreviation of operations. It refers to a range of agencies such as station operations, flight operations, or central operations.

PA: Public Address – verbal broadcasts to the passenger cabin made from a flight attendant station or from the flightdeck.

Pairing: A sequence of assigned flying that starts and ends at the pilot's domicile. They range from a one-day pairing (often called a "turnaround" or "turn") to multiday pairings with hotel overnights. The term is also used when referring to a scheduled flight between two airports. For example, a JFK-LAX flight is "city pair". Pilots assigned to the flight would "fly that pairing".

PC: Proficiency Checkride – typically a yearly simulator event (often 6 months from the last procedural training event – PT) where pilots must demonstrate proficiency to performance standards. The PC program has been replaced by the AQP program at many airlines.

PF: Pilot Flying – The pilot directly responsible for maneuvering the aircraft or managing the autopilot.

PIREP: Pilot Report – a report of the actual weather conditions observed. PIREPs are typically transmitted over radio or via ACARS.

Placard: Prominently posted flightdeck notices displaying system limits. For example, maximum flap extension speed limits for each flap position and maximum landing gear extension/retraction speeds. Pilots are directed to ensure compliance with placard limits before changing system settings.

PM: Pilot Monitoring – The pilot responsible to monitor the accuracy and quality of the PF's flying. The PM makes deviation callouts, trend callouts, and interventions as required to keep the aircraft within procedural and safety limits.

Pressing the field: The voluntary technique when a pilot delays their typical slowing and configuring of the aircraft. It decreases flight time, improves efficiency, or increases the challenge of managing aircraft energy state. ATC vectors can also induce this situation while trying to compress aircraft spacing. Mismanaging aircraft energy while pressing the field can result in an unstabilized approach.

PT: Proficiency Trainer – typically a yearly simulator training event (usually 6 months from the last proficiency checkride – PC) where an instructor trains designated events. The PT program has been replaced by the AQP program at many airlines.

Push: Both an action (aircraft pushback off of the gate) and a time (push time). When crew scheduling assigns a flight with minimal notice, they may ask, "Can you can make push?"

PWS: Predictive Windshear System – an aircraft system that detects approaching or current windshear conditions that may exceed aircraft performance capabilities. PWS displays and audio warnings alert pilots to delay takeoff if hazardous conditions are detected along the departure path or to break off an approach (either execute a go around or a maximum performance windshear recovery maneuver) if windshear is predicted or detected along the approach path.

QRC: Quick Reaction Card – an immediately accessible flightdeck card that includes a number of non-normal procedures that warrant quick response.

QRH: Quick Reaction Handbook – a detailed flightdeck guide with checklists and reference tables needed to handle aircraft and system malfunctions.

RA: Resolution Advisory – an alert from the TCAS system that indicates an inflight conflict with another aircraft. The conflict is highlighted on aircraft displays. The system often provides directive audio to resolve the flightpath conflict.

RCAM: Runway Condition Assessment Matrix – a system for rating reduced braking and stopping effectiveness of runways and taxiways. Reports are made using Runway Condition Codes (RCC) criteria. RCAMs uses the following Runway Condition Codes (RCC) values: dry-GOOD (RCC 6), wet-GOOD (RCC 5), GOOD TO MEDIUM (RCC 4), MEDIUM (RCC 3), MEDIUM TO POOR (RCC 2), POOR (RCC 1), and NIL (RCC 0).

RCC: See RCAM.

Reserve line: A section of a pilot's schedule where they are assigned specific blocks of days of duty eligibility. Pilots must be ready to fly during assigned reserve days. They may be assigned a pairing in advance, wait at home, or wait nearby the airport for possible assignment. See also Hard Lines.

RNAV: Area Navigation (previously Random Navigation) – a navigation option where aircraft use onboard GPS and FMS systems to proceed directly from their present position to a distant fix. It smooths the zigs and zags of flying along established airways or over published fixes.

RNP: Required Navigation Performance – a classification of highly precise navigation used for point-to-point movement and within a class of non-precision RNP approaches. The system requires FMS navigation using GPS accuracy. Aircraft are coupled to the autopilot.

RPDM: Recognition Primed Decision Making – a decision-making model from Gary Klein that describes the decision-making process where cues lead to recognized patterns, patterns activate action scripts, action scripts are assessed through mental simulation, and mental simulation is driven by mental models.[2]

RRM: Risk and Resource Management model – a risk management strategy from Volant Systems, LLC that promotes individual and crew awareness of rising risk factors and then guides the application of resources to manage the situation and reduce overall risk.

RTO: Rejected takeoff – anytime a takeoff is initiated and is then aborted for any reason. Events range from immediate low-speed rejects (from configuration warning horns) to high-speed rejects (for unsafe flight conditions).

SA: Situational Awareness – awareness of the current status of the aircraft, surrounding conditions, and aircraft trajectory. See Chapters 5 and 12.

SID: Standard Instrument Departure – charted instrument procedures designed to sequence aircraft from takeoff and local ATC control (TRACON) until reaching high-altitude ATC Center sectors.

SMS: Safety Management System – a comprehensive safety management structure that combines top-down, organization-wide risk management programs with bottom-up reporting of risk factors from frontline operators to achieve responsive and resilient management of risk factors. It includes a range of programs, departments, policies, procedures, and practices to identify, analyze, and manage risk.

SOP: Standard Operating Procedures - procedures written and used within a company for line operations.

STAR: Standard Arrival Routes – charted instrument procedures designed to sequence aircraft from high-altitude ATC Center sectors to local airport arrival controllers (TRACON).

Sterile flightdeck: Also commonly called "sterile cockpit" – it refers to a series of protocols intended to limit unnecessary conversation and distraction during critical flight phases. A common application is the prohibition against unnecessary conversation during dynamic ground movement until climbing through 10,000′ MSL.

TAF: Terminal Aerodrome Forecast – The FAA forecast of expected weather conditions at an airport. They are published with specific time periods. TAFs may also include intermittent weather conditions that are more restrictive, such as reduced ceiling and visibility from convective weather activity.

TCAS: Traffic Alert and Collision Avoidance System – a system that analyzes the movement of nearby aircraft to provide automated, directive collision avoidance instructions using both flightdeck displays and audio warnings/directives. See also RA.

Threats: Factors or situations that increase operational risk. This concept is a cornerstone of the Threat and Error Management (TEM) safety program used by many airlines since the 1990s.

TOGA: Takeoff/Go Around – this refers to both a flightdeck switch used to engage TOGA logic using flight director guidance and to the flight procedure employed when transitioning from an approach to a missed approach or a go around.

TRACON: Terminal Radar Approach Control Facilities, also known as Traffic Control – an ATC agency that handles aircraft movement between the high-altitude Center sectors and local Tower controllers. They direct and monitor aircraft movement during departures (SIDs – Standard Instrument Departures) and arrivals (STARs – Standard Arrival Routes). They manage course and speed assignment to maintain required spacing between aircraft on IFR clearances for visual and instrument approach procedures. As workload allows, they also provide flight following and traffic advisories for VFR aircraft.

Trip: A common airline reference for a pairing (one-day to multi-day). Uses include checking in for a trip, flying a trip, finishing a trip, and trip trade.

Tunnel vision: Tunnel vision and tunnel focus are terms used to describe how our minds narrow their focus when we become task-overloaded. Unable to perceive and process the deluge of information, we limit our attention toward a few selected parameters or indications. Consequently, we miss other parameters and indications. This strategy can either be successful when we accurately select relevant parameters, or unsuccessful when we miss relevant parameters. This effect is often associated with task overload and event quickening.

V-speeds: A series of speeds that govern aircraft operating limits. Some of the more commonly used V-speeds are:

V_1 – decision airspeed that separates the decision to either reject or continue a takeoff. It is typically associated with an engine failure or fire and is trained using a simulator event known as a V_1 cut.

V_R – airspeed to begin aircraft pitch rotation for takeoff.

V_2 – minimum maneuvering airspeed from shortly after takeoff until 400'.

V_{TARGET} – intended speed to hold while flying down final approach until transitioning to flare and landing. V_{TARGET} is determined by computing additional knots based on local conditions to the computed V_{REF} airspeed.

V_{REF} – minimum speed allowed on final approach. It is based on aircraft weight and configuration.

VMC: Visual Meteorological Conditions – weather conditions that allow pilots to maintain their own spacing with nearby aircraft and clouds.

VSI: Vertical Speed Indicator – a flightdeck instrument that displays the rate that the aircraft is either climbing or descending. The flightdeck instrument is also called the VVI (Vertical Velocity Indicator).

VVI: See VSI.

NOTES

1 List is from the Human Factors and Ergonomics Society website: hfes.org/About-HFES/What-is-Human-Factors-and-Ergonomics

2 Klein (2003, p. 28).

REFERENCE

Klein, G. (2003). *The Power of Intuition: How to Use Your Gut Feelings to Make Better Decisions at Work.* New York, NY: Currency Books.

Bibliography

Active Pilot Monitoring Working Group. (2014). *A Practical Guide for Improving Flightpath Monitoring*. Alexandria, VA: Flight Safety Foundation.

Alter, A. (2017). *Irresistible: The Rise of Addictive Technology and the Business of Keeping Us Hooked*. New York, NY: Penguin Press.

ASRS – Aviation Safety and Reporting System – National Aeronautics and Space Administration (NASA) Online Database. (n.d.). Retrieved from Aviation Safety and Reporting System: https://asrs.arc.nasa.gov/search/database.html.

Buck, C. R. (1994). *The Pilot's Burden*. Ames: Iowa State University.

Cook, R. I., & Woods, D. D. (2006). Distancing through Differencing: An Obstacle to Organizational Learning Following Accident. In E. Hollnagel, D. D. Woods, & N. Leveson, *Resilience Engineering: Concepts and Precepts* (pp. 329–338). Burlington, NJ: Ashgate Publishing Company.

Cushing, S. (1994). *Fatal Words: Communication Clashes and Aircraft Crashes*. Chicago: The University of Chicago Press.

Dekker, S. (2002). *The Field Guide to Human Error Investigations*. Burington, NJ: Ashgate Publishing Company.

Dekker, S. (2006). *The Field Guide to Understanding Human Error*. Burlington, NJ: Ashgate Publishing Company.

Dekker, S. (2007). *Just Culture: Balancing Safety and Accountability*. Burlington, NJ: Ashgate Publishing Company.

Dekker, S. (2011). *Drift into Failure: From Hunting Broken Components to Understanding Complex Systems*. Burlington, NJ: Ashgate Publishing Company.

Dekker, S. (2015). *Safety Differently: Human Factors for a New Era*. Boca Raton, FL: CRC Press.

Dismukes, R. K., Berman, B. A., & Loukopoulos, L. D. (2007). *The Limits of Expertise: Rethinking Pilot Error and the Causes of Airline Accidents*. Burlington, NJ: Ashgate Publishing Company.

Endsley, M. R. (1995). Toward a theory of situation awareness in dynamic systems. *Human Factors*, 1(37), 32–64.

Endsley, M. R. (2022). *SA Technologies - Publications*. Retrieved from SA Technologies: https://satechnologies.com/publications/.

Ericsson, A., & Pool, R. (2016). *Peak: Secrets From the New Science of Expertise*. Boston, MA: Houghton Mifflin Harcourt.

FAA. (2014, September 17). *91-79A - Mitigating the Risks of a Runway Overrun Upon Landing*. Retrieved from FAA Advisory Circulars: https://www.faa.gov/documentLibrary/media/Advisory_Circular/AC_91-79A_Chg_2.pdf.

Fallucco, S. J. (2002). *Aircraft Command Techniques: Gaining Leadership Skills to Fly the Left Seat*. Burlington, NJ: Ashgate Publishing Company.

Field, J., Boland, E., van Rooij, J., Mohrmann, J., & Smeltink, J. (2018). *Startle Effect Management*. Amsterdam: Netherlands Aerospace Centre: EASA - European Aviation Safety Agency.

Flin, O., O'Connor, P., & Crichton, M. (2008). *Safety at the Sharp End: A Guide to Non-Technical Skills*. Burlington, NJ: Ashgate Publishing Company.

Gladwell, M. (2008). *Outliers: The Story of Success*. New York, NY: Little, Brown and Company.

Heath, C., & Heath, D. (2010). *Switch: How to Change Things When Change Is Hard.* New York, NY: Broadway Books.

Hoffer, W., & Hoffer, M. M. (1989). *Freefall: A True Story.* New York, NY: St. Martin's Press.

Kern, T. (2009). *Blue Threat: Why to Err Is Inhuman.* Lewiston: Pygmy Books, LLC.

Kern, T. (2011). *Going Pro: The Deliberate Practice of Professionalism.* Lewiston: Pygmy Books, LLC.

Klein, G. (1999). *Sources of Power: How People Make Decisions.* Cambridge, MA: The MIT Press.

Klein, G. (2003). *The Power of Intuition: How to Use Your Gut Feelings to Make Better Decisions at Work.* New York, NY: Currency Books.

Lanfermeijer, S. (2021). *19 February 1989 – Flying Tiger 66.* Retrieved from Tailstrike.com: https://tailstrike.com/database/19-february-1989-flying-tiger–66/

Loudon, R., & Moriarty, D. (2017, August 1). *Better Briefing.* Retrieved from Royal Aeronautical Society: https://www.aerosociety.com/news/briefing-better/

Loukopoulos, L., Dismukes, R. K., & Barshi, I. (2009). *The Multitasking Myth: Handling Complexity in Real-World Operations.* Burlington, NJ: Ashgate Publishing Company.

Martin, W., Murray, P., & Bates, P. (2012). The Effects of Startle on Pilots During Critical Events: A Case. *30th EAAP Conference: Aviation Psychology & Applied Human Factors - Working Towards Zero* (pp. 388–394). Queensland: Griffith University.

Mosier, K. (2010). The Human in Flight: From Keinesthetic Sense to Cognitive Sensibility. In E. Salas, & D. Maurino, *Human Factors in Aviation – 2nd Edition* (pp. 147–174). Burlington, NJ: Academic Press.

Mosier, K. L. (1991). Expert Decision Making Strategies. *Proceedings of the Sixth International Symposium on Aviation Psychology*, (pp. 266–271). Columbus, OH.

Nowik, O. (2021, February 3). *12 Powerful Habits of People Dedicated to Lifelong Learning.* Retrieved from Lifehack.org: https://www.lifehack.org/articles/communication/12-signs-you-are-lifelong-learner.html

NTSB. (1997). *NTSB/AAR-97/01 and PB97–910401: Wheels-up Landing – Continental Airlines Flight 1943 – Douglas DC-9 N10556 – Houston, TX, February 19, 1997.* Washington D.C.: National Transportation Safety Board.

NTSB. (2000). *PB2001–910402, NTSB/AAR-01/02, DCA99MA060 – Runway Overrun During Landing – American Airlines Flight 1420 – McDonnell Douglas MD-82, N215AA – Little Rock, Arkansas – June 1, 1999.* Washington D.C.: National Transportation Safety Board.

NTSB. (2006). *LAX05IA312: Jet Blue Airways Flight 292 – Airbus A320, N536JB – Nose Wheels Cocked 90 degrees – 21 September, 2005.* Washington D.C.: National Transportation Safety Board.

NTSB. (2010). *NTSB/AAR-10/03: Loss of Thrust in Both Engines After Encountering a Flock of Birds and Subsequent Ditching on the Hudson River – US Airways Flight 1549 – Airbus A320-214, N106US – Weehawken, New Jersey – January 15, 2009.* Washington D.C.: National Transportation Safety Board.

NTSB. (2017, September 22). *NTSB Issues Final Report for Oct. 2016 Mike Pence Boeing 737-700 LaGuardia Runway Excursion.* Retrieved from Aviation Safety Network: An Exclusive Service of Flight Safety Foundation: https://reports.aviation-safety.net/2016/20161027_B737_N278EA.pdf

Reason, J. (1990). *Human Error.* New York, NY: Cambridge University Press.

Reason, J. (2008). *The Human Contribution: Unsafe Acts, Accidents, and Heroic Recoveries.* Burlington, NJ: Ashgate Publishing Company.

Regan, M., Lee, J. D., & Young, K. L. (Eds.). (2009). *Driver Distraction: Theory, Effects, and Mitigation.* Boca Raton, FL: CRC Press.

Sheridan, T. B. (2010). The System Perspective on Human Factors in Aviation. In E. Salas, & D. Maruino, *Human Factors in Aviation – 2nd Edition* (pp. 23–64). Burlington, NJ: Academic Press.

Tremblay, S., & Banbury, S. (Eds.). (2004). *A Cognitive Approach to Situation Awareness: Theory and Application*. London: Routledge.

TSB. (1999). *A98H0003: In-Flight Fire Leading to Collision with Water – Swissair Transport Limited – McDonnell Douglas MD-11 HB-IWF – Peggy's Cove, Nova Scotia 5 nm SW – 2 September 1998*. Gatineau: Transportation Safety Board of Canada.

Vadulich, M. A., Wickens, C. D., Tsang, P. S., & Flach, J. M. (2010). Information Processing in Aviation. In E. Salas, & D. Maurino, *Human Factors in Aviation – 2nd Edition* (pp. 175–216). Burlington, NJ: Academic Press.

Wischmeyer, E. (2005). Why'd They Do That? Analyzing Pilot Mindset in Accidents and Incidents. *Proceedings of the International Symposium on Aviation Psychology* (pp. 639–644). Dayton, OH: ISAP.

Woods, D., & Cook, R. (1999). Perspectives on Human Error: Hindsight Bias and Local Rationality. In D. e. (eds), *Handbook of Applied Cognition* (pp. 141–171). New York, NY: Wiley.

Woods, J., Dekker, S., Cook, R., Johannesen, L., & Sarter, N. (2010). *Beyond Human Error – 2nd Edition*. Burlington, NJ: Ashgate Publishing Company.

Index

Note: **Bold** page numbers refer to tables and page numbers followed by "n" denote endnotes.

Taylor & Francis Group
an **informa** business

Taylor & Francis eBooks

www.taylorfrancis.com

A single destination for eBooks from Taylor & Francis with increased functionality and an improved user experience to meet the needs of our customers.

90,000+ eBooks of award-winning academic content in Humanities, Social Science, Science, Technology, Engineering, and Medical written by a global network of editors and authors.

TAYLOR & FRANCIS EBOOKS OFFERS:

- A streamlined experience for our library customers
- A single point of discovery for all of our eBook content
- Improved search and discovery of content at both book and chapter level

REQUEST A FREE TRIAL

support@taylorfrancis.com

Routledge
Taylor & Francis Group

CRC Press
Taylor & Francis Group